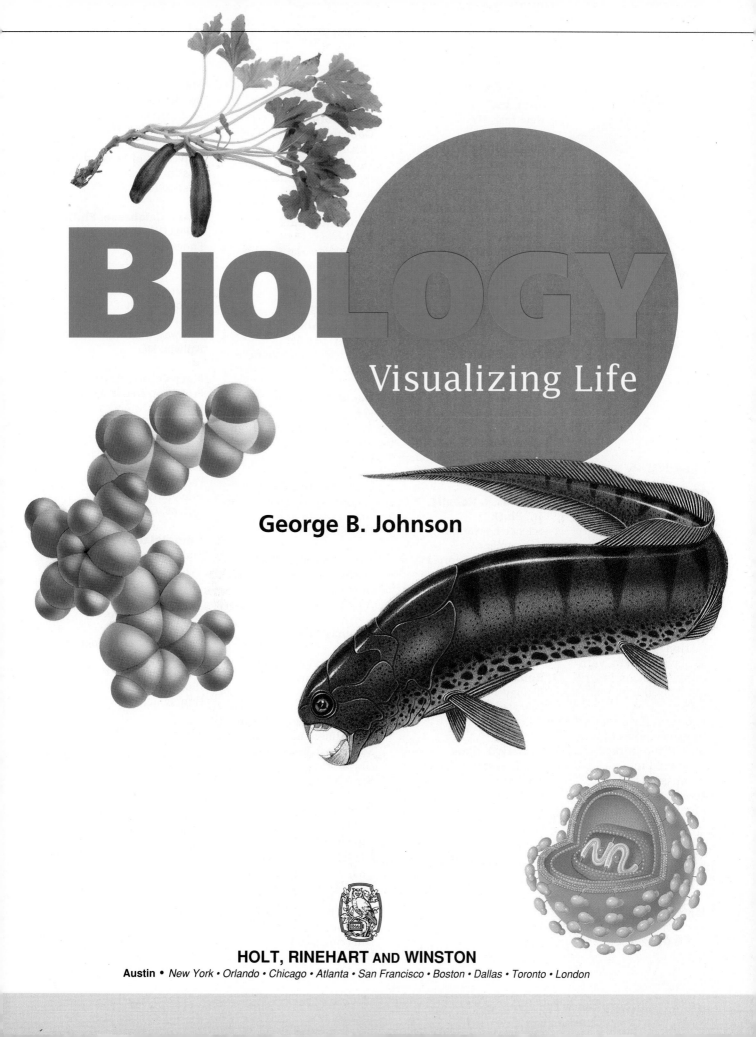

BIOLOGY

Visualizing Life

George B. Johnson

HOLT, RINEHART AND WINSTON

Austin • *New York • Orlando • Chicago • Atlanta • San Francisco • Boston • Dallas • Toronto • London*

Acknowledgments

To learn about the biology of this morning glory flower, see page 407.

Design Development
Foca, Inc.
New York, NY

Contributing Writers
Jo Arnett
San Jose, CA

Gary J. Brusca, Ph.D.
Department of Biological Sciences
Humboldt State University
Arcata, CA

Tracey Cohen
Science Writer
Highland Park, NJ

Rhoda Elovitz
Writer
Austin, TX

Jacquelyn Jarzem, Ph.D.
Instructor
Austin Community College
Austin, TX

Thomas R. Koballa, Jr., Ph.D.
Science Education Department
University of Georgia
Athens, GA

Glenn Leto
Biology Teacher
Barrington High School
Barrington, IL

Ronnee Yashon
Biology Teacher
Lakeview High School
Chicago, IL

Christina Zikos
WARD'S Natural Science Establishment, Inc.
Rochester, NY

Lab Reviewers
Don Chmielowiec
Alex Molinich
George Nassis
Laboratory Investigations
WARD'S Natural Science Establishment, Inc.
Rochester, NY

Kenneth Rainis
Safety
WARD'S Natural Science Establishment, Inc.
Rochester, NY

Reviewers
Hugh C. Allen
Biology Teacher
American High School
Dade County, FL

David Anderson
Biology Teacher
Miami Northwestern High School
Miami, FL

Carol Baskin, Ph.D.
School of Biological Science
University of Kentucky
Lexington, KY

Lowell Bethel, Ph.D.
Science Education Center
University of Texas
Austin, TX

Thomas G. Betz, M.D., M.P.H.
Austin/Travis County Health Department
Austin, TX

Mark W. Bierner, Ph.D.
Department of Biology
Southwest Texas State University
San Marcos, TX

Barry Bogin, Ph.D.
Professor of Anthropology
University of Michigan
Dearborn, MI

Beverly J. Bradley, Ph.D., R.N., C.H.E.S.
Supervisor of School Health Programs
San Francisco Unified School District
San Francisco, CA

Steve Bratteng
Austin, TX

Linda Butler, Ph.D.
Lecturer
University of Texas at Austin
Austin, TX

Diane Calabrese, Ph.D.
PAPILLONS: Diversified Endeavours
Columbia, MO

Thomas J. Conley
Biology Teacher/Science Chairman
Parkway West High School
Ballwin, MO

Mary Coyne, Ph.D.
Professor, Department of Biological Sciences
Wellesley College
Wellesley, MA

Joe Crim, Ph.D.
Professor of Zoology
University of Georgia
Athens, GA

John Delaney
Biology Teacher
Harlandale High School
San Antonio, TX

Andrew A. Dewees, Ph.D.
Department of Biological Sciences
Sam Houston State University
Huntsville, TX

Claudia Dickerson
Biology Teacher
Oliver Wendell Holmes High School
San Antonio, TX

Robert Dimmick
Biology Teacher
Harlandale High School
San Antonio, TX

Tommy C. Douglas, Ph.D.
Graduate School of Biomedical Sciences
University of Texas Health Science Center at Houston
Houston, TX

Marvin Druger, Ph.D.
Departments of Biology and Science Teaching
Syracuse University
Syracuse, NY

John Edwards, Ph.D.
Department of Zoology
University of Washington
Seattle, WA

Richard Farrar
Lead Science Teacher
Northern High School
Accident, MD

Carl Gans, Ph.D.
Department of Biology
The University of Michigan
Ann Arbor, MI

Cynthia Hanes R.D., L.D.
Austin Regional Clinic
Austin, TX

Denny O. Harris, Ph.D.
School of Biological Sciences
University of Kentucky
Lexington, KY

Joseph M. Hepfinger
Biology Teacher
Webster Groves High School
Webster Groves, MO

Agnes Higbie
Biology Teacher
Fulton Junior High School
Indianapolis, IN

Vivian Huang
Biology Teacher
Skyline High School
Oakland, CA

Andrea Huvard, Ph.D.
Department of Biology
California Lutheran University
Thousand Oaks, CA

Duane E. Jeffery
Department of Zoology
Brigham Young University
Provo, UT

Deborah L. Jensen
Biology Teacher/department Chairperson
Oak Ridge High School
Conroe, TX

Janet Jones
Biology Teacher
Sullivan High School
Chicago, IL

Gayle Karriker
Biology Teacher/Science Chairperson
Olympic High School
Charlotte, NC

Karen Martin, Ph.D.
Natural Sciences Division
Pepperdine University
Malibu, CA

Emily Mims
Biology Teacher
Edison High School
San Antonio, TX

Patricia Mokry
Biology Teacher
Westlake High School
Austin, TX

Ted Molskness, Ph.D.
Oregon Regional Primate Research Center
Beaverton, OR

Jane Moncure
Biology Teacher
East Mecklenburg High
Charlotte, NC

Betty K. Moore
Biology Teacher
East Mecklenburg High
Charlotte, NC

David Moury, Ph.D.
Department of EPO Biology
University of Colorado
Boulder, CO

Dorothy Ngongang
Biology Teacher
Providence Senior High School
Charlotte, NC

Martin Nickels, Ph.D.
Anthropology Program
Illinois State University
Normal, IL

Celia Rainwater
Biology/Anatomy Teacher
Clark High School
San Antonio, TX

Peter Raven, Ph.D.
Director
Missouri Botanical Garden
St. Louis, MO

Patricia Recker
Biology Teacher
John Jay High School
San Antonio, TX

Linda Fox Simmons
Nutrition & Health
Training Alternatives
Austin, TX

Marian Smith, Ph.D.
Biology Department
Southern Illinois University
Edwardsville, IL

Scott Spear, M.D.
UT Student Health Center
Austin, TX

George F. Spiegel, Jr.
Biology Instructor
Austin Community College
Austin, TX

Susan Talkmitt
Biology Teacher
Monterey High School
Lubbock, TX

William Thwaites, Ph.D.
Biology Department
San Diego State University
San Diego, CA

David Zeigler, Ph.D.
Pembroke State University
Pembroke, NC

Page 561 contains information on the biology of this salamander and other amphibians.

The leg bone's connected to the hip bone? Check it out on page 624.

A Message From the Author

You are about to have a lot of fun. You might doubt this because you think science is supposed to be both hard and dull, right? It's not. Especially not biology! Biology is the study of the living world, of dinosaurs, of AIDS and acid rain, of how bees fly and what tigers eat and why people grow old. Remember when you were young and tried to capture lightning bugs or touch snowflakes—the sheer wonder of looking at things to see how nature works? That's what biology is, looking at nature and asking questions about it. Each of us was acting as a biologist when we were kids, and there was nothing dull about the excitement we felt then. And biology doesn't have to be dull now. The questions are every bit as interesting as they were, and seeking the answers is every bit as much fun. The secret is not to get bogged down in details. Scientists have learned a lot of information—after all, they have been studying biology for hundreds of years—and wading through this sea of facts can be discouraging. People who find science hard usually get bogged down learning information. So don't do it! Learn ideas instead. The heart of biology is a set of simple ideas that explain why things work the way they do and how they got that way. Focus on understanding these ideas and you won't lose sight of the questions, and the fun.

Of course, some of the information in this book *is* important to you, personally. You need to know how to avoid catching AIDS, why smoking cigarettes will give you cancer, what you can do to help save the environment. Also, in a broader sense, biology is important to you because you are going to have to live the rest of your life in a world very different from today, a rapidly changing world crowded with people in which biology will play a very important role.

In writing this book, I have tried to practice what I preach, to focus on ideas rather than information, and not to lose sight of the fun of science. You will notice that there are a lot of pictures in this book—I believe learning is visual and that pictures help. So take it seriously, keep track of the ideas—and have fun!

Dr. George B. Johnson
Washington University
St. Louis, MO

Biology
Visualizing Life

Table of Contents

Unit 1 Study of Life

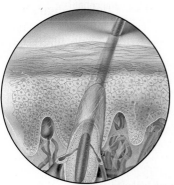

For the role of cells in living things, including your skin, see page 46.

Energy is a unifying scientific theme. For others, see page 16.

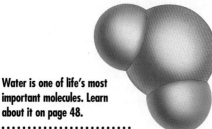

Water is one of life's most important molecules. Learn about it on page 48.

How will the loss of water from a cell affect this sunflower? See page 75.

What do enzymes have to do with the color of this cat's fur? For an answer, see page 93.

Unit 2 Continuity of Life

To learn about Mendel's studies of garden peas, see page 119.

How does DNA duplicate itself? See page 140.

See page 157 to learn about how the *Xenopus* frog is used in genetic engineering studies.

How does natural selection result in the various forms of life on Earth? See page 186.

How did Miller's apparatus help us understand the origin of life? See page 200.

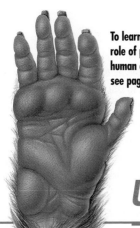

To learn about the role of primates in human evolution, see page 221.

Unit 3 The Environment

How do organisms interact in an ecosystem? See page 248.

See page 275 to read about how these mussels were affected by the removal of their major predator.

To learn how you can help an ecosystem, see page 307.

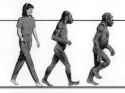

Unit 4 Diversity of Life

See page 325 to learn about how living things are classified.

To learn about organisms such as these blue-green bacteria, see page 342.

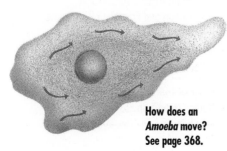

How does an *Amoeba* move? See page 368.

Study page 385 to learn about the biology of fungi such as this mushroom.

How do water and nutrients move through plants? See page 410.

Unit 5 Animal Kingdom

Wheat is one of our most important grains. To learn more about it, see page 428.

How do complex animals develop from a single egg? See page 459.

See page 482 to learn how this animal played an important role in the movement of animals to land.

To study the major groups of animals and to learn about this snail, see page 512.

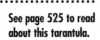

See page 525 to read about this tarantula.

What's this? See page 552.

To learn about the biology of this lizard, see page 573.

Unit 6 Human Life

See page 613 for an introduction to the systems of the human body.

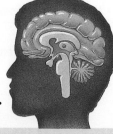

What role does the brain play in the body's hormone system? See page 671.

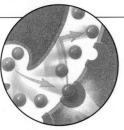

How do drugs affect the nervous system? See page 692.

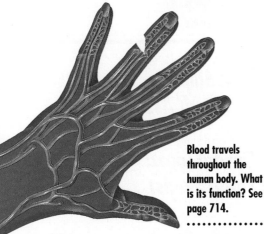

Blood travels throughout the human body. What is its function? See page 714.

See page 744 to learn how antibodies help protect body cells against viral infection.

A taco tastes great, but now what happens? Learn about digestion starting on page 768.

Study page 783 to begin learning about human reproduction.

Biology
Visualizing Life

Features

Science in Action

To learn what inspired Diana Punales-Morejon to become a genetic counselor see page 146.
....

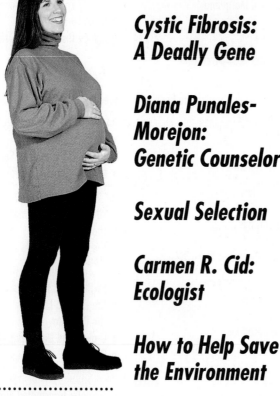

On page 792, you can find out how this woman is helping to ensure that her baby will be healthy.
....

These high school students are helping preserve the environment. To find out what you can do, see page 304.

Journeys

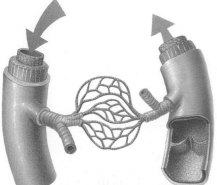

See page 371 for a close-up look at this aquatic organism.

On page 715, you can study the details of how blood flows from arteries to veins.

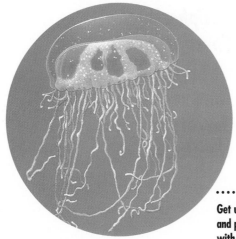

Get up-close and personal with a jellyfish on page 506.

Page 587 explores the biology of the cedar waxwing.

Science, Technology, and Society

Do you understand the issues involved in logging the world's forests? Examine the controversy on page 290.

Discoveries in Science

The quest to feed our world's population has deep roots in all cultures. See page 314 to learn how agriculture and technology have attempted to solve our food production problems.

Investigations

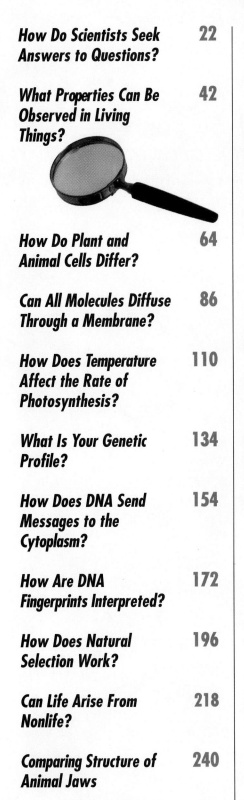

Unit 1

Study of Life

Biology is the science devoted to the study of life. Some biologists study creatures too small to see with the naked eye. Others study molecules, like the delicate threads of DNA that are found in genes. Since you were a child, you probably have been curious about the animals and plants with which you share the world. In this unit you will discover that biologists work like detectives, trying to figure out how the living world works by testing their guesses with experiments. Of all the sciences, biology is one of the most interesting—and fun.

The Science of Biology

Biology is the study of the living world and its organisms, like this *Anopheles* mosquito. It was the study of this type of mosquito that led to the discovery of how malaria, a deadly disease, is transmitted.

YOU ARE ABOUT TO EMBARK ON THE STUDY OF BIOLOGY. IT IS THE SCIENCE OF LIFE, OF OURSELVES, AND OF THE LIVING WORLD AROUND US—BUTTERFLIES, VIRUSES, HOW A CHILD GROWS AND WHAT DINOSAURS WERE LIKE. PEOPLE HAVE PUZZLED ABOUT LIFE FOR MANY CENTURIES, BUT ONLY IN THE LAST FEW HAVE WE REALLY BEGUN TO UNDERSTAND HOW THE WORLD AROUND US CAME TO BE THE WAY IT IS.

1.1 Biology Today

Objectives

1 Describe the research for the cure and prevention of disease.

2 List the ways a knowledge of biology will affect the future of life on Earth.

3 List the ways biology affects you now and will continue to affect you in the future.

Studying Life

Science is a way of investigating the world, of observing nature in order to form general rules about what causes things to happen. A scientist is an observer who searches for knowledge to find solutions to problems. Science has changed the world very rapidly in modern times. What scientists have learned about life and how they learned it are the subjects of this text.

You are about to embark on the study of biology, the science of life. **Biology** is the study of living things, of ourselves, and of the living world. While biological scientists, called biologists, continue to learn new things all the time, they now have a pretty clear picture of what living things are like and how they function. Indeed, biologists have learned so much that they have begun to fashion tools to change the living world in very important ways—an exciting and sobering prospect. Biology's impact by the year 2000 will be so great that every person will need to know about biology to make many personal decisions.

Figure 1.1
Exciting advances in science, biology in particular, make the headlines of many newspapers and are the subjects of many fascinating articles read by millions of people each day.

Biology and Medicine

"What do I need to know about AIDS?"

Stop for a moment and think about the headlines you read in the newspaper this week, the news stories you heard on the radio or TV. Mixed in with bad news and politics is news that modern science is exploring new ways to cure genetically inherited disorders like cystic fibrosis and muscular dystrophy. The fight against infectious diseases like AIDS and malaria, and ways to protect the body from cancer and heart disease, affects everyone, as shown in **Figure 1.2**. The most direct impact of biology on our lives is in medicine, where scientific advances are improving health and health care every day.

Biological knowledge is used in stopping infectious diseases

At the beginning of this century, the infectious diseases flu, tuberculosis, and pneumonia were the top three causes of death in the United States. The flu epidemic of 1918 killed 22 million people worldwide in 18 months! Now, thanks to intense biological research, deaths from these three diseases are far less common. Because of the discovery of antibiotics like penicillin and the development of vaccines that prevent infection, these diseases will probably not be major causes of death in the United States in coming years.

Figure 1.2
AIDS is one of the most devastating diseases of the 1990s. A knowledge of biology helps us to understand what causes AIDS and how the virus that causes it is spread.

a **The virus that causes AIDS is small but deadly. Because it may take years for the symptoms of infection to appear, people can spread the virus unknowingly.**

b **Anyone can get AIDS. Knowing how the virus is transmitted can be the first step in controlling its spread.**

The battle against disease is not over. More than 1 million people are likely to die of malaria this year alone. Spread by mosquitoes, this disease is prevalent in tropic regions—more than 250 million people suffer from malaria at any one time. Almost every child under the age of five who contracts malaria dies. The organism that causes malaria has a very complex life cycle, and is therefore difficult to eliminate. Using modern genetic engineering techniques, scientists are trying to design a vaccine that will attack at a critical stage of the organism's life cycle. Also, new ways to control the large mosquito population are being researched.

New strains of bacteria causing tuberculosis, the fatal lung disease, have recently arisen that are resistant to today's antibiotics.

Other scientists are working to find a cure for AIDS, a fatal disease caused by a virus that destroys the body's ability to defend itself from infections. Because the virus changes so quickly, normal vaccines don't work. No one yet knows what the solutions to these problems will be, but many approaches are being explored.

Biological knowledge is used in curing genetic disorders

Among the most disheartening medical problems are fatal disorders that result from inherited defective genes. Two common fatal human genetic disorders are cystic fibrosis and muscular dystrophy. Cystic fibrosis kills 1 in 1,800 Caucasian children. Their lungs are clogged with mucus because of a defect in a single gene. Muscular dystrophy kills about 1 in 10,000 humans. Their muscles waste away because of a defect in another gene.

Until recently, such genetic disorders were almost always fatal, and no cure was known. In 1990, biologists trying a new approach used genetic engineering techniques to transfer copies of a normally functioning gene from a healthy individual into a patient with defective copies of that gene. Gene transfer therapy offers the first hope that cures for previously fatal genetic disorders will be found. Progress with the new therapy is reported almost daily. A major effort is now underway to catalog every gene in the human body, a program that may soon open the avenue to curing many other genetic disorders.

c **This child has AIDS, which she acquired before her birth. Scientists predict that 40 million people world-wide will be infected with the virus that causes AIDS by the year 2000.**

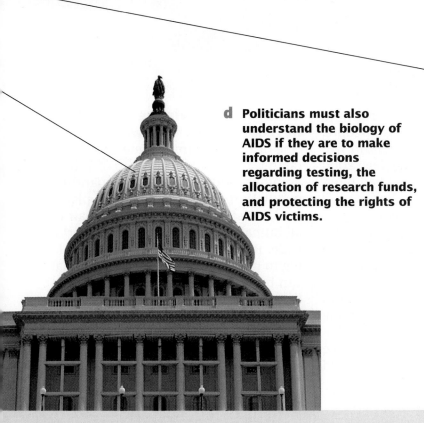

d **Politicians must also understand the biology of AIDS if they are to make informed decisions regarding testing, the allocation of research funds, and protecting the rights of AIDS victims.**

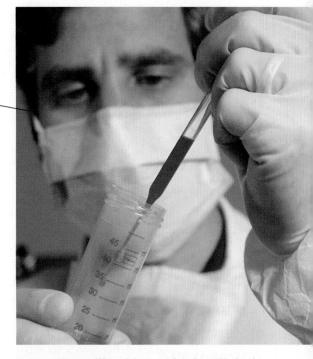

e **Researchers, like this one in the Clinical Immunology Lab at East Tennessee State University, use their knowledge of viruses to understand the biology of the AIDS virus.**

Figure 1.3
Scientists throughout the world work to create solutions to global problems like toxic waste, destruction of the rain forests, and hunger.

a Cleaning toxic spills is one way to preserve the planet for future generations. These researchers are testing contaminated soil in New Jersey.

b Destruction of the Amazonian rain forest must be avoided if we are to preserve its precious resources. This scientist is studying the bird life in the Cuyabeno Nature Reserve in the Amazon Basin.

Biology and the Environment

"What can I do to help preserve the environment?"

When you were born, the world held somewhat more than 4 billion people. This year the world's population will pass 5.5 billion, and by the year 2000 it is estimated that it will exceed 6 billion. This exploding population is placing great stress on the planet, using a lot more energy, consuming more resources, and producing more waste than ever before in the history of the planet. One of the greatest challenges we face entering the next century is to find ways to support so many people without harming the Earth. The scientists in **Figure 1.3** use their knowledge and skill to seek solutions for our world and its future health.

Biological knowledge is used in feeding a hungry world

One of the most immediate challenges facing today's world is to produce enough food to feed its expanding population. Only three kinds of plants (rice, wheat, and corn) provide half of all human energy requirements worldwide. Researchers are presently working to increase the amount of food that can be obtained from a farm without demanding heavy use of fertilizers, pesticides, and equipment that consumes large amounts of energy. New experimental plants are being developed that are more resistant to disease and more tolerant of poor growing conditions.

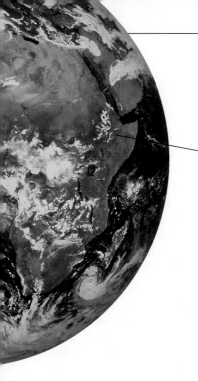

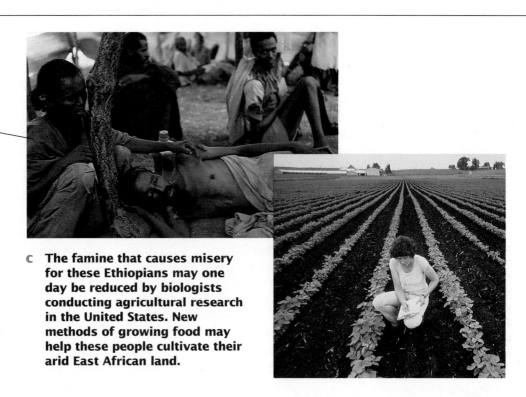

c The famine that causes misery for these Ethiopians may one day be reduced by biologists conducting agricultural research in the United States. New methods of growing food may help these people cultivate their arid East African land.

Genetic engineers are also trying to improve the productivity of existing crops by adding genes to them that increase their resistance to pests, or increase growth rates.

Our impact on the environment is often negative

Think for a moment about the materials you consume and discard in a day—the plastic, paper, glass, and metal. We are beginning to exceed the capacity of the Earth to absorb the waste we generate. Dumping pollutants into the atmosphere, for example, produces "acid rain" that kills forests and poisons lakes. Other industrial wastes are destroying the ozone in the atmosphere that shields us from the sun's harmful rays. Indeed, the sheer volume of carbon dioxide released into the atmosphere by the burning of gasoline, oil, and coal is causing the Earth's temperature to rise alarmingly because carbon dioxide traps the sun's heat.

Preserving a world for your children

An old saying, attributed to many different sources from Native Americans to Amish farmers, states that "We do not inherit the world from our parents, we borrow it from our children." To ensure that our children have the resources they need tomorrow, we must stop wasting precious resources that cannot be replaced, like topsoil and groundwater.

Most important, we must not destroy the world's biological diversity. Unfortunately, we are doing just that, and on a very large scale. The world's tropical rain forests are being destroyed at a breathtaking rate, along with much of the world's biological richness. About 1.3 acres of rain forest per second are being cut for lumber or burned to create pastures for grazing cattle. If this practice continues at such an alarming rate, little or no rain forest will remain in 30 years, and the creatures that lived there, fully one fifth of the world's species, will be gone forever. Because of this incredible destruction, more animals and plants are expected to become extinct in your lifetime than at any time during the last several hundred million years—more even than during the great extinction 65 million years ago when the dinosaurs disappeared from the planet. Biologists and others are working to save as much as possible by creating preserves, and by educating the public about the need to save some of the world's biological richness for future generations.

Biology and You

Much of what you will learn in this course will give you information you will need in making critical personal decisions like the ones contemplated in **Figure 1.4**.

Should I smoke? Biologists now know how cigarette smoking causes lung cancer—chemicals in the smoke enter the cells of your lungs and damage them, and cancer results. Most people who die of lung cancer today are smokers. Smoking cigarettes is really a form of suicide. You'll learn a great deal about the dangers of smoking in Chapter 30.

How can I avoid developing heart disease? Biologists have learned that diet and exercise have a major influence over whether you are likely to have a heart attack or stroke. You will learn about what you can do to prevent heart disease in Chapter 31.

What are the risks in taking drugs? Most mind-altering drugs are addictive and thus very harmful. This includes the nicotine in cigarettes and the alcohol in wine, beer, and spirits. Biologists have learned much about the physical basis of drug dependency, and about the far-reaching effects drugs can have. You will learn about drug abuse and safe drug use in Chapter 30.

How does our species reproduce? The biology of reproduction is now well known, and because you may someday plan to raise a family of your own, you need to gain a clear understanding of these critical processes. You will learn about human reproduction in Chapter 34.

Figure 1.4
Asking questions is the first step to finding answers. A knowledge of biology will help answer many of the questions that these young adults have.

"Many of my friends smoke and they say smoking is all right as long as you quit when you're older. I wonder if that is true?"

"My uncle died three years ago at 45 of a heart attack. Will I have heart problems too?"

"I see other kids using drugs. They say they can stop whenever they want to. I wonder if that's true?"

"My sister is going to have a baby. What's going to happen to her?"

Section Review

❶ **List three decisions that you will be likely to make in the next 10 years that are related to biology.**

❷ **Why is research in the field of genetic engineering considered so important?**

❸ **How does a knowledge of biology help you in playing a role to preserve the Earth?**

SCIENCE IS A WAY OF INVESTIGATING THE WORLD. A SCIENTIST IS AN
OBSERVER WHO IS DRIVEN BY THE SEARCH FOR NEW KNOWLEDGE USING
METHODS THAT DIFFER FROM THOSE USED BY WRITERS OR
PHILOSOPHERS. PHILOSOPHERS MAY MAKE HYPOTHESES AFTER THINKING
THROUGH A PROBLEM. HOWEVER, SCIENTISTS MAKE HYPOTHESES AFTER
OBSERVING A NUMBER OF SPECIFIC CASES.

1.2 Science Is a Search for Knowledge

Objectives

❶ Describe how a hypothesis is formed and the role of testing to verify hypotheses.

❷ Describe the importance of controls in testing hypotheses.

❸ Compare the scientific definition of a theory to the use of theory in language.

A Case Study in Science

Perhaps the best way to see how science works is to look at a real case where science has been used to solve a problem and improve human health. We will study malaria, a disease that kills more humans than any other. The man you see in **Figure 1.5** contracted malaria and is now being treated with life-saving drugs. In 1941, more than 4,000 Americans died of malaria. In the year 2000, by contrast, fewer than five people are likely to die in the United States of malaria! This disease has been virtually eliminated from the United States as the result of one man's work. His story is a very real example of how science works. His investigations were simple. They involved careful observation and the formulation of clear questions. Finding the cause of malaria is one of the greatest medical advances of all time.

Observations suggest questions to investigate

In the summer of 1897 an English physician, Ronald Ross, worked in a remote field hospital in Secunderabad, India. Ross set out to find the cause of malaria. Of all tropical diseases, malaria was the greatest killer, taking more than a million lives a year in India alone. No one knew what caused the disease—most doctors thought it was brought about by poisonous mists or vapors. Working alone, Ross discovered the pattern by which the disease spread.

Ross observed that patients in the field hospital who did not have malaria were more likely to develop the deadly disease in the open wards (those without screens or netting) than in wards with closed windows or screens. Ross wondered why people in open wards were much more likely to get malaria than those in closed wards.

Figure 1.5
This man contracted malaria in Thailand. He is being treated with an intravenous solution of quinine. Worldwide, over one million people die of malaria each year.

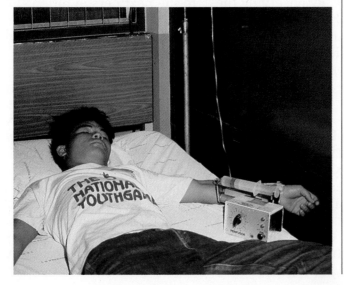

Scientific Method

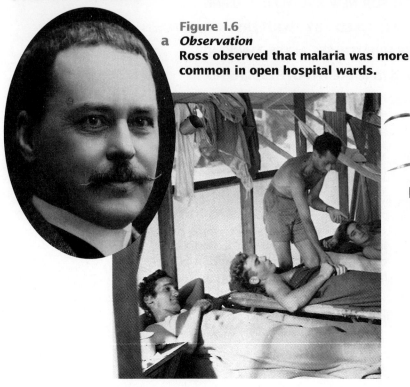

Figure 1.6

a *Observation*
Ross observed that malaria was more common in open hospital wards.

b *Hypothesis*
He formed the hypothesis that mosquitoes were the mode of transmission.

Hypothesis: The basis of further investigation

Observing this pattern, Ross suggested an explanation for why people in open wards were much more likely to get malaria, as shown in **Figure 1.6**. We call such an explanation a hypothesis. A **hypothesis** is a testable explanation for an observation. Ross proposed that mosquitoes in the open wards might be spreading the disease from patients with malaria to patients who did not have the disease. By observing the mosquitoes closely, Ross noted they were *Anopheles*. Using this fact, Ross formulated a hypothesis. Ross's hypothesis was that *Anopheles* mosquitoes were spreading the disease from one patient to another.

Predictions: The framework for testing hypotheses

If Ross's hypothesis was correct, then several consequences could reasonably be expected. We call these expected consequences predictions. A **prediction** is what you expect to happen if a hypothesis is accurate. Ross predicted that *if* the *Anopheles* mosquitoes were

spreading malaria (hypothesis), *then* mosquitoes that had bitten malaria patients and sucked up some of their blood should have picked up the parasite *Plasmodium* (prediction), which is always present in the blood of malaria victims. Ross also predicted that parasites should be alive within the mosquito. Somehow the parasites make their way from the mosquito's stomach to its saliva so that the parasites are transferred with the mosquito's saliva to the next person bitten.

Testing under controlled conditions can verify predictions

The controlled test of a hypothesis is called an experiment. Ross did two types of experiments. He looked for living malaria parasites in *Anopheles* mosquitoes that had bitten malaria patients. He carefully dissected the mosquito's stomach and found the live parasites. He then located the mosquito's salivary gland and by careful dissection showed that the parasite spreads throughout an infected mosquito's body and was indeed present in the salivary gland.

c *Predictions*
He predicted that only mosquitoes that had bitten malaria patients would carry the parasite.

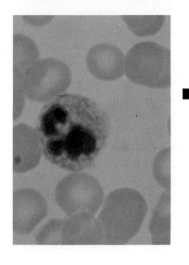

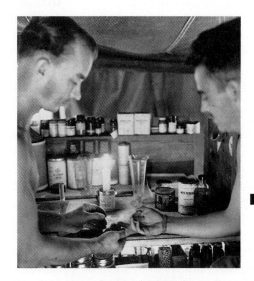

d *Control experiments*
As a control experiment, Ross checked for the parasite in mosquitoes that had never been near malaria patients.

e *Theory*
After his theory was proposed, other scientists set out to verify his results to support or refute his theory.

If a person is bitten by a malaria-carrying mosquito, that person will receive a dose of the parasite in the saliva left behind by the mosquito. To test this prediction, Ross carried out a control experiment. A **control experiment** is one in which the condition suspected to cause the effect is compared to the same situation without the suspected condition (a control group). Nothing else is changed or altered in any way. In Ross's experiment, the suspected condition was mosquitoes feeding on malaria victims. As a control, Ross checked mosquitoes that had not bitten someone with the disease to see if they also contained the parasites. If they did, then malaria patients could not possibly be the source of the parasites in the mosquitoes, and Ross's prediction must be wrong. Gathering newly hatched mosquitoes which had not yet fed, he allowed them to feed on malaria-free blood, and then he examined them as adults. Their stomachs and salivary glands lacked the parasite. The control group of mosquitoes did not contain malaria parasites.

Theories: Explanations for observations

A collection of related hypotheses that have been tested and supported is called a theory. A **theory** is a unifying explanation for a broad range of observations. Theories can have a major impact on science when they tie many accepted and proven hypotheses together into a unified concept. Ross's theory that malaria is transmitted by *Anopheles* mosquitoes carrying it from one person to another was an important milestone in medicine. The idea that malarial epidemics could be prevented by combating mosquitoes was first put forth in a letter written by Ross to the government of India in 1901. Before the end of that year, American army doctors had eliminated almost all malaria from Havana, Cuba, where malaria had reached an epidemic stage. The success of eliminating the deadly disease in these areas was brought about by the reduction of the mosquito population. There have been few advances in the history of medicine more dramatic than the discovery of the cause of malaria.

Science requires continued verification of hypotheses

The essence of science is to reject any hypothesis not supported by observations and the results of control experiments. A new hypothesis is examined very closely to see what it predicts, and the predictions are then rigidly tested. If the predictions are not supported, the hypothesis is rejected. If they are confirmed, the hypothesis is subjected to further verification. One very critical aspect of science is that a scientist's work is held up for review by other scientists. The validity of one's hypothesis is questioned by others until similar results are obtained from similar control experiments. This system of checking and rechecking hypotheses ensures that most, if not all, scientific information is factual information.

Hypotheses that do not explain observations are rejected

A scientist works by systematically showing that certain hypotheses are invalid, that is, they are not consistent with the results of experiments. The results of all experiments are used to evaluate alternative hypotheses. An experiment is successful when it shows that one or more of the alternative hypotheses are inconsistent with observations. By conducting experiments, Ross was able to eliminate the hypothesis that mosquitoes *could* transmit the malaria parasite without biting malaria victims. He retained the alternative hypothesis that if mosquitoes did bite malaria victims, then the mosquitoes could not transmit the parasite. Scientific progress is often made the same way a marble statue is, by chipping away the unwanted bits.

Theories Have Limited Certainty

Theories are the solid foundation of science, that of which we are most certain. There is no absolute certainty, however, no scientific "truth." The possibility always remains that future evidence will cause a theory to be revised or discarded. A scientist's acceptance of a theory is always provisional. See **Figure 1.7**.

The word "theory" is used differently by scientists and the general public. To a scientist, a theory represents that of which he or she is most certain; to the general public the word "theory" implies a lack of knowledge, a guess. How often have you heard someone say "It's only a theory," to imply lack of certainty? As you can imagine, confusion often results. In this text, the word theory will always be used in its scientific sense, as a generally accepted scientific principle.

Some theories, like the theory illustrated in **Figure 1.8**, are so strongly supported that the likelihood of their being rejected in the future is very small. Most of us would be willing to bet that the sun will rise in the east tomorrow, or that if an apple dropped, it would fall. In physics, the theory of the atom is universally accepted, although until recently no one had ever seen one. In biology, the theory of evolution by natural selection is so broadly supported by evidence that biologists accept it with as much certainty as they do the theory of gravity. We will examine the theory of evolution in Chapter 9. It is a very important theory to biologists because the theory of evolution provides the framework that unifies biology as a science.

IF NEWTON WAS A BOTANIST...

Ah, ha— the stem of an apple is not permanently fixed to the tree.

J. harris

Figure 1.7
Newton could have interpreted the falling apple from a different perspective.

Alfred Wegener thought the continents on Earth were once one giant continent. His idea was ridiculed by other scientists who did not think that the continents could move. After many years and much research, enough evidence was gathered to show his theory partially correct. New data showed the sea floor was spreading along with the continents. Continental drift theory was replaced by the theory of plate tectonics.

The continents were once part of a larger continent called Pangaea.

Heat and pressure beneath the Earth's mantle caused the continents to drift apart.

Continental shapes, and fossil and rock evidence indicate that Pangaea existed.

The theory of plate tectonics resolved some of the controversy with the theory of continental drift. As you can see in this illustration, all the continents rest on giant plates, which are indicated by the different shades of color. The arrows indicate their direction of movement. During a human lifetime, the plates do not seem to move a great deal. On a geologic time scale, however, the plates are very mobile.

Figure 1.8
The theory of continental drift was tested many times before it became widely accepted.

Scientific Method: The Systematic Study of a Question or Problem

It was once fashionable to claim that scientific progress was the result of applying a series of steps called the "scientific method." In this view, science is a sequence of logical "either/or" steps, each step rejecting one of two incompatible alternatives. Trial-and-error testing could inevitably lead one through a maze of uncertainty. If this view were indeed true, a computer could be programmed to be a good scientist—but science is not done this way. If you ask successful scientists how they do their work, you will find that without exception they design experiments with a good idea of how their experiments are going to come out. Not just any hypothesis is tested, but rather a hunch or educated guess based on all the scientist knows and that allows his or her imagination full play. Because insight and imagination are so important in scientific progress, some scientists are better than others.

Section Review

❶ Under what conditions is a hypothesis supported?

❷ In the Ross case study, what was the relationship between his experiments and his hypothesis?

❸ How does the scientist's use of the word "theory" differ from the way the term is used by the general public?

SCIENTISTS HAVE BEEN STUDYING LIVING THINGS FOR SEVERAL HUNDRED YEARS. FROM THE MOUNTAIN OF INFORMATION COLLECTED, GENERAL PRINCIPLES OF PARTICULAR IMPORTANCE HAVE EMERGED. YOU WILL ENCOUNTER THESE PRINCIPLES REPEATEDLY AS YOU EXPLORE THE IDEAS IN THIS TEXT—THEY ARE THE FRAMEWORK OF BIOLOGY, THE SKELETON THAT SUPPORTS ALL YOU WILL LEARN.

1.3 Studying Biology

Objectives

❶ Describe the role of energy in life processes.

❷ List the characteristic properties of cells.

❸ Describe the role of genes in controlling cell functions and development.

❹ Relate evolution to natural selection.

Themes Unify Ideas

In this text, we have selected six unifying principles or themes to use as a framework for your study. These themes are introduced here, and you will see them repeatedly throughout the text. Your goal should be to understand how the many topics you study in biology are examples of themes.

Theme 1: Energy and Life

All organisms require energy to carry out life processes. Energy is used to grow and do work. Without it, life soon stops. Almost all the energy that drives life on Earth is obtained from the sun. Plants capture the energy of sunlight and use it to make complex molecules in a process called photosynthesis. These molecules then serve as the source of fuel for animals that eat them. The flow of energy among organisms, including the teenager shown in **Figure 1.9**, helps determine how organisms interact within their environment, which is another important concept in biology.

Figure 1.9
The teenager to the right is using the *energy* that he gets from the food he eats and, indirectly, from the sun.

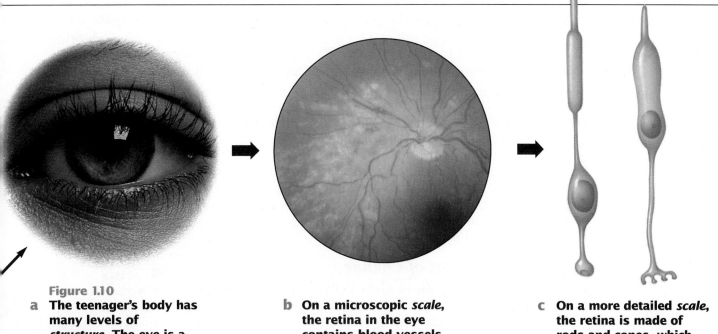

Figure 1.10

a The teenager's body has many levels of *structure*. The eye is a complex tool for sight.

b On a microscopic *scale*, the retina in the eye contains blood vessels. Light enters the eye and lands on the retina.

c On a more detailed *scale*, the retina is made of rods and cones, which send information to the brain.

Theme 2: Scale and Structure

Living things are made of the same materials as the rest of the universe, of atoms assembled into molecules. All the physical principles that apply to stars and home computers also apply to you and to every other living creature. Living things differ from nonliving ones only in their degree of organization.

All organisms are composed of cells, tiny compartments surrounded by membranes. Your body has trillions of cells. Cells form the structures of your eye shown in **Figure 1.10**. The complex chemical processes that occur within cells are much the same in all organisms, and all cells have the same basic structure: a covering called a membrane that surrounds the cell and controls what information and materials enter and leave it; an internal fluid and skeleton that gives shape to the cell and supports the other things within it; and a central zone or nucleus which contains the cell's genes, the hereditary instructions coded within long complex molecules called DNA. Many cells have specialized structures within their cell walls called organelles that carry out some of the cell's activities. You will learn much more about cell processes in Chapters 3, 4, and 5. Cell activities are often influenced by outside molecules that attach to special proteins in the cell's membrane. Much of a cell's biology is determined by the nature of its membrane.

Theme 3: Stability

Control is an essential aspect of life. *Humans and other organisms must maintain a constant internal environment in order to function properly;* for example, your body temperature must not vary by more than a few degrees. Your body has many mechanisms for maintaining **homeostasis**, a word that means keeping things the same.

Theme 4: Evolution

Biologists have long suspected that life on Earth is the result of evolution. Life forms are slowly changing and have apparently been changing since Earth formed. Charles Darwin proposed the hypothesis that this change is the result of a long process called natural selection. Natural selection proposes that those organisms with more favorable mutations will be more likely to survive and reproduce. These favorable mutations better enable an organism to overcome the many challenges presented by its environment. Darwin's theory provides biology with a basis for understanding the diversity of life on Earth. *Evolution results from a long history of organisms adjusting to a diverse and changing environment.* Evolution provides the vast diversity of species that exist on Earth. A **species** is a group of organisms that look similar and can produce fertile offspring in their environment. *Natural selection leads to changes in species over time.*

Theme 5: Patterns of Change

In organisms composed of many cells, such as you, most of the cells are specialized, each performing distinct functions. For example, gland cells secrete hormones, muscle cells contract, and nerve cells conduct electrical signals. *All of the many different kinds of specialized cells, however, are descended from the same single fertilized egg cell—as the cells grow and divide, their genes manage an orderly process of change called development.* The process is controlled by genes, and is the same in all humans.

Children resemble their parents because instructions for development are passed from parents to offspring. These instructions are in the form of **genes**, which are segments of long DNA molecules. Sometimes damage to the DNA occurs. These changes, called mutations, are usually harmful, but sometimes mutations help an organism to better survive.

Theme 6: Interacting Systems

Living things interact with each other and with their environment in complex ways, like the hoverfly in **Figure 1.11**. Ecology is the study of complex communities of organisms in relation to their environment. *A living community is highly structured and interdependent. This interdependence is the result of a long process of evolution in which selection has favored cooperation.* This complex web of interactions is easily disrupted when the environment is polluted and individual species become extinct, as is happening in much of the world today.

Figure 1.11
This hoverfly is gathering nectar from the flower. Without the flower, the fly would not have food. Without the fly, the flower would not be able to pollinate and reproduce.

Section Review

❶ Why would energy be considered a theme in biology?

❷ What properties do all cells share?

❸ What is the role of a gene in the development of an organism?

❹ How is evolution influenced by natural selection?

Highlights

Life is plentiful. In this rain forest alone, approximately 30 million different species of plants, animals and insects can be found.

	Key Terms	**Summary**

1.1 Biology Today

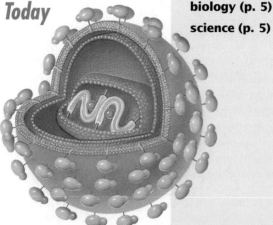

The AIDS virus causes one of the most devastating diseases of the 1990s.

Key Terms

biology (p. 5)

science (p. 5)

Summary

• Biologists are combating infectious diseases. Smallpox, tuberculosis, and pneumonia have been largely conquered. AIDS and malaria are still being studied.

• Genetic disorders may be eliminated by gene therapy involving the transfer of normally functioning genes to affected individuals.

• As the population increases, Earth's ability to sustain the human population is being strained.

1.2 Science Is a Search for Knowledge

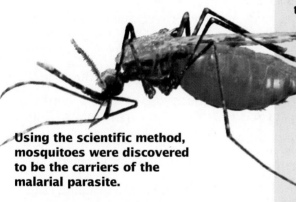

Using the scientific method, mosquitoes were discovered to be the carriers of the malarial parasite.

Key Terms

hypothesis (p. 12)

prediction (p. 12)

control experiment (p. 13)

theory (p. 13)

Summary

• Scientists, by asking questions and making observations, determine scientific principles.

• Scientific progress is made by posing hypotheses and testing predictions. Verification of hypotheses is required before they are widely accepted.

• Controlled experiments are important in testing hypotheses. Control groups ensure that only one variable in an experiment has been changed.

• A theory links well-supported hypotheses together as one concept.

1.3 Studying Biology

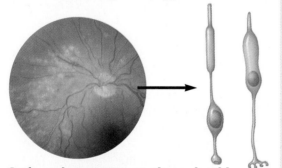

Scale and structure, as shown by this retina and the rods and cones, is one of the themes of biology.

Key Terms

development (p. 17)

homeostasis (p. 17)

genes (p. 17)

species (p. 18)

Summary

• Energy in biological systems plays a role in organization—from cells to ecosystems.

• Structure dictates function. Genes are the foundation of this concept.

• Turning genes "on" and "off" produces different kinds of cells.

• Mutations produce new versions of genes.

• Individuals with favorable mutations are more likely to survive and reproduce.

• Darwin proposed that evolution was the result of natural selection.

Understanding Vocabulary

1. For each pair of terms, explain the differences in their meanings.
   a. ecology, biology
   b. hypothesis, theory
   c. homeostasis, development

Relating Concepts

2. Copy the unfinished concept map below onto a sheet of paper. Then complete the concept map by writing the correct word or phrase in each oval containing a question mark.

Understanding Concepts

Multiple Choice

3. Science is mainly concerned with
   a. asking questions and seeking answers.
   b. test tubes and beakers.
   c. making life easier for people.
   d. maintaining tropical rain forests.

4. Your knowledge of the relationships between such things as cigarette smoking and lung cancer and between diet and heart disease is important to you because
   a. biologists are learning more about lung cancer and heart disease.
   b. this knowledge is likely to affect personal decisions you make in the future.
   c. it will prepare you to be a health care professional in the next century.
   d. lung cancer and heart disease are typically associated with the use of drugs.

5. Gene transfer therapy
   a. involves transferring a defective copy of a gene into a healthy person.
   b. will enable biologists to cure some genetic disorders.
   c. can begin when every gene in the human body is cataloged.
   d. has produced a cure for AIDS.

6. Scientific principles are generated from
   a. tests.
   b. predictions.
   c. observations.
   d. variables.

7. In Ross's experiment, the control group was
   a. yellow fever mosquitoes.
   b. mosquitoes that had not bitten a person with malaria.
   c. stomachs and salivary glands of blood sucking mosquitoes.
   d. people living in places other than India.

8. The scientific method is
   a. a series of logical steps that leads to a final solution.
   b. trial-and-error tests of scientific problems.
   c. a process of investigation influenced by insight and imagination.
   d. a hunch or guess tested with the aid of computers.

9. Theories
   a. are always subject to revision.
   b. reflect scientific truths.
   c. always result in major scientific breakthroughs.
   d. are tested but never rejected.

10. The energy used by living things on Earth comes from
    a. burning fossil fuels.
    b. the sun.
    c. waste products of photosynthesis.
    d. DNA in plant and animal cells.

11. As proposed by Charles Darwin, life on Earth evolves by means of
    a. speciation.
    b. mutation.
    c. homeostasis.
    d. natural selection.

12. What biological theme is highlighted when studying the cell, its membrane and its organelles?
    a. energy and life
    b. scale and structure
    c. stability
    d. all of the above

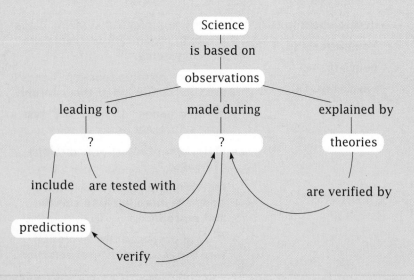

Completion

13. Mainly due to the work of _____ , malaria has been virtually eliminated from the United States. He _____ that malaria is spread from one person to another by the mosquitoes.

14. The world's population today is more than _____ people. By the year 2000, experts predict that the world's population will be about _____ .

15. When a _____ explains a set of observations it is called a _____ .

16. Segments of DNA molecules are called _____ .

17. Mutations are caused by changes to _____ .

Short Answer

18. What are scientists doing to help farmers produce enough food for the world's expanding population?

19. Why is it important that the world's tropical rain forests not be destroyed?

20. It has been said that medicine is the area of modern biology that has most affected our lives. What evidence supports this statement?

21. Describe the relationship between hypothesis and prediction in Ross's experiment.

Interpreting Graphics

22. Look at the photograph of a scientist at work. Based on your examination of the photo, write a hypothesis about what you think the scientist could be investigating. Then, write a prediction that you expect to occur if the hypothesis is true.

Reviewing Themes

23. *Interacting Systems*
How might the extinction of one species in an environment affect other organisms with which it interacts?

24. *Stability*
You perspire when you get too hot, and shiver when you get too cold. How do these actions help you maintain a constant body temperature?

25. *Evolution*
A politician who opposes the teaching of biological evolution in schools said her reason is "It's only a theory." Why are scientists likely to be upset by the politician's statement about biological evolution?

Thinking Critically

26. *Inferring Conclusions*
Scientific information is always changing, yet you are embarking on a yearlong study of scientific information. Why should you learn it?

27. *Compare and Contrast*
How is scientific progress a lot like constructing a stone statue by chipping away at the unwanted bits?

Cross-Discipline Connections

28. *Biology and History*
Edward Jenner conducted a daring experiment, the results of which led to the eradication of smallpox in many parts of the world. What was his hypothesis? What experience led him to formulate this hypothesis?

Discovering Through Reading

29. Read the article "The War Among the Greens," in *Newsweek*, May 4, 1992, page 78. What is the mission of the Sierra Club? How is the willingness of the Sierra Club's leadership to compromise on issues like clearcutting in national forests likely to affect the future of life on Earth?

Investigation

How Do Scientists Seek Answers to Questions?

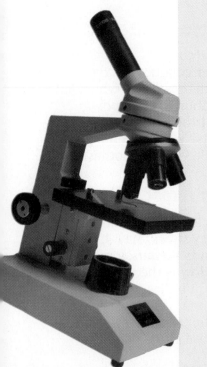

Objectives

In this Investigation, you will:
- *formulate* a hypothesis
- *design* an experiment
- *test* a hypothesis
- *average* data
- *compare* experimental results
- *display* experimental results in graph form

Materials

- watch with second hand
- paper

Prelab Preparation

1. Form a cooperative group of four students. Work with one member of your group to complete steps 3–10.

2. In this investigation, you will explore the scientific methods used to answer a question. The question that you will attempt to answer is:
   What effect will holding your breath have on your pulse rate?

Procedure: Testing a Hypothesis

1. To find the pulse rate, have your teammate sit quietly at your desk. Find your partner's pulse in either wrist using your middle finger. Count the number of beats during a 60-second period. This number is the pulse rate. For example, a person might have a pulse rate of 70 beats per minute.

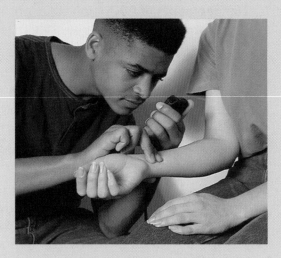

2. *Why would counting for a fraction of a minute and multiplying the result increase the error in your answer?*

3. To find the average pulse rate repeat step 1 twice. Average your three answers and record your result. *Why is it better to work with the average pulse rate than one single measurement?*

4. Hold your breath for 45 seconds. Record your pulse rate immediately after holding your breath.

5. Compare your average pulse rate with the rate after holding your breath. *Do you think your data matches those of your classmates?* Formulate a hypothesis that answers this question. Use Ross's hypothesis on page 12 as a guide.

6. Design an experiment to test your hypothesis. In designing your

experiment, one factor is varied and the response of another factor is measured. The factor varied is the independent variable. The factor that responds to the independent variable is called the dependent variable. *Identify the variables in your experimental design. What variables will you keep constant?*

7. Write out your experimental plan and have your teacher approve your experimental plan. Run the experiment and record your data in a table.

8. Compare your results with the other team in your group and with the other teams in your class. Record their data in your table.

9. Make a graph like the one below and show the data collected from each group in the class.

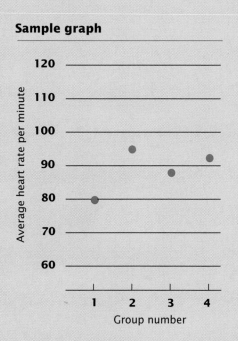

Sample graph

Average heart rate per minute (y-axis, values 60, 70, 80, 90, 100, 110, 120)
Group number (x-axis, values 1, 2, 3, 4)

10. Clean up your materials and wash your hands before leaving the lab.

Analysis

1. *Analyzing Data*
   What conclusions can be drawn from your data? Explain how the data support or refute your hypothesis.

2. *Analyzing Methods*
   Why must you know the average pulse rate during normal breathing?

3. *Evaluating Methods*
   What are some possible sources of error in this experiment?

4. *Evaluating Methods*
   What is the value of comparing results?

5. *Applying Concepts*
   Experiments are used to collect data from two or more groups of subjects or from the same subjects at two different times. The group exposed to changes in the independent variable is called the experimental group. The other group is not exposed to changes in the independent variable. This group is called the control group. Identify the experimental group and control group in your experiment.

Thinking Critically

How would the consumption of different kinds of foods affect your pulse rate? How would you form a hypothesis and test it? How do sugars act in the body? Would proteins or fats affect your pulse rate the same way? Is there an accurate way to predict the effect other than by direct experimentation?

Discovering Life

Review

- **biological themes (Section 1.3)**

- **the term *atom* (Glossary)**

- **the term *cell* (Glossary)**

Are all the things you see when you walk along the seashore alive? How do you know? In order to decide, you will have to examine them more closely. In this chapter you will find out what to look for to define life.

IMAGINE THAT YOU ARE WALKING IN THE WOODS AND YOU UNEXPECTEDLY ENCOUNTER A LARGE FORMLESS BLOB LYING STILL ON THE FOREST FLOOR. IS THE BLOB ALIVE? HOW WOULD YOU DECIDE? MAKE A LIST OF THE THINGS YOU MIGHT DO TO DETERMINE WHETHER THE BLOB IS ALIVE. IN THIS SECTION YOU'LL FIND OUT IF YOUR LIST HAS THE SAME THINGS THAT A SCIENTIST WOULD DO.

2.1 What Is Life?

Objectives

❶ Explain the difficulty in defining life using visually observable properties.

❷ Describe the five properties shared by all living organisms.

❸ Relate the properties of life to the biological themes in Chapter 1.

First Guesses at Defining Life

Figure 2.1
If you found this blob on the ground, would you think it was a living thing? How would you be able to tell? What are the characteristics of living things?

In making your list about the blob shown in **Figure 2.1**, the first thing you would probably look for would be movement.

Movement Almost all animals move around. Squirrels dash along tree branches, sharks knife through the water, humans ride bicycles and run.

Everywhere you look, the world is alive with movement. However, movement from one place to another is not in itself a sure sign of life. A tree does not move about, for example, but it is alive. A cloud does move about, but it is not alive. Even if you could see the blob move on the ground in front of you, that would not be enough to tell you that the blob is alive.

Sensitivity So what should you do next? One thing you might do is poke the blob to see whether it responds. If you did that, you would be checking its **sensitivity**, its ability to respond to the stimulus of being poked.

Almost all organisms respond to stimuli. Deer flee from sounds they sense as dangerous, and plants grow towards light. Air movement, sound, light, and temperature are all stimuli. But not every stimulus produces a response. Can you imagine getting a response from kicking a redwood tree? If the blob just sits there after you poke it, and the tree doesn't move after you kick it, that doesn't mean they are not alive. Sensitivity, while a better criterion than movement, is still not a good single characteristic to define life.

Microscopy: *The Invisible World*

Why Are Microscopes Used?

If you hold a leaf out at arm's length, you can see the pattern of the veins and perhaps a few marks made by insects. As you bring the leaf closer you see more details of the structures, maybe a few cracks. But just when you are beginning to see a lot of detail, everything starts to blur. How can you get a closer look?

Scientists in the mid-1400s realized that they needed more than the human eye to study objects. As microscopy evolved, scientists have learned more than they could ever have imagined about plant and animal life.

Two important concepts relating to microscopes are magnification and resolution. Magnification is the ability of a microscope to make an image appear larger. Resolution is the ability to distinguish small, close objects. These concepts are equally important. If the details of a large image are unclear, the viewer would see only a fuzzy blur.

Kinds of Microscopes

Each type of microscope has its own strengths and limitations. Scientists have learned which microscopes can give the most information about whatever they are trying to see.

A compound microscope uses two lenses

Microscopes that use two lenses are called compound microscopes. A typical compound microscope, such as the one you use in biology class, has a light bulb or mirror in the base that shines light upward through the specimen. Light rays pass through the objective lens and then through the lens in the eyepiece. The image you see is magnified by both lenses. Total magnification is determined by multiplying the magnifications of the two lenses. If your microscope has a 10X eyepiece lens, and the 40X objective lens is in place, the object you are looking at appears 400 times larger than it actually is.

A biologist can use a compound microscope to study living cells. Cells appear to be essentially transparent, although there are small variations in thickness and density. As a result, the cell and some structures inside are visible, but the image is not very distinct. More details of the structures inside cells can be seen by thinly slicing cells and dyeing them with stains. Looking at a cell this way has obvious disadvantages—only one thin slice of cell is seen, and, of course, the cell is dead. However, sectioning and staining cells enables biologists to see many structures not visible in living cells.

These human cheek cells have been magnified 225X using a light microscope. They are seen using an ordinary bright field.

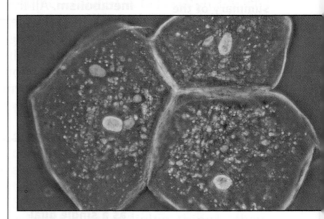

These human cheek cells have also been magnified 225X using a light microscope. The cells are unstained and are seen using phase contrast, which alters light waves so that details can be seen more easily.

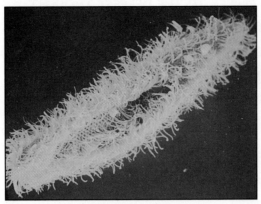

This *Paramecium* has been magnified 1,000X using a scanning electron microscope (SEM). Note the numerous cilia it uses for locomotion.

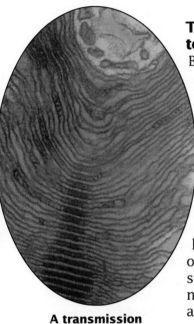

A transmission electron microscope (TEM) enables biologists to see details of the structure of cell organelles such as this Golgi body, magnified 13,000X.

TEMs and SEMs cause electrons to magnify objects

Because light has wavelike characteristics, there is a limit to the size of an object that can be viewed as a sharp, focused image. Practically speaking, bacteria with a diameter of 0.5 µm are about the smallest living things that can be distinguished using a good mass-produced light microscope. In the early part of this century, physicists showed that an accelerated stream of electrons also had wavelike properties similar to those of light. Microscopes using electrons instead of light to form images can magnify images at least 100 times as much as the light microscope.

In a transmission electron microscope (TEM), a stream of electrons passes through the specimen and strikes a fluorescent screen. By replacing the fluorescent screen with a piece of photographic film, a photograph called a transmission electron micrograph can be made. Sections of specimens viewed with a TEM are sliced much more thinly than sections prepared for the light microscope. These sections are treated with stains that block electrons, causing details to appear dark.

The scanning electron microscope (SEM) enables biologists to see detailed three-dimensional images of the surfaces of cells. Specimens are not sliced, but are placed on a small metal cylinder and coated with a very thin layer of metal. Like the picture on a television set, the image is formed one line at a time as the beam of electrons scans the specimen from side to side. The electrons that bounce off the specimen form an image that can be viewed on a video screen, or a scanning electron micrograph can be made. Because electrons would bounce off of the gas molecules in air, the stream of electrons and the specimen to be viewed must be placed in a vacuum chamber. Therefore, living cells cannot be viewed with an electron microscope.

The micrographs made with electron microscopes are always black-and-white, never colored. This is because the stream of electrons is only one wavelength. The fluorescent screen, the video screen, and the photographic film simply detect the presence or absence of electrons—light or dark. However, electron micrographs often have color added in the darkroom to make certain structures stand out in the micrograph.

New ways to look at cells

New video and computer techniques are extending the resolution and level of detail that can be detected by microscopes. The scanning tunneling electron microscope (STM) uses a needle-like probe to measure differences in voltage due to electrons on the surface of an object. A computer tracks the movement of the probe across the object, creating an image of the cell surface. The STM is used to view living cells.

This representation of DNA has been captured using a scanning tunneling electron microscope (STM).

YOUR BODY IS A CHEMICAL MACHINE. EVERYTHING IT DOES, FROM THE SMASHING STROKE OF A TENNIS PLAYER TO THE DEEPEST THOUGHT OF A SCIENTIST OR POET, CAN BE UNDERSTOOD AS A CHEMICAL PROCESS. BECAUSE CHEMISTRY AND BIOLOGY ARE CLOSELY RELATED, A BRIEF INTRODUCTION TO CHEMISTRY WILL HELP YOU BETTER UNDERSTAND HOW LIVING ORGANISMS FUNCTION.

2.2 Basic Chemistry

Objectives

❶ Relate atoms, elements, ions, and molecules to each other.

❷ Describe the structural features of an atom.

❸ Distinguish between covalent and ionic bonds and how they are formed.

Atoms: The Basic Structural Units of Matter

All matter in the universe is composed of tiny particles called atoms. An **atom** is the smallest particle of matter that can retain its chemical properties. For example, the smallest particle of carbon that still has all the chemical properties of carbon is a carbon atom.

Figure 2.4
An atom consists of a nucleus surrounded by electrons that move about the nucleus at high speeds. This model shows regions of space outside the nucleus called energy levels.

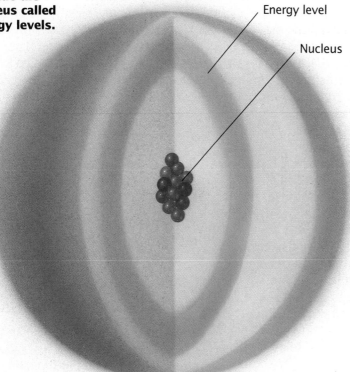

Energy level

Nucleus

An **element** is a substance composed of only one type of atom. There are 92 kinds of elements found in nature, but only 11 elements are common in living things. Over 99 percent of the atoms in your body are either nitrogen (N), oxygen (O), carbon (C), or hydrogen (H).

Atoms are composed of electrons, protons, and neutrons

An atom consists of a dense core called a nucleus surrounded by tiny moving particles called **electrons**. Electrons move about the nucleus in various energy levels. The nucleus contains two other kinds of particles, protons and neutrons. Most of the interior of an atom is empty space. If the nucleus of an atom were the size of an apple, the nearest electron could be more than a mile away. **Figure 2.4** shows a model for the structure of an atom.

Each proton in the nucleus has a positive charge, while the neutrons have no charge. Each electron has a negative charge. Atoms contain equal numbers of electrons and protons, therefore they have no charge.

Atoms can gain or lose energy

You may wonder why electrons are not pulled into the nucleus. It takes energy to overcome attractive forces and keep electrons moving about the nucleus. The energy needed to do this is similar to the energy it takes to hold an apple in your hand when gravity is pulling it toward the ground. The apple in your hand and electrons moving around the nucleus both possess energy. If you were to release the apple, it would fall to the ground. Electrons have energy of position, or potential energy, just like the apple. Electrons also have energy due to their motion, or kinetic energy.

Electrons move about the nucleus in different energy levels. The farther an electron is from the nucleus, the more energetic it is. An atom may have several energy levels stacked one on top of another, like the skin of an onion.

The sun's energy can be transferred to electrons by light photons. If a photon of light crashes into an atom, the transfer of energy jolts electrons to an energy level that is farther from the nucleus of the atom. The electron is able to maintain its new high energy level because it stores energy obtained from the photon.

Formation of Bonds Stabilizes Atoms

When an atom reacts, it gains, loses, or shares electrons. An atom that gains or loses electrons is called an **ion**. If the ion has more protons than electrons, it is said to be positively charged. If the ion has more electrons than protons, it is said to be negatively charged. Ions with positive charges are electrically attracted to ions with negative charges.

The force holding two atoms or ions together is called a **chemical bond**. All chemical bonds involve interactions between electrons in the energy levels farthest from the nucleus. These outer electrons determine an atom's chemical behavior. In other words, the atom's outermost electrons determine its ability to react with other atoms. Atoms bond to fill their outer energy levels with electrons. A full outer energy level makes an atom chemically stable. Carbon, nitrogen, and oxygen require eight electrons to fill their outer energy levels. Since these atoms all have fewer than eight electrons in their outer energy levels, they form bonds to gain, lose, or share electrons.

An **ionic bond** is the force of attraction between oppositely charged ions. To form an ionic bond, an atom loses one or more electrons so that it has a full outer energy level. The other atom involved in bonding gains one or more electrons so that it too has a full outer energy level. **Figure 2.5** shows a model for the formation of salt.

Figure 2.5

a **Sodium (Na) and chlorine (Cl) atoms are reactive. The sodium atom has one extra electron in its outer energy level. The chlorine atom is missing one electron in its outer energy level.**

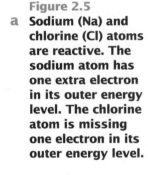

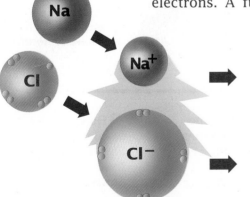

b **When a sodium atom loses an electron to a chlorine atom, two ions form. The force of attraction between the ions is an ionic bond.**

c **Oppositely charged sodium and chloride ions cluster to form salt crystals.**

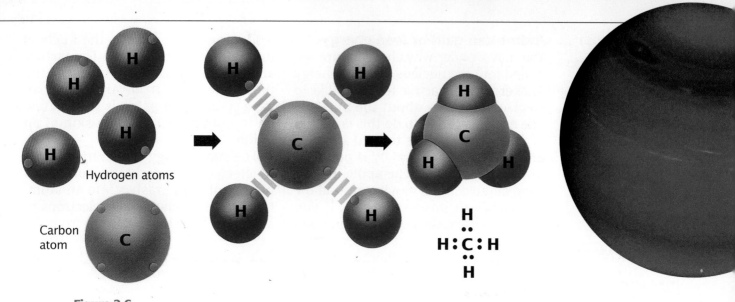

Figure 2.6

a Hydrogen (H) and carbon (C) atoms are unstable.

b This carbon atom is electrically attracted to four hydrogen atoms.

c Methane forms when the carbon and hydrogen atoms share electrons.

d Methane gives Neptune's atmosphere its bluish-green appearance.

Molecules result from the formation of covalent bonds

A second type of chemical bond results when two atoms share one or more electrons. A **covalent bond** is different from an ionic bond because electrons are *shared*, rather than *lost* or *gained*. **Figure 2.6** shows a model for covalent bonding between an atom of carbon and four atoms of hydrogen. The result is a molecule of methane.

A **molecule** is a group of atoms held together by covalent bonds. The force that holds these atoms together comes from sharing electrons. Most molecules in your body are made of more than two atoms, because most atoms must share electrons with more than one other atom to fill their outer energy levels. Oxygen atoms are able to form covalent bonds with two other atoms. Nitrogen atoms can form covalent bonds with up to three other atoms. Carbon atoms can form covalent bonds with as many as four different atoms.

Hydrogen requires just two electrons to fill its outer energy level. Each hydrogen atom is composed of a proton and a single electron. When two hydrogen atoms are close enough to each other, the two electrons can be shared equally by the two nuclei. The result is the diatomic (two atom) molecule of hydrogen gas (H_2), shown in **Figure 2.7**.

Figure 2.7
Two hydrogen atoms can form a covalent bond.

Section Review

❶ How do atoms differ from ions?

❷ How are elements related to molecules?

❸ Describe the structure of an atom.

❹ How do covalent bonds differ from ionic bonds?

YOU HAVE JUST LEARNED HOW ATOMS COMBINE TO FORM BONDS.
MOLECULES ALSO REACT TO FORM MACROMOLECULES. THERE ARE
FOUR TYPES OF MACROMOLECULES IN YOUR BODY. THEY ARE
COMPOSED ALMOST ENTIRELY OF CARBON, HYDROGEN, AND OXYGEN.
IN THIS SECTION, YOU WILL LEARN ABOUT THE BASIC STRUCTURE
AND FUNCTION OF EACH MACROMOLECULE.

2.3 Molecules of Life

Objectives

❶ **Define sugar and describe the process that occurs in the formation of polysaccharides.**

❷ **Describe the solubility and energy storage properties of lipids.**

❸ **Explain the factors that affect the three-dimensional structure of proteins.**

❹ **Define nucleic acids and describe their functions.**

Organic Compounds Are Derived From Carbon

Just as atoms can be joined to form molecules, molecules can be joined to build **macromolecules**. All organisms are composed of four major classes of macromolecules: proteins, lipids, carbohydrates, and nucleic acids. Each kind of macromolecule has different subunits. The primary component of all macromolecules is carbon.

The properties of carbon are important to biological systems, including those living in the coral reef shown in **Figure 2.8**. In the last section you read that carbon has four electrons in its outer energy level, and that a carbon atom seeks to fill that energy level by sharing its electrons with other atoms. Carbon atoms form long chains that are the backbone of many different kinds of molecules. Molecules with carbon-carbon bonds are called **organic compounds**.

Figure 2.8
Diamonds are pure carbon. Carbon compounds are the primary components of living things, including plants and animals that live in this coral reef.

Carbohydrates Are Energy Sources

A **carbohydrate** is composed of carbon, hydrogen, and oxygen in a ratio of one carbon atom to two hydrogen atoms to one oxygen atom. Some carbohydrates, such as table sugar, are simple, small molecules. Other carbohydrates, like the starch in potatoes, exist as chains hundreds of subunits long. Carbohydrates contain many carbon-hydrogen bonds. They are well suited to be energy sources because their bonds store considerable energy.

Carbohydrates can be either simple or complex molecules

Among the simplest carbohydrates are sugars, small molecules that taste sweet. While sugars may have as few as three carbon atoms, the sugars involved in energy storage, like glucose, have six. These sugars have the formula $C_6H_{12}O_6$. Carbohydrates are made by linking individual sugars together to form long chains called polysaccharides. Polysaccharides are insoluble in water. They can be deposited in specific storage areas in a cell. This ability to store energy in the form of polysaccharides lets organisms build up energy reserves.

Starch and glycogen are both complex carbohydrates

Starch is a polysaccharide composed of glucose subunits. Amylose is the simplest kind of starch. It exists as a long, unbranched chain of glucose molecules. Baking or boiling starchy plants such as potatoes and corn breaks these chains into shorter fragments that are soluble and can be used by the cell.

Humans consume a great deal of carbohydrates; the seeds of rice, wheat, and corn supply about two-thirds of all the calories used by people. Animals store glucose in the form of long, branched chains, called glycogen. **Figure 2.9** shows the structural differences of starch and glycogen.

Cellulose provides structural support

Many organisms use polysaccharides as structural molecules as well as for energy storage. Plants manufacture a polysaccharide called cellulose. Cellulose consists of glucose subunits linked in a way that most animals cannot break down. Cellulose forms in the cell walls of plants. When you eat plants containing cellulose, it passes through your body undigested. This undigested cellulose is called dietary fiber and is an important component of your diet. In contrast, cows and horses are able to graze on grass. These animals have bacteria that digest or break down cellulose in their intestines. You lack these bacteria and would starve on a diet of grass.

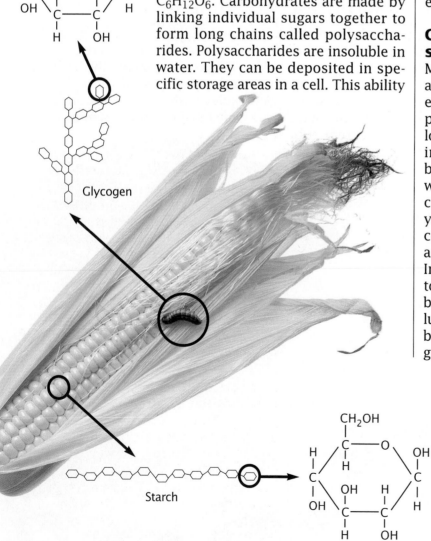

Glycogen

Starch

Figure 2.9
Starch, the carbohydrate found in corn, is chopped into glucose subunits and converted to glycogen when it is eaten by this worm. The glucose subunits are linked in different ways.

Table 2.1 Types of Fats

Type	Found in
Saturated Fat 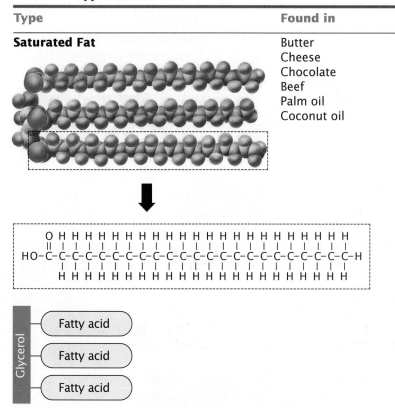	Butter Cheese Chocolate Beef Palm oil Coconut oil

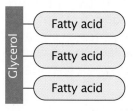

Type	Found in
Unsaturated Fat 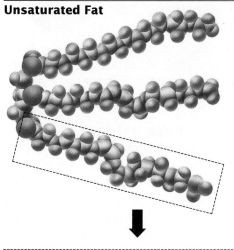	Avocado Olives Olive oil Peanuts Peanut oil Almonds Corn oil Fish Mayonnaise Safflower oil Sunflower oil

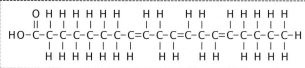

Lipids Store Energy

A **lipid** is not soluble in water, but is soluble in oil. The most important kind of lipid is fat, an energy storage molecule. Fats have more carbon-hydrogen bonds than carbohydrates and, therefore, can store more energy. They have three very long chains of CH_2 units that join at one end like the letter "E." The backbone of the structure is a glycerol molecule, and the branches are fatty acids made from CH_2 units.

There are two major categories of fats: saturated fats and unsaturated fats. In addition to fats, there are two other types of lipids: steroids and waxes. Steroids include the sex hormones, cholesterol, chlorophyll (an important pigment found in plants), and retinol (the vision pigment of your eyes). Earwax and beeswax are examples of lipids that are waxes.

Fats can be saturated or unsaturated

When the three chains of fatty acid molecules line up neatly side by side, the fat is said to be saturated. As you can see in **Table 2.1**, most carbon atoms in a saturated fat are bonded to two hydrogen atoms. The carbon atoms are saturated with hydrogen atoms. If some units of the three fatty acid chains are linked by double bonds, kinks occur in the chain. This type of fat is called an unsaturated fat. The carbon atoms could bond to additional hydrogen atoms if the double bonds between adjacent carbon atoms were broken. Unsaturated fats usually exist as liquids called oils at room temperature. In contrast, most saturated fats are solid at room temperature. Two exceptions are palm oil and coconut oil. Although both are liquid at room temperature, they are classified as saturated fats because they lack double bonds.

It is possible to make an oil into a solid fat by adding hydrogen. The double bonds become single bonds and the chains can then line up. Peanut butter is usually hydrogenated. The peanut fats are converted to saturated fats, so they don't separate as oils while the jar sits on the store shelf.

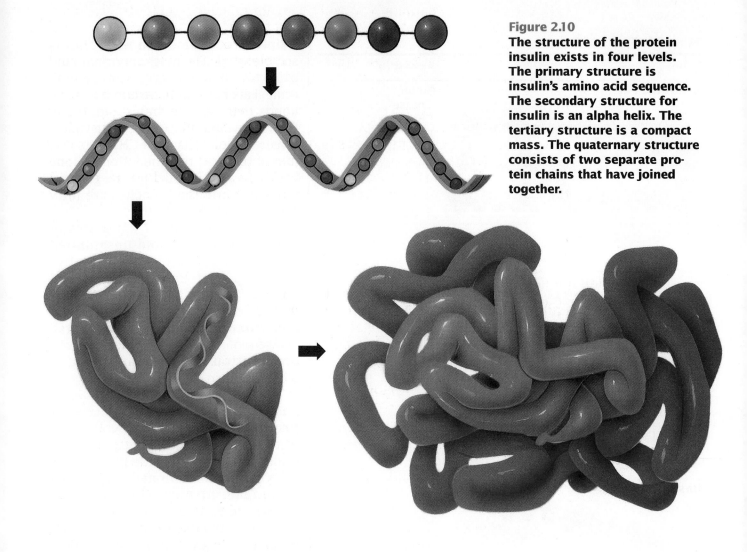

Figure 2.10
The structure of the protein insulin exists in four levels. The primary structure is insulin's amino acid sequence. The secondary structure for insulin is an alpha helix. The tertiary structure is a compact mass. The quaternary structure consists of two separate protein chains that have joined together.

Proteins Provide Structure and Increase Reaction Rate

A **protein** is composed of long chains of subunits called amino acids. There are 20 different kinds of amino acids that humans use. In a particular type of protein the positions of the component amino acids can vary along the chain. Thus, there is an almost endless variety of possible proteins. Think of a type of protein as a paragraph in a book made from 20 letters in the English alphabet. Any letter may be present at any position as the words of the paragraph are composed. A typical protein has approximately 100 amino acids linked together in its chain. There are many millions (actually 20^{100}) of different possible amino acid sequences for such a protein.

Proteins have a three-dimensional structure

The chemical properties of a particular protein depend on its structure. The actual sequence of amino acids in the protein is called its primary structure. Because amino acids interact with their neighbors, parts of the chain coil or bend. This coiling and bending determines the protein's secondary structure. In most proteins, the entire chain folds into a compact mass called its tertiary structure. When two or more folded proteins combine to form clusters, the mix of proteins forms a quaternary structure. **Figure 2.10** shows the four levels of structural arrangements for a protein.

In the next few chapters, you will study various types of proteins. Pay particular attention to the shapes of these macromolecules, because shape determines a protein's biological function.

Proteins function as structural molecules and enzymes

Proteins often play structural roles in organisms. Cartilage and tendons are made of a protein called collagen, as is the matrix of your skin and bone. A protein called keratin forms the horns of a rhinoceros and the feathers of a bird, as well as your own hair.

Proteins play a second very important role in organisms: they act as enzymes. Enzymes increase the rate at which chemical reactions occur such as those that take place during metabolism. Most chemical reactions necessary for growth, movement, and other body activities would not take place without enzymes.

Nucleic Acids Contain Genetic Information

The fourth major class of macromolecules is called **nucleic acids**. There are two types of nucleic acids: DNA (deoxyribonucleic acid) and RNA (ribonucleic acid). Subunits of DNA and RNA are called **nucleotides**. These nucleotides are grouped into units called genes, which encode information concerning how a given organism will grow and develop. RNA is involved in making working copies of genes. These copies are used in assembling amino acids to make proteins.

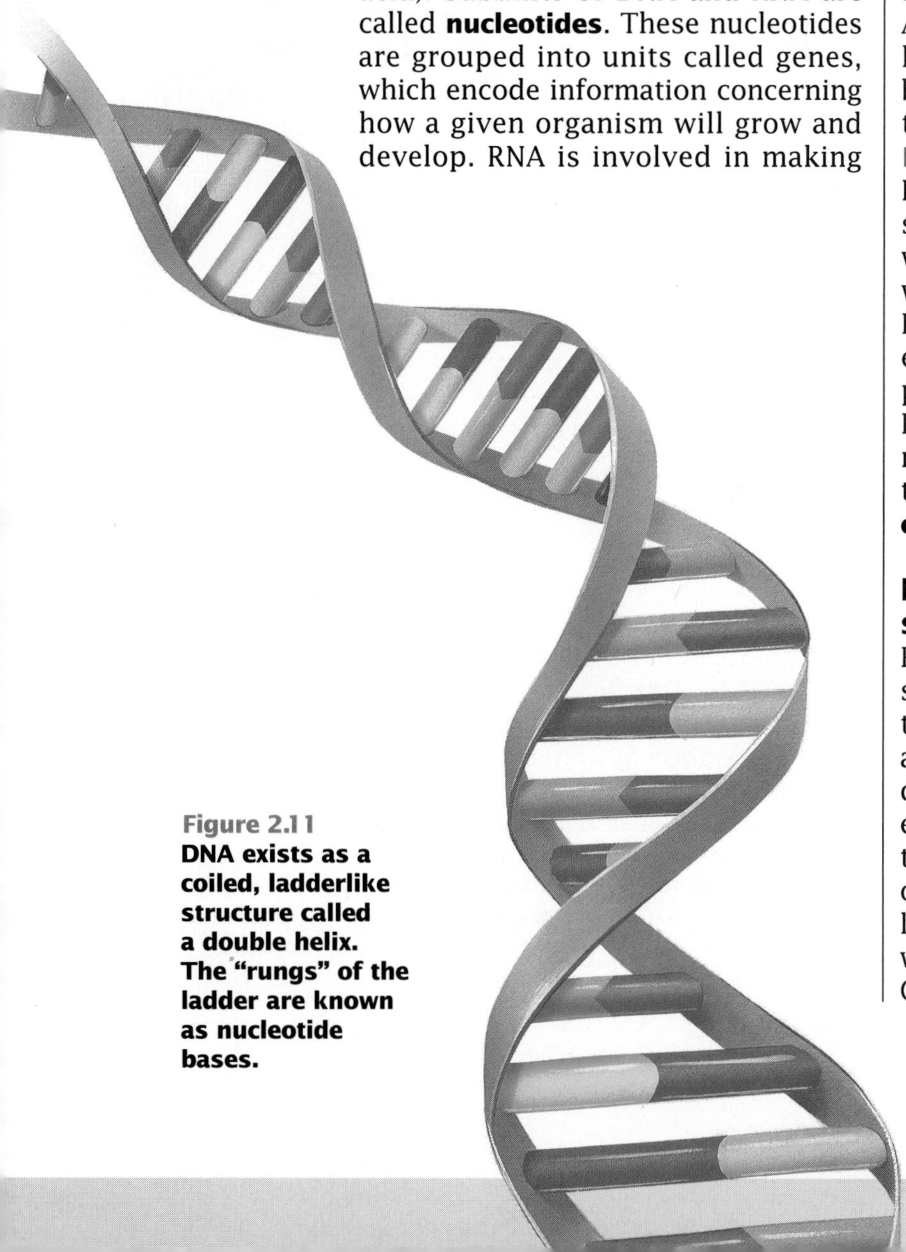

DNA is a double helix

A DNA molecule consists of two interlocking coil-shaped strands that resemble a spiral staircase. Chemists refer to this coiling structure as a double helix. **Figure 2.11** shows the structure of a DNA double helix. DNA encodes the sequences of all the cell's proteins, as well as information that determines when each protein is to be produced. In simple cells like bacteria, the DNA exists as a long molecule. But in complex cells like those of your body, the DNA exists in numerous different segments. These segments, along with proteins, form compact bodies called **chromosomes**.

RNA helps in the synthesis of proteins

RNA molecules have a variety of shapes, depending on their function in the cell. Specific RNA molecules serve as scaffolds for the assembly of all the different proteins in the cell. Others exist in the cell as long, single-stranded threads that carry DNA's message from one part of the cell to another. You will learn more about how DNA and RNA work together to build proteins in Chapter 7.

Figure 2.11
DNA exists as a coiled, ladderlike structure called a double helix. The "rungs" of the ladder are known as nucleotide bases.

Macromolecule Summary

The structures of the four basic macromolecules (carbohydrates, lipids, proteins, and nucleic acids) have the same building blocks. All four are composed of long chains of similar subunits. Using **Table 2.2** you can compare and contrast the structures and functions of macromolecules.

Table 2.2 Classes of Macromolecules

Class	Subunit	Function	Example	
Carbohydrates	Sugar	Stores energy	Starch, glycogen	Starch granules
		Structural component	Cellulose, chitin	
Lipids	Fatty acid	Stores energy	Body fat	Human fat cells
		Membrane bilayer	Plasma membrane	
		Steroid hormones	Testosterone	
		Pigments	Chlorophyll	
Proteins	Amino acid	Catalysis by means of enzymes	Lactase	Hair
		Structural component	Hair, cartilage	
		Peptide hormone	Insulin	
Nucleic Acids	Nucleotide	Stores genetic information	DNA	Chromosomes
		Makes proteins	RNA	

Section Review

❶ How is a sugar related to polysaccharides and starch?

❷ How do lipids react in water?

❸ What causes a protein to have a tertiary structure?

❹ What is the function of DNA in cells?

Chapter 2 Highlights

These seeds are the start of new life, but is a seed itself alive? Use the five properties of life to decide.

	Key Terms	Summary
2.1 What Is Life?  Is this blob alive? How would you go about answering this question?	sensitivity (p. 25) metabolism (p. 27) homeostasis (p. 27) heredity (p. 27)	• Some of the most obvious properties of life cannot be used alone to decide whether something is alive. • The characteristic properties of living things can also appear in nonliving things: batteries use energy, clouds move, oceans maintain constant temperatures.
2.2 Basic Chemistry Atoms are the basic structural units of matter.	atom (p. 30) element (p. 30) electron (p. 30) ion (p. 31) chemical bond (p. 31) ionic bond (p. 31) covalent bond (p.32) molecule (p. 32)	• Each of the 92 elements found on Earth is made of atoms. Nitrogen, oxygen, carbon, and hydrogen make up more than 99 percent of the atoms in your body. • When electrons absorb energy, they move to higher energy levels farther from the nucleus. When they release energy, they fall to lower energy levels. • Bonds form when atoms lose, gain, or share electrons. Molecules are groups of atoms held together by covalent bonds. Some compounds are held together by the force of attraction among ions.
2.3 Molecules of Life  Carbon is the primary chemical component of the macromolecules found in living things, including the plants and animals shown in this coral reef.	macromolecule (p. 33) organic compound (p. 33) carbohydrate (p. 34) lipid (p. 35) protein (p. 36) nucleic acid (p. 37) nucleotide (p. 37) chromosome (p. 37)	• Carbon is the most abundant element in living things. Molecules with carbon-carbon bonds are called organic compounds. • Organisms use carbohydrates to store energy and provide structural support. • Lipids are not soluble in water. Fats store energy more efficiently than carbohydrates because they have more carbon-hydrogen bonds. • The sequence of amino acids in a particular protein determines its shape and chemical properties. The shape of a particular protein determines its biological activity or function. • Nucleic acids contain genetic information and direct protein production.

Understanding Vocabulary

1. From each group of terms, select the one that does not fit the pattern and explain why it does not fit.
   a. homeostasis, movement, reproduction, heredity
   b. electrons, protons, compound, neutron
   c. protein, lipid, element, nucleic acid
   d. nitrogen, carbon, hydrogen, tin

Relating Concepts

2. Copy the unfinished concept map below onto a sheet of paper. Then complete the concept map by writing the correct word or phrase in each oval containing a question mark.

Understanding Concepts

Multiple Choice

3. Living things are made mostly of compounds that contain
   a. carbohydrates.   c. carbon.
   b. blood.   d. scandium.

4. Sugar, starch, and cellulose are examples of
   a. carbohydrates.   c. lipids.
   b. monosaccharides.   d. proteins.

5. A protein's secondary structure is caused by
   a. neighboring amino acids interacting.
   b. the amino acid sequence.
   c. folding of the total chain of amino acids.
   d. its clustering with other proteins.

6. Polysaccharides are formed
   a. from amino acids and nucleotides.
   b. when sugars and proteins react.
   c. when animal fats are hydrogenated.
   d. by linking sugars together.

7. Which class of macromolecules stores more energy than carbohydrates?
   a. proteins   c. lipids
   b. polysaccharides   d. nucleic acids

8. Molecules are formed
   a. from ionic bonds.
   b. by atoms carrying electrons.
   c. when electrons are shared.
   d. when protons gain or lose energy.

9. An electrically charged atom is a(n)
   a. element.   c. ion.
   b. organic   d. molecule.
      compound.

10. Which of the following is not a characteristic property of life?
    a. maintains homeostasis
    b. uses energy
    c. undergoes development
    d. made of cells

11. Chemical bonds help atoms achieve chemical stability because bonding
    a. forms ions.
    b. fills an atom's outer energy level.
    c. prevents electrons from moving about the atom's nucleus.
    d. causes DNA to release energy.

Completion

12. Chemical bonds that result from the attraction of oppositely charged ions are called _____ , while those that result from atoms sharing electrons are called _____ .

13. The smallest particle of matter is a(n) _____ ; when it is electrically charged it is called a(n) _____ .

14. The force that holds two atoms together is called a(n) _____ .

Short Answer

15. Describe how the properties of carbon enable it to serve as the "backbone" of so many different kinds of molecules.

16. "Nucleic acids are biologically important molecules." Do you agree or disagree with this statement? Explain.

17. Name the four classes of biologically important macromolecules and the subunit of each.

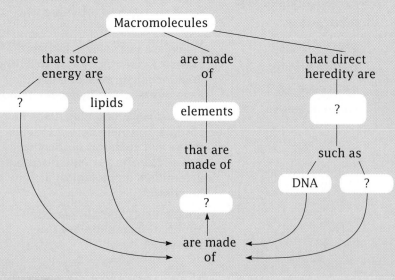

Interpreting Graphics

18. Label the protons, neutrons, and electrons of the carbon atom shown above. Recall that a carbon atom requires four electrons to fill its outer energy level and become stable. How many electrons does a bonded carbon atom have moving around its nucleus?

19. Use a reference on food groups to help you answer the following. Name the carbohydrates you see in the picture. Name the proteins you see in the picture. Name the lipids you see in the picture.

Reviewing Themes

20. *Patterns of Change*
A pattern exists in the way electrons behave when covalent bonds are formed. Describe the pattern. How does this pattern differ from the pattern of electron behavior in ionic bonds?

21. *Energy and Life*
All living things store energy but they often do it in different ways. What are two forms of energy storage found in humans?

Thinking Critically

22. *Inferring Conclusions*
Why do football players and other athletes often consume a complex carbohydrate diet while training?

23. *Inferring Conclusions*
Butter and margarine are similar in appearance, but are chemically different. Margarine is actually an oil, sometimes called an unsaturated fat. During the manufacturing process, what is done to margarine to convert it to a hard fat?

24. *Analyzing Concepts*
Contrast the functions of proteins and carbohydrates in the human body.

Cross-Discipline Connections

25. *Biology and Health*
Dietary fiber is a carbohydrate that cannot be digested by the human body. Why do nutrition experts recommend that high fiber foods like whole-grain cereals and raw fruit be eaten regularly?

Discovering Through Reading

26. Read pages 115–119 of the article "Eloquent Remains," in *Scientific American*, May 1992. What is paleo-DNA? Of what value is paleo-DNA to molecular archaeologists?

Investigation

What Properties Can Be Observed in Living Things?

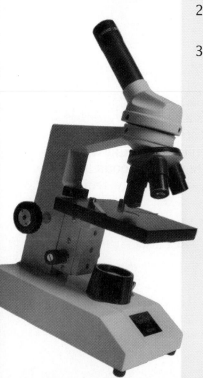

Objectives

In this investigation you will:

- *observe* and *identify* the characteristics of living things

Materials

- *Detain* slowing agent
- forceps
- microscope slide (with concave impression)
- coverslip
- compound light microscope
- medicine dropper
- pond water
- toothpick

Prelab Preparation

1. Review what you have learned about the characteristics of life by answering the following:
   - What are the characteristics of living things?
   - What characteristics might be easily observed in microscopic organisms?
   - What characteristics might be difficult to observe?

2. Explain why some characteristics are difficult to observe.

3. Read the information about the use and care of the compound light microscope in the Appendix.
   - What power objective must be used to make your first observations?
   - State the reasons why this objective must be used first.
   - Name the adjustment control that should never be used when you focus with the high power objective.

Procedure: Observing Characteristics of Living Things

1. Form a cooperative group of four students. Work with one member of your group to complete steps 2–8.

2. Using the medicine dropper, place several drops of pond water and a drop of *Detain* inside the depression on the slide. Stir with a toothpick.

3. Carefully place the coverslip on top of the pond water.

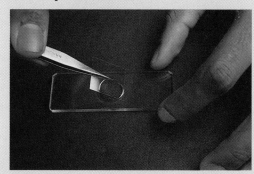

4. Make sure the microscope is level. Then, place the prepared slide on the microscope stage. *What might happen if the microscope were tilted at a sharp angle?*

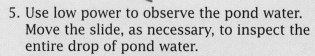

5. Use low power to observe the pond water. Move the slide, as necessary, to inspect the entire drop of pond water.

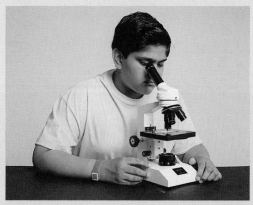

6. Make a drawing of the first thing you observe that appears to be alive. List the characteristics shown by this thing.

7. Repeat step 6 for each object you think might be classified as living.

8. Clean up your work area. Wash your hands before leaving the lab.

9. Combine your work with that of the other team in your group. *What purpose does a comparison of results serve in a scientific investigation?*

10. Pool your group's data with that of the rest of the class. Prepare a composite list of the characteristics you observed in the pond water.

Analysis

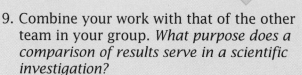

1. *Identifying Relationships* What characteristics can be used to distinguish between microscopic living and nonliving things?

2. *Analyzing Observations* List the properties of life that could not be observed in the pond water organisms. Explain why these characteristics cannot be seen with a microscope.

3. *Making Predictions* If you were to observe a drop of pond water for an extended period of time, what changes might occur?

4. *Making Inferences* Is movement an indication that an organism is alive? Why could movement be a misleading indicator for defining life?

Thinking Critically

1. What limitations does microscopic observation present in determining whether something is alive?

2. You see an object that you think is alive in the pond water. Devise a plan to test your prediction.

Cells

Review

3.1 World of the Cell

3.2 Membrane Architecture

3.3 Inside the Cell

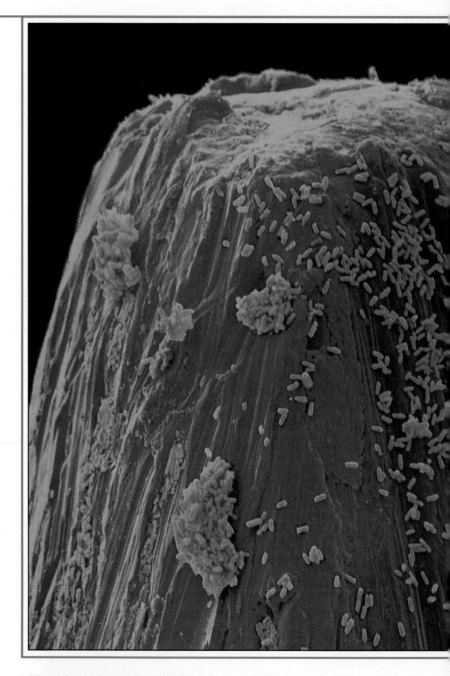

This color-enhanced scanning electron micrograph shows bacteria, one-celled organisms, on the point of a pin.

YOUR BODY IS COMPOSED OF MORE THAN 100 TRILLION CELLS. AS LONG AS CELLS CARRY OUT THEIR FUNCTIONS NORMALLY, YOU ARE GENERALLY UNAWARE OF THEM. BUT IN SOME DISEASES, SUCH AS CANCER, CELLS BEHAVE ABNORMALLY. IN THE SEARCH FOR THE CAUSES AND POSSIBLE CURES FOR CANCER, SCIENTISTS FIRST EXAMINE THE WORLD OF THE CELL TO SEE HOW CELLS WORK NORMALLY.

3.1 *World of the Cell*

Objectives

❶ **Explain why cells must be small if they are to function efficiently.**

❷ **Contrast the behavior of polar and nonpolar molecules.**

❸ **Describe how the structure of water shapes the membrane of the cell.**

At the Edge of the Cell

Plants and animals are made up of a maze of tiny compartments. Each compartment is called a **cell**. Since the mid-1800s we have known that all living things are composed of cells. Most microscopic creatures like the one in **Figure 3.1** are single cells. A cell is the smallest unit that can carry on all of the activities of life.

In some ways, a cell is like a submarine. A submarine has a tough outer surface, called a hull, which wraps around the complex machinery that makes the submarine function. A cell is also filled with complex machinery and has a tough outer surface, which is called the **cell membrane**. The cell membrane serves the same purpose as the hull of a submarine. It separates what is inside from what is outside. Nothing gets into or out of the sub except through the hatches in the hull, and nothing gets into or out of a cell except through "gates" in its cell membrane.

Imagine that the submarine's hull suddenly disappears. The machinery would be scattered all over the sea floor. Without the cell membrane to hold together the machinery and the substances the cell needs for life, the cell would die. However, a cell membrane is much more than a simple container. The complex machinery that is inside the cell cannot function unless raw materials continuously enter into the cell and destructive waste products promptly leave the cell. By regulating what goes into and out of a cell, the cell membrane helps to maintain the internal environment of the cell.

Figure 3.1
This microscopic protozoan, *Paramecium* (133X), is a single cell.

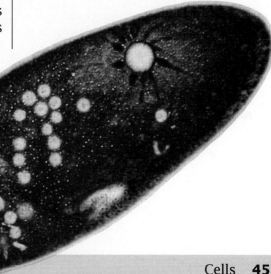

What Limits the Size of a Cell?

What is the most obvious thing that can be said about cells? It is that they are very small. Your body has about 100 trillion cells. If these cells were each the size of a shoe box and were lined up end to end, they would stretch in a line about 30 billion km (18.6 billion mi.)— to the sun and back 100 times! So cells must be pretty small to cram that many into your body. Study **Figure 3.2**. What factors limit the size of a cell?

Cell parts cannot be too far from the cell membrane

Every bit of food and information needed by the cell must enter through the cell membrane. When cells are small, no part of their complex machinery lies too far from the area outside the cell. If a cell were larger, fewer of its interior structures could be near the cell membrane. That is bad for just the same reason that long supply lines are bad for an army—too many things could go wrong and responses to information would be too slow. Thus, small cells work more efficiently because their supply lines are short.

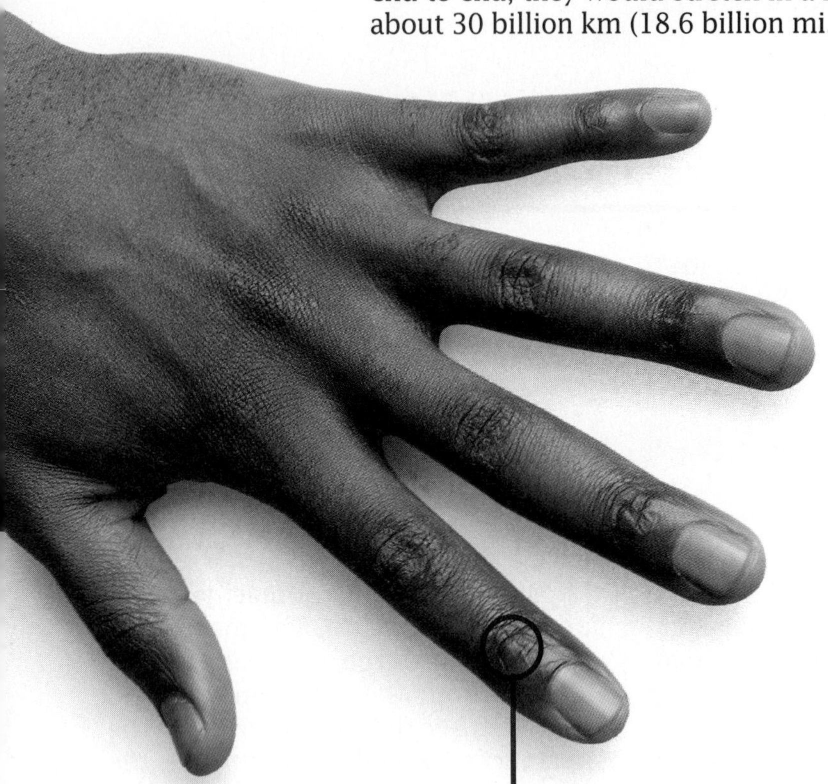

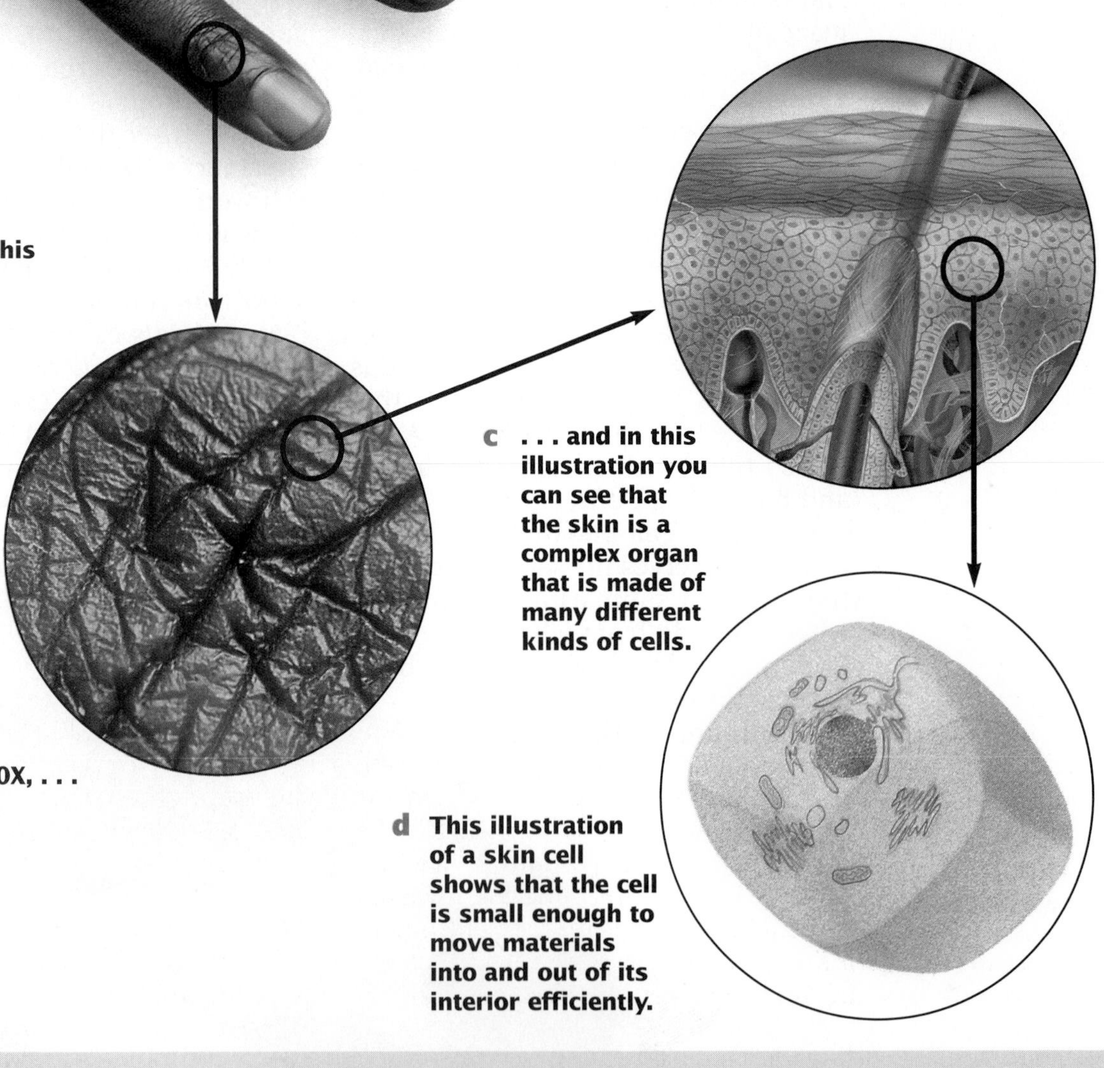

Figure 3.2

a **The skin of this finger . . .**

b **. . . is now magnified 10X, . . .**

c **. . . and in this illustration you can see that the skin is a complex organ that is made of many different kinds of cells.**

d **This illustration of a skin cell shows that the cell is small enough to move materials into and out of its interior efficiently.**

The Ratio of Surface Area to Volume in Cells

The surface area of a cell is a measurement of the exterior of the cell. The volume is a measurement of the internal contents of the cell.

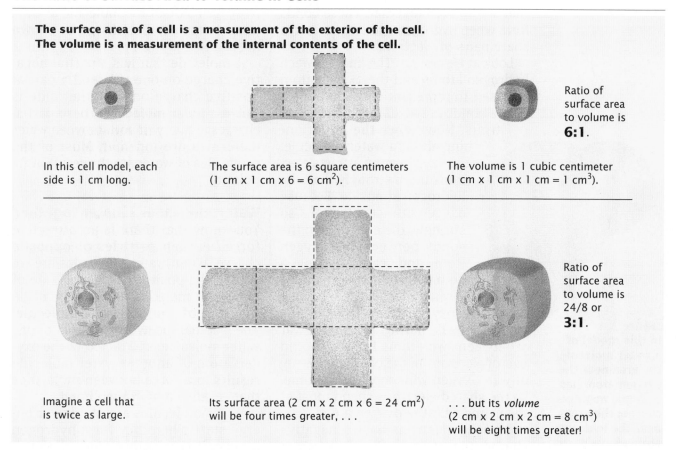

In this cell model, each side is 1 cm long.

The surface area is 6 square centimeters (1 cm x 1 cm x 6 = 6 cm²).

The volume is 1 cubic centimeter (1 cm x 1 cm x 1 cm = 1 cm³).

Ratio of surface area to volume is **6:1**.

Imagine a cell that is twice as large.

Its surface area (2 cm x 2 cm x 6 = 24 cm²) will be four times greater, . . .

. . . but its *volume* (2 cm x 2 cm x 2 cm = 8 cm³) will be eight times greater!

Ratio of surface area to volume is 24/8 or **3:1**.

Figure 3.3
As a cell gets larger, its volume increases at a faster rate than its surface area. A cell's surface area must be large enough to meet the needs of its volume.

A cell's volume increases faster than its surface area

As a cell grows, it takes in more food and creates more wastes. Since these substances must pass into and out of the "gates" in the cell membrane, the membrane must be large enough to service the cell's needs. As the cell grows, so does its membrane. But cells cannot grow indefinitely. So what limits cell size?

One factor is the relationship between the surface area and the volume of the cell. As a cell grows, its volume increases at a much faster rate than its surface area. A small cell, such as the one shown in the top row of **Figure 3.3**, has enough surface area to meet its needs. But a cell as large as the one shown in the bottom row might not. The ratio of a cell's surface area to its volume ultimately limits how large that cell can become. Cells cannot grow so large that their surface areas become too small to take in enough food and to remove enough wastes.

Water and the Cell

All cells are surrounded by water. Single-celled creatures swim in small ponds and also in vast oceans. Even the cells of your body, such as blood cells or skin cells, are surrounded by a thin film of water. Water is present inside the cell too. All the complex machinery inside the cell performs its functions in water. The cell membrane is shaped by the water found inside and outside of the cell. To understand how water can shape a cell membrane, you must first look closely at the structure of a water molecule.

Water is a polar molecule
The chemical formula for water is H_2O. A water molecule is made of two

hydrogen atoms and one oxygen atom that are bonded together. These bonds form when hydrogen and oxygen atoms share pairs of electrons.

Look at **Figure 3.4**. The lines between hydrogen atoms and the oxygen atom are used to represent covalent bonds. These bonds are actually pairs of electrons. Now here's the important thing about a water molecule: the oxygen atom attracts electrons more strongly than the hydrogen atoms do. Because oxygen attracts electrons so strongly, the electrons in the bonds between the oxygen atom and each hydrogen atom are not shared equally. They are more likely to be near the oxygen atom. Think of this unequal sharing as a tug of war between two atoms for the shared pair of electrons in the bond. In this tug of war, oxygen wins most of the time, so the shared electrons spend most of their time near the oxygen atom.

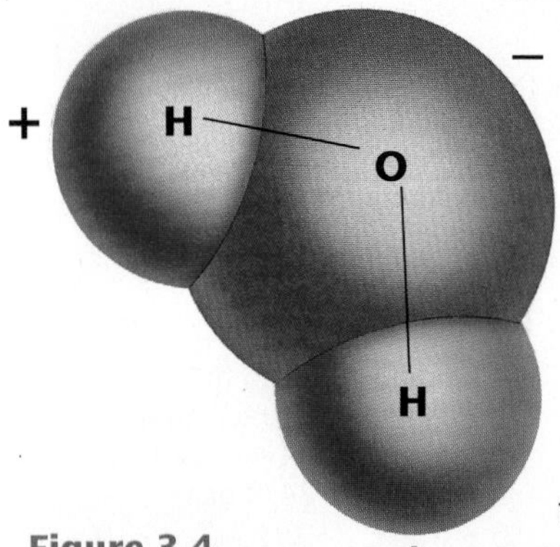

Because electrons have a negative charge, the oxygen part of the water molecule is slightly negative. The hydrogen atoms in the water molecule have slightly positive charges because electrons rarely spend their time near the positive hydrogen nuclei. We think of the oxygen side of the molecule as having a partial negative charge and the hydrogen side of the molecule as having a partial positive charge, as shown in **Figure 3.4**.

A molecule that has a partial negative charge on one side and a partial positive charge on the other side is called a **polar molecule**. These partial charges are very important when water molecules are together. Most of the properties of water are the result of its polarity.

Water molecules cluster together

You know that there is an attractive force between particles of opposite charge. When water molecules are together, the positively charged side of one water molecule attracts the negative side of another water molecule. The fact that the hydrogen atom of one water molecule is attracted to the oxygen atom of another water molecule results in a force between molecules that is called a **hydrogen bond**.

Water molecules are at a lower energy state when they form hydrogen bonds with each other. Since all things tend toward lower energy, there is a natural tendency for water molecules to form hydrogen bonds. When forming hydrogen bonds, the water molecules will cluster together as illustrated in **Figure 3.5**.

Figure 3.4
In this model of a water molecule, the area near the oxygen atom has a partial negative charge; the areas near the hydrogen atoms have partial positive charges.

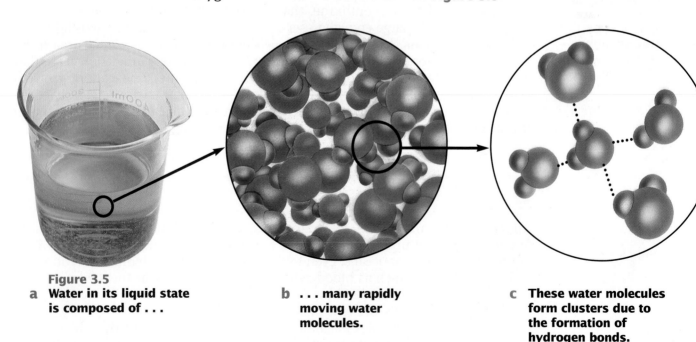

Figure 3.5
a **Water in its liquid state is composed of . . .**

b **. . . many rapidly moving water molecules.**

c **These water molecules form clusters due to the formation of hydrogen bonds.**

Water and the Cell Membrane

Now that you know something about the nature of water molecules, you can look more closely at how water shapes the cell membrane. The basic plan of the cell membrane begins with a sheet of lipids. You learned in Chapter 2 that lipids are molecules such as fats and oils. The interaction between water and lipids is what shapes the cell membrane.

Figure 3.6 shows what happens when oil is poured into a beaker of water and thoroughly mixed. Soon after mixing, small beads of oil form. Eventually the water and oil separate into two distinct layers. The water and oil won't stay mixed. Why? Water molecules start to cluster together because they form hydrogen bonds with each other. Unlike water, which is polar, lipids are **nonpolar molecules** because they have no negative and positive poles. Polar and nonpolar substances like oil and water will separate after being mixed. The oil, which is a nonpolar lipid, is not attracted to the water. Because the water molecules attract one another, the oil is pushed away. In this way, lipids only interact with each other. You can see the lipid and water layers in **Figure 3.7**.

Figure 3.6

a **If you pour oil into a beaker of water . . .**

b **. . . and stir it thoroughly, . . .**

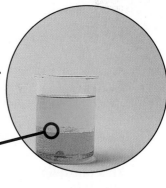

c **. . . the oil and water will separate because water is polar and oil is nonpolar.**

Figure 3.7
The force of water molecules pushing a sheet of lipids out of the way is the force that shapes the cell membrane.

Section Review

❶ **What are the advantages of being composed of many small cells instead of a few large ones?**

❷ **Describe how a cell might be affected if it doubled in size.**

❸ **How does a polar molecule differ from a nonpolar molecule?**

❹ **Compare the interaction of water molecules with each other to that of water molecules with lipid molecules.**

Cells **49**

YOU CANNOT JUDGE A BOOK BY ITS COVER, BUT A CELL IS SHAPED LARGELY BY THE KINDS OF PROTEINS IN ITS CELL MEMBRANE. SCIENTISTS STUDYING THE CELLS OF CYSTIC FIBROSIS PATIENTS HAVE FOUND A SINGLE DEFECTIVE PROTEIN IN THEIR CELL MEMBRANES THAT IS RESPONSIBLE FOR THE DISEASE. IF SCIENTISTS CAN CORRECT THE PROTEIN, THEY CAN CURE THE DISEASE.

3.2 Membrane Architecture

Objectives

❶ **Describe how phospholipids are organized to form a fluid cell membrane.**

❷ **Describe the functions of proteins in the cell membrane.**

❸ **Explain how cell membrane proteins interact with the lipid bilayer.**

Structure of the Lipid Bilayer

A cell membrane's framework consists of molecules called **phospholipids**. A phospholipid, as shown in **Figure 3.8a**, is a lipid in which the long chains or "tails" are joined to a short polar "head" that contains phosphorus. The long tails are nonpolar, so water molecules will tend to push them away. The heads are attracted to water because they form hydrogen bonds with the water molecules.

Water can interact with the polar heads and repel the nonpolar lipid tails best if the phospholipids are aligned into two layers. The polar heads of the phospholipids point toward the water that is inside and outside of the cell, and the tails point inward toward each other, as illustrated in **Figure 3.8b**. This double layer of phospholipids forms a tough yet flexible **lipid bilayer**, which is shown in **Figure 3.8c**.

Figure 3.8

a **This phospholipid molecule . . .**

b **. . . is part of a lipid bilayer.**

c **The lipid bilayer forms the framework of the cell membrane.**

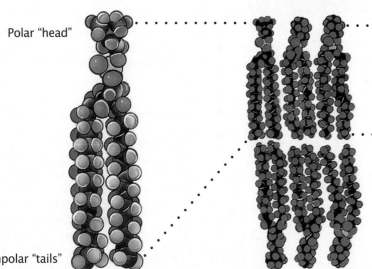

Polar "head"

Nonpolar "tails"

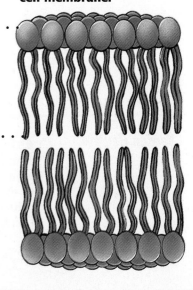

Characteristics and Functions of the Lipid Bilayer

Lipid bilayers stop polar molecules

The lipid bilayers that are found in cell membranes have two important characteristics. One important characteristic of the lipid bilayer is that most polar molecules cannot go across it. Polar molecules are attracted to the water inside the cell or to the water outside the cell. However, they cannot interact with the nonpolar tails of the phospholipids within the lipid bilayer. The result is that the interior part of the cell membrane forms a nonpolar zone that acts as a barrier to polar molecules. But most food molecules and other substances needed by the cell are polar. So the cell must have a way of allowing these molecules to cross the barrier. If the cell membrane were made only of a lipid bilayer, there would be no way for food and other things to pass in and out of the cell.

The solution is to have passageways through the barrier. The cell membrane has passageways made of proteins, such as those shown in **Figure 3.8d**. By making "gates" that can open and shut in the lipid bilayer, proteins enable the passageways to regulate precisely the substances that go into and out of cells. You will see later that cell membrane proteins have other roles, too.

Lipid bilayers are fluid

A second important characteristic of lipid bilayers is that their phospholipid and protein molecules are not rigidly fixed in place. These molecules move about like rubber life preservers floating on the surface of a swimming pool. The lipid bilayer is fluid.

Because they are not fixed in place, the phospholipid and protein molecules that make up the cell membrane can shift from one region of the cell membrane to another. This is important because cell membranes can be structured to fit the needs of different cell types.

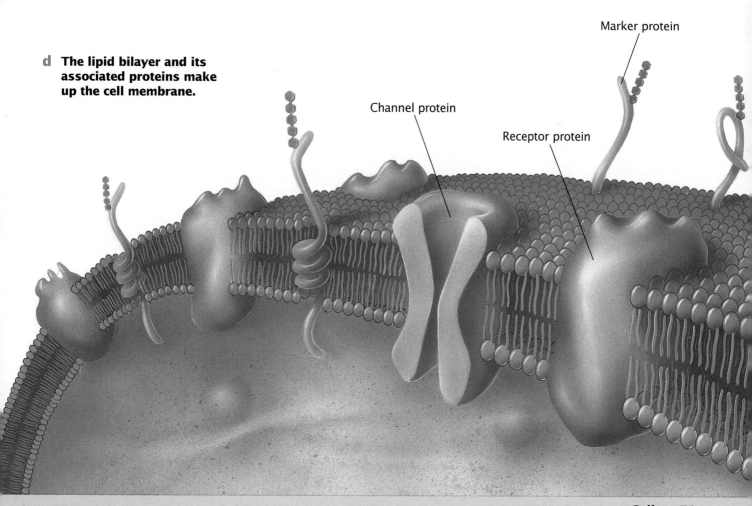

d **The lipid bilayer and its associated proteins make up the cell membrane.**

Marker protein

Channel protein

Receptor protein

Roles of Cell Membrane Proteins

If you were able to peer at the surface of a cell, it would look like a smooth sea of phospholipids interrupted by proteins sticking out from the surface, some like boulders, others like tall trees. Proteins that protrude from the cell membrane may serve as channels, receptors, or markers.

Channels allow some molecules to pass through the membrane

Proteins arranged into a shape like a doughnut form channels through the cell membrane. See the channel model in **Figure 3.9**. Many molecules and ions that are needed by the cell cross the membrane through these passageways. However, these channels are like locked doors—only people with a key can enter. In the same way, each channel will admit only certain molecules.

Receptors transfer information across the membrane

Receptor proteins in the cell membrane are shaped like boulders, as shown in **Figure 3.10**. Receptors convey information from the world outside the cell to the inside of the cell. The end of the receptor that sticks out from the cell surface has a special shape that holds only a particular type of molecule. When a molecule of the right shape fits into the receptor, its presence causes a change at the other end of the receptor. This change, in turn, triggers responses inside the cell.

Markers aid cell identification

Cell surface markers are elongated proteins that often have carbohydrates on their surface, as shown in **Figure 3.11**. These protein markers are the "name tags" of the cell. Every cell of your body has markers on its surface saying that it is part of you and not of some other individual. These markers organized the first tissues of your body before you were born. As new tissues and organs formed, cell surface markers told your cells where to go and with which other cells to join. You will learn more in Chapter 4 about how cell surface markers help the cell to interact with its environment.

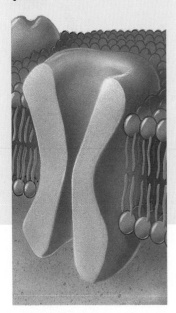

Figure 3.9
Channels
These proteins act like passageways through which only certain molecules pass.

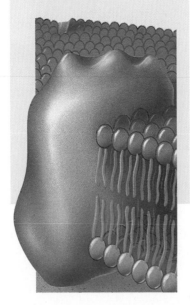

Figure 3.10
Receptors
These proteins transmit information into the cell by reacting to certain other molecules.

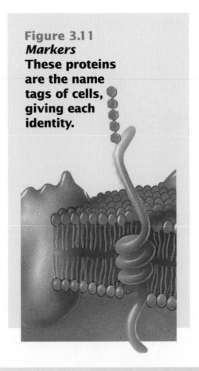

Figure 3.11
Markers
These proteins are the name tags of cells, giving each identity.

Proteins: A Limitless Variety

You remember from Chapter 2 that proteins are molecules made of subunits called amino acids. Proteins in the cell membrane can serve as channels, receptors, or markers. How can proteins play such different roles? And how do proteins, which can be polar, fit into the inner nonpolar region of the membrane bilayer? To answer these questions, take a closer look at the structure of protein.

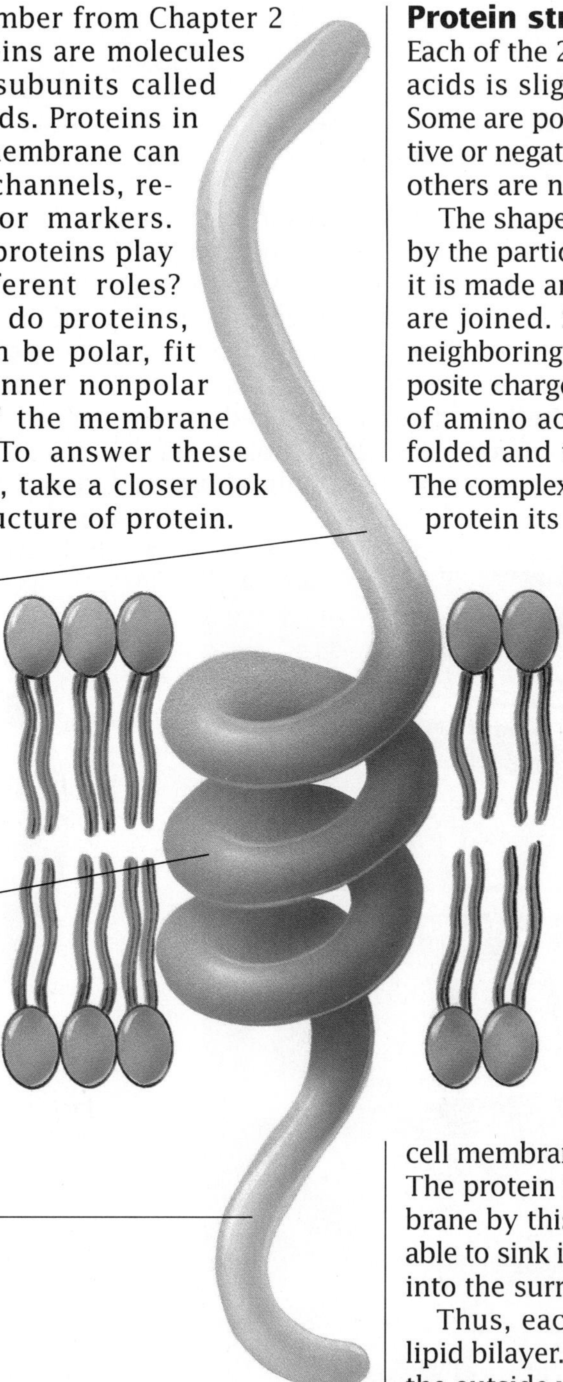

Protein structure is variable

Each of the 20 different kinds of amino acids is slightly different chemically. Some are polar; that is, they have positive or negative electrical charges, while others are nonpolar.

The shape of a protein is determined by the particular amino acids of which it is made and the order in which they are joined. Some amino acids attract neighboring amino acids that have opposite charges. As a result, the long line of amino acids in a protein becomes folded and twisted as in **Figure 3.12**. The complexity of the shape gives each protein its unique function.

Some proteins fit into a cell membrane

A protein that fits into the cell membrane has three sections. The two end sections, **Figure 3.12a** and **Figure 3.12c**, contain many polar amino acids that form hydrogen bonds with water. However, the middle section, **Figure 3.12b**, is made of many nonpolar amino acids. This nonpolar coil fits into the nonpolar interior of the lipid bilayer, allowing the protein to float in the cell membrane like a ship on the ocean. The protein is anchored into the membrane by this nonpolar region. It is unable to sink inside the cell or to float off into the surrounding water.

Thus, each cell is a prisoner of its lipid bilayer. A cell communicates with the outside world by means of the proteins embedded in its lipid shell.

Figure 3.12

a The amino acids in this region of the protein are mostly polar. Therefore, this polar region is compatible with the water outside the cell.

b The amino acids in this region of the protein are mostly nonpolar. Therefore, this region is compatible with the nonpolar area in the center of the lipid bilayer.

c This region is also polar and is compatible with the water inside of the cell.

Section Review

❶ How do phospholipids interact with water to form lipid bilayers?

❷ Why are lipid bilayers said to be fluid?

❸ Explain three roles proteins play in cell membranes.

❹ Explain how the structure of a protein enables it to fit in the cell membrane.

Cells   **53**

Jack Wang: Protein Chemist

Why I Became a Protein Chemist

"I came to the United States from China when I was six years old. Although I don't recall much about moving to this country, I do remember that my parents expected me to excel in school. My parents were not scientists, but they were well educated and raised me to respect people who are scholars. As a result, I grew up with a natural desire for learning.

"In my sophomore year of high school, I took biology and learned about genetics. The teacher talked about the inheritance of traits. I enjoyed learning how features such as hair and eye color are passed on from generation to generation. When the time came for college, I already knew that I wanted to learn more about biology and how nature worked."

Jack spends his spare time with his wife and children. His two sons are in Little League. "We love baseball. During the Little League season, I coach the teams my sons play on. We also closely follow our favorite major league team—the New York Mets."

Name:	**Jack Wang**
Home:	**Lexington, Massachusetts**
Employer:	**Genetics Institute, Cambridge, MA**
Personal Traits:	• **Patience**
	• **Honesty**
	• **Perseverance**
	• **Detail Oriented**
	• **Friendliness**
	• **Inquisitiveness**

The Fascination of Protein Chemistry

Career Path

High School:
- Biology
- Chemistry
- Physics

College:
- Cell Biology
- Organic Chemistry
- Genetics

Graduate School:
- Biochemistry

"I purify and characterize proteins. Purification involves passing a sample through a set of columns. This begins separating proteins found within the sample. Every protein has its own characteristics—different masses, shapes, and charges. By putting a protein mixture through a column with a particular matrix that selects for certain properties, you can separate the proteins. Some of the proteins will bind to the column more than others. You actually use a wide variety of columns, each with its own ability to select proteins. By running a mixture through several columns in a row, you end up eventually purifying the single protein you're after."

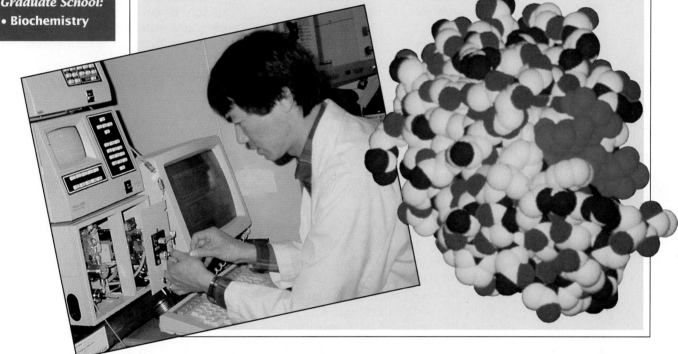

Research Focus

Dr. Wang works in the Protein Chemistry Department at the Genetics Institute in Cambridge, Massachusetts. He has worked on two projects. The first project developed a synthetic form of a protein called factor VIII. Factor VIII is found in the blood and is required for normal blood clotting. When this protein is missing from a person's blood, a severe bleeding disorder called hemophilia A results. People with this type of hemophilia often can receive the missing protein through concentrated blood products. Unfortunately, for years the blood supply was relatively impure and some hemophiliacs became infected with forms of hepatitis and the human immunodeficiency virus. Now, with the synthetic form of factor VIII that Dr. Wang helped develop, many hemophiliacs are safe from such dangers.

Dr. Wang's second project involves researching a human growth factor that stimulates the production of bone. "This protein is actually quite revolutionary," he says. "It's especially useful for someone who has a badly broken bone. The growth factor is undergoing clinical trials and probably won't be on the market for another three or four years."

WHEN YOU HAVE A CASE OF STREP THROAT, YOUR CELLS ARE UNDER ATTACK BY BACTERIA. BACTERIA ARE SINGLE-CELLED ORGANISMS. BACTERIAL CELLS ARE MUCH LESS COMPLICATED THAN YOUR CELLS. THEY ARE ALSO THE ANCESTORS OF YOUR CELLS. OVER THE COURSE OF TIME, BACTERIAL CELLS EVOLVED. THE RESULT WAS THE KINDS OF CELLS THAT MAKE UP YOUR BODY.

3.3 Inside the Cell

Objectives

❶ **Name two characteristics that distinguish eukaryotes from prokaryotes.**

❷ **Relate each organelle to a task essential to the life of the cell.**

❸ **Contrast the structure of an animal cell with that of a plant cell.**

❹ **Describe how eukaryotes evolved from prokaryotes.**

Two Types of Cells

All cells can be divided into two large categories. A eukaryotic cell, or **eukaryote** (*yoo KAR ee oht*), is a large, complex cell that contains a membrane-bound compartment called a **nucleus**. The nucleus contains DNA within chromosomes. A prokaryotic cell, or **prokaryote** (*pro KAR ee oht*), is a very small, simple cell that lacks a nucleus. Its DNA is a single, circular molecule that is not enclosed in a membrane-bound compartment. Fossils of the first cells on Earth, which existed about 3.5 billion years ago, reveal that these ancient cells were prokaryotes. Today, the only living prokaryotes are the bacteria. Compare the bacterium to the eukaryote in **Table 3.1**.

Eukaryotes and prokaryotes share several characteristics. For example, both kinds of cells have a cell membrane. They also contain **cytoplasm** (*SYT uh plaz uhm*). Cytoplasm is everything inside the cell membrane except the nucleus in eukaryotes, and the DNA in prokaryotes. These cells also contain **ribosomes** (*RY buh sohmz*), structures on which proteins are made.

Table 3.1

Prokaryote	Eukaryote
No nucleus	Nucleus
No membrane-bound organelles	Many organelles
1–10 µm in size	2–1,000 µm in size
Evolved 3.5 billion years ago	Evolved 1.5 billion years ago
Only bacteria	All other cells

This bacterium (13,600X) is a prokaryotic cell.

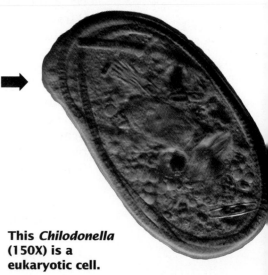

This *Chilodonella* (150X) is a eukaryotic cell.

Eukaryotic Cells Have Compartments

The nucleus is not the only compartment inside eukaryotic cells. The cytoplasm of eukaryotes contains many specialized parts called **organelles**, as shown in **Figure 3.13**. Each organelle is a specialized compartment that carries out a specific function. Examples of organelles include the nucleus and ribosomes. Many organelles, such as the nucleus, are membrane-bound.

Organelles isolate cell activities

A eukaryotic cell is like a building with many rooms. Imagine that you had to attend a class in a building where the gymnasium, the cafeteria, the library, and the boiler room were all located within the same room. It sounds confusing. Having separate compartments allows special activities to be restricted to particular places. This type of organization has many advantages over that of prokaryotic cells.

Figure 3.13
This is a diagram of a cell from a human liver.

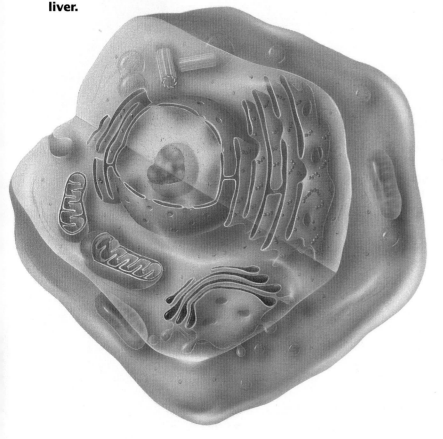

Organelles contribute to the specialization of eukaryotic cells

The organelles of a living eukaryotic cell are constantly growing, moving, reproducing, appearing, and disappearing. Multiple arrangements of these active, minute structures enable eukaryotic cells to specialize—that is, to perform special functions.

For example, a muscle cell in your leg, which must contract quickly, contains a greater number of energy-producing organelles than a bone cell does. Likewise, the striped, pulsating cells of heart muscle are far different from the boxlike cells that make up the growing layers of the skin on your hand. And skin cells differ greatly from the nerve cells that send electrical signals from your brain throughout your body. Other examples of specialized eukaryotic cells include gland cells that secrete hormones, and sperm cells that are capable of movement.

Keep in mind that all the cells of plants and animals originate from a single fertilized egg cell. As these cells divide and form new cells, they change in slightly different ways. At maturity, they become specialized. Single-celled eukaryotes evolved into complex, multicellular organisms because they had the ability to become specialized. These specialized cells make eukaryotes capable of survival in complex environments.

"Any living cell carries with it the experience of a billion years of experimentation by its ancestors."

Max Delbrück
Cell Biologist
Nobel Prize Winner

Cells Perform Basic Functions of Life

As you learned in Chapter 2, one of the five characteristics of living things is that they are composed of one or more cells. Like all living things, cells like the human liver cell in **Figure 3.14** display other properties shared by all organisms. Cells use energy. Cells maintain homeostasis. And cells reproduce.

You are already familiar with some of the ways in which cells exhibit the last two properties listed above. For example, you learned earlier that the cell membrane helps to maintain a constant internal environment. You also learned that DNA occurs in prokaryotic cells, and in eukaryotic cells where it is found in the nucleus.

Like the cell's nucleus, the other organelles in a eukaryotic cell are membrane-bound, specialized compartments whose contents are separated from the cytoplasm. Isolated from the hustle and bustle of the rest of the cell, each organelle of the cell stands ready to perform a specific task. The coordinated activities of the different organelles make it possible for cells to go about the business of living.

Cells reproduce

When a growing cell approaches a particular size, certain organelles, such as the nucleus shown in **Figure 3.15**, help it prepare to divide. Cell division enables single-celled organisms to reproduce by dividing in two. It also enables multicellular organisms to grow. In Chapter 4, you will learn how cells divide.

Cells manufacture and release energy

Two kinds of organelles act as cellular powerhouses, playing essential roles in food manufacture and energy release. **Mitochondria** (*myt uh KAHN dree uh*) are found in all eukaryotic cells where they release the energy stored in food. A mitochondrion is shown in **Figure 3.16**. Your cells carry from 10 to several hundred of these mitochondria, each a tiny chemical factory breaking down food molecules to produce energy.

Chloroplasts, on the other hand, are organelles that have the amazing ability to make food in the form of sugars using air, water, and the energy from sunlight. This process is called photosynthesis. You will learn more about photosynthesis in Chapter 5. Chloroplasts are found only in algae, such as seaweed, and in green plants.

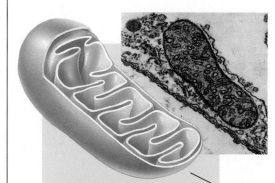

Figure 3.16
Mitochondrion
Without mitochondria your cells would be unable to produce the energy on which all of your activities depend.

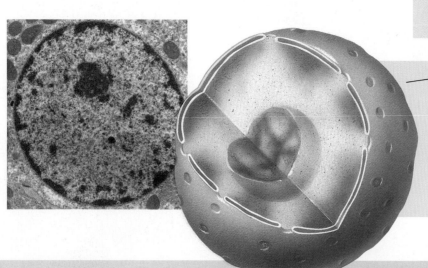

Figure 3.15
Nucleus
The DNA inside the nucleus directs the cell's growth and development.

Cells maintain homeostasis

As you learned in Chapter 2, every living thing, no matter how big or how small, must be able to maintain a constant internal environment. Whether it is a one-celled organism or part of a larger living thing, a cell must be able to monitor and adjust internal conditions. You learned earlier in this chapter how the cell membrane helps to maintain homeostasis by controlling what can enter or exit a cell.

Some organelles help cells to maintain homeostasis by moving supplies from one part of a eukaryotic cell to another. In the small cell of a bacterium, a molecule can go from one place to another fairly quickly. But in large eukaryotic cells with intricately arranged membranes, molecular traffic needs to be directed more precisely. To accomplish this, eukaryotic cells have a system of membranes called the **endoplasmic reticulum** (*ehn duh PLAZ mihk rih TIHK yuh luhm*), or ER, as shown in **Figure 3.17**. In addition to transporting proteins, other parts of the ER help make lipids the cell needs.

The proteins and the lipids made by the ER are then transported to another organelle called the **Golgi** (*GOHL jee*) **body**. The Golgi body puts the finishing touches on these molecules and then releases them in membrane-wrapped bubbles called vesicles. Substances packaged in this way can be sent to particular places in the cell. Some types of vesicles fuse with the cell membrane, releasing their contents outside the cell. An illustration and photograph of a Golgi body are shown in **Figure 3.18**.

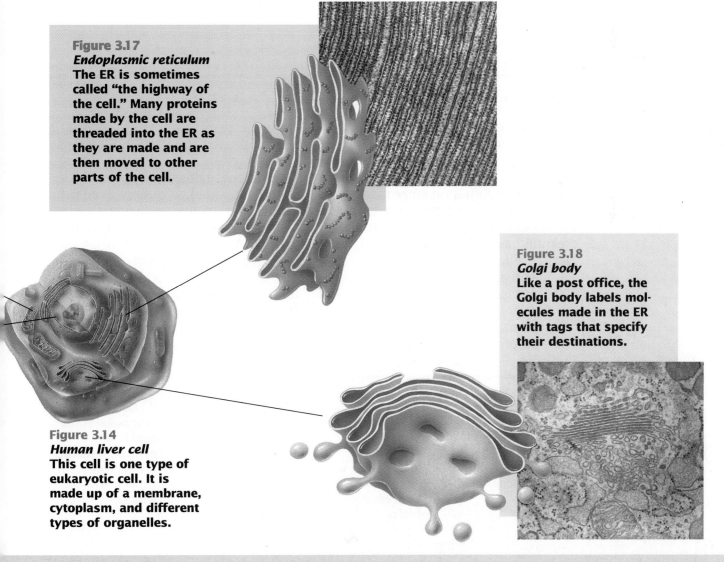

Figure 3.17
Endoplasmic reticulum
The ER is sometimes called "the highway of the cell." Many proteins made by the cell are threaded into the ER as they are made and are then moved to other parts of the cell.

Figure 3.18
Golgi body
Like a post office, the Golgi body labels molecules made in the ER with tags that specify their destinations.

Figure 3.14
Human liver cell
This cell is one type of eukaryotic cell. It is made up of a membrane, cytoplasm, and different types of organelles.

Kinds of Eukaryotic Cells

Figure 3.19
This illustration of a plant cell from timothy grass contains a cell membrane, a large central vacuole, chloroplasts that produce sugar, and a strong cell wall.

Although all eukaryotic cells contain nuclei and organelles, they differ from one another with regard to the presence or absence of certain structures and organelles. Some of the most dramatic differences, for example, are those that distinguish animal cells from those of plants.

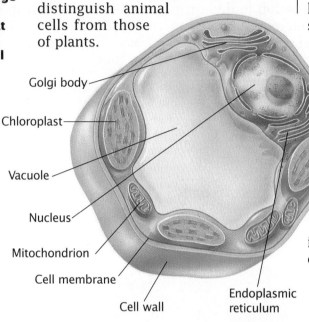

Golgi body

Chloroplast

Vacuole

Nucleus

Mitochondrion

Cell membrane

Cell wall

Endoplasmic reticulum

Plant cells have cell walls and chloroplasts

Plant cells possess a **cell wall** in addition to a cell membrane. As **Figure 3.19** shows, the cell wall lies outside the cell membrane. Plant cell walls contain cellulose, a polysaccharide that gives strength and rigidity to the plant cell. The cells of algae, fungi, and some bacteria also have cell walls. Plant cells store waste products, nutrients, and water in large, centrally located membrane-bound spaces called **vacuoles**. The pressure exerted by the stored water enables the plant to stand upright. When its vacuoles lack water, a plant will become limp.

Chloroplasts are also found in plant cells but not in animal cells. During photosynthesis, chloroplasts use energy of light to make sugars.

How Did Eukaryotes Evolve?

As biologists studied eukaryotes they were surprised to see that many organelles resembled bacteria. Some organelles have a double set of membranes. The interior membrane of such an organelle is like the cell membrane of a bacterium. It looks as if a bacterium has been engulfed by a much larger cell.

Scientists hypothesize that bacterial "trespassers" remained inside cells, gradually losing their ability to live independently. These invading bacteria became organelles, and eukaryotic cells were the result. The first eukaryotic cells appeared about 1.5 billion years ago.

Additional evidence supports the hypothesis that eukaryotes evolved from prokaryotes. Some eukaryotic organelles such as mitochondria have their own DNA, which suggests that they once lived as independent cells. Eukaryotic organelles sometimes also have their own ribosomes, which are very similar to the ribosomes of bacteria. Also, some organelles divide in a manner similar to that of bacteria.

Section Review

❶ **What characteristics would help you distinguish a bacterium from a *Chilodonella*?**

❷ **Name five organelles found in cells and describe how each enables the cell to display the properties of life.**

❸ **In what ways does a plant cell differ from an animal cell?**

❹ **Do you think that eukaryotes could have evolved without prokaryotes? Explain why or why not.**

This child, who is 19 months old, has grown from a single cell to nearly 100 trillion cells.

Key Terms

Summary

3.1 World of the Cell

The cell, like this *Paramecium,* is the fundamental unit of life.

cell (p. 45)

cell membrane (p. 45)

polar molecule (p. 48)

hydrogen bond (p. 48)

nonpolar molecule (p. 49)

- A cell is the smallest unit of life.
- Small cells can take in food and communicate with each other more efficiently than large cells.
- Water molecules attract other water molecules and shape the cell membrane by repelling lipid molecules.

3.2 Membrane Architecture

A cell membrane is made of a lipid bilayer, its associated proteins, and other molecules. The membrane is the interface between a cell and its environment.

phospholipid (p. 50)

lipid bilayer (p. 50)

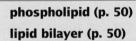

- The cell membrane is a tough, flexible lipid bilayer made of phospholipids, protein and other molecules.
- Proteins in the cell membrane permit certain substances to go into and out of cells.

3.3 Inside the Cell

The eukaryotic cell is composed of cytoplasm and compartments called organelles. Compartments isolate cell activities.

eukaryote (p. 56)

nucleus (p. 56)

prokaryote (p. 56)

cytoplasm (p. 56)

ribosome (p. 56)

organelle (p. 57)

mitochondrion (p. 58)

chloroplast (p. 58)

endoplasmic reticulum (p. 59)

Golgi body (p. 59)

cell wall (p. 60)

vacuole (p. 60)

- All cells have DNA and cytoplasm.
- A eukaryotic cell has a nucleus and other membrane-bound organelles. A prokaryotic cell does not.
- Organelles are compartments that enable a cell to function by making and releasing energy, helping the cell to maintain homeostasis, and enabling the cell to reproduce.
- Unlike animal cells, plant cells have cell walls and chloroplasts.

Understanding Vocabulary

1. For each pair of terms, explain the differences in their meanings.
   a. cell wall, cell membrane
   b. polar molecule, nonpolar molecule
   c. prokaryote, eukaryote
   d. chloroplast, mitochondrion
   e. cytoplasm, organelle

Relating Concepts

2. Copy the unfinished concept map below onto a sheet of paper. Then complete the concept map by writing the correct word or phrase in each oval containing a question mark.

Understanding Concepts

Multiple Choice

3. The flexible lipid outer surface of a cell is called the
   a. cell membrane.   c. hull.
   b. phospholipid.   d. cytoplasm.

4. The growth of cells is limited by the ratio of
   a. surface area to volume.
   b. organelles to surface area.
   c. organelles to cytoplasm.
   d. nucleus to cytoplasm.

5. A cell membrane is composed of
   a. phospholipids.
   b. water molecules.
   c. proteins.
   d. phospholipids and proteins.

6. Which of the following does not enable the cell to communicate with its environment?
   a. channels   c. receptors
   b. phospholipids   d. markers

7. Besides the cell membrane, two components of all cells are
   a. cytoplasm, cell wall.
   b. DNA, cytoplasm.
   c. organelles, nucleus.
   d. DNA, mitochondria.

8. A part of the cell that functions to maintain homeostasis is the
   a. cytoplasm.   c. nucleus.
   b. cell wall.   d. cell membrane.

9. Single-celled organisms that do not have nuclei are called
   a. prokaryotes.   c. organelles.
   b. eukaryotes.   d. mitochondria.

10. Which of the following releases energy from nutrients?
    a. mitochondrion   c. Golgi body
    b. endoplasmic reticulum   d. protein channel

11. Lipids and proteins are transported through the cell by the
    a. nucleus.
    b. chloroplast.
    c. mitochondrion.
    d. endoplasmic reticulum.

Completion

12. Water molecules attract each other because they are _____ . Lipids are not attracted to water because they are _____ .

13. The double layer of phospholipids surrounding the cytoplasm is called a(n) _____ _____ .

14. Structures that perform specialized functions within the cell are called _____ .

15. Plant cells differ from animal cells in that they possess a(n) _____ , which is a thick layer of cellulose surrounding the cell membrane.

Short Answer

16. Explain how water shapes the cell membrane.

17. What advantage do small cells have over large cells?

18. List two kinds of evidence that suggest eukaryotes evolved from prokaryotes.

19. What are the names and functions of three cell membrane proteins?

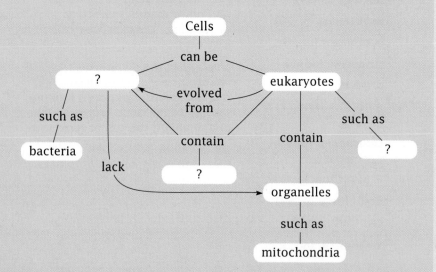

Interpreting Graphics

20. Look at the figure below (right) to answer the following questions. Identify the polar heads and the nonpolar tails in the lipid bilayer. How many layers make up the cell membrane? Why have the layers aligned themselves in the manner shown? Describe the pathway that molecules and ions would take to get inside the cell.

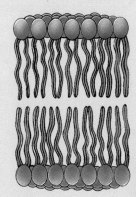

21. Compare the structure of each protein in the figure below to its function. Explain how the structure of each protein enables it to accomplish its specific task.

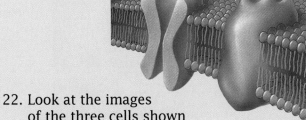

22. Look at the images of the three cells shown below. For each cell, place an X under each term that appropriately describes it.

Prokaryote Eukaryote Animal Plant

Reviewing Themes

23. *Stability*
    How does a cell membrane help a cell maintain homeostasis?

24. *Energy*
    How do chloroplasts and mitochondria function in the capture and release of energy within the cell?

Thinking Critically

25. *Inferring Conclusions*
    How would a cell be affected if the cell membrane were completely solid and watertight?

26. *Comparing and Contrasting*
    How does a bacterium swimming around in a pond differ from a cell in your body? How is it similar?

27. *Inferring Conclusions*
    To solve a hit-and-run case, police want to know if a microscopic sample of material scraped from a car bumper is animal or plant matter. What evidence should the police look for in the sample in order to eliminate the driver of the car as a suspect?

28. *Building on What You Have Learned*
    You learned in Chapter 2 that movement is a characteristic exhibited by most animals. What features of animal cells make movement possible?

Cross-Discipline Connection

29. *Biology and Art*
    Before photographs were commonly used in biology, drawings of cells and organisms were the most accurate way to share information. Describe the benefits and disadvantages of using art instead of a photograph.

Discovering Through Reading

30. Read the article "Cell Videos Catch Asbestos in the Act," in *Science News*, September 21, 1991, page 180. How do asbestos fibers get into cells? How do asbestos fibers cause cancer?

Investigation

How Do Plant and Animal Cells Differ?

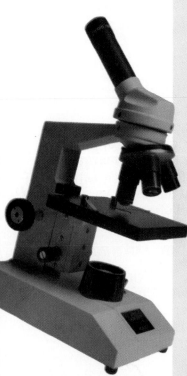

Objectives

In this investigation you will:
- *observe* cell structures
- *compare* and *contrast* animal and plant cells
- *relate* the structure of a cell to its function

Materials

- lens paper
- compound light microscope
- *Elodea* plant
- forceps
- glass slides
- water
- coverslips
- medicine dropper
- prepared slide of human cheek cells
- paper towels

Prelab Preparation

1. Review what you have learned about cells by answering the following questions:
   - How does a cell wall differ from a cell membrane?
   - How is plant cell structure related to the ability of plants to make food from sunlight?
   - What organelles are found in both plant and animal cells?
2. Review the procedures for proper use of the microscope in the Appendix. Summarize the procedure for observing a specimen first at low power and then at high power.
3. Review the procedures in the Appendix for making a wet mount slide. Summarize the steps.

Procedure: Comparing Cells of Plants and Animals

1. Use lens paper to clean the microscope lenses.
2. Make a wet mount slide of a young *Elodea* leaf taken from the growing tip of the plant.

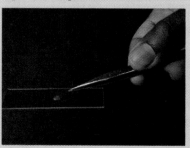

3. Observe your specimen under low power. *What structures do you recognize?*

4. Observe the specimen using high power. *What structures do you see that were not visible under low power?*
5. Draw your specimen as it appears at each magnification. Label the cell structures and note the magnification represented by each drawing.

6. Write a brief description of your observations next to each drawing.

7. Obtain a prepared slide of human cheek cells.

8. Repeat steps 3 through 6.

9. Make a table similar to the one shown.

10. Use the table to record evidence suggesting that the cell structures listed in the table are present in your specimens. Use the line labeled "Other" to record observations of structures not listed in the table.

	Elodea leaf	Cheek cells
Cell membrane		
Cell wall		
Cytoplasm		
Nucleus		
Chloroplast		
Other		

11. Use the information in your table to write a summary that describes the differences between plant cells and animal cells.

12. Dispose of your *Elodea* specimen as directed by your teacher. Carefully clean and dry the slide and coverslip.

13. Return your laboratory materials to their proper places and wash your hands thoroughly before leaving the laboratory.

Analysis

1. *Comparing Observations*
Compare the sizes and shapes of animal and plant cells.

2. *Comparing Observations*
Compare the structures observed in animal and plant cells.

3. *Making Generalizations*
Based on your observations, describe a generalized plant cell and a generalized animal cell.

4. *Inferring Relationships*
What is the relationship between plant cell structure and the ability of plants to stand upright?

5. *Evaluating Methods*
Describe some problems that you had when preparing slides for viewing. How would you attempt to solve these problems the next time you make a slide?

Thinking Critically

1. Compare the organelles found in the plant and animal cells shown in **Figures 3.14** and **3.19**. What can you infer about the evolution of some of the organelles that are found both in plant cells and in animal cells?

2. For each of the organelles listed in your table, describe what function the cell would lose if the organelle were damaged or suddenly lost.

DO YOUR CELLS BELONG TO YOU?

CASE #: S0069817
Moore vs. the Regents
Moore vs. the Univ. of California

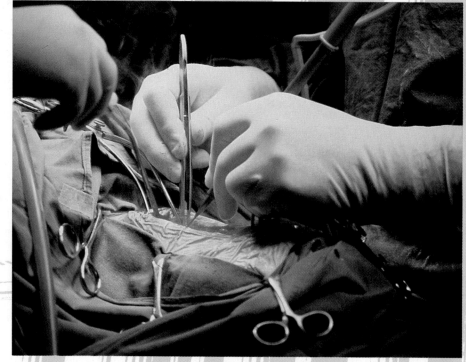

These doctors are performing spleen surgery.

Do you own your body? Are the cells of your arm, your liver, or your heart really yours? The answer might seem obvious: Yes! Of course! But that might not be so. Consider the highly controversial case of Moore v. the Regents of the University of California. The court's decision might startle you.

John Moore was thrilled with his cure. His leukemia had not come back since his spleen was removed. But now the doctors at the University of California wanted him to come back for some additional tests. Was something wrong?

Moore gave blood and skin samples to his doctor. The doctor said he should not be worried. Moore did not have leukemia; these were only follow-up tests.

But Moore soon found out that when the doctors who treated him realized that his cells were of a rare type, they applied for and were issued a patent on the cells they had removed from Moore. A patent allows the holder to own the product. The doctors now owned Moore's cells!

After the doctors had removed his spleen, they took the cells, worked with them, and sold their DNA and the DNA's products to a

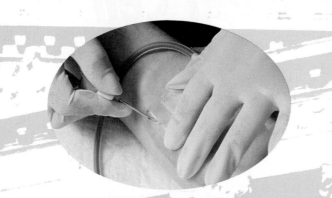

A nurse is taking a sample of blood.

biotechnology company. The company was going to market the cells and their products and make a tremendous profit. Didn't Moore have any right to the profits or at least a right to say, "Don't use my cells—I own them"?

The Case Goes to Court

John Moore took his case to court, and *Moore* v. *the Regents of the University of California* became one of the most publicized cases in medical history.

Moore's cells were special because their DNA produced a type of protein called lymphokine. Lymphokines are rare, and have been used to treat patients with cancer and blood diseases. One example of a lymphokine is interferon. Interferon is a chemical that blocks certain viruses from attacking cells within the body. If interferon is given to patients with cancer, the drug sometimes helps stop the spread of the

disease. Moore's lymphokine was much rarer than interferon and had the possibility of use as a treatment for other rare leukemias, which are forms of cancer. Because cells can be kept alive, growing, and also reproducing for years, doctors wanted to use Moore's cells for research and making money.

The Verdict Is In

After an extended hearing, the Supreme Court of California ruled in favor of the doctors. They said that once cells are removed from your body they are not considered yours anymore. They can be used by the doctors any way they like. The court said John Moore never expected to have his cells back when he gave permission to remove his spleen.

The court was afraid that scientific experimentation, a very necessary thing, would become more difficult if they didn't allow doctors to work with human cells freely. The work with human cells is extremely important because testing the effects of drugs and other treatments on living human cells must occur before they are used on human beings. If the scientists had to check into the ownership of the cells they use, it would take a great deal of time and study would be slowed.

Although the court did not allow John Moore to recover any money, they said he could go back to court and sue the doctors because they did not inform him that his cells were being used for this purpose. The court said Moore had a right to consent to this use.

Thinking Critically

❶ Do you agree with the court's decision not to allow John Moore the ownership of his cells? Why or why not?

❷ What do you think John Moore might have done if he had known beforehand that the doctors were going to experiment with and sell his cells' products?

❸ What arguments could John Moore's attorney have made to prove Moore's ownership of the cells?

❹ How much do you think the doctor should have told John Moore before his spleen was removed? How much after?

❺ Recently, a couple in California decided to have a baby so that the baby could donate bone marrow cells to their older child who suffered from leukemia. According to the Moore case, did those cells belong to the baby? Why or why not?

Acting on the Issue

❶ Find out what legislation may be under consideration in your state related to the sale of tissue.

❷ Find out if biotechnology companies now require consent forms before purchasing human tissue.

The Living Cell

Review

- structure of cell membrane (Section 3.2)

- characteristics of prokaryotes and eukaryotes (Section 3.3)

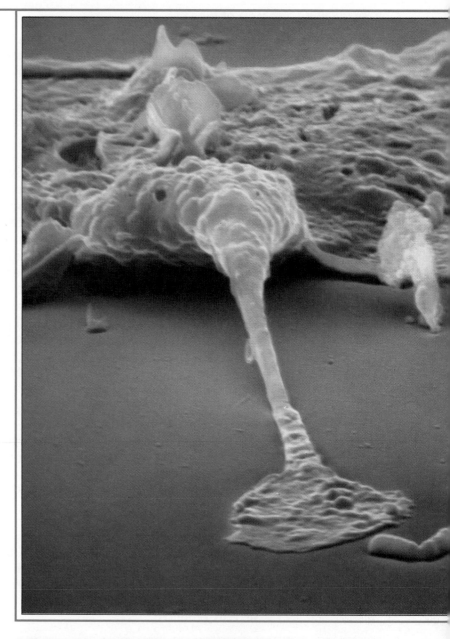

Cells are very much alive. A macrophage (top), a cell that attacks foreign intruders in the bloodstream, is about to engulf a harmful bacterium (lower right).

LIKE YOU, CELLS NEED INFORMATION ABOUT THEIR ENVIRONMENTS. IN THE SAME WAY THAT YOUR SENSES—SUCH AS SIGHT, HEARING, AND TOUCH—ENABLE YOU TO GATHER INFORMATION, PROTEINS IN THE CELL MEMBRANE CONNECT THE CELL TO ITS ENVIRONMENT. THOSE PROTEINS ARE THE CELL'S AVENUES OF COMMUNICATION WITH THE WORLD AROUND IT.

4.1 How Cells Receive Information

Objectives

❶ Explain how electrical currents affect voltage-sensitive channels.

❷ Describe how receptor proteins enable cells to sense chemical signals.

❸ List three different tasks performed by cell surface markers.

Sensing Electrical Signals

Figure 4.1
Actions of the human body are controlled by a network of nerves.

a **Nerves run throughout the body.**

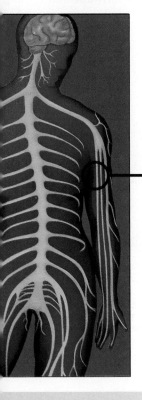

All living cells need to "keep in touch" with their environments. They do so in several ways, such as by sensing chemical signals or by marking their surfaces for identification. Some cells have protein channels that are sensitive to electricity. Because these channels respond to electrical signals, they are known as **voltage-sensitive channels**. Voltage-sensitive channels are important because they allow electrical signals to pass along nerves. Without them, your brain and nerves couldn't function, as you'll see in Chapter 29.

It is not difficult to understand how these channels work. They are like little doors with magnets on them. The doors usually remain closed. But they flip open or shut in response to electrical signals, as shown in **Figure 4.1c–d**. The center of each channel is occupied by a protein containing many charged amino acids. When an electrical signal reaches the channel, the door flips out of the way, and the channel opens to allow ions to enter the cell. You will see a voltage-sensitive channel in greater detail in Section 4.2.

b **A nerve is made of individual nerve cells.**

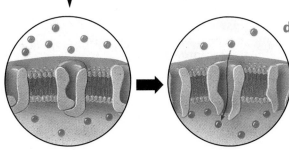

c **Along the cell membrane, the concentrations of ions are balanced. The voltage-sensitive channels remain closed.**

d **But when additional ions upset the balance, the voltage-sensitive channels spring open and ions pass through.**

Sensing Chemical Signals

Only a very few specialized cells have "fingers" that are sensitive to pressure or "eyes" that are sensitive to light. However, almost all cells can sense chemical signals by means of **receptor proteins** protruding through their membranes.

Receptor proteins send signals into the cell

Notice in **Figure 4.2b** that the end of the receptor protein has a unique shape that will hold only a particular type of molecule. Proteins like this one act like television antennas: they capture signals and deliver information. When the right molecule binds to the receptor protein, changes are triggered in the protein that are detected within the cell. Thus, the "signal" picked up by a receptor protein is a specific molecule outside the cell, a molecule that does not even enter the cell. Instead, what passes into the cell is information.

Hormones provide a good example of how receptor proteins carry information into a cell. A hormone is a chemical substance that acts as a messenger. Insulin, the hormone your body uses to regulate the level of sugar in your blood, has a shape that fits a specific receptor protein. Most of your cells have only a few insulin receptor proteins, but each of your liver cells possesses as many as 100,000! When an insulin molecule encounters an insulin receptor protein, the molecule binds to that protein, as shown in **Figure 4.2c**. The opposite end of the receptor protein changes shape, initiating a change in cell activity that ultimately causes levels of sugar in the blood to fall.

If blood sugar levels fall too low, however, the body responds by producing less insulin. Therefore, fewer receptor proteins are activated, and blood sugar levels gradually rise. Insulin and its receptor proteins thus help to keep blood sugar at a constant level. If cells have too few insulin receptors, one form of a disease called diabetes (*deye uh BEET eez*) results. In this type of diabetes, which occurs mainly among adults over 40 years old, insulin levels in the bloodstream are above normal but cells are unable to take in enough sugar.

Figure 4.2

a After you eat a muffin, the level of sugar in your blood rises. Insulin lowers this level by causing your cells to take up and store sugar. Here's how it works.

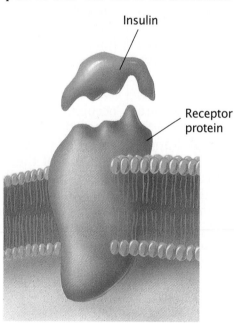

Insulin

Receptor protein

b An insulin molecule has a specific shape that fits into a receptor protein. Like a jigsaw-puzzle piece, the insulin molecule joins with the receptor protein.

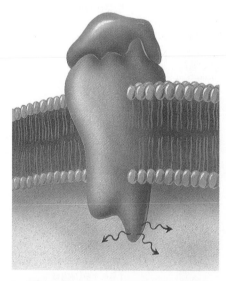

c The receptor protein changes shape, sending a message into the cell to take in and store excess sugar.

Sensing Cellular Identity

Just as a football jersey bears a number and last name, cells carry markers for identification. Every cell in your body contains a unique set of membrane proteins called **cell surface markers**. Your immune system uses these markers to recognize your cells and to distinguish them from damaging invaders, such as harmful bacteria. Cell surface markers also indicate cell type, such as whether a cell is a blood cell, a muscle cell, or a liver cell.

The uniqueness of an individual's cell surface markers explains why organ donors and recipients must be carefully matched. Notice the cell surface marker on the heart cell in **Figure 4.3d**. To ensure that the heart is a "good match," the donor and recipient must share many of the same identifying proteins on their cell surfaces. As with all proteins, the structure of cell surface markers is determined by genes. If the donor and recipient are genetically related, there is a greater probability of a match. Cell surface markers are so unique that a perfect match never occurs, except between identical twins. The organ recipient, therefore, must take drugs to suppress the immune system, which could reject the transplanted organ.

Stability

Why is it necessary that a cell be able to gather information about its environment?

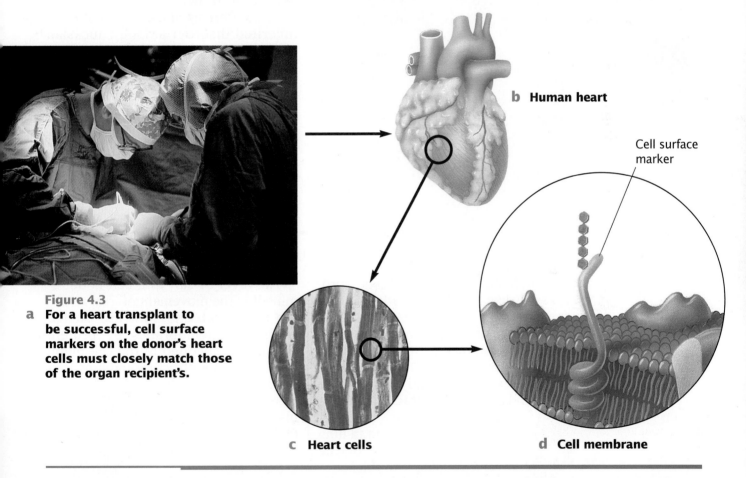

b **Human heart**

Cell surface marker

Figure 4.3

a **For a heart transplant to be successful, cell surface markers on the donor's heart cells must closely match those of the organ recipient's.**

c **Heart cells**

d **Cell membrane**

Section Review

❶ How do voltage-sensitive channel proteins respond to electrical signals?

❷ How do receptor proteins pass information into cells?

❸ Why are protein markers important in matching organ donors to recipients?

Cystic Fibrosis: A Deadly Gene

A Rare but Deadly Disease

The cystic fibrosis patient below is breathing into a vitalograph, a device that measures lung function.

For a child with cystic fibrosis, life is mostly a series of respiratory infections, doctors' visits, and medications. Cystic fibrosis causes thick mucus to build up in the lungs, making breathing difficult.

The thick mucus also coats the hairlike projections lining air passages, weakening the body's immune system. Excess mucus interferes with the functioning of other organs too. In the liver and pancreas, mucus blocks the flow of digestive enzymes in the intestine, so food is not digested properly. Worn down by repeated bouts of illness, a cystic fibrosis patient rarely lives beyond his or her twenties.

Cystic fibrosis is the most common inherited disorder among Caucasian people. The disease, which begins in infancy, afflicts more than 25,000 Americans and causes 500 deaths every year. Forty years ago, the average life span of a cystic fibrosis patient was five years. Today, improved medical therapies and nutrient-rich diets have enabled cystic fibrosis sufferers to survive into adulthood.

Salt Imbalance Is a Clue

Research into the cause of cystic fibrosis reads like a detective story. One clue is that cystic fibrosis patients have excess amounts of sodium and chloride in their sweat, making it very salty. At the University of North Carolina, researchers found that salt imbalance causes thick mucus to accumulate in the patients' lungs. High levels of salt in lung cells draw water out of the mucus, causing it to thicken.

The level of salt in a cell is determined by the movement of ions across the cell membrane. Ions are carried across cell membranes by protein channels embedded in the cell membrane. Since the structure and functions of proteins are determined by genes, problems with ion transport can be assumed to have a genetic basis.

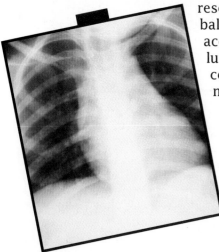

The X ray at left shows the chest of a healthy person. The white area in the center is the shadow of the heart.

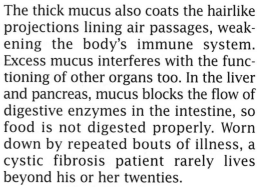

In this X ray of a 20-year-old person with cystic fibrosis, the airways are partially blocked with a thick buildup of mucus.

CFTR: The Defective Gene

The CFTR gene is located near the center of the long arm of chromosome 7.

In 1989, an American-Canadian research team found the defect in the gene that causes cystic fibrosis. The gene, called the cystic fibrosis transmembrane conductance regulator (CFTR) gene, produces a protein that usually helps maintain normal levels of chloride. In about 60 percent of cystic fibrosis patients, the protein made by the CFTR gene is missing an amino acid called phenylalanine. When this amino acid is missing, the protein doesn't fold into its correct shape and loses its function.

Researchers are not certain whether the defective protein is actually a transport molecule. Some research data indicate that the protein might instead be involved in a regulatory network that controls ion transport. Although most cystic fibrosis patients have the mutation causing the defective protein described above, another 61 mutations in the CFTR gene have been identified.

Cystic fibrosis (CFTR) gene

Treatment and a Possible Cure

Locating the cystic fibrosis gene gives researchers an opportunity to develop more effective treatments and possibly a cure for the disorder. In one experiment, viruses that cause the common cold were used to insert corrected CFTR genes into the lung cells of live laboratory rats. The cold viruses were first disabled, so they could not cause infection. Then they were modified to carry healthy CFTR genes. In tests on hundreds of rats, the altered viruses inserted the healthy CFTR gene into the cells lining the animals' lungs. Once inside the cells, the genes began to produce normal protein. Researchers hope to develop a spray that will deliver a normal CFTR gene into the lungs of cystic fibrosis patients and possibly cure the disease. But since the cells that line the lungs are replaced every few weeks, the gene therapy would have to be repeated on a regular basis.

Another possible therapy involves the defective protein that the CFTR gene produces. Using information from the CFTR gene, scientists have figured out the structure of the protein. With this knowledge, the scientists might be able to develop medications that could force the protein to work normally.

Another possibility is the use of live animals to produce large quantities of normal protein that could be incorporated into a lung spray for cystic fibrosis patients. Healthy CFTR genes would be inserted into the embryos of mice or goats. As adults, the females would produce the protein in their milk. The protein could then be harvested and sprayed directly into patients' lungs.

In 1991, Jeff Pinard, who suffers from a relatively mild form of cystic fibrosis, began to conduct genetic research at the University of Michigan. He hopes to isolate some of the DNA mutations that cause cystic fibrosis.

ON A HOT DAY OR AFTER A HARD WORKOUT, YOU MIGHT REACH FOR A COLD SPORTS DRINK. THE SODIUM IONS IN THE DRINK REPLENISH THE ONES YOUR BODY LOST THROUGH PERSPIRATION. THESE IONS ENTER YOUR CELLS THROUGH PROTEIN CHANNELS IN THE CELL MEMBRANE. IN THIS SECTION YOU WILL SEE HOW CELLS TAKE IN NEEDED SUBSTANCES, SUCH AS SODIUM, AND DISCARD WASTES.

4.2 Moving Into and Out of Cells

Objectives

❶ Describe the difference between diffusion and osmosis.

❷ Distinguish facilitated diffusion from active transport.

❸ Explain how the sodium-potassium pump and the proton pump transport ions.

❹ Describe how large substances can enter and exit cells.

Diffusion and Osmosis

In addition to receiving information, cells need a way to move water molecules, food particles, and other ions through their membranes. Some molecules, such as water, pass through freely. Others must be carried through channels.

Diffusion mixes different molecules

Look at **Figure 4.4**, in which a student places ink droplets into a beaker of water. The ink slowly spreads throughout the water. What's happening? As the particles of ink dissolve in the water, their random motion soon carries them farther and farther out into the water until they are evenly distributed.

The mixing of two substances by the random motion of molecules is called **diffusion** (*dih FYOO zhuhn*). As shown in **Figure 4.4**, molecules diffuse from a region where their concentration is high (the ink) to a region where their concentration is lower (the water), until they are evenly dispersed.

Figure 4.4
a **When you place a few droplets of ink into a beaker of water, . . .**

b **. . . the ink will settle to the bottom . . .**

c **. . . and diffuse randomly through the water until the ink particles are evenly dispersed.**

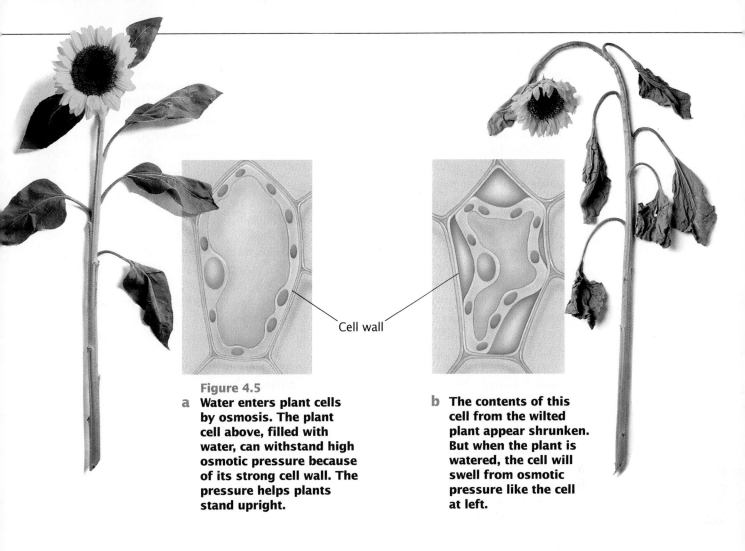

Figure 4.5

a Water enters plant cells by osmosis. The plant cell above, filled with water, can withstand high osmotic pressure because of its strong cell wall. The pressure helps plants stand upright.

Cell wall

b The contents of this cell from the wilted plant appear shrunken. But when the plant is watered, the cell will swell from osmotic pressure like the cell at left.

Patterns of Change

How does osmosis enable a cell to respond to changes within an organism?

Water travels through membranes by osmosis

Looking at **Figure 4.4**, you might think that only the ink diffuses through the water. But water molecules can also diffuse. When water molecules diffuse through a cell membrane the process is called **osmosis** (*ahz MOH sihs*). Since water molecules are small, they can slip freely through the gaps between the phospholipids in the cell membrane. As a result, water molecules constantly move back and forth through the cell membrane. This free movement of water has a very important function: It enables cells to absorb water.

Unlike the water molecules randomly bustling about outside the cell, many of the water molecules inside the cell are busy interacting with sugars, proteins, and other polar molecules. As a result, more water molecules flow into the cell by osmosis than flow out of the cell. Therefore, there is often a net movement of water into the cell.

Water in a cell creates osmotic pressure

When water enters a cell, it creates pressure. This pressure is called **osmotic** (*ahz MAH tihk*) **pressure**. If osmotic pressure is very high, it can cause a cell to burst. Most cells cannot withstand high osmotic pressure unless their membranes are braced to resist the swelling. Many kinds of organisms, such as the plants in **Figure 4.5**, have cell walls to protect and support their cells.

What about organisms without cell walls? Many single-celled organisms have specialized organelles to pump water out. In multicellular animals, cells do not burst because a balance exists between the concentration of fluids inside and outside the cells. In your body, for example, the concentrations of salts, sugars, and other ions are about the same inside cells as in the fluids surrounding them. Thus, water enters the cells of the body at the same rate that it leaves. The cells do not burst.

Selective Transport of Substances

Unlike water molecules, many substances cannot easily pass through the cell membrane. For example, molecules such as sugars and proteins are often too large to slip through the gaps in the cell membrane. In addition, these molecules are often polar and cannot pass through the nonpolar region of the lipid bilayer. Polar molecules and ions, therefore, use channels made of proteins to move in or out of the cell. A protein channel is shown in **Figure 4.6b**.

The transport of substances through membrane protein channels is called **selective transport**. This form of transport is said to be "selective" because each kind of channel will allow only a particular type of molecule or ion to pass through the membrane. Thus, the cell can control the substances that enter and leave. There are two modes of selective transport: facilitated diffusion and active transport.

Facilitated diffusion works in two directions

Some channels are like open doors. As long as a molecule or ion fits into the channel, it is free to pass through in either direction. Each kind of molecule or ion diffuses toward the side where it is least concentrated. Eventually, diffusion balances the concentrations of that molecule or ion on both sides of the membrane. Because some channels assist, or facilitate, the transport of substances, this form of diffusion is called **facilitated diffusion**. Sugar molecules and ions such as chloride ions enter cells by facilitated diffusion.

Active transport moves molecules in one direction

If all channels worked like open doors, many molecules and ions would simply flood out by facilitated diffusion. But some channels work only in one

Figure 4.6

a *Ahhh!* There's nothing like the taste of a good thirst quencher after a tough workout. Have you ever wondered how the cells of your body replenish nutrients?

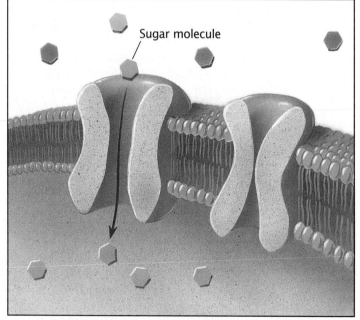

Sugar molecule

b Your bloodstream carries minerals and sugars to your body's cells, which pull them in through protein channels.

direction. Like turnstiles at a subway station, these channels let certain molecules and ions into the cell, but do not let them out. Active transport enables a cell to stockpile certain substances in far greater concentrations than they occur outside the cell. Follow the steps in **Figure 4.7** to see how a channel can be one-way.

The cell must use some of its energy to change the shape of the channel protein and to move substances across the membrane to a region of higher concentration. The operation of these one-way channels, therefore, is called **active transport**. Active transport plays a critical role in acquiring sugars and other molecules, even when more of these molecules already exist inside the cell. Surprisingly, almost all the active transport in cells is carried out by only two kinds of channels: the sodium-potassium pump and the proton pump.

The sodium-potassium pump transports ions

The **sodium-potassium pump** is an active transport system that enables the cell to admit ions needed for important biological processes, such as the conduction of nerve impulses through the body. Uniquely shaped receptor sites on protein channels enable the ions to move in one direction only.

As you can see in **Figure 4.7**, this mechanism works by actively pumping sodium ions out of cells and potassium ions into cells. In just one second, each channel can move more than 300 sodium ions out of the cell. More than one-third of all the energy expended by your body's cells is spent driving the sodium-potassium pump. When there are very few sodium ions in the cell, facilitated diffusion channels enable sodium ions to rush back into the cell.

There is one catch, however: The channels must be opened. Some are opened by electrical currents. Others are opened only when sodium ions are paired to partner molecules such as sugar or amino acids. Because so many sodium ions rush back in through these channels, large numbers of partner molecules are pulled through as well, even if they are already present in high concentrations within the cell. The protein channels that admit sodium ions and their partners are called coupled channels, as shown in **Figure 4.7d**.

Figure 4.7
The sodium-potassium pump, an active transport system, actively pumps specific ions across the cell membrane in one direction.

a **Sodium ions within the cell fit precisely into receptor sites on the channel protein.**

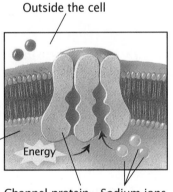

Outside the cell

Potassium ions

Inside the cell   Energy

Channel protein   Sodium ions

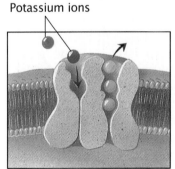

Sodium ions

b **The channel changes shape, pumping the sodium ions across the membrane. Potassium ions outside the cell move into receptor sites.**

c **The sodium ions are released and cannot reenter through this channel. At the same time, potassium ions are pumped across the channel into the cell.**

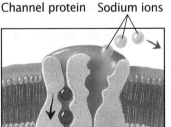

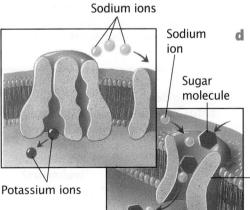

Sodium ion

Sugar molecule

d **Potassium ions enter the cell. Sodium ions outside the cell, along with sugar molecules, reenter the cell through a coupled channel.**

Coupled channel

Potassium ions

How Eukaryotic Cells Divide

Evolution

Why do you think

a more complex

type of cell division

evolved among

eukaryotes?

Figure 4.10
A chromosome and its copy may look something like an "X" because they generally are connected near the middle.

Eukaryotic cells carry far more DNA than bacteria. During cell division, the DNA in eukaryotic cells is packaged into tightly wound structures that are called **chromosomes**, as shown in **Figure 4.10**. Cell division plays a major role in the development of eukaryotic organisms. Just by looking at your hand, you can imagine how many millions of cells must have divided to form each crease on every finger. The magnified hand of the human fetus shown in **Figure 4.11b** highlights the importance of eukaryotic cell division.

A typical human cell contains 46 chromosomes. Instead of being attached to the cell membrane as with bacterial DNA, eukaryotic chromosomes are contained within the nucleus. Because eukaryotic cells have more DNA than do prokaryotic cells, and because the DNA is confined within a nucleus, eukaryotic cell division is more complex than bacterial cell division. First, each chromosome must be copied exactly. Next, the chromosomes must be sorted out precisely so that each new cell gets a complete set. Finally, the cell itself divides in half.

Two processes have evolved that enable eukaryotic cells to divide successfully: mitosis and cell division. **Mitosis** (*meye TOH sihs*) is the process by which the nucleus of a eukaryotic cell divides into two nuclei, each containing a complete set of the cell's chromosomes. Mitosis can be broken down into four distinct phases: prophase, metaphase, anaphase, and telophase. These phases are summarized in **Figure 4.12**.

In many cells, mitosis is followed by cell division, also called **cytokinesis** (*syt oh kuh NEE sihs*). During cell division, the cell divides into two cells, each with its own nucleus. Plant cells—which have strong cell walls that cannot be pinched like the cell membranes of animal cells—form a new cell wall in the center of the cell. This new cell wall divides the cell in half, like a partition dividing a room.

The end result of mitosis and cell division is two cells with the same genetic information, where only one cell existed before. To follow the events of eukaryotic cell division in an animal cell, see the steps shown in **Figure 4.12**.

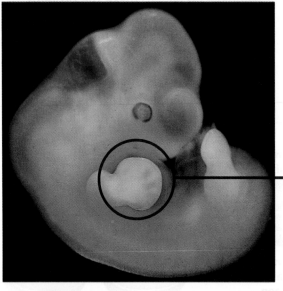
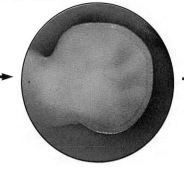
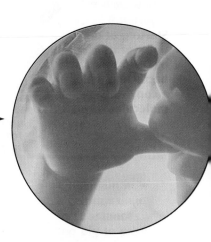

Figure 4.11
a **Cell division plays an essential role in the growth of an organism.**

b **For example, the hand of this human fetus is made of millions of cells. Some cells die, while others continue to divide until . . .**

c **. . . the hand is fully formed. After birth, skin cells will keep dividing to replace those that wear out.**

A Generalized Picture of Mitosis and Cell Division

Figure 4.12

a *DNA replication*
This animal cell is ready to divide. Its chromosomes are not yet visible because they are extended and uncoiled. The DNA of each chromosome is copied. Each chromosome consists of two identical strands.

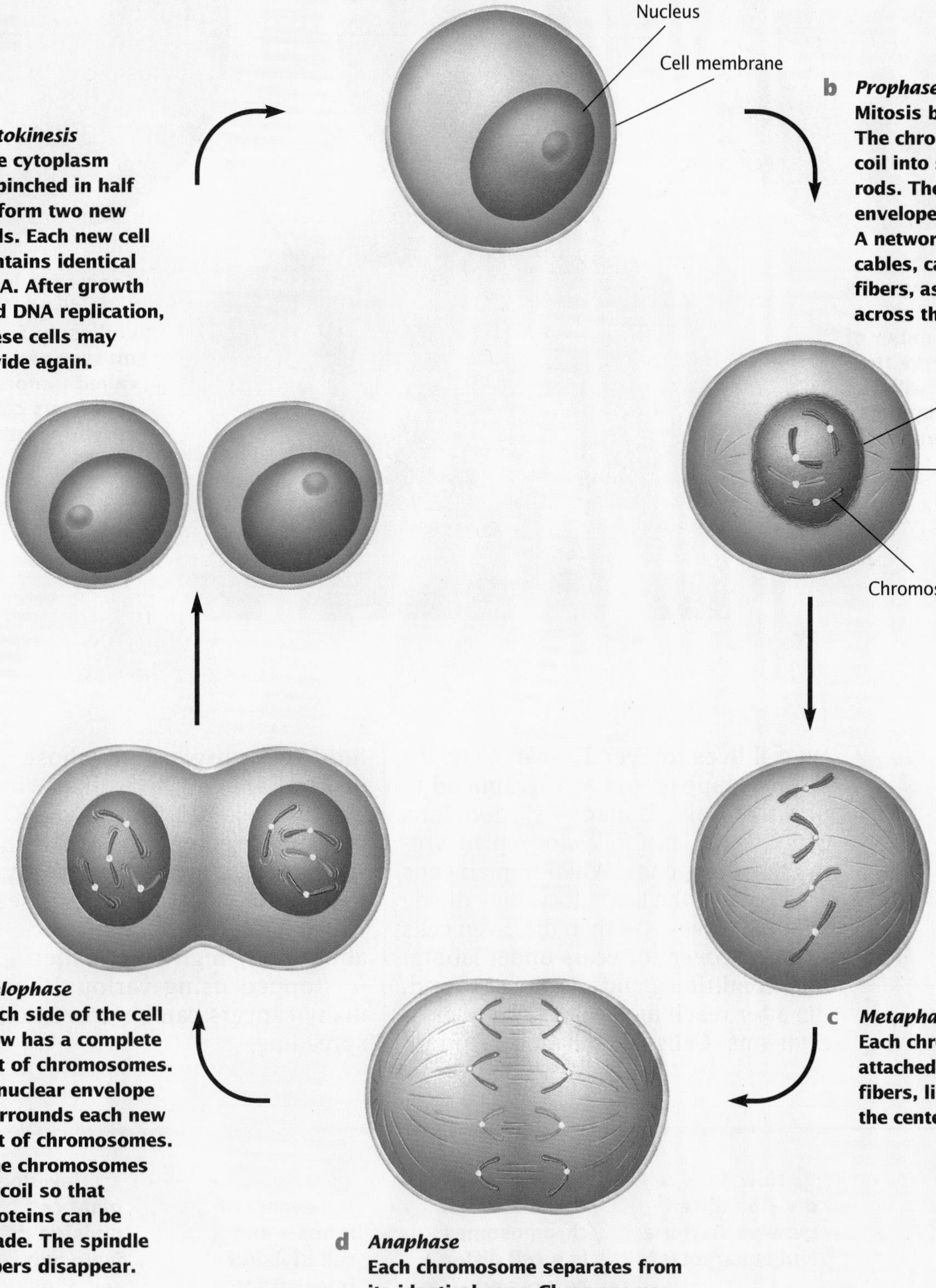

Nucleus

Cell membrane

b *Prophase*
Mitosis begins. The chromosomes coil into short, fat rods. The nuclear envelope breaks up. A network of protein cables, called spindle fibers, assembles across the cell.

f *Cytokinesis*
The cytoplasm is pinched in half to form two new cells. Each new cell contains identical DNA. After growth and DNA replication, these cells may divide again.

Nuclear envelope

Spindle fibers

Chromosomes

e *Telophase*
Each side of the cell now has a complete set of chromosomes. A nuclear envelope surrounds each new set of chromosomes. The chromosomes uncoil so that proteins can be made. The spindle fibers disappear.

c *Metaphase*
Each chromosome, attached to spindle fibers, lines up in the center of the cell.

d *Anaphase*
Each chromosome separates from its identical copy. Chromosomes are reeled to opposite sides of the cell. The spindle fibers start to break down.

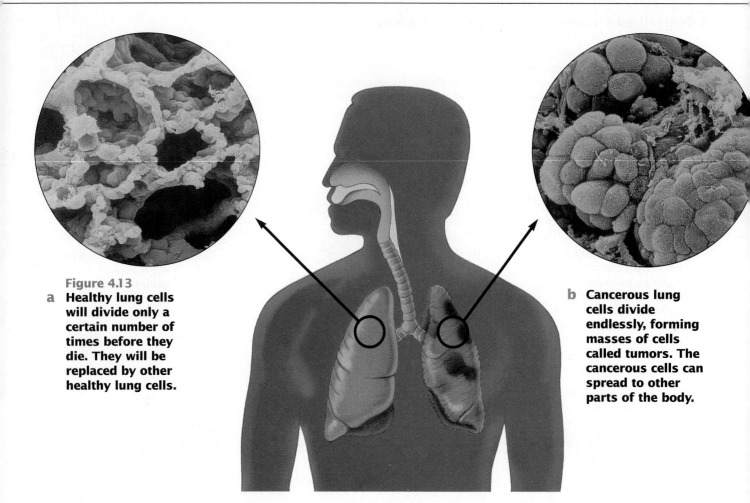

Figure 4.13

a Healthy lung cells will divide only a certain number of times before they die. They will be replaced by other healthy lung cells.

b Cancerous lung cells divide endlessly, forming masses of cells called tumors. The cancerous cells can spread to other parts of the body.

Life Span of Cells

No cell lives forever. Human cells, for example, appear to be programmed to undergo only so many cell divisions and then die, as if following a plan written into the genes. When human cells are grown in the laboratory, they divide about 50 times and then die. Even cells that are frozen for years under laboratory conditions, and are then thawed, die after reaching a certain number of divisions. Cells in your body contain hidden "hourglasses" whose grains of sand are cell divisions. When the "sand" runs out, the cells die. On rare occasions, some cells, such as those in **Figure 4.13b**, appear to disobey these instructions. **Cancer** is a disease in which cells grow and divide at an abnormally high rate. If their growth is stopped using various therapies, many cancers can be prevented from spreading.

Section Review

❶ How does cell division differ between bacteria and eukaryotes?

❷ How does the number of chromosomes in a cell differ before and after cell division?

❸ Summarize the events of mitosis and cell division in eukaryotes.

❹ How does plant cell division differ from animal cell division?

Chapter 4 *Highlights*

How do you smell perfume? Perfume molecules diffuse through the air, just like ink diffuses through water (p. 74).

	Key Terms	Summary
4.1 How Cells Receive Information **Insulin joins with a receptor protein to send chemical messages into the cell.**	**voltage-sensitive channel (p. 69)** **receptor protein (p. 70)** **cell surface marker (p. 71)**	• **Many proteins embedded in the cell membrane transmit information into cells.** • **Some proteins are cell surface markers that identify cells. Markers are unique to each individual.**
4.2 Moving Into and Out of Cells 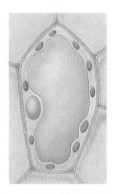 **A strong cell wall prevents this plant cell from being burst by osmotic pressure.**	**diffusion (p. 74)** **osmosis (p. 75)** **osmotic pressure (p. 75)** **selective transport (p. 76)** **facilitated diffusion (p. 76)** **active transport (p. 77)** **sodium-potassium pump (p. 77)** **proton pump (p. 78)** **chemiosmosis (p. 78)** **endocytosis (p. 78)** **exocytosis (p. 78)**	• **Homeostasis is maintained in cells by the cell membrane, which controls movement of substances into and out of cells.** • **Some substances pass through the cell membrane by diffusion, the movement of molecules from a higher to lower concentration.** • **Water enters or leaves a cell by osmosis, the diffusion of water across a cell membrane.** • **Selective transport and facilitated diffusion do not require energy to move substances into cells. Active transport requires energy, and can move substances to a region of higher concentration.** • **Larger particles are moved into or out of the cell by endocytosis and exocytosis.**
4.3 How Cells Divide 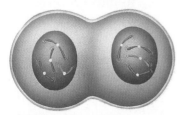 **When cells reach a certain size, they either stop growing or divide. This cell is in the final stage of mitosis and will soon divide into two cells, each with identical DNA.**	**chromosome (p. 80)** **mitosis (p. 80)** **cytokinesis (p. 80)** **cancer (p. 82)**	• **Eukaryotic cell division consists of two processes: mitosis, or division of the nucleus, and cytokinesis, division of the cytoplasm.** • **In mitosis, the nucleus of a cell is divided into two nuclei, each with the same number of chromosomes as the parent cell.** • **In cytokinesis, the cytoplasm divides to form two distinct cells.** • **Most cells undergo a certain number of divisions and then die. Cancer cells, however, continuously grow and divide at an abnormally high rate.**

Understanding Vocabulary

1. For each pair of terms, explain the differences in their meanings.
   a. diffusion, osmosis
   b. cell division, mitosis
   c. facilitated diffusion, active transport
   d. endocytosis, exocytosis

Relating Concepts

2. Copy the unfinished concept map below onto a sheet of paper. Then complete the concept map by writing the correct word or phrase in each oval containing a question mark.

Understanding Concepts

Multiple Choice

3. Voltage-sensitive channels help cells communicate by
   a. electrical signals.
   b. chemical stimuli.
   c. binding to molecules.
   d. insulin molecules.

4. Cells sense chemical signals by using
   a. hormones.
   b. receptor proteins.
   c. signaling messengers.
   d. surface markers.

5. The process by which water moves into and out of the cell is
   a. facilitated diffusion.
   b. osmosis.
   c. active transport.
   d. diffusion.

6. A cell uses some of its energy to move molecules by
   a. osmotic pressure.
   b. active transport.
   c. diffusion.
   d. osmosis.

7. The sodium-potassium pump
   a. requires no energy.
   b. moves potassium out of the cell.
   c. enables sugars to enter cells.
   d. works independently of channels.

8. Which is an example of active transport?
   a. sodium-potassium pump
   b. electron pump
   c. endocytosis
   d. facilitated diffusion

9. Particles too large to pass through protein channels in the cell membrane may enter the cell by
   a. exocytosis.
   b. selective transport.
   c. endocytosis.
   d. osmotic pressure.

10. During mitosis
    a. chromosomes are copied.
    b. chromosomes move to opposite sides of the cell.
    c. cytoplasm divides in half.
    d. a new cell wall forms in the center of the cell.

11. If a cell has 8 chromosomes before cell division, how many chromosomes will each of the two new cells have at the end of cell division?
    a. 16
    b. 8
    c. 4
    d. 32

12. During cell division in bacteria
    a. a circle of DNA is copied.
    b. chromosomes coil and move.
    c. the cell splits into three parts.
    d. two new nuclei are formed.

Completion

13. After mitosis, the cytoplasm of cells is pinched in half. In cells of _____ , cell walls are formed.

14. If placed in water, a cell would swell and possibly burst due to _____ .

15. A type of active transport system that uses light or chemical energy to move molecules that are required for cell metabolism is called the _____ .

Short Answer

16. What two helpful jobs are performed by the surface markers on cells?

17. Explain why animal cells do not burst due to osmotic pressure.

18. Describe the events that result in the formation of two complete nuclei in a eukaryotic cell.

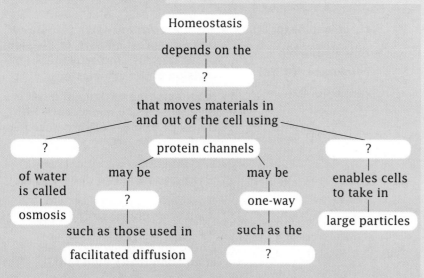

Homeostasis

depends on the

?

that moves materials in and out of the cell using

? — protein channels — ?

of water is called — may be

osmosis — ? — may be — enables cells to take in

such as those used in — one-way — large particles

facilitated diffusion — such as the — ?

Interpreting Graphics

19. Look at the drawings of the five cells at different stages of cell division. Write the letters identifying each cell in the order that indicates the correct sequence of the events of cell division.

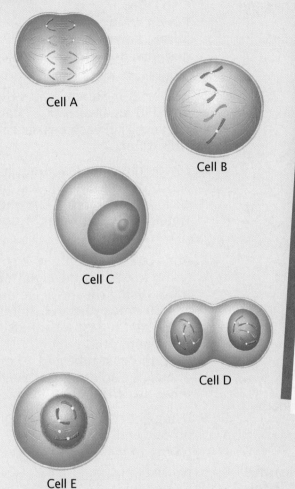

Cell A

Cell B

Cell C

Cell D

Cell E

20. Looking at the drawing below, explain how each molecule passed through the cell membrane.

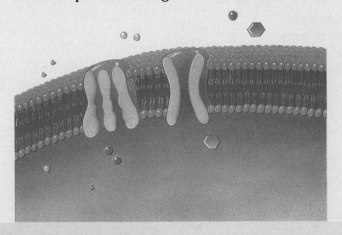

Reviewing Themes

21. *Interacting Systems*
    How do cell surface markers affect the success of organ transplants? What can be done to increase the success of organ transplants?

22. *Patterns of Change*
    How does mitosis promote genetic consistency?

23. *Evolution*
    Bacteria simply split, but eukaryotic cells reproduce by mitosis. How has the evolution of mitosis in eukaryotic cells aided their reproduction?

24. *Energy and Life*
    How is the cell's energy used in an active transport system like the sodium-potassium pump?

Thinking Critically

25. *Inferring Conclusions*
    On the basis of your understanding of osmosis, explain what would happen to a marine clam placed in a freshwater aquarium.

26. *Building on What You Have Learned*
    How does the structure of water enable it to interact with sugars, proteins, and other polar molecules within the cell?

Cross-Discipline Connection

27. *Biology and Drama*
    William Shakespeare said that we are all actors and the world is our stage. With two students, write a skit showing the events of mitosis and cell division. Then, act out your play and judge its accuracy and drama.

Discovering Through Reading

28. Read "A Herpes Key," in *Discover*, November 1990, page 22. How does the herpes virus infect cells? What effect does covering cells with fibroblast growth-factor have on the rate of infection?

Investigation

Can All Molecules Diffuse Through a Membrane?

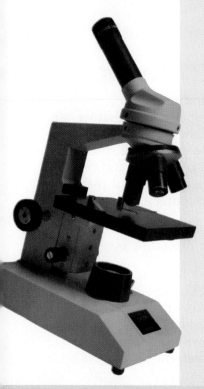

Objectives

In this investigation you will:
- *demonstrate* the movement of molecules through a selectively permeable membrane
- *perform* chemical tests for the presence of starches and sugars
- *collect* and *interpret* data

Materials

- lab apron
- safety goggles
- disposable gloves
- wax pencil
- two 250-mL beakers
- 8 test tubes
- water
- medicine dropper
- starch solution
- enzyme solution
- test-tube rack
- two 15-cm pieces of dialysis tubing
- heavy-duty thread
- rubber gloves
- Lugol's iodine solution
- Benedict's solution
- hot plate
- 400-mL beaker water bath

Prelab Preparation

Review what you have learned about sugars, starches, and diffusion by answering the following questions:
- How does starch differ from a simple sugar?
- How does a cell membrane control the passage of molecules into and out of a cell?

Procedure: Investigating Diffusion

1. Form a cooperative group of four students. Work with a member of your team to complete steps 2–16.

2. **CAUTION: Put on a lab apron, safety goggles, and disposable gloves.**

3. With a wax pencil, label one 250-mL beaker "B-1: Starch." Label the second 250-mL beaker "B-2: Starch and Enzyme." Fill each beaker half way with water.

4. Divide the eight test tubes into two sets of four. Label one test tube in each set as "Starch," "Enzyme," "B-1: Water," and "B-2: Water," respectively.

5. Add water to a depth of 2 cm (1 in.) to each of the eight test tubes. Using the medicine dropper, add 20 drops of starch solution to each "Starch" test tube. Rinse out the dropper thoroughly. To each "Enzyme" test tube add 20 drops of enzyme solution. Set the test tubes aside in a test-tube rack.

6. Using heavy-duty thread, tightly tie off one end of each piece of dialysis tubing to make two "bags."

7. Pour starch solution into the first bag until it is two-thirds full. Tightly tie the bag's opening with thread. Rinse the outside of the bag and place it in the "B-1: Starch" beaker.

8. Pour starch solution into the second bag until it is two-thirds full. Using a clean medicine dropper, add 20 drops of enzyme solution into the bag. Tie it tightly with thread. Rinse the outside of the bag and place it in the "B-2: Starch and Enzyme" beaker. *What is the purpose of rinsing the outside of each bag before continuing?*

9. Add 20 drops of water from the "B-1: Starch" beaker to each test tube labeled "B-1: Water." Similarly, add 20 drops of water from the "B-2: Starch and Enzyme" beaker to each test tube labeled "B-2: Water."

10. Set the beakers aside for 15 minutes. Meanwhile, copy the table below and proceed to step 11.

		Presence of starch	Presence of sugar
Initial test (steps 4–13)	Starch solution		
	Enzyme solution		
	B-1 Water		
	B-2 Water		
Final test (step 15)	B-1 Water		
	B-2 Water		

11. **CAUTION: Lugol's iodine solution is poisonous; avoid skin/eye contact.**
Using one set of four test tubes, add two drops of Lugol's iodine solution to each test tube. If the liquid turns dark blue, starch is present. Record the results in your table.

12. **CAUTION: Use care when working with the hot plate and hot liquids.**
Fill a 400-mL beaker half way with water. Place it on a hot plate and heat until boiling. Heat the second set of test tubes in the boiling water bath for 5 minutes.

13. **CAUTION: Benedict's solution is an irritant; avoid skin/eye contact.**
Add 10 drops of Benedict's solution to each test tube. If the liquid turns orange, sugar is present. Record your results.

14. Empty and clean all the test tubes before continuing with the investigation.

15. After 15 minutes, use two test tubes to separately test the water in beaker B-1 for the presence of starch and sugar. Use another two test tubes to test the water in beaker B-2 for the presence of starch and sugar. Record your results.

16. Clean up your materials and wash your hands before leaving the lab.

Analysis

1. *Evaluating Methods*
Why was it necessary to make initial tests of the starch solution, the enzyme solution, and the water in each beaker?

2. *Analyzing Data*
Describe the contents of each dialysis bag, and discuss the evidence that supports your description.

3. *Analyzing Data*
Was starch or sugar present in the water of either beaker immediately after each bag was placed in the water? What evidence supports your answer?

4. *Analyzing Data*
What do your results suggest about the contents of the water in beakers B-1 and B-2 at the end of the investigation?

5. *Making Inferences*
Explain how the dialysis tubing demonstrates selective permeability.

6. *Making Inferences*
Why are some molecules unable to pass through a membrane while others move through freely?

Thinking Critically

Why would leaving out any of the procedural steps in this investigation affect your ability to draw valid conclusions from your results?

Energy and Life

Review

5.1 Cells and Chemistry

- **Chemical Reactions
  in Living Things**
- **Actions of Biological Catalysts**

5.2 Cells and Energy

- **How Cells Use Energy**
- **Energy Flow in the Living World**

5.3 Photosynthesis

- **Harnessing the Sun's Energy**
- **Stage 1: Capturing Light Energy**
- **Stage 2: Using Light Energy to
  Make ATP and NADPH**
- **Stage 3: Building Carbohydrates**

5.4 Cellular Respiration

- **Releasing Energy
  From Organic Molecules**
- **Regulating Cellular Respiration**

**Running, jumping, and leaping over hurdles require lots
of energy. In this chapter, you will learn that the energy
needed for life comes ultimately from the sun.**

JUST AS A RACE CAR DRIVER LEARNS HOW A CAR WORKS BY STUDYING ITS ENGINE, A SCIENTIST UNDERSTANDS HOW LIVING THINGS WORK BY STUDYING CELL CHEMISTRY. LIKE CARS, CELLS ARE COMPLEX MACHINES, FULL OF DELICATE DETAIL AND POWERED BY CHEMICAL ENERGY. TO UNDERSTAND HOW YOUR BODY WORKS, YOU MUST LOOK "UNDER THE HOOD" AT THE CHEMICAL MACHINERY IN YOUR CELLS.

5.1 Cells and Chemistry

Objectives

❶ **Distinguish between energy-storing and energy-releasing chemical reactions.**

❷ **Define activation energy and describe its role in chemical reactions.**

❸ **Describe the interaction between an enzyme and its substrate.**

❹ **Describe the role of enzymes in cells.**

Chemical Reactions in Living Things

Chemical reactions do not take place only in fizzing test tubes. Within cells, all the activities associated with life also are driven by chemical reactions.

A chemical reaction is the process of making or breaking the chemical bonds that link atoms. In plants, chemical reactions use light energy to form the chemical bonds of sugars such as glucose. The energy in these chemical bonds can then be used by cells to power their lives. A potato like the one in **Figure 5.1** is an excellent food source because it is crammed with glucose, linked in a thicket of long chains called starch. When the bonds of the glucose are broken, the stored energy is released.

Figure 5.1

a **When you eat a baked potato, starch is broken down into sugar in your mouth and small intestine. The energy in these sugar molecules will be used by your body to do work.**

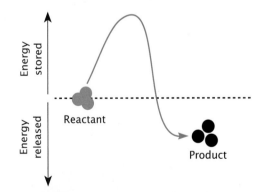

b **Notice that the product contains less energy than the reactant. Reactions that release energy are called exergonic reactions.**

Any chemical reaction that results in a net release of energy is **exergonic**. Your cells put this released energy to work making proteins and other molecules of which your body is built. Building these molecules uses a lot of energy because many new chemical bonds must be made. Extra energy must be supplied to cause any chemical reaction in which the chemical bonds of the products have more energy than the bonds of the reactants. Any chemical reaction that requires a net input of energy is called **endergonic**. In the potato plant in **Figure 5.2**, the formation of glucose using the energy in light is an example of an endergonic reaction.

Thousands of chemical reactions are going on at any moment in each cell of a living plant and in every cell of your body, a symphony of living chemistry. All of these chemical reactions, taken together, are called **metabolism** (*muh TAB uh liz uhm*).

Figure 5.2

a **This potato plant uses light energy to form the chemical bonds of glucose.**

Chemical reactions need energy to get started

The heat from a flame ignites the twigs in a campfire. The spark from a spark plug causes gasoline in an engine to ignite. In both cases, an input of energy is used to start the chemical reaction. The amount of energy needed to cause a chemical reaction to start is called **activation energy**. Think of a boulder you must move up and over the top of a hill. To get it rolling downhill, you must first push it to the top. Activation energy is simply a chemical push that gets a reaction going.

Even if the product contains less energy than the reactants, activation energy must still be supplied before that reaction can occur. For example, the combustion of gasoline provides the energy needed to power an automobile. But only after the key is turned in the ignition will a spark from each spark plug ignite the gasoline in the engine's cylinders. The sparks provide the activation energy needed to trigger the burning of gasoline.

Reactions in cells must occur quickly

The burning of gasoline fuel to power an engine requires a spark or high temperature to get a reaction going. Cells also "burn" fuels. Like an engine, most cellular reactions would require very

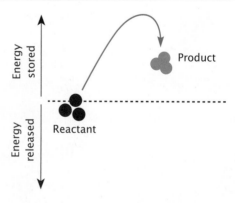

b **Reactions that store energy are endergonic reactions.**

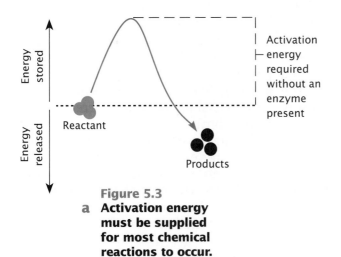

Figure 5.3

a Activation energy must be supplied for most chemical reactions to occur.

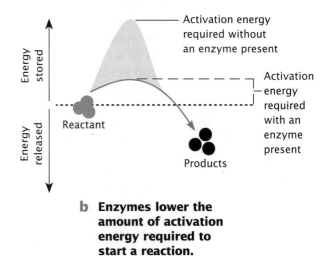

b Enzymes lower the amount of activation energy required to start a reaction.

high temperatures to proceed quickly enough to keep the cell functioning. But high temperatures would kill the cell. Fortunately, cellular reactions can occur quickly at relatively low temperatures through the action of enzymes. You read in Chapter 2 that enzymes are proteins that can hasten a chemical reaction.

How does an enzyme increase the speed of a reaction? A reaction could proceed faster if less energy were needed to get each molecule started. Think about the activation energy needed to move a heavy boulder over a hill. One way to reduce the amount of

energy necessary would be to reduce the hill's size. Digging away some of the ground in front of the boulder would reduce the amount of energy needed to send the rock rolling over the hill. Enzymes cause reactions to occur at a lower activation energy, as shown in **Figure 5.3**. Using an enzyme in a reaction is like lowering the top of the hill.

Enzymes are biological catalysts. **Catalysts** make a reaction proceed faster without themselves being used up during the reaction. An enzyme-catalyzed reaction is faster because it has a lower activation energy than does an uncatalyzed reaction.

Actions of Biological Catalysts

To understand the importance of enzymes, consider how red blood cells pick up carbon dioxide and deliver it to the lungs to be exhaled. Carbon dioxide is a cellular waste product that will poison the body if not

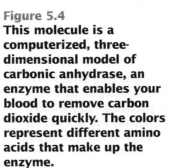

Figure 5.4
This molecule is a computerized, three-dimensional model of carbonic anhydrase, an enzyme that enables your blood to remove carbon dioxide quickly. The colors represent different amino acids that make up the enzyme.

continuously removed. When carbon dioxide is converted to carbonic acid, it can be easily carried within the bloodstream. However, the chemical reaction that converts carbon dioxide to carbonic acid is very slow. Only 200 molecules of carbonic acid form in an hour. Fortunately, an enzyme present in the blood called carbonic anhydrase increases the rate of this reaction to 600,000 molecules of carbonic acid formed every second. The enzyme, shown in **Figure 5.4**, accelerates the reaction rate about 10 million times! Without carbonic anhydrase, your blood would quickly become poisoned with carbon dioxide.

Energy Flow in the Living World

Almost all the energy needed for life comes ultimately from the sun, which shines continuously on Earth. Using a process called **photosynthesis** (*foh toh SIHN thuh sihs*), green plants capture sunlight and use the energy to convert carbon dioxide and water into energy-storing carbohydrates. Oxygen is produced as a waste product. Photosynthesis is also performed by algae and some bacteria. Organisms that eat plants—and those that eat plant-eaters—use the energy in carbohydrates to fuel their own life processes. All living things use a process called **cellular respiration** to obtain energy from the bonds of food molecules. For humans, food molecules are the foods we eat, which include carbohydrates. For plants, the food molecules are carbohydrates made by photosynthesis.

As you can see in **Figure 5.9**, energy flows in connected pathways throughout the living world. The energy-requiring reactions of photosynthesis transform sunlight energy into chemical energy. And the energy-releasing reactions of cellular respiration enable living things to use chemical energy to do work. You will learn more about the processes of photosynthesis and cellular respiration in Sections 5.3 and 5.4.

Figure 5.9

a **Light energy streaming down from the sun . . .**

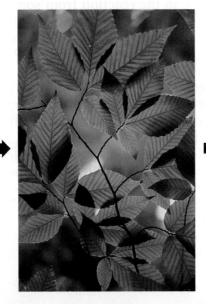

b **. . . is converted by green plants into carbohydrates.**

c **When an animal eats plants, it uses energy within carbohydrates in the plants.**

d **And when that animal is eaten by another animal, energy is transferred.**

Section Review

❶ What life activities performed by cells require energy?

❷ What are biochemical pathways? How do enzymes interact with biochemical pathways?

❸ How does ATP supply energy for the cell?

❹ Compare and contrast the roles of plants and animals in the flow of energy in the living world.

WHAT DID YOU HAVE FOR LUNCH YESTERDAY? BEEF ENCHILADAS? THE BEEF CAME FROM A COW THAT ATE GREEN GRASS. NO MATTER WHAT YOU ATE, IF YOU TRACE THE FOOD BACK TO ITS ORIGIN YOU END UP WITH A GREEN PLANT. CLEARLY, YOU DEPEND ON GREEN PLANTS FOR ENERGY. PLANTS DEPEND ON SUNLIGHT TO MAKE FOOD, BECAUSE LIFE ITSELF IS POWERED BY THE SUN.

5.3 Photosynthesis

Objectives

❶ **Summarize the evolution of photosynthesis.**

❷ **Describe how green plants and algae capture energy from sunlight.**

❸ **Explain how a plant cell uses light to make ATP and NADPH.**

❹ **Describe how photosynthesis provides the energy needed by all living things.**

Harnessing the Sun's Energy

As you read in Section 5.2, the energy used by most living things comes ultimately from the sun. Each day, light energy that reaches Earth equals the energy of about 1 million atomic bombs. About 1 percent of that light energy is captured by photosynthesis and used to make energy-rich carbohydrates. Almost all living things, such as the cow in **Figure 5.10**, depend on the products of photosynthesis to survive.

Figure 5.10
The green plants that this cow is eating provide it with the energy it needs to carry out its life activities.

How did photosynthesis evolve?

Photosynthesis evolved billions of years ago among bacteria. Early forms of bacteria captured the electrons they needed for photosynthesis by stripping hydrogen atoms from hydrogen sulfide (H_2S), generating sulfur as a byproduct. For about 1 billion years, this early type of photosynthesis was the only kind of photosynthesis that occurred. In all that time, no oxygen gas existed in Earth's atmosphere. Approximately 3 billion years ago, a second method of photosynthesis evolved. In this new type of photosynthesis, electrons were obtained by removing hydrogen atoms from water (H_2O). Oxygen gas (O_2) was released into the atmosphere as a waste product. As a result of this type of photosynthesis, Earth's atmosphere is now rich in oxygen gas.

Photosynthesis takes place in three stages. In the first stage, energy is captured from light. In the second stage, the energy is used to make ATP and a high-energy compound called NADPH. During the third stage, the ATP and NADPH are used to power the manufacture of energy-rich carbohydrates from CO_2 in the air.

Stage 1: Capturing Light Energy

If plants use light energy for photosynthesis, our first question might be: Where is the energy in light? Twentieth-Century physics has taught us that light actually consists of tiny packets of energy called **photons** (*FOH tahnz*). When light shines down on you, your body is being bombarded by a stream of photons smashing onto its surface. Some of these photons, such as X rays and ultraviolet light, carry a greater amount of energy, while others, such as radio waves, carry very little energy. Our eyes absorb only photons carrying intermediate amounts of energy, which is why we see only "visible" light, as shown in **Figure 5.11**. Plants are even choosier, absorbing mostly blue and red light. They reflect back what is left of the visible light, which is why most plants appear green.

Molecules called pigments absorb light

How can the molecules of a leaf or a human eye "choose" which photons to absorb? Molecules that absorb light are called **pigments**. In your eyes, the pigment is called retinal. The eyes of insects contain pigments that absorb photons of a higher energy level than those absorbed by retinal, so insects can see violet light that humans cannot.

But the insect pigment does not absorb low energy photons as well as retinal does, so insects cannot see the red light that humans can.

Plants capture light energy in chlorophyll

The major light-absorbing pigment in plants is chlorophyll. While it absorbs fewer kinds of photons than retinal, chlorophyll is much more efficient at capturing these photons. Where is chlorophyll located? Look at the structure of the leaf in **Figure 5.12**. In green plants, chlorophyll is found in chloroplasts within plant cells.

When atoms in a pigment absorb light, electrons are boosted to higher energy levels. The energy in the photons is transferred to the electrons, causing the move. Boosting an electron requires an exact amount of energy, just as when climbing a ladder you must

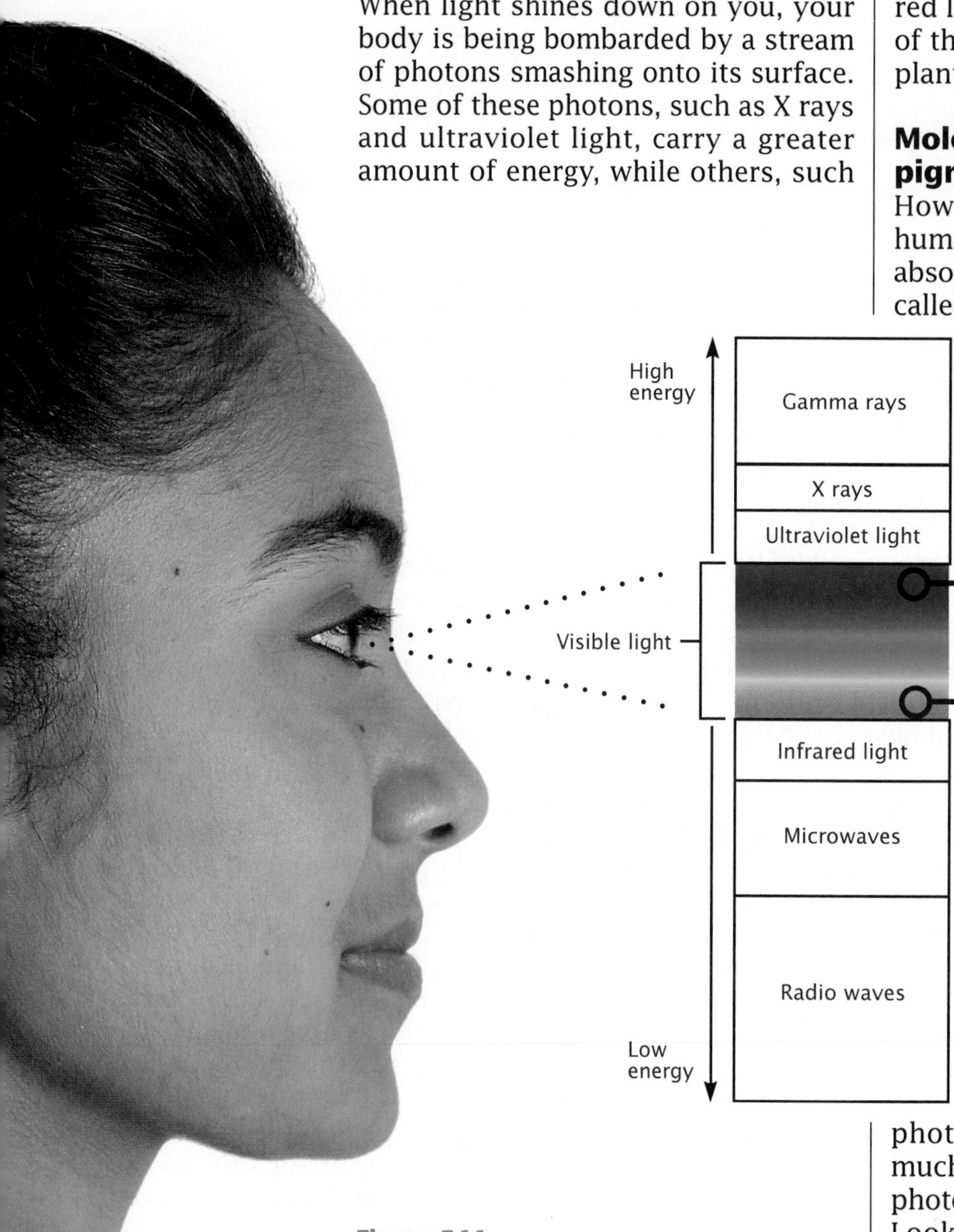

High energy

Gamma rays

X rays

Ultraviolet light

Visible light

Infrared light

Microwaves

Radio waves

Low energy

Figure 5.11

Visible light is that portion of the spectrum that humans can see. The spectrum consists of a range of waves from low-energy radio waves to high-energy gamma rays. These waves also range in wavelength. Green plants absorb red and blue wavelengths and reflect green.

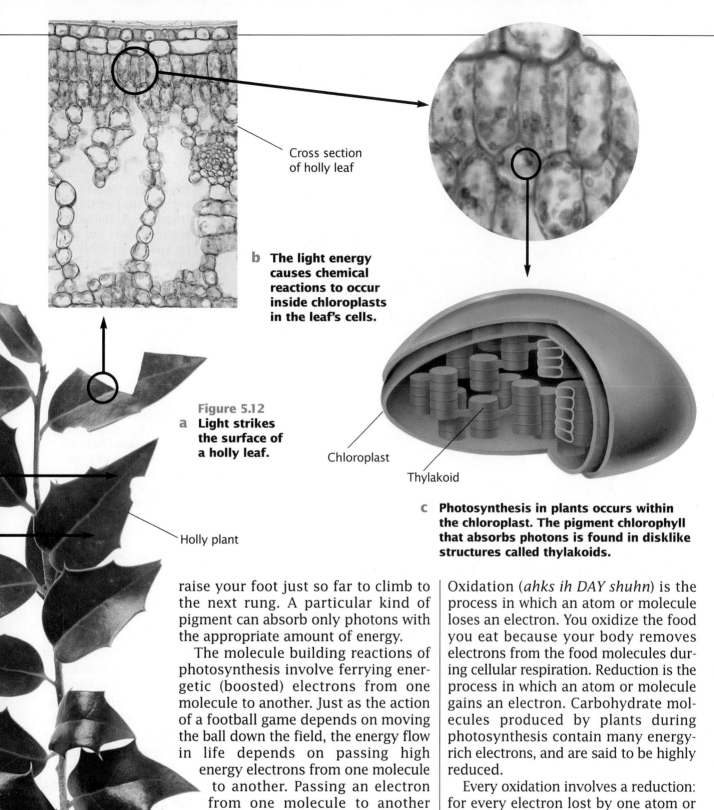

Cross section
of holly leaf

b **The light energy
causes chemical
reactions to occur
inside chloroplasts
in the leaf's cells.**

Figure 5.12
a **Light strikes
the surface of
a holly leaf.**

Chloroplast

Thylakoid

Holly plant

c **Photosynthesis in plants occurs within
the chloroplast. The pigment chlorophyll
that absorbs photons is found in disklike
structures called thylakoids.**

raise your foot just so far to climb to the next rung. A particular kind of pigment can absorb only photons with the appropriate amount of energy.

The molecule building reactions of photosynthesis involve ferrying energetic (boosted) electrons from one molecule to another. Just as the action of a football game depends on moving the ball down the field, the energy flow in life depends on passing high energy electrons from one molecule to another. Passing an electron from one molecule to another transfers the energy contained in that electron from molecule to molecule as well.

Chemical reactions that involve the transfer of electrons from one atom or molecule to another are called **oxidation-reduction reactions**. Many of the chemical reactions of photosynthesis and cellular respiration are examples of oxidation-reduction reactions.

Oxidation (*ahks ih DAY shuhn*) is the process in which an atom or molecule loses an electron. You oxidize the food you eat because your body removes electrons from the food molecules during cellular respiration. Reduction is the process in which an atom or molecule gains an electron. Carbohydrate molecules produced by plants during photosynthesis contain many energy-rich electrons, and are said to be highly reduced.

Every oxidation involves a reduction: for every electron lost by one atom or molecule (oxidation), an electron is gained by another atom or molecule (reduction). In cells, electrons do not travel alone, but in the company of a proton. Recall that a proton and an electron together make a hydrogen atom. Thus, oxidation-reduction reactions usually involve the loss of hydrogen atoms from one molecule and the gain of hydrogen atoms by another molecule.

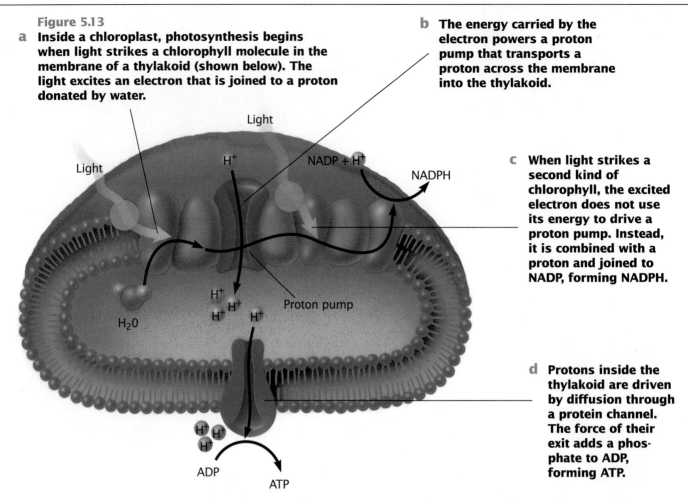

Figure 5.13

a Inside a chloroplast, photosynthesis begins when light strikes a chlorophyll molecule in the membrane of a thylakoid (shown below). The light excites an electron that is joined to a proton donated by water.

b The energy carried by the electron powers a proton pump that transports a proton across the membrane into the thylakoid.

c When light strikes a second kind of chlorophyll, the excited electron does not use its energy to drive a proton pump. Instead, it is combined with a proton and joined to NADP, forming NADPH.

d Protons inside the thylakoid are driven by diffusion through a protein channel. The force of their exit adds a phosphate to ADP, forming ATP.

Stage 2: Using Light Energy to Make ATP and NADPH

Scale and Structure

How does the

structure of the

thylakoid enable

green plants to

capture light

energy and

transform it into

chemical energy?

Inside the chloroplasts, chlorophyll molecules are contained within disk-like structures called thylakoids, as shown in **Figure 5.13**. When photons of light strike the chloroplasts, electrons are boosted within chlorophyll molecules to higher energy levels, as shown in **Figure 5.13a**. Each excited electron, traveling as part of a hydrogen atom, leaves the chlorophyll and jumps to a nearby protein in the membrane of the thylakoid. The electron is then passed from protein to protein, like a ball being passed down a line of people.

Soon the hydrogen atom carrying the light-excited electron arrives at its destination, a proton pump. The proton pump knocks the excited electron down to its original energy level, releasing energy in the process. This energy powers the pumping of a proton across the thylakoid membrane into the interior of the thylakoid. These events are shown in **Figure 5.13b**.

ATP is made when protons are forced through a protein channel

The payoff lies in what happens to protons pumped into the thylakoid. More protons are pumped inward, until the interior of the thylakoid is bursting at the seams. Straining to escape, the protons are driven by diffusion through the only exit available, a protein channel. As shown in **Figure 5.13d**, these protein channels use the force of the exiting protons to add a phosphate group to a molecule of ADP, making ATP.

A second kind of chlorophyll absorbs photons of higher energy than those absorbed by the ATP-making chlorophyll. The light-excited electrons of this second chlorophyll are carried by a hydrogen atom and attached to an electron carrier called NADP, forming NADPH. These steps are shown in **Figure 5.13c**. The ATP and NADPH will be used to help power the last stage of photosynthesis, the building of new carbohydrates.

Stage 3: Building Carbohydrates

The ultimate goal of photosynthesis is to capture carbon atoms from carbon dioxide in the air and use them to make carbohydrates that store energy. In a series of reactions, plants produce a number of carbon-containing molecules. From these molecules, plants can then assemble more complex carbohydrates such as glucose and other compounds needed for energy and growth. This series of reactions is called the Calvin cycle, for its discoverer Melvin Calvin of the University of California. The energy to fuel the Calvin cycle comes from ATP made during the second stage of photosynthesis. Follow the Calvin cycle in **Figure 5.14**.

The overall process of photosynthesis can be summarized as:

$$6CO_2 + 6H_2O \overset{\text{(light)}}{\rightarrow} C_6H_{12}O_6 + 6O_2$$

This equation indicates that carbon dioxide and water, in the presence of sunlight, will react to form sugars and oxygen gas.

Many plants store some of the sugars they produce by linking molecules to form complex carbohydrates such as starch. Starches can be stored either in the cells that formed them or in plant storage areas. For example, sugar made in the leaves of a potato plant is stored as starch in the potato tuber. The plant may later break down these starches to make the ATP needed by the cell, as you will see in Section 5.4.

Figure 5.14

a **Using ATP generated from the second stage of photosynthesis, . . .**

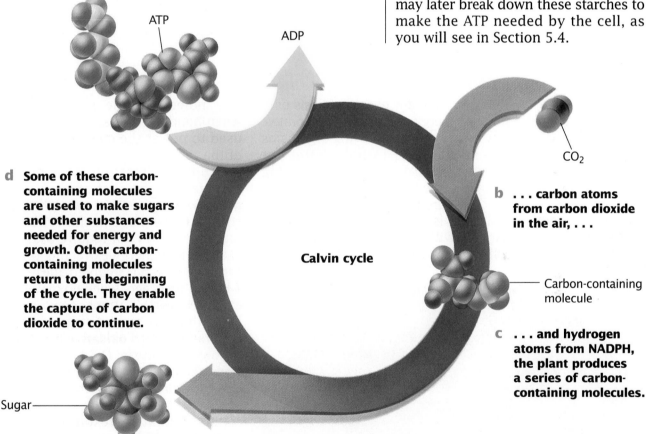

ATP

ADP

CO₂

d **Some of these carbon-containing molecules are used to make sugars and other substances needed for energy and growth. Other carbon-containing molecules return to the beginning of the cycle. They enable the capture of carbon dioxide to continue.**

Calvin cycle

b **. . . carbon atoms from carbon dioxide in the air, . . .**

Carbon-containing molecule

c **. . . and hydrogen atoms from NADPH, the plant produces a series of carbon-containing molecules.**

Sugar

Section Review

❶ **How did the appearance of a new kind of photosynthesis about 3 billion years ago change Earth's atmosphere?**

❷ **How do green plants and algae capture energy from sunlight?**

❸ **What happens to the ATP and NADPH made during the second stage of photosynthesis?**

❹ **List the products of photosynthesis.**

ALTHOUGH ONLY PLANT CELLS WITH CHLOROPLASTS PRODUCE CARBOHYDRATES, ALL CELLS OF A PLANT USE THESE CARBOHYDRATES FOR ENERGY. IN BOTH PLANTS AND ANIMALS—INDEED, IN ALMOST ALL ORGANISMS—THE ENERGY FOR LIVING IS OBTAINED BY RECYCLING THE SUGARS PRODUCED BY PHOTOSYNTHESIS. THE ENERGY IN THESE MOLECULES IS RELEASED IN CELLULAR RESPIRATION.

5.4 Cellular Respiration

Objectives

1. **Explain the importance of cellular respiration to living things.**

2. **Summarize the process of glycolysis.**

3. **Contrast fermentation with oxidative respiration.**

4. **State the role of oxygen in oxidative respiration.**

Releasing Energy From Organic Molecules

You have seen how photosynthesis uses sunlight energy to make carbohydrates. You also know that all living organisms depend on these carbohydrates for energy. The process by which living things release the energy stored in the bonds of carbohydrates and other food molecules is called cellular respiration. As you will see, the first result of cellular respiration is the formation of ATP molecules. The energy released when bonds in ATP are broken is then used to power the chemical reactions of the cell.

Cellular respiration takes place in two stages. The first stage is called **glycolysis** (*gly KAHL uh sihs*). Glycolysis takes place in the cell's cytoplasm and does not require oxygen. It is an ancient energy-extracting process thought to have evolved more than 3 billion years ago, when no oxygen gas existed in Earth's atmosphere. In most living things, a second stage of cellular respiration called **oxidative respiration** follows glycolysis. Oxidative respiration takes place within mitochondria. It is far more effective than glycolysis at recovering energy from food molecules. Oxidative respiration is the method by which plant and animal cells get the majority of their energy.

Glycolysis breaks down glucose into two pyruvate molecules

Glycolysis is one of the most ancient biological processes we know. It evolved among bacteria, the first life forms on Earth. Bacteria, like those in **Figure 5.15**, relied upon glycolysis to make the ATP needed to drive

Figure 5.15
Glycolysis evolved among ancient bacteria that were similar to the photosynthetic bacteria (160X), below.

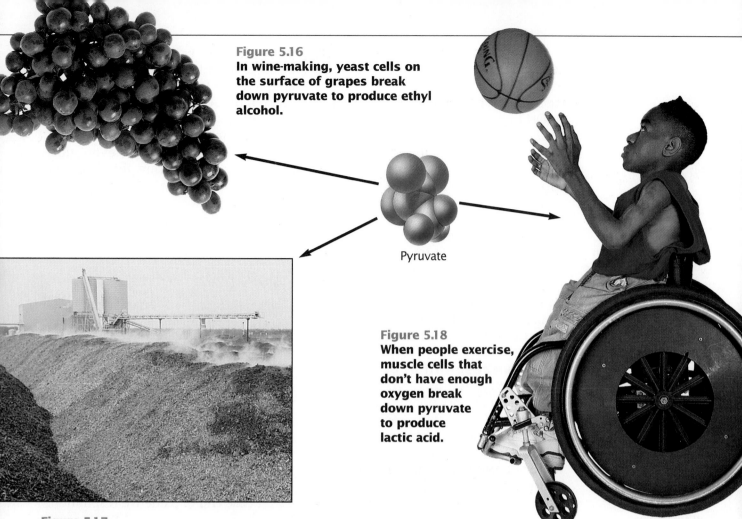

Figure 5.16
In wine-making, yeast cells on the surface of grapes break down pyruvate to produce ethyl alcohol.

Pyruvate

Figure 5.18
When people exercise, muscle cells that don't have enough oxygen break down pyruvate to produce lactic acid.

Figure 5.17
Bacteria act upon pyruvate to convert organic plant and animal wastes into "biogas"—a methane fuel used for heating, cooking, and, in some countries, for transportation. The biogas power-plant pictured above is in California.

chemical reactions within their cells. Today virtually every living organism, including you, still uses glycolysis.

The word "glycolysis" means "the splitting of glucose." In a series of 10 reactions, a molecule of glucose is split into two identical smaller molecules, each called **pyruvate** (*py ROO vayt*). Although the cell must use some ATP to begin glycolysis, the overall process produces more ATP than was used to initiate it. For each molecule of glucose that enters glycolysis, the cell harvests two molecules of ATP.

During glycolysis, an electron carried on a hydrogen atom is stripped from glucose. This electron is donated to an electron carrier molecule called NAD, forming NADH. For glycolysis to keep going, however, the electrons stripped away from the glucose molecules must be donated to some other organic molecule. This frees the NAD to go back and accept more hydrogens from glycolysis. Thus, glycolysis is followed by either fermentation or oxidative respiration.

Fermentation takes place in the absence of oxygen

Glycolysis evolved before the Earth's atmosphere contained oxygen. Consequently, the earliest energy-harvesting pathway did not require oxygen. The breakdown of organic compounds such as glucose in the absence of oxygen is called **fermentation**. During fermentation the hydrogen from NADH is attached to the pyruvate, forming lactic acid or ethyl alcohol (ethanol), the alcohol in beer and wine.

In the conversion to alcohol, pyruvate loses a molecule of carbon dioxide as it accepts an electron from NADH. This process regenerates NAD, which enables glycolysis to continue. Many microorganisms that live in the absence of oxygen use fermentation to produce small amounts of ATP. Some pathways of fermentation are shown in **Figures 5.16–5.18.**

Fermentation occurs in your muscle cells when they are forced to operate without enough oxygen. Electrons freed by glycolysis are donated from NADH to pyruvate without the release of carbon dioxide, forming lactic acid. This process allows glycolysis to continue in your muscles as long as the supply of glucose holds out. Blood circulation removes excess lactic acid from muscles—but when lactic acid cannot be removed as fast as it is produced, your muscles cease to work well. Try raising and lowering your arms rapidly 100 times, for example. Because of this limit in removing lactic acid, the world record for running one mile is slightly under four minutes and not less.

Oxidative respiration occurs in the presence of oxygen

When Earth's atmosphere became rich in oxygen, an alternative to fermentation became possible. Instead of using the hydrogen atoms freed by glycolysis to form ethanol or lactic acid, the hydrogen atoms could now be attached to oxygen atoms, forming water. This pathway is a wonderful alternative because the attachment process can be coupled to a proton pump and used to make more ATP. Not only that, the end product of glycolysis undergoes steps not possible in the absence of oxygen. Pyruvate is used to make even more ATP than is made during glycolysis and fermentation. **Figure 5.19** summarizes these processes.

The equation for the breakdown of glucose by oxidative respiration is:

$$C_6H_{12}O_6 + 6O_2 + ADP + P \rightarrow$$
$$6CO_2 + 6H_2O + ATP$$

This equation indicates that glucose and oxygen react to form carbon dioxide, water, and ATP molecules. Inside mitochondria, oxidative respiration picks up where glycolysis left off. Each of the two pyruvate molecules produced by glycolysis is oxidized, freeing a high-energy electron and a carbon, which is released as CO_2. The electron is donated to NAD, forming NADH, which will be used at the end of oxidative respiration.

Figure 5.19
Glucose is broken down into pyruvate during glycolysis. Two molecules of ATP are released. Without oxygen present, fermentation can follow, resulting in the formation of lactic acid or ethyl alcohol and carbon dioxide gas. When oxygen is present, oxidative respiration can occur. This process yields as many as 36 ATP molecules.

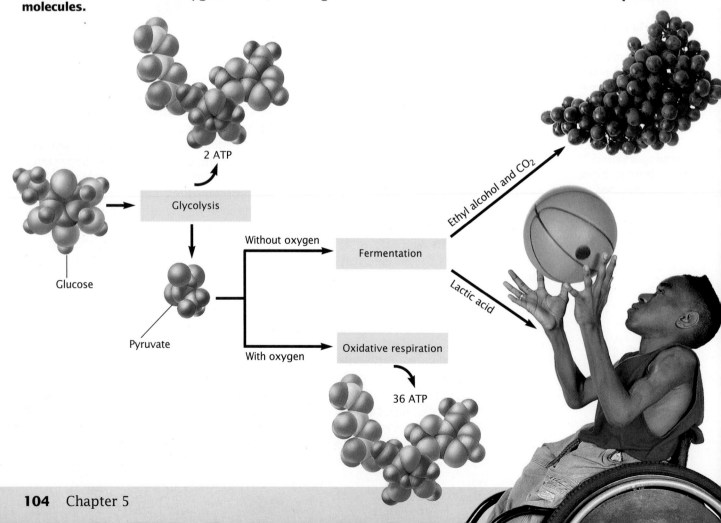

Glucose

Glycolysis

2 ATP

Pyruvate

Without oxygen

With oxygen

Fermentation

Oxidative respiration

Ethyl alcohol and CO₂

Lactic acid

36 ATP

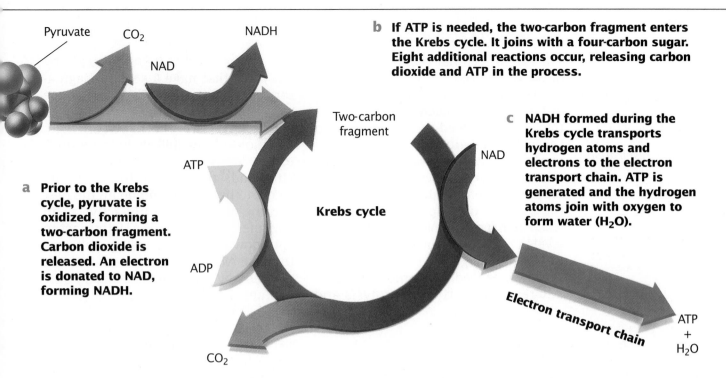

Figure 5.20
Oxidative respiration occurs in two major steps: the Krebs cycle and the electron transport chain.

After each pyruvate is oxidized, only a two-carbon fragment remains. If the cell has enough ATP, the two-carbon fragment is funneled into fat synthesis and its energetic electrons are stored. If the cell needs ATP now, the two-carbon fragment will continue on to the next steps of oxidative respiration.

The Krebs cycle yields ATP and carbon dioxide

The two-carbon fragment, left over after the oxidation of pyruvate, joins with a four-carbon sugar. Then, in rapid-fire order, eight additional reactions occur. When the two-carbon fragment is used up, its two carbon atoms are expelled as two molecules of carbon dioxide. In addition, one ATP molecule has been made and four more energetic electrons have been harvested. All that remains is the original four-carbon sugar, now free to join with another two-carbon fragment. This cycle of nine reactions is known as the Krebs cycle, after Sir Hans Krebs, the biochemist whose work in the 1930s revealed how these reactions work. **Figure 5.20** summarizes the major events of the Krebs cycle.

The electron transport chain makes more ATP

The energetic electrons in the molecules of NADH that formed during the Krebs cycle are used to make ATP in a series of reactions known as the electron transport chain. The membranes of a mitochondrion contain proteins that serve as proton pumps. Using these proton pumps, a mitochondrion pumps protons outward. Driven by diffusion, the protons then pass back into the interior of the mitochondrion. The energy of the reentering protons is used by the mitochondrion to attach a phosphate group onto ADP, making new molecules of ATP. A living cell is never without a supply of ATP.

What happens to the electrons after the proton pumps have used their energy? The hydrogen atoms carrying them are joined to oxygen gas to form water. Because the electrons stripped from pyruvate need to find a final home, cellular respiration requires oxygen. The energy cannot be extracted from pyruvate without oxygen to siphon off the spent electrons. Otherwise, the proton pumps and other electron-ferrying components of the mitochondrion would soon become clogged with used electrons.

Oxidative respiration is very efficient. The breakdown of a molecule of glucose to pyruvate during glycolysis yields a net of only two ATP molecules, but oxidative respiration can yield as many as 36 additional molecules of ATP!

Regulating Cellular Respiration

Patterns of Change

How does the regulation of cellular respiration resemble a thermostat controlling the temperature in your home?

The rate of cellular respiration slows down when your body's cells already have enough ATP. This is very sensible, but what signals each mitochondrion that it is supposed to slow down and cease ATP production? The control works through a system of feedback inhibition in which excess product shuts off the reaction. **Feedback inhibition** is the slowing or stopping of an early reaction in a biochemical pathway when levels of the end product of the pathway become high.

How does feedback inhibition work? Key reactions early in glycolysis and the Krebs cycle are catalyzed by enzymes that have a second "regulatory" site. This site is the same shape as ATP, as shown in **Figure 5.21**.

When ATP levels in the cell are high, ATP molecules will likely become stuck to the site. The binding of ATP to the site goads the protein to change its shape to better accommodate the fit—and the new shape is not active as an enzyme. High levels of ATP thus act to shut down the processes the cell uses to make ATP. Like a well-designed car, your energy-producing machinery only operates when you step on the gas.

Figure 5.21
Feedback inhibition slows or stops an early reaction in a biochemical pathway when levels of the end product are high. Molecules of the end product bind to an enzyme's regulatory site.

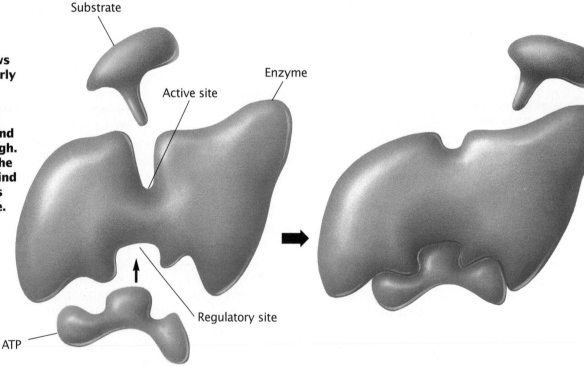

Substrate

Enzyme

Active site

Regulatory site

ATP

a In this reaction, ATP binds with the regulatory site.

b The enzyme then changes shape. The active site is no longer able to catalyze its usual reaction, and no more ATP is made.

Section Review

❶ **How is cellular respiration important for living cells?**

❷ **Why do you think glycolysis might be particularly advantageous for certain organisms?**

❸ **How do fermentation and oxidative respiration differ?**

❹ **Explain how feedback inhibition enables cellular respiration to slow down when supplies of ATP are sufficient.**

Green grass is the mainstay of this cow's diet. Using what you've learned, can you explain the link between sunlight and dairy products?

	Key Terms	Summary

5.1 Cells and Chemistry

Enzymes, such as carbonic anhydrase, hasten chemical reactions.

exergonic (p. 90)

endergonic (p. 90)

metabolism (p. 90)

activation energy (p. 90)

catalyst (p. 91)

substrate (p. 92)

active site (p. 92)

- Chemical reactions store or release energy needed for life.
- Enzymes make reactions occur rapidly enough to sustain life.
- Enzymes act on specific substrates, and are affected by heat, acidity, and substrate concentration.

5.2 Cells and Energy

Living things depend on the energy stored in green plants.

biochemical pathway (p. 95)

ATP (p. 95)

photosynthesis (p. 96)

cellular respiration (p. 96)

- In a biochemical pathway, energy is released by the breakdown of chemical compounds.
- The energy released is stored temporarily in a molecule called adenosine triphosphate, or ATP.
- Photosynthesis captures the light energy available for life.
- Cellular respiration releases energy stored in food molecules.

5.3 Photosynthesis

Photosynthesis begins when light strikes a chloroplast.

photon (p. 98)

pigment (p. 98)

oxidation-reduction reaction (p. 99)

- Photosynthesis evolved billions of years ago among bacteria.
- Green plants and algae contain the pigment chlorophyll.
- Light striking chlorophyll boosts electrons to higher energy levels. As electrons pass through a series of oxidation-reduction reactions, their energy is used to produce carbohydrates.

5.4 Cellular Respiration

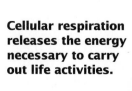

Cellular respiration releases the energy necessary to carry out life activities.

glycolysis (p. 102)

oxidative respiration (p. 102)

pyruvate (p. 103)

fermentation (p. 103)

feedback inhibition (p. 106)

- Living things obtain energy from food through cellular respiration.
- In glycolysis, glucose splits into two pyruvate molecules.
- In the absence of oxygen, fermentation converts pyruvate to lactic acid or ethyl alcohol. With oxygen present, the Krebs cycle and the electron transport chain produce most of the ATP needed for life.
- Feedback inhibition controls reactions of respiration.

Understanding Vocabulary

1. For each set of terms, complete the analogy.
   a. reactants:change ::enzyme:_____
   b. fermentation: anaerobic:: oxidative respiration:_____
   c. oxidation:lose an electron:: reduction:_____
   d. photosynthesis: stores energy:: cellular respiration:_____

Relating Concepts

2. Copy the unfinished concept map below onto a sheet of paper. Then complete the concept map by writing the correct word or phrase in each oval containing a question mark.

Understanding Concepts

Multiple Choice

3. Enzymes in the human body
   a. are changed during the reaction they catalyze.
   b. decrease the activation energy of a chemical reaction.
   c. are completely burned in the reactions.
   d. increase the speed of proton pumps.

4. The sum of all chemical reactions in the body is called
   a. meiosis.
   b. metabolism.
   c. respiration.
   d. activation energy.

5. Chemical reactions require energy to get started. What is this energy called?
   a. metabolism
   b. chemical energy
   c. activation energy
   d. potential pressure

6. Which is an example of a biochemical pathway?
   a. glycolysis   c. ATP
   b. pyruvate   d. catalyst

7. The formation of ATP involves the
   a. addition of a phosphate to ADP.
   b. reaction of chemical stimuli.
   c. bonding of carbon molecules.
   d. release of energy and ADP.

8. The energy supplied by the oxidation of glucose is used to
   a. break down ATP.
   b. make ATP from ADP.
   c. produce starch.
   d. change carbon dioxide to oxygen.

9. Products of glycolysis include
   a. chemical pathways.
   b. ATP and alcohol.
   c. pyruvate and ATP.
   d. glucose and energy.

10. When oxygen is unavailable, muscle tissue converts pyruvate to
    a. alcohol.   c. enzymes.
    b. ATP.   d. lactic acid.

11. The ultimate goal of photosynthesis is to
    a. make ATP from carbon dioxide.
    b. construct carbon-containing molecules that serve as an energy source.
    c. convert ADP to ATP by using energy from the sun.
    d. use enzymes to speed up chemical reactions.

Completion

12. Cells need _____ in order to move, transport food, and get rid of waste.

13. In oxidation-reduction reactions, the molecule that gains the electron is _____ , while the molecule that loses the electron is _____ .

14. Inside the cell, glycolysis occurs in the _____ , but oxidative respiration occurs in the _____ .

15. The sun's energy is used to produce food during _____ , while energy stored in food is released during _____ .

16. The energy needed by the human body to carry out the processes of life comes from _____ .

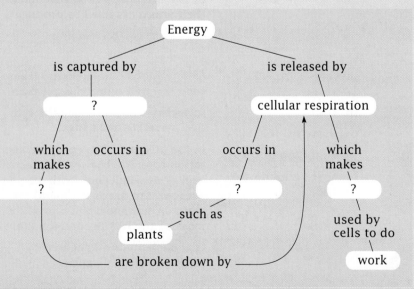

Short Answer

17. Do plants perform cellular respiration? Support your answer.

18. If you tried to grow a plant in a dark closet, what would happen? Explain your answer.

19. What is a green plant's source of carbon for photosynthesis?

20. Where does the glucose come from that the cell oxidizes or breaks down during respiration?

Interpreting Graphics

21. Look at the processes below.

    Photosynthesis:

    $$6CO_2 + 6H_2O \xrightarrow{\text{(light)}} C_6H_{12}O_6 + 6O_2$$

    Oxidative respiration:

    $$C_6H_{12}O_6 + 6O_2 + ADP + P \rightarrow 6CO_2 + 6H_2O + ATP$$

    • What are the products of photosynthesis?

    • What are the reactants for oxidative respiration?

    • How are photosynthesis and respiration related?

22. Match the photograph below with one of these three pathways of cellular respiration:

    a. Fermentation → lactic acid
    b. Fermentation → ethyl alcohol and CO_2
    c. Glycolysis

Reviewing Themes

23. *Interacting Systems*
    If all the plants on Earth died, what would happen to the animals?

24. *Interacting Systems*
    How does the feedback system of cellular respiration operate?

25. *Evolution*
    What evidence suggests that glycolysis is a more ancient process than oxidative respiration?

Thinking Critically

26. *Inferring Conclusions*
    Temperature affects the rate of enzyme reactions. How does freezing food affect the enzyme activity of organisms that cause decay?

27. *Comparing and Contrasting*
    How is the interaction of a lock and key analogous to the interaction of an enzyme and its substrate?

28. *Inferring Conclusions*
    What happens to the carbon dioxide gas produced by yeast cells when bread is baked?

29. *Building on What You Have Learned*
    In Chapter 3 you read about the proton pump. How does this pump function during photosynthesis?

Cross-Discipline Connection

30. *Biology and Social Studies*
    Yogurt is a fermented dairy product with an unusual history. Use library resources to find out how fermentation has been used in the production of yogurt, cheese, soybean products, wine, and beer.

Discovering Through Reading

31. Read the article "Oxygen, the Great Destroyer," in *Natural History*, August, 1992, pages 46–53. How do plants protect themselves from oxidation?

Investigation

How Does Temperature Affect the Rate of Photosynthesis?

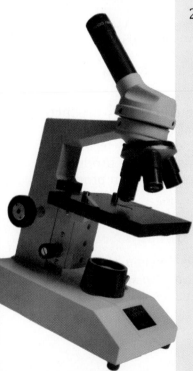

Objectives

In this investigation you will:
- *measure* the rate of photosynthesis
- *evaluate* the effect of temperature on the rate of photosynthesis

Materials

- water
- 600-mL beaker
- sodium bicarbonate ($NaHCO_3$)
- *Elodea* or other aquatic plant
- glass funnel
- test tube
- thermometer
- watch or clock with second hand

Prelab Preparation

1. Review what you have learned about a scientific method by answering the following questions:
   - What is a hypothesis?
   - What is a control experiment?

2. Review what you have learned about photosynthesis by answering the following questions:
   - Summarize the process of photosynthesis. List the reactants and the products.
   - What is meant by rate of photosynthesis?
   - Which product of photosynthesis would be easiest to measure in an aquatic plant? Explain your reasoning.
   - List the physical factors that might affect the rate of photosynthesis.

Procedure: Testing a Hypothesis

1. Form a cooperative team with a classmate to complete steps 2–10.

2. You and your partner are to design an experiment that demonstrates the effects of temperature on the rate of photosynthesis. With your partner, construct a hypothesis that you will test in your experiment. *Explain the reasoning for your hypothesis.*

3. Design a control experiment that tests your hypothesis. *What variable is being tested in your experiment? What variable is being measured in your experiment?* Be sure to include a table to contain your observations and data. Proceed with your experiment only after your experimental design and your table have been approved by your teacher.

4. Pour water into the 600-mL beaker until it is half full. Dissolve 3 g of sodium bicarbonate in the water.

5. Place a 6–10 cm (2–4 in.) length of *Elodea* in the bottom of the beaker. Place the glass funnel over the plant.

6. Fill a test tube with water. Placing your thumb tightly over the mouth of the test tube, invert the test tube, and place it over the end of the funnel.

7. Use a thermometer to find the temperature of the water in the beaker.

8. Set the beaker in direct sunlight or under a bright light. After two minutes observe the funnel. Record your observations.

9. Count the bubbles as they rise in the funnel's tube for a period of 210 seconds, recording the total number of bubbles at 30-second intervals.

10. Measure the temperature of the water again. If the temperature has changed by more than three degrees, find the average of the two temperatures. Record this average in your table as the water's temperature.

11. Pour out the water from the beaker and replace it with water differing in temperature by five to eight degrees. Add 3 g of sodium bicarbonate to the water and repeat steps 5–10.

12. On a separate sheet of paper, draw a graph like the one shown below. Graph your data. Use a separate line for data collected at each temperature.

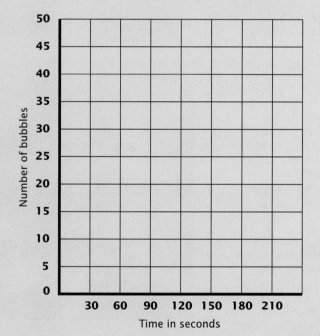

13. Clean up your materials and wash your hands before leaving the lab.

Analysis

1. *Summarizing Data*
Summarize your results by describing the data illustrated in your graph.

2. *Analyzing Data*
State a conclusion that relates to your hypothesis. Explain how your data support your conclusion.

3. *Making Inferences*
Explain why the data give an indication of the rate of photosynthesis.

4. *Analyzing Relationships*
Can you assume that raising the temperature of the water to 50°C (122°F) would continue the trend indicated by your data? Explain your answer.

Thinking Critically

1. Suggest a reason for adding sodium bicarbonate to the water in the beaker. Describe a control experiment that would determine the effects of sodium bicarbonate on the rate of photosynthesis.

2. What other physical factors might be affecting your experimental results?

Energy and Life   **111**

WHAT IS LIFE? SEARCHING FOR ANSWERS

1500

1500 Observations of fishes in dried ponds, and snakes and frogs emerging from the mud of river-banks lead many people to believe that life **generates spontaneously** from non-living matter.

1600 Belgian physician **Jan Baptist van Helmont** places wheat grains in a sweaty shirt and returns 21 days later to find mice. He concludes that human sweat changes wheat grains into mice.

Anton van Leeuwenhoek

1668 The first to observe animalcules (bacteria) through a lens, **Anton van Leeuwenhoek**, a Dutch amateur scientist, reports his findings to scientists of The Royal Society of London. They conclude that larger animals may not be able to arise from nonliving matter, but microscopic organisms can.

Redi's experiment

1668 Observing maggots and flies developing on decaying meat, Italian physician **Francesco Redi** questions spontaneous generation. He concludes that if flies are kept away from rotten meat, maggots do not develop.

1920

1924 Soviet biochemist **A.I. Oparin** and British geneticist **J.B.S. Haldane** suggest that life may have come from nonliving matter on primitive Earth, when the atmosphere consisted of methane, ammonia, water vapor, and hydrogen. Energy supplied by electrical storms and ultraviolet light may have broken down atmospheric gases into elements, which reacted to form amino acids, the building blocks of proteins.

1953 American scientists **Stanley L. Miller** and **Harold C. Urey** of the University of Chicago test Oparin's hypothesis. Their experiment generates four amino acids, among other molecules.

1960 Exobiology, a term coined by American geneticist Joshua Lederberg, is the study of extraterrestrial life origins on Earth, or the existence of life outside Earth or its atmosphere.

Carl Sagan

1974 American astronomer **Carl Sagan** and other scientists from Cornell University beam a signal containing coded messages into outer space, with hopes of hearing from extra-terrestrial life forms.

1860

1750 The English Catholic priest **John Needham**, in support of his belief in spontaneous generation, reenacts Redi's experiments but concludes that life generates spontaneously.

Pasteur's swan-necked flasks

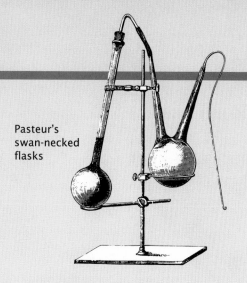

1860 French chemist **Louis Pasteur** ends the spirited controversy over spontaneous generation. In a series of experiments, Pasteur proves that microorganisms appear only as contaminants from the air and not "spontaneously."

1775 Italian biologist **Abbe Lazzaro Spallanzani** challenges the findings of Needham. He repeats Redi's experiments and concludes that life does not generate spontaneously.

1855 German pathologist **Rudolf Virchow** proposes that "all living things arise from preexisting living cells."

Louis Pasteur

2004

1976 Despite data collected on **Viking Missions 1 and 2**, scientists still question whether life exists on Mars.

1992 Program Sentinel (a radio telescope with 131,072 channels)—a project of **SETI** (Search for Extraterrestrial Intelligence)—sifts through a shower of radio waves falling on Earth for signals indicating extraterrestrial intelligence.

1985 Alexander Graham Cairns-Smith proposes that naturally occurring clay crystals may have served as the original templates for assembling amino acids into protein chains, providing the fundamental materials from which life originated.

2000 An **antenna** is proposed to be set up on the far side of the Moon to evaluate radio signals for possible intelligent extra-terrestrial life.

Human expedition to Mars

Moon

2004 The focus of **future NASA missions**: settling the issue of life on Mars, past or present.

Continuity of Life

O f all life's mysteries, one of the most puzzling has been heredity. Why do we resemble our parents? What mechanism told the tiny embryo that became you how to develop? We are now able to answer that question, for biologists have largely solved the mystery of heredity. In this unit, you will discover that the mechanism consists of chromosomes, which pass instructions from parent to child about how to grow and develop. Learning how heredity works has been one of the greatest triumphs of science.

Genetics and Inheritance

Review

- **heredity (Section 1.3)**
- **chromosome (Section 4.3)**
- **mitosis (Section 4.4)**
- **the term *gene* (Glossary)**

No two people are exactly alike. What you look like depends on what combination of genes you have inherited from your parents.

THE TRANSMISSION OF TRAITS FROM PARENTS TO THEIR OFFSPRING IS CALLED HEREDITY. FOR HUNDREDS OF YEARS, PEOPLE WERE PUZZLED BY THE PATTERNS OF HEREDITY. ONLY IN THIS CENTURY HAVE SCIENTISTS BEGUN TO UNDERSTAND THE MOLECULAR BASIS OF HEREDITY, A DISCOVERY THAT HAS ITS ROOTS IN THE ELEGANT EXPERIMENTS OF GREGOR MENDEL.

6.1 *The Puzzle of Heredity*

Objectives

❶ Explain how Mendel discovered the laws of heredity using the information available to him.

❷ Outline the garden pea experiments of Gregor Mendel.

❸ Define the terms heterozygous, homozygous, dominant, and recessive.

❹ Compare and contrast each of Mendel's laws of heredity.

Gregor Mendel and the Garden Pea

Gregor Mendel

Figure 6.1
Mendel studied different traits of the garden pea, *Pisum sativum*, in an attempt to understand heredity.

The scientific study of heredity is called **genetics** (*juh NEHT ihks*). The basis for this science began when scientists crossed, or bred, varieties of animals and plants in an attempt to understand heredity. For example, in the 1790s the British farmer T.A. Knight crossed a type of garden pea plant that had purple flowers with one that had white flowers. What color flowers do you think the offspring had? People at that time thought traits blended the way paint colors mix. If heredity is a blending process, wouldn't all the flowers be lavender? But all of the offspring of Knight's cross had purple flowers! If two of these purple-flowering offspring were crossed, however, some of their offspring had purple flowers and some had white. Why was this so? Knight's studies puzzled people for many years.

Some of the most important pieces in the complex puzzle of heredity were assembled in the garden of an Austrian monastery in the 1860s. There lived Gregor Mendel, shown in **Figure 6.1**, a monk who was interested in both science and mathematics. Curious about the patterns of heredity, Mendel repeated the garden pea plant crosses done by Knight and others. The difference between Mendel's experiments and those done by earlier researchers was that Mendel counted the pea plants resulting from each cross. Mendel worked diligently to express the results of his experiments in terms of numbers. This quantitative approach to science, which counted and measured data, was just becoming fashionable in Europe. Mendel hoped that this detailed and numerical procedure would give some hint of how hereditary processes work. The results of his experiments changed the course of biology.

Mendel conducted his experiments in three steps

Mendel chose pea plants for his experiments because they are small, easy to grow, produce large numbers of offspring, and mature quickly. Mendel also took advantage of the reproductive characteristics of pea plants. He could either allow the plants to self-fertilize, or he could cross-fertilize them. Self-fertilization occurs when the sperm (located inside a pollen grain) of one flower fertilizes the egg (located in the base of the flower) of the same flower. Cross-fertilization occurs when pollen is transferred by hand from the flower of one plant to a flower of another.

Also, Mendel took advantage of the fact that many varieties of garden pea are available and can be easily grown. He selected seven pea plant traits for study. These seven traits were: flower color, seed color, seed shape, pod color, pod shape, flower position on the stem, and plant height. Each of these traits had two contrasting forms. For example, flower color could be purple or it could be white. In this text, we will follow only Mendel's experiments with flower color, although he studied the inheritance of all seven traits. The three steps that Mendel followed when he designed his experiments are shown in **Table 6.1**.

Table 6.1 How Mendel Designed His Experiments

	What Mendel Did	What Mendel Found
Step 1 ***Parental Generation*** First Mendel produced a parental (P) generation.	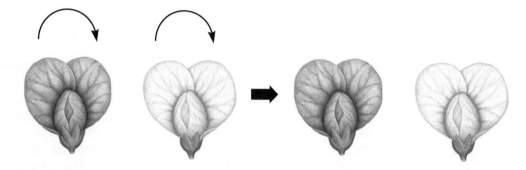 Mendel allowed the pea plants to self-fertilize for many generations. He then collected and grew the seeds from these plants. This would ensure that the plants he used for his research were "true-breeding," meaning that their offspring would produce only one form of a particular trait.	Mendel called the offspring the parental generation, or P generation for short. These plants were true-breeding, that is the white-flowering variety would only produce white flowers and purple-flowering variety would produce only purple flowers.
Step 2 ***F_1 Generation*** Next Mendel produced the F_1 generation.	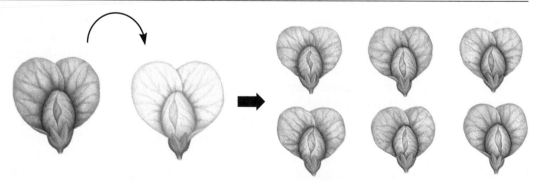 Mendel cross-fertilized the two different parental (P) varieties. He took pollen from a plant with white flowers and placed it on a flower that was purple. He then collected and grew the resulting seeds.	Mendel called the offspring the F_1 generation (F_1 for "first filial" generation, from the Latin word for son or daughter). Mendel found that all the F_1 plants had only purple flowers. No F_1 plants had white flowers. *(See Step 3 on the next page.)*

Step 3
F₂ Generation
Then Mendel produced the F₂ generation.

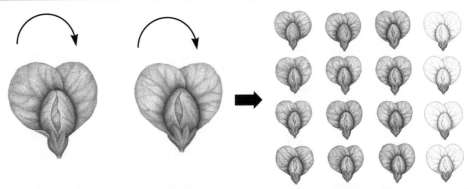

Mendel allowed the F₁ generation plants to self-fertilize. As before, he grew the seeds from these plants.

Mendel called the offspring the F₂ (second filial) generation. White flowers reappeared among the offspring. All together, Mendel grew 929 F₂ individuals. There were 705 plants with purple flowers and 224 plants with white flowers.

Patterns of Change

Explain how a form of a trait that appears in both parents could be absent from their children.

Mendel developed a model to explain his results

Mendel realized the ratio of purple-flowering plants to white-flowering plants in the F₂ generation was just about 3:1. Mendel obtained this same result for each cross. Surely this ratio must mean something important. To explain his results, Mendel came up with a simple model, a set of rules that could be used to accurately predict patterns of heredity. These rules summarize Mendel's ideas about inheritance.

1. Parents transmit information about traits to their offspring. Mendel called this information "factors."

2. Each individual has two factors for each trait, one from each parent. The two factors may or may not have the same information. If the two factors do have the same information (for example, if both have information for purple flowers), the individual is said to be **homozygous** (*hoh muh ZY guhs*) for the trait. If the two factors have different information (for example, one factor codes for purple flowers and the other for white flowers), the individual is said to be **heterozygous** (*heht uh roh ZY guhs*).

3. The alternative forms of a factor are called **alleles** (*uh LEELS*). The many alleles that an organism possesses make up its **genotype** (*JEE nuh typ*). An organism's physical appearance, which is determined by its alleles, is called its **phenotype** (*FEE nuh typ*).

4. An individual possesses two alleles for each trait. One allele is contributed by the female parent, and the other is contributed by the male parent. The two alleles in each pair are not affected by each other. They are passed on when an individual matures and produces gametes (eggs and sperm). During the formation of gametes, the paired alleles segregate (separate) randomly so that a gamete receives a copy of one allele or the other.

5. The presence of an allele does not ensure that the trait will be expressed in the individual that carries it. In heterozygous individuals, only the **dominant** allele achieves expression. The **recessive** allele is present but remains unexpressed. In Mendel's F₁ generation, purple flower color was caused by a dominant allele. For every pair of contrasting forms of a trait—tall versus short, or green seeds versus yellow seeds—the allele for one form of the trait was always dominant and the allele for the other form of the trait was always recessive.

Visualizing Mendel's Model

Mendel's model can be understood easily by diagraming Mendel's crosses. For example, consider Mendel's cross of parental (P) purple-flowering pea plants with parental white-flowering pea plants. For the sake of convenience, letters are used to represent alleles. The recessive allele is represented by the lower case letter "w" (for "white"), and the dominant allele is represented by the corresponding upper case letter "W." A plant that is true-breeding for the recessive white flower would be designated "ww," and a plant that is true-breeding for the dominant purple flower would be designated "WW." A heterozygote would be designated "Ww."

A simple diagram called a Punnett square can help you visualize crosses. Named for the British geneticist Reginald Punnett, a Punnett square is also a handy device for predicting the results of a cross.

In a Punnett square, the symbols for all the possible alleles carried by male gametes (sperm) are arranged along the top of the square, while all the possible alleles carried by the female gametes (eggs) are shown along the left side. By combining the symbol for an allele carried by a male gamete with the symbol for the allele carried by the female gamete, all the possible gamete combinations can be predicted, as shown in **Figure 6.2**.

Figure 6.2
This Punnett square illustrates a cross between two true-breeding varieties of garden pea plants. This was Step 2 of Mendel's experiment.

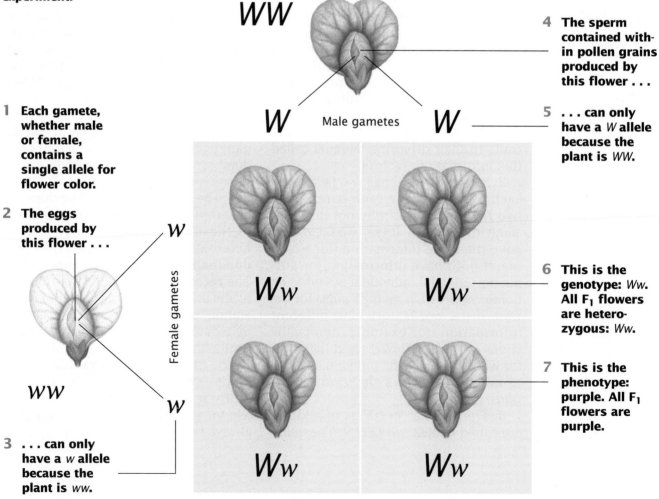

1 Each gamete, whether male or female, contains a single allele for flower color.

2 The eggs produced by this flower . . .

3 . . . can only have a *w* allele because the plant is *ww*.

Male gametes

Female gametes

4 The sperm contained within pollen grains produced by this flower . . .

5 . . . can only have a *W* allele because the plant is *WW*.

6 This is the genotype: *Ww*. All F$_1$ flowers are heterozygous: *Ww*.

7 This is the phenotype: purple. All F$_1$ flowers are purple.

Crossing two heterozygous plants

When heterozygous (Ww) F_1 individuals with purple flowers are allowed to self-fertilize, what will the resulting F_2 individuals be like? Let's predict what we would expect to find using Mendel's rules. To make a prediction, consider what scientists call probability. Probability is simply the likelihood that something will happen. For example, when you toss a coin, the probability that the coin will land with "heads" up is 50 percent, or one-half, because the coin is just as likely to fall showing "tails." Similarly, because of the segregation of alleles, the probability that an egg or sperm cell in one of Mendel's heterozygous F_1 pea plants (Ww) will contain the allele W is 50 percent, or one-half, just as in a coin toss. Obviously, the reverse is also true: the probability that a gamete will contain the allele w is also 50 percent or one-half.

Study the Punnett square for a cross involving two purple-flowered heterozygous plants (Ww) shown in **Figure 6.3**. In this particular cross, one-half the male and female gametes carry the allele for purple flowers (W). The other one-half of the gametes carry the allele for white flowers (w). You can see that Mendel's model clearly predicts that 75 percent of the F_2 generation will have purple flowers and 25 percent will have white flowers, a 3:1 ratio. The Punnett square also shows the genotypes found in the F_2 generation. Twenty-five percent of the F_2 is homozygous (ww) with white flowers, 50 percent heterozygous (Ww) with purple flowers, and the remaining 25 percent is homozygous (WW) with purple flowers. The 3:1 ratio Mendel had repeatedly observed is the expression of an underlying 1:2:1 ratio of genotypes in which the heterozygotes look like one of the homozygotes.

Figure 6.3
This Punnett square illustrates a cross between two F_1 individuals. This was Step 3 of Mendel's experiment.

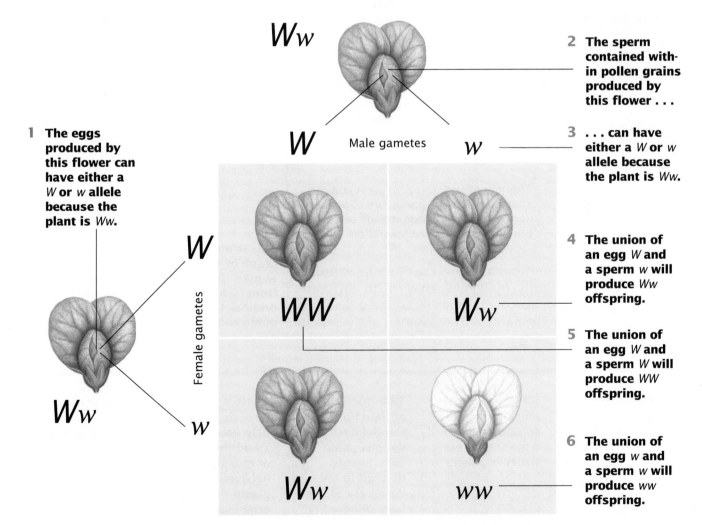

1 The eggs produced by this flower can have either a W or w allele because the plant is Ww.

2 The sperm contained within pollen grains produced by this flower . . .

3 . . . can have either a W or w allele because the plant is Ww.

4 The union of an egg W and a sperm w will produce Ww offspring.

5 The union of an egg W and a sperm W will produce WW offspring.

6 The union of an egg w and a sperm w will produce ww offspring.

Male gametes

Female gametes

Mendel's Laws of Heredity

As you have just seen, Mendel's model analyzes the 3:1 ratio Mendel saw in the F_2 generation in a neat and very satisfying way. Similar patterns of heredity have been observed in countless other organisms. Mendel summarized his results by formulating two laws.

To describe how traits can disappear and reappear in a certain pattern from generation to generation, Mendel proposed what is now called the law or principle of segregation.

> **The law of segregation states that the members of each pair of alleles separate when gametes are formed. A gamete will receive one allele or the other.**

You saw this law at work in the Punnett squares shown on pages 120 and 121.

Mendel went on to study the inheritance of two or more pairs of traits. He crossed pea plants with contrasting forms of two traits, such as flower color and plant height. He found that the inheritance of one trait did not influence the inheritance of the other trait. This idea is the basis for Mendel's second law, the law or principle of independent assortment.

> **The law of independent assortment states that two or more pairs of alleles segregate independently of one another during gamete formation.**

We now know that this law applies only for genes located on different pairs of chromosomes or far apart on the same chromosome pair. (Keep in mind that chromosomes had not yet been discovered in Mendel's time.)

Mendel's laws were not discovered for more than thirty years

Mendel was a member of a local science society. Each member tackled a scientific investigation that focused on an area of personal interest. The research would later be discussed at meetings. Mendel's paper describing his results was published in the science society journal in 1866. Unfortunately, his paper failed to arouse much interest, and the importance of his work was not recognized. In 1900, sixteen years after Mendel's death, several investigators independently rediscovered Mendel's pioneering paper. Scientists soon realized that chromosomes, discovered not long before Mendel's death, are the carriers of heredity.

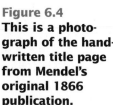

Figure 6.4
This is a photograph of the handwritten title page from Mendel's original 1866 publication.

Evolution

How does the law of independent assortment provide a source of variation from one generation to the next?

Section Review

❶ How did Mendel's approach to science differ from that of earlier researchers?

❷ In three steps, explain how Mendel designed the pea plant experiments that led to his discovery of the laws of heredity.

❸ How could a Punnett square help you understand the results of crosses between pea plants or other organisms?

❹ How does the law of segregation differ from the law of independent assortment?

MENDEL'S THEORY DISPELLED THE MYSTERY OF WHY TRAITS SEEMED TO APPEAR AND DISAPPEAR MAGICALLY FROM ONE GENERATION TO THE NEXT. STRIPPED OF RATIOS AND SYMBOLS, MENDEL'S WORK SHOWS US THAT PATTERNS OF HEREDITY REFLECT THE TRANSMISSION OF ENCODED INFORMATION FROM PARENTS TO OFFSPRING. THIS INFORMATION IS LOCATED ON CHROMOSOMES.

6.2 Chromosomes

Objectives

1. **Explain the relationship between genes and chromosomes.**

2. **Summarize the events that occur during meiosis.**

3. **Define crossing over and explain its role in evolution.**

4. **Describe how sex chromosomes determine the sex of humans.**

Genes and Chromosomes

Mendel concluded that factors containing information about traits are transmitted from parents to offspring. But what exactly are these factors? This question dominated biology for more than half a century after Mendel's work was rediscovered in 1900.

In 1909, a Danish biologist named Wilhelm Johannsen first used the word "gene" to describe the physical units of heredity that Mendel had called factors. A **gene** is a segment of the DNA molecule that carries the instructions for producing a specific trait. Recall that the alternative forms of a factor, or gene, are called alleles.

In Mendel's time, no one knew of chromosomes or genes. Chromosomes were first observed in 1879 by the German scientist Walther Flemming. In 1902, the American biologist Walter Sutton gave a convincing argument supporting an earlier notion that genes are located on chromosomes. Ten years later, Thomas Hunt Morgan, shown in **Figure 6.5**, verified Sutton's idea. Morgan presented clear-cut evidence that the presence of white eye color in fruit flies, which usually have red eyes, is associated with a particular gene on a particular chromosome. Biologists around the world soon accepted the chromosomal theory of inheritance—the genes on chromosomes are the units of inheritance. In Chapter 7 you will learn how the information in a gene results in the formation of individual traits.

Figure 6.5
The American biologist Thomas Hunt Morgan showed that the genes for specific traits of a fruit fly (inset) are located on the fruit fly's chromosomes.

Humans have forty-six chromosomes in most cells

If you could look at almost any cell in your body, you would find 46 chromosomes that are similar to the one shown in **Figure 6.6**. If you sorted the chromosomes by size and shape, you would find that they exist in 23 pairs. The members of almost every pair look similar, and contain genes that affect the same characteristics. Cells in which chromosomes occur in pairs are said to be **diploid** (2n) cells.

The chromosomes in your cells are about 40 percent DNA and 60 percent protein. The protein serves as a scaffold around which the slender thread of DNA is tightly wound. If the single DNA strand were stretched out in a straight line, it would be about 50 mm (2 in.) long.

The typical chromosome in your body contains thousands of genes. The information contained in each gene on every chromosome was used by your body to enable you to grow and develop, and now it helps run your body every day. If you have children someday, you will pass some of this information to them.

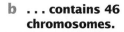

b ... contains 46 chromosomes.

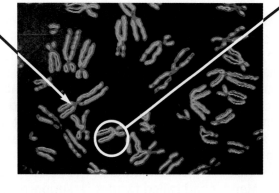

c Each chromosome is formed from DNA and its associated proteins.

Figure 6.6
a **Almost every cell in the human body . . .**

How Gametes Form: Meiosis

Stability

Why is meiosis essential for the survival of a species?

In the previous section, you learned that almost all your cells have 46 chromosomes. Which cells are the exceptions? Your gametes—egg or sperm cells—have only 23 chromosomes. Cells such as gametes that contain only one chromosome of each pair are called **haploid** (n) cells.

Why must gametes have only 23 chromosomes? Consider this: If gametes had the same number of chromosomes as other body cells, when egg and sperm unite during fertilization the new individual would have twice as many chromosomes as its parents. Imagine how this number would soon become impossibly large with each new generation. However, the chromosome number does not double with each generation because eggs and sperm are formed by a special form of nuclear division called meiosis. **Meiosis** (*my OH sihs*) is a type of nuclear division in which the chromosome number is halved. Like mitosis, meiosis is followed by cell division. In humans, specialized reproductive cells with 46 chromosomes undergo meiosis and cell division to give rise to egg or sperm cells that have only 23 chromosomes each. Study this process in **Table 6.2** located on the next page.

Table 6.2

Event	Sperm Formation	Egg Formation

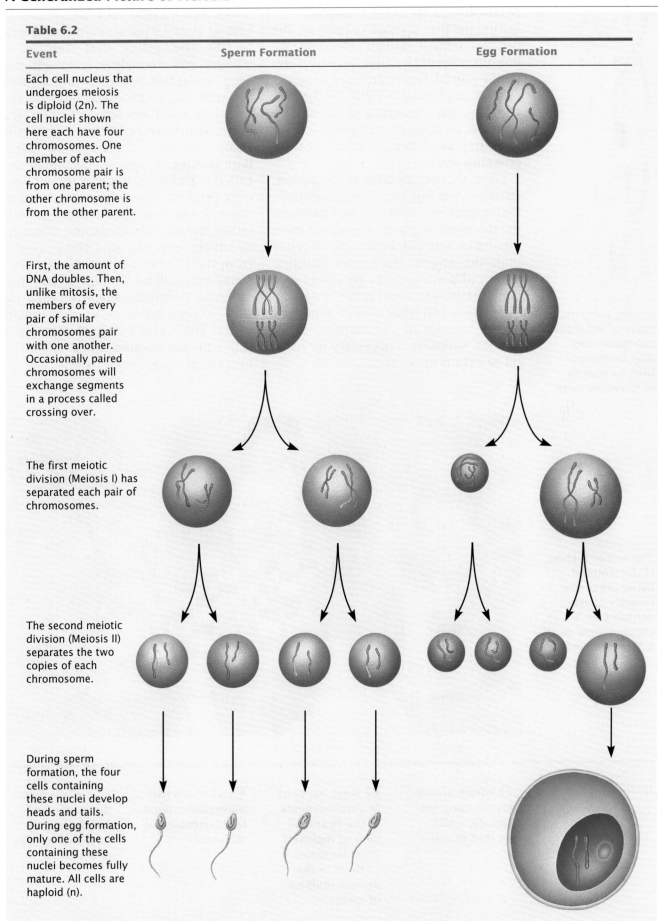

Each cell nucleus that undergoes meiosis is diploid (2n). The cell nuclei shown here each have four chromosomes. One member of each chromosome pair is from one parent; the other chromosome is from the other parent.

First, the amount of DNA doubles. Then, unlike mitosis, the members of every pair of similar chromosomes pair with one another. Occasionally paired chromosomes will exchange segments in a process called crossing over.

The first meiotic division (Meiosis I) has separated each pair of chromosomes.

The second meiotic division (Meiosis II) separates the two copies of each chromosome.

During sperm formation, the four cells containing these nuclei develop heads and tails. During egg formation, only one of the cells containing these nuclei becomes fully mature. All cells are haploid (n).

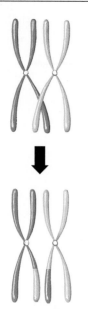

Figure 6.7
During meiosis, parts of adjacent chromosomes may exchange DNA by means of crossing over.

Crossing over during meiosis recombines genetic material

During the first division of meiosis, the two members of each chromosome pair line up with each other side by side. While paired together, they may exchange segments of DNA, as the two chromosomes are doing in **Figure 6.7**. This reciprocal exchange of corresponding segments of DNA is called **crossing over**.

From the perspective of evolution, crossing over has an enormous impact. Exchanging segments of DNA between the members of a pair of chromosomes results in new combinations of genes in particular gametes, just as shuffling a deck of playing cards generates new combinations of cards dealt in a hand. These new combinations of genes act as one source of variation within a species. Variation is necessary for natural selection to occur.

Sex chromosomes determine what sex a child will be

One of your 23 pairs of chromosomes carries the genes that determine whether you are male or female. If you are female, this pair of chromosomes consists of two chromosomes designated as X chromosomes. If you are male, you have only one X chromosome. The other chromosome in the pair is a much smaller chromosome called a Y chromosome. A female can only produce eggs with an X chromosome. A male can produce sperm with either an X or Y chromosome. Thus individuals receiving an X chromosome from their father become females because they will be XX. Individuals that receive a Y chromosome from their father become males because they will be XY. The X and Y chromosomes are called the **sex chromosomes**. Study the process in **Figure 6.8**.

Figure 6.8
The sex of a child is determined by whether the father's sperm contains an X chromosome or a Y chromosome.

Section Review

❶ **Where are the instructions for specific traits located in cells?**

❷ **What happens to chromosomes in the first division of meiosis? What happens to them in the second division of meiosis?**

❸ **What are two biologically important outcomes of meiosis?**

❹ **Explain how the sperm and not the egg determines the sex of a child.**

MANY GENES PRESENT IN HUMAN POPULATIONS CAN CODE FOR CHARACTERISTICS FAR MORE IMPORTANT THAN HAIR COLOR OR EYE COLOR. SOME OF THE MOST DEVASTATING HUMAN DISORDERS RESULT FROM ALLELES THAT LEAD TO MALFUNCTIONING OF IMPORTANT PROCESSES IN THE BODY. WHILE ALLELES ARE PASSED FROM PARENT TO CHILD, THE EXPRESSION OF A DEFECT FOLLOWS MENDEL'S SIMPLE IDEAS.

6.3 Human Genetic Disorders

Objectives

❶ **Recognize the relationship between mutation and human genetic disorders.**

❷ **Explain the patterns of inheritance of cystic fibrosis and sickle cell anemia.**

❸ **Explain the purpose of genetic counseling.**

❹ **Describe some of the benefits that gene technology might offer.**

Mutations Are Changes in Genes

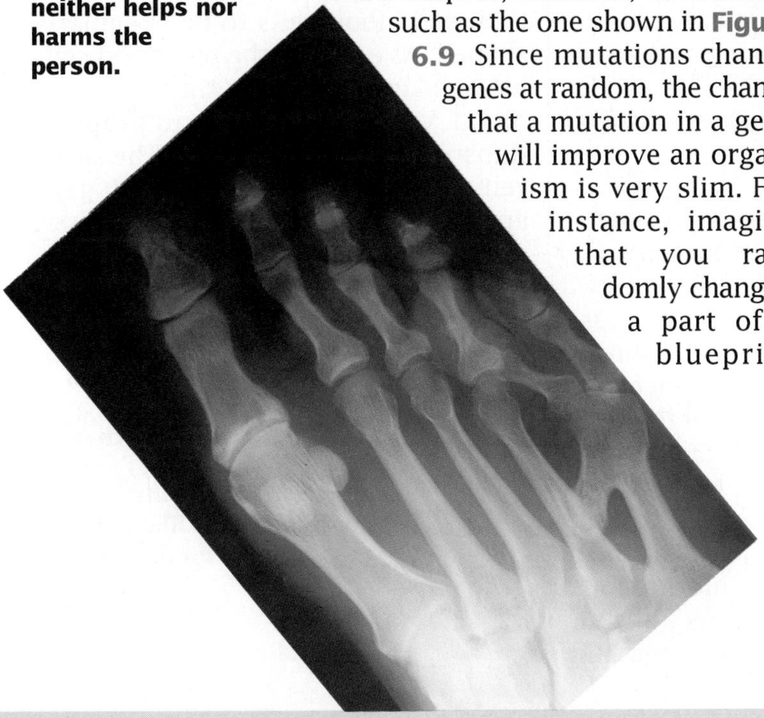

Figure 6.9
A genetic mutation causes some humans to have six toes. In this case, the mutation neither helps nor harms the person.

For you to develop and function properly, your genes must code for certain proteins. Unfortunately, genes are sometimes damaged or copied incorrectly. A change in the gene is called a **mutation** (*myoo TAY shuhn*).

Mutations occur only rarely, but there are so many genes in our chromosomes that each of us carries dozens of mutations. The effects of a mutation can be helpful, harmful, or neutral, such as the one shown in Figure 6.9. Since mutations change genes at random, the chance that a mutation in a gene will improve an organism is very slim. For instance, imagine that you randomly changed a part of a blueprint for construction of an airplane. Some changes would not matter, but others would. Occasionally however, a mutation may have a beneficial effect. Mutations act as a source of the variation that is needed for a species to adapt to changing conditions or a new environment, and, thus, evolve over time.

Mutations in humans can cause genetic disorders

Most mutations are rare in human populations. Almost all mutations occur in recessive alleles and, therefore, are not expressed in heterozygous individuals. Because mutations are rare, it is unlikely that a person carrying a copy of the recessive allele caused by a mutation will marry someone who carries the same mutation. Instead, he or she will most likely marry a person who is homozygous dominant for that allele, a person who does not have the mutation. Therefore, their children will not be homozygous for the mutant allele and will not have the genetic disorder.

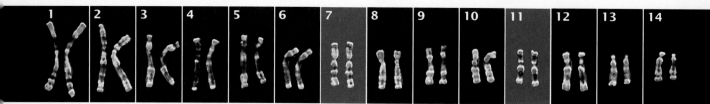

| 1 | 2 | 3 | 4 | 5 | 6 | 7 | 8 | 9 | 10 | 11 | 12 | 13 | 14 |

Four Genetic Disorders

In some cases, particular mutant alleles have become common in human populations. The harmful effects that they produce are called genetic disorders. Four genetic disorders are described below. Not all genetic disorders are alike, however. Cystic fibrosis and sickle cell anemia are caused by mutations that result in harmful recessive genes. Down syndrome is caused when chromosomes fail to separate properly during meiosis. The individual is born with 47 chromosomes, as in **Figure 6.10**. Hemophilia is caused by a recessive gene, but the defective gene is on the X chromosome.

In the past, people with genetic disorders could do little to alleviate the harmful effects of their mutations. However, modern research into the causes and cures of such disorders is offering new encouragement to afflicted individuals. New techniques for identifying harmful genes can help couples determine any risks their children might face. New forms of gene therapy that are now being developed offer hope to sufferers of genetic abnormalities. As more information is learned about genes, scientists can begin to apply their new knowledge toward developing cures for genetic disorders.

	F	f
F	FF	Ff
f	Ff	ff

Figure 6.11
If both parents carry a defective copy of the CF gene (f), their child has a one-in-four chance of developing the disease.

Cystic fibrosis is caused by a recessive gene on chromosome 7

Cystic fibrosis (CF) is the most common fatal genetic disorder among people of European ancestry. The disease is carried by a recessive allele on chromosome 7. See **Figure 6.11**. In the United States, cystic fibrosis strikes one child in every 2,000. In these individuals, membrane channels that normally transport chloride ions into cells do not function. As a result, mucus accumulates in the lungs and pancreas, clogging important ducts in these organs. Cystic fibrosis patients have difficulty breathing and cannot properly digest their food. Most of the current treatments for CF patients can help relieve the symptoms of the disease, although they do not cure it.

Researchers have recently isolated the defective gene that causes cystic fibrosis. The discovery has enabled scientists to devise tests to identify people who carry the gene and who run the risk of having children with the disease. Isolation of the gene has led to the possibility of gene transfer therapy, in which a healthy copy of the CF gene is transferred to the lungs of CF patients to cure the disease.

Sickle cell anemia is caused by a recessive gene on chromosome 11

Sickle cell anemia is a recessive genetic disorder of a gene located on chromosome 11. See **Figure 6.12**. It is particularly common in African populations. People with sickle cell anemia have defective hemoglobin proteins that cause their red blood cells to be irregularly shaped. As a result, the red blood cells have difficulty moving through small blood vessels and cannot properly transport oxygen to tissues. The sickle cell mutation in the hemoglobin gene apparently first arose in Central Africa

	S	s
S	SS	Ss
s	Ss	ss

Figure 6.12
If each parent carries a copy of the sickle cell gene, there is a one-in-four chance that a child's red blood cells will sickle like the cell on the left.

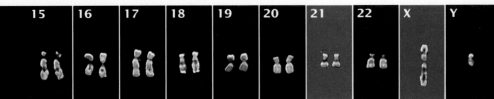

centuries ago. There, up to 45 percent of the population is heterozygous for the sickle cell gene. Evolution has favored the sickle cell allele in Central Africa because heterozygous people are more resistant to malaria, a leading cause of illness and death in the tropics.

Down syndrome is caused by an extra copy of chromosome 21

The developmental features produced by having an extra copy of chromosome 21 were first described in 1866 by J. Langdon Down and are now known as Down syndrome. See **Figure 6.13**. The features that characterize Down syndrome include extra folds in the upper eyelids, a broad and somewhat flattened nose, short stature, and, most importantly, varying degrees of mental retardation. Although people with Down syndrome are physically challenged, they are able to lead active lives and make positive contributions to society.

Down syndrome occurs in about one out of every 1,000 children. This syndrome is much more common among children born to older mothers. In mothers older than 45 the risk is as high as one in sixteen births. Pregnant women over the age of 35 are usually advised to have a medical procedure called amniocentesis performed by their physician. Amniocentesis is a technique in which a small amount of the fluid surrounding the fetus is removed and analyzed. Fetal cells in this fluid are then examined for defects and abnormalities in chromosome number or structure. Results from amniocentesis can reveal whether or not the fetus has a genetic disorder. Along with Down syndrome, amniocentesis can detect other types of disorders such as hemophilia, sickle cell anemia and cystic fibrosis.

Egg   Sperm

Zygote

Figure 6.13
When Chromosome 21 does not separate properly during meiosis in one parent, a child may receive three copies of the chromosome.

Hemophilia is the result of a recessive gene on the X chromosome

Hemophilia is a recessive genetic disorder in which the blood is slow to clot or does not clot at all. When you cut yourself, the blood in the immediate area of the cut solidifies into a clot that seals the wound. The blood clot is formed by several kinds of protein fibers that circulate in the blood. A mutation causing one of these proteins to be defective leads to hemophilia. See **Figure 6.14**. In afflicted individuals, small cuts are difficult to heal and internal bleeding can be fatal. However, treatments are available, such as injections of genetically engineered clotting factors that are lacking in the blood.

A dozen genes encode proteins involved in blood clotting, and mutations can occur in any of them. Two of the genes are located on the X chromosome. Any male who inherits a mutant copy of these alleles will develop hemophilia because his other sex chromosome is a Y chromosome, which lacks an allele of the gene. This pattern of heredity is called sex linkage. Traits that are determined by genes located on the X chromosome are said to be **sex-linked traits**.

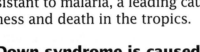

	X	Y
X	XX	XY
X^H	$X^H X$	$X^H Y$

Figure 6.14
A female who is heterozygous for hemophilia can pass the gene for hemophilia to sons and daughters, but only the sons can have the disorder. With proper treatment, hemophiliacs, like this boy, can lead active lives.

Genetic Counseling and Technology

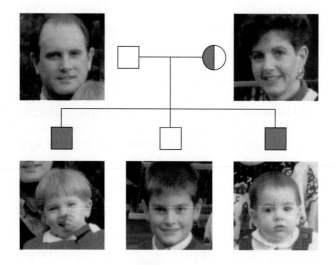

Figure 6.15

a Two boys in this family have hemophilia, but no one else does. How can this happen?

b This pedigree shows that the mother carries the gene (shaded) for hemophilia, and that two of her sons have received this gene.

Most genetic disorders cannot be cured. However, researchers are learning much about them and making progress towards successful therapies. In the absence of a cure, sometimes the only thing to do is avoid having children who could possibly be affected by a genetic disorder. Genetic counseling is a process that helps identify parents at risk for having children with genetic defects. With such information, parents can decide whether or not to have children.

You might ask, "If most genetic disorders are caused by mutations that affect recessive alleles, how do parents know if they are heterozygous for a damaging trait?" Usually it is difficult to be sure, but if one family member is affected by a genetic disorder such as cystic fibrosis or hemophilia, there is a possibility that other family members could be heterozygous for the disorder. These heterozygous individuals are said to be carriers of the trait. In such cases, genetic counselors can prepare a family **pedigree** *(PEHD uh gree)*, a record that shows how a trait is inherited over several generations, as shown in **Figure 6.15**. They will look for the presence of any relatives who have a genetic disorder, and assess the likelihood that a particular individual could be a carrier.

In some cases, therapy is available for genetic disorders if they are diagnosed early enough. For example, phenylketonuria (PKU) is a disorder in which affected individuals lack an enzyme responsible for converting the amino acid phenylalanine into another amino acid, tyrosine. As a result, phenylalanine builds up in the body. People with this disorder suffer severe mental retardation. If PKU is diagnosed shortly after birth, the newborn can be placed on a low phenylalanine diet. Such a diet ensures that the baby gets enough phenylalanine to make proteins, but not enough to do any damage. The child maintains the low phenylalanine diet until six years of age. At this age, the child's brain is fully developed and PKU is usually no longer a problem. Because this disease can be easily diagnosed after birth by inexpensive laboratory tests, many states require testing of all newborns for PKU.

Gene technology is making it possible to correct genetic disorders by replacing copies of defective genes with copies of healthy ones. In 1990, this approach was tried for the first time on human patients. In one case, a healthy copy of a gene encoding an enzyme was successfully transferred. In another case, the transferred gene was a potent cancer-fighting gene.

Section Review

❶ Describe how a mutation could result in a genetic disorder.

❷ Compare the pattern of inheritance of a dominant genetic disorder with that of a recessive genetic disorder.

❸ How is a pedigree used to trace the inheritance of a genetic disorder?

❹ Describe how gene technology could be used to cure a genetic disorder.

Highlights

These children are brothers. Even though they have the same parents, they both look very different.

	Key Terms	Summary
6.1 The Puzzle of Heredity  **Mendel used pea plants in his attempts to understand patterns of heredity.**	genetics (p. 117) homozygous (p. 119) heterozygous (p. 119) allele (p. 119) genotype (p. 119) phenotype (p. 119) dominant (p. 119) recessive (p. 119)	• Mendel's experiments with pea plants marked the beginning of genetics, the scientific study of heredity. • Mendel noted that factors now called genes transmit information about traits from parents to off-spring. The different forms of genes are called alleles. • Organisms that have two identical alleles for a particular trait are said to be homozygous for that trait. Organisms that have two different alleles for that same trait are said to be heterozygous.
6.2 Chromosomes  **This chromosome formed from DNA and its associated proteins.**	gene (p. 123) diploid (p. 124) haploid (p. 124) meiosis (p. 124) crossing over (p. 126) sex chromosome (p. 126)	• Chromosomes contain DNA and proteins. They carry genetic information from one generation to the next. • Meiosis is a type of nuclear division that results in the formation of haploid gametes. • A pair of chromosomes known as sex chromosomes carries the genes that determine whether an individual is male or female.
6.3 Human Genetic Disorders **Occasionally, a harmless genetic mutation causes a person to have six toes.**	mutation (p. 127) sex-linked trait (p. 129) pedigree (p. 130)	• A mutation is a change in a gene. • Cystic fibrosis, sickle cell anemia, Down syndrome, and hemophilia are four genetic disorders. • Genetic counseling can help identify parents at risk for having children with genetic defects. • Therapy is available for some genetic disorders if they are diagnosed early enough. • Gene technology offers hope for correcting genetic disorders by replacing copies of defective genes with copies of healthy ones.

Understanding Vocabulary

1. For each set of terms, complete the verbal analogy by filling in the blank.
   a. seen in F_1 generation:dominant::not seen in F_1 generation: _____
   b. homozygous: BB::heterozygous: _____
   c. 23 chromosomes: haploid::46 chromosomes: _____
   d. XX chromosomes: female::XY chromosomes: _____

Relating Concepts

2. Copy the unfinished concept map below onto a sheet of paper. Then complete the concept map by writing the correct word or phrase in each oval containing a question mark.

Understanding Concepts

Multiple Choice

3. The scientific study of the inheritance of traits is called
   a. genetics.   c. heredity.
   b. genotype.   d. osmosis.

4. Mendel studied contrasting traits of _____ pea plants in his experiments.
   a. dominant
   b. mutant
   c. white-flowering
   d. true-breeding

5. The crossing of a white-flowering plant with a purple-flowering plant (ww x WW) results in plants with the genotype
   a. Ww.   c. WW or ww.
   b. purple.   d. purple and pink.

6. The formation of an equal number of W and w gametes from a Ww individual demonstrates
   a. independent assortment.
   b. segregation.
   c. dominance.
   d. recessiveness.

7. The phenotypic ratio resulting from the cross Ww x Ww is
   a. 1:1.   c. 3:1.
   b. 1:2:1.   d. 2:1.

8. A gene is
   a. a complete molecule of DNA.
   b. a short segment of DNA.
   c. made of many chromosomes.
   d. a tiny particle first seen by Mendel in the 1860s.

9. Crossing over
   a. produces variation in the chromosomes.
   b. enables the second division to occur.
   c. causes the number of chromosomes in a cell to be reduced by half.
   d. produces XY but not XX chromosomes.

10. Cell division that halves the number of chromosomes is called
    a. mitosis.   c. meiosis.
    b. endocytosis.   d. zygote.

11. The purpose of genetic counseling is to
    a. repair defective genes.
    b. cure genetic defects that occur in unborn children.
    c. tell parents that they cannot have children.
    d. help parents understand the nature and risk of genetic disorders.

Completion

12. An organism showing the dominant trait may be homozygous or _____ , but an organism showing the recessive trait must be _____ .

13. An organism's _____ is the physical expression of its genotype.

14. Different forms of the same gene are called _____ .

15. A human baby with the sex chromosomes XY is a _____ .

16. A trait is _____ if a male has only one allele of the gene for the trait, as in the case of hemophilia.

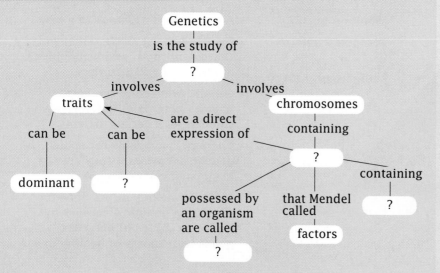

Short Answer

17. Mendel was a very meticulous and diligent experimenter. What else is known about Mendel that might help explain the success of his experiments?

18. To begin his experiments, Mendel used true-breeding pea plants. How might the results of his experiments have been different if he had started with pea plants that were not true-breeding?

19. During meiosis, the number of chromosomes is cut in half. What would happen when a sperm fertilizes an egg if the number of chromosomes was not reduced by one-half during meiosis?

Interpreting Graphics

20. A Punnett square is a diagram used to show the results of a cross. The Punnett square below shows the results of a cross between parents with cleft chins. Study the Punnett square carefully.

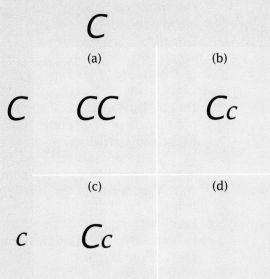

What is the genotype of the father?
What is the phenotype of the offspring in block a?
What is the genotype of the offspring in block d?
What is the phenotypic ratio for the offspring produced by the cross?

Reviewing Themes

21. *Stability*
Two tall pea plants are crossed. How does knowing that both parent plants are heterozygous (Tt) for this trait and that only the dominant trait is expressed help you predict the genotypic and phenotypic ratios for the offspring?

22. *Evolution*
How does crossing-over contribute to genetic diversity?

Thinking Critically

23. *Inferring Conclusions*
When Mendel crossed true-breeding purple flowers (WW x ww), he found that only purple flowers appeared in the F_1 generation. A ratio of 3 purple flowers to 1 white flower appeared in the F_2 generation. Suppose that Mendel found only pink flowers in the F_1 generation and a ratio of 1 purple flower to 2 pink flowers to 1 white flower in the F_2 generation. How would you explain these results?

24. *Inferring Conclusions*
A husband and wife have three children, two girls and a boy. The couple is expecting a fourth child. What is the likelihood that the child will be a boy?

Cross-Discipline Connection

25. *Biology and History*
Identify the contributions of these people to today's understanding of genetics.

1900	Hugo deVries
1901–1903	W. S. Sutton
1909	W. Johannsen
1910–1916	T. H. Morgan, A. H. Sturtevant, H. J. Muller

Discovering Through Reading

26. Read the article "Hand-Me-Down Genes," in *Newsweek*, January 27, 1992, page 53. What are the benefits of knowing your family's medical history?

Investigation

What Is Your Genetic Profile?

Objectives

In this investigation you will:
- *observe* a variety of human traits
- *collect* and *record* data about your phenotype
- *infer* possible genotypes from your data
- *calculate* percentages

Materials

- mirror

Prelab Preparation

Review what you have learned about genetics by answering the following questions:
- Define the terms "phenotype," "genotype," "dominant," and "recessive."
- What is a pedigree? Why are pedigrees often necessary when studying human hereditary traits?
- When reading a pedigree, how do you distinguish males from females and children from parents?

Procedure: Making Your Genetic Profile

1. Form a cooperative group of two students.

2. Make a table similar to the one shown at the top of the next page for recording your observations.

3. Working with a partner, check for each trait described below. Record your phenotype in the appropriate column in your table.

   a. **Mid-digital hair**
      Each segment of a finger is called a digit. *Is hair present on the middle digit of any of your fingers?*

   b. **Tongue rolling**
      Look in the mirror. *When you stick out your tongue, can you roll up the edges on each side?*

   c. **Dimples**
      When you smile, are there small indentations in your cheeks?

   d. **Cleft chin**
      Do you have an indentation in the middle of your chin?

   e. **Attached earlobes**
      Do the tips of your earlobes hang partially free, or are they completely attached to the side of your head?

   f. **Freckles**
      Do you have small reddish-brown spots on your skin?

   g. **Widow's peak**
      Pull your hair back over your forehead. *Does the hairline come down to a short point in the middle of your forehead or does it go straight across?*

   h. **Five fingers**
      Were you born with five fingers on each hand?

Trait	Phenotype	Percent of Class	Dominant or Recessive	Genotype
Mid-digital hair				
Tongue rolling				
Dimples				
Cleft chin				
Attached earlobes				
Freckles				
Widow's peak				
Five fingers				

4. Compare your data with that of your classmates. Calculate the percentage of the class that exhibits each trait.

5. Look at the pedigree below. Colored figures represent individuals that possess the trait. Explain why individual IV-8 and her parents provide the evidence that attached earlobes is a recessive trait.

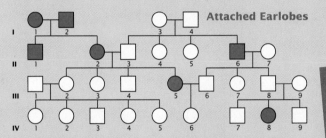

6. Analyze the pedigrees for tongue rolling and five fingers. Explain whether each trait is dominant or recessive. Each of the other traits listed in your table is dominant.

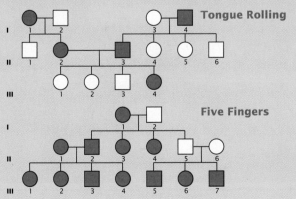

7. Based on the available information, use appropriate symbols for each trait to record your possible genotypes in your table.

Analysis

1. *Analyzing Data*
   Does the information you collected and studied during this investigation indicate that dominant traits are the most common? Explain.

2. *Evaluating Methods*
   Look at the pedigree for five fingers. Explain why individuals II-5, II-6, and their children are the most important for analyzing whether this trait is dominant or recessive.

Thinking Critically

What would happen to your percentages if you were to perform this investigation with five other classes and were to record their data?

How Genes Work

Review

- **DNA and proteins (Section 2.3)**
- **mitosis (Section 4.3)**
- **genes and chromosomes (Section 6.2)**

The genes that determine physical traits and regulate body functions are passed from parent to child.

ONE OF THE MOST IMPORTANT ADVANCES IN HUMAN KNOWLEDGE WAS THE REALIZATION THAT OUR UNIQUE CHARACTERISTICS ARE ENCODED WITHIN MOLECULES OF DNA. THE APPLICATIONS OF THIS KNOWLEDGE HAVE PROFOUNDLY INFLUENCED THE WAY PEOPLE THINK ABOUT THEMSELVES, THEREBY MAKING THE BIOLOGICAL NATURE OF HUMAN BEINGS SEEM APPROACHABLE.

7.1 *Understanding DNA*

Objectives

❶ Explain how researchers concluded that DNA is the genetic material.

❷ Summarize how scientists determined the structure of DNA.

❸ Describe the structure of the DNA molecule.

❹ Summarize the process of DNA replication.

How Scientists Discovered That DNA Is the Genetic Material

As you learned in Chapter 6, chromosomes are made of DNA and protein. Because chromosomes and heredity are linked, biologists hypothesized that either DNA or protein was the genetic material. But which molecule was it?

Griffith discovers the process of transformation

In 1928 the British microbiologist Frederick Griffith was using mice to study the bacterium that causes pneumonia. One type, or strain, of bacteria was enclosed in coats made of complex sugar molecules. When Griffith infected mice with this strain, the mice died. In the other strain of bacteria, the coat was absent. Surprisingly, this strain did not kill mice. Griffith found that if he first killed the coated bacteria by boiling them, they were then harmless to mice. When Griffith mixed the dead coated bacteria (harmless) with live uncoated bacteria (harmless), the mixture killed mice! He examined the blood of the dead mice and found that the live uncoated bacteria had grown coats. Somehow the live uncoated bacteria had acquired the ability to make coats from the dead coated strain. Griffith called this process **transformation.** His experiment is summarized in **Figure 7.1**.

Figure 7.1
Griffith's experiment showed that live uncoated bacteria acquired the ability to make coats from dead coated bacteria.

a When Griffith infected mice with live coated bacteria, the mice died.

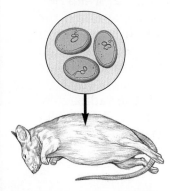

b When Griffith infected mice with live uncoated bacteria, they lived.

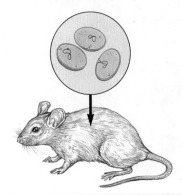

c When Griffith infected mice with dead coated bacteria, the mice lived.

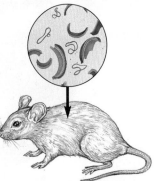

d When Griffith infected mice with a mixture of dead coated and live uncoated bacteria, they died.

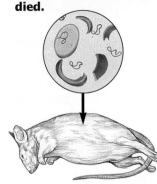

Avery shows that DNA is the transforming principle

What was the material that transformed the bacteria from harmless (without coats) to deadly (with coats)? Was this "transforming principle" made of DNA or of protein? In 1944 biologist Oswald Avery and two colleagues at Rockefeller University in New York showed that DNA is the transforming principle. He extracted DNA from bacteria with coats and added it to a population of bacteria without coats. Some of the bacteria that grew from this mixture now had coats.

In a separate experiment, Avery found that when he added protein-destroying enzymes to bacteria, transformation still occurred. However, when DNA-destroying enzymes were added to bacteria, transformation did not occur. Avery's work provided clear evidence that DNA was the genetic material in these bacteria.

Hershey and Chase confirm that DNA is the genetic material

At first, Avery's results were not widely appreciated. Many biologists were reluctant to give up the idea that protein was the genetic material. In 1952, however, Alfred Hershey and Martha Chase, two scientists at Cold Spring Harbor Laboratory on Long Island, New York, performed an experiment using viruses that infect bacteria. The viruses attach to the surfaces of bacteria and inject their hereditary information into the cells like tiny hypodermic needles. Once inside the bacteria, this hereditary information directs the production of hundreds of new viruses. When the new viruses are mature, they burst out of the infected bacteria and attack new cells. These bacteria-infecting viruses have a very simple structure: a core of DNA surrounded by a protein coat.

To identify the hereditary material, Hershey and Chase used radioactive phosphorus (^{32}P) to label the DNA and radioactive sulfur (^{35}S) to label the protein coats in the viruses. The Hershey-Chase experiment is summarized in **Figure 7.2**. When Hershey and Chase examined the new viruses that burst out of the bacteria, they found that the viruses contained the radioactive phosphorus (^{32}P) label, but not the radioactive sulfur (^{35}S) label. The conclusion was undeniable—DNA is the hereditary material.

Figure 7.2

Hershey and Chase used radioactive labeling to identify the DNA and the protein coats of viruses. Their experiment showed that DNA, not protein, is the hereditary material of viruses.

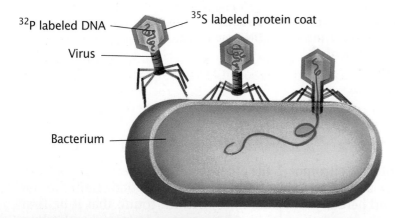

^{32}P labeled DNA

^{35}S labeled protein coat

Virus

Bacterium

a The viral DNA (red) is injected into the bacterium, where it will direct the production of new viruses. The protein coat (green) remains outside the bacterium.

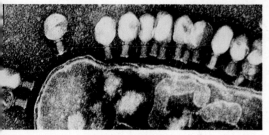

b The bacterium bursts, releasing the newly made viruses. The new viruses contain DNA labeled with ^{32}P.

How Scientists Determined the Structure of DNA

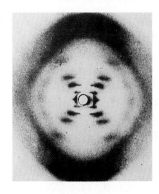

Figure 7.3
This photograph produced by Rosalind Franklin suggested that DNA was helical—the form indicated by the pattern of X-ray reflections in the photograph.

After the experiments of Hershey and Chase, scientists began to study DNA intensely. Scientists already knew that DNA was composed of subunits called **nucleotides** (*NOO klee oh tydz*). Every nucleotide has three parts: a sugar, a phosphate group, and a base. The sugar and phosphate group are the same in every nucleotide, but there are four different bases. The two larger bases, adenine and guanine, are called **purines** (*PYUR eenz*). The two smaller bases, cytosine and thymine, are called **pyrimidines** (*py RIHM uh deenz*).

In 1949 Erwin Chargaff, a biochemist working at Columbia University in New York City, made a key discovery about the chemical structure of DNA by studying the DNA from different organisms. Chargaff found that the amount of adenine in a DNA molecule always equals the amount of thymine (A=T). Likewise, the amount of guanine always equals the amount of cytosine (G=C). These observations, now known as Chargaff's rules, suggested that DNA has a regular structure.

The value of Chargaff's rules became clear when Rosalind Franklin, a chemist working at King's College in London, began studying the structure of DNA using X-ray diffraction. Franklin's X-ray diffraction images, one of which is shown in **Figure 7.3**, suggested that the DNA molecule resembled a tightly coiled spring, a shape called a helix.

Watson and Crick build a model showing DNA's structure

In the early 1950s a young American scientist, James Watson, went to Cambridge, England, on a research fellowship. At Cavendish Laboratories he met Francis Crick, a British physicist interested in DNA. Together, Watson and Crick attempted to construct a model of DNA. They applied the clues provided by Chargaff's rules and Franklin's X-ray diffraction studies. Using tin and wire models of the bases, sugars, and phosphate groups, Watson and Crick deduced that the structure of the DNA molecule is a **double helix**, a spiral staircase of two strands of nucleotides whose bases face each other. The double helix is held together by weak hydrogen bonds between the bases. Adenine can only form hydrogen bonds with thymine, and guanine can form hydrogen bonds only with cytosine. **Figure 7.4** shows the structure of DNA. Francis Crick and James Watson were awarded the Nobel Prize in 1962 for their work in formulating this brilliant model.

Figure 7.4

a **Inside the nucleus of a cell are chromosomes, which contain long strands of DNA.**

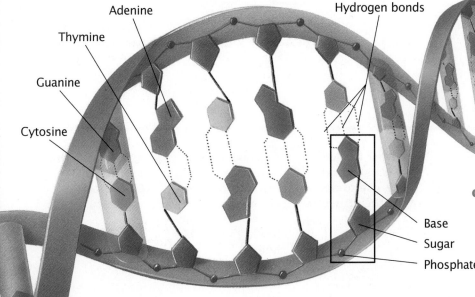

Adenine

Thymine

Guanine

Cytosine

Hydrogen bonds

b **DNA consists of two strands of nucleotides, joined by hydrogen bonds and twisted into a double helix.**

c **Every DNA nucleotide contains a sugar, a phosphate group, and a base.**

Base

Sugar

Phosphate group

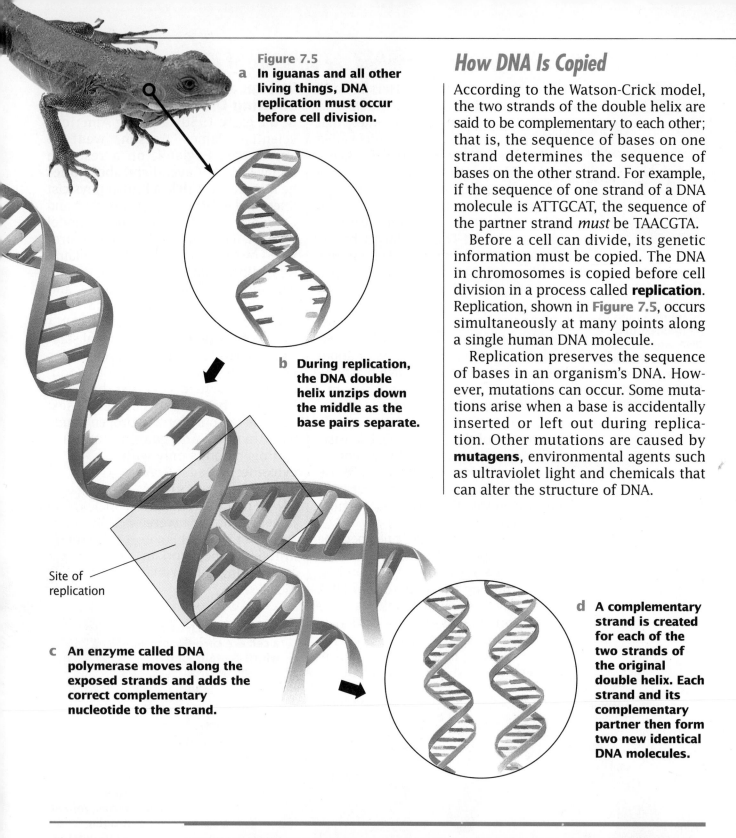

Figure 7.5

a In iguanas and all other living things, DNA replication must occur before cell division.

b During replication, the DNA double helix unzips down the middle as the base pairs separate.

Site of replication

c An enzyme called DNA polymerase moves along the exposed strands and adds the correct complementary nucleotide to the strand.

d A complementary strand is created for each of the two strands of the original double helix. Each strand and its complementary partner then form two new identical DNA molecules.

How DNA Is Copied

According to the Watson-Crick model, the two strands of the double helix are said to be complementary to each other; that is, the sequence of bases on one strand determines the sequence of bases on the other strand. For example, if the sequence of one strand of a DNA molecule is ATTGCAT, the sequence of the partner strand *must* be TAACGTA.

Before a cell can divide, its genetic information must be copied. The DNA in chromosomes is copied before cell division in a process called **replication**. Replication, shown in **Figure 7.5**, occurs simultaneously at many points along a single human DNA molecule.

Replication preserves the sequence of bases in an organism's DNA. However, mutations can occur. Some mutations arise when a base is accidentally inserted or left out during replication. Other mutations are caused by **mutagens**, environmental agents such as ultraviolet light and chemicals that can alter the structure of DNA.

Section Review

1 What did Griffith, Avery, and Hershey and Chase each contribute to identifying the genetic material?

2 What evidence did Watson and Crick use to deduce the structure of DNA?

3 What are the restrictions on the four different nucleotide bases when pairing in the double helix?

4 Describe the process by which DNA is copied.

THE DISCOVERY THAT **DNA** IS THE GENETIC MATERIAL LEFT STILL MORE QUESTIONS UNANSWERED. HOW IS THE INFORMATION IN **DNA** USED? SCIENTISTS NOW KNOW THAT **DNA** DIRECTS THE CONSTRUCTION OF PROTEINS. PROTEINS DETERMINE THE SHAPES OF CELLS AND SPEED THE RATES OF CHEMICAL REACTIONS SUCH AS THOSE THAT OCCUR DURING METABOLISM AND PHOTOSYNTHESIS.

7.2 How Proteins Are Made

Objectives

❶ Identify and describe the two stages of gene expression.

❷ Explain why the genetic code is said to be universal.

❸ Compare and contrast the roles of the three types of RNA.

❹ Explain the relationships among codons, anticodons, and amino acids.

The Transfer of Genetic Information

Figure 7.6
During gene expression, the information in DNA is used to assemble proteins.

Once scientists understood the structure of DNA, they were able to learn how a specific protein is built from information found in the DNA of one gene. Scientists now know that DNA is used as a blueprint to make a similar molecule called **ribonucleic acid**, or RNA for short. This RNA then directs the formation of proteins. The use of genetic information in DNA to make proteins is called **gene expression**. Gene expression takes place in two stages. The first stage is called **transcription**. During transcription, an RNA copy of a gene is made. During **translation**, the second stage of gene expression, three different kinds of RNA work together to assemble amino acids into a protein molecule. Gene expression is summarized in **Figure 7.6**.

Through gene expression, the messages encoded in DNA direct all cellular activities. For example, when you eat carbohydrates such as those found in a bowl of cereal or a slice of bread, certain genes direct the production of a protein called insulin. Insulin helps your body maintain its blood sugar level. Other genes direct the production of hundreds of other structural proteins and enzymes.

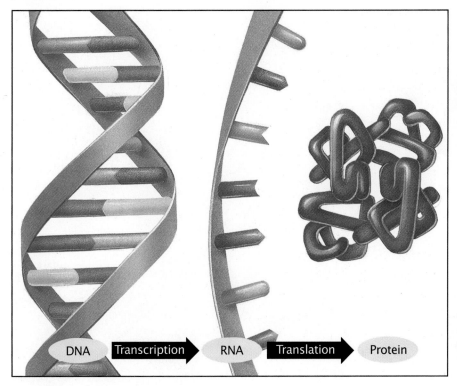

DNA → Transcription → RNA → Translation → Protein

How DNA Makes RNA

Transcription is the process by which genetic information encoded in DNA is transferred to an RNA molecule. Genetic information must be copied because DNA cannot leave the nucleus. Just as an architect protects building plans from loss or damage by keeping them in a central place, your cells protect genetic information by keeping the DNA safe within the nucleus. Instead of sending out the DNA, copies of genes are sent out into the cell to direct the assembly of proteins. These working copies of genes are made of a single strand of RNA.

RNA is chemically similar to DNA except that its sugars have an additional oxygen atom, and the base thymine (T) is replaced by a structurally similar base called uracil (U). RNA occurs in three different forms. **Figure 7.7** shows transcription of one form of RNA, messenger RNA (mRNA). Just as monks once copied manuscripts by faithfully transcribing each letter, so enzymes in your cells' nuclei make mRNA copies of your genes by copying their nucleotides.

During transcription, each gene is copied from a fixed starting position called a promoter site. Here, an enzyme called RNA polymerase binds to one strand of the DNA double helix and moves along the DNA strand like a train on a track. As it passes over each nucleotide, the RNA polymerase pairs the base in that nucleotide with its complementary RNA base. Cytosine is paired with guanine. But because RNA contains uracil instead of thymine, adenine is now paired with uracil. In this way, a complementary strand of mRNA is gradually built. Nucleotide sequences located at the end of the genes tell the RNA polymerase where to stop. In eukaryotes, after transcription of a gene is finished, the mRNA passes out of the nucleus through pores in the nuclear membrane and into the cytoplasm. There the second stage of gene expression, translation, takes place.

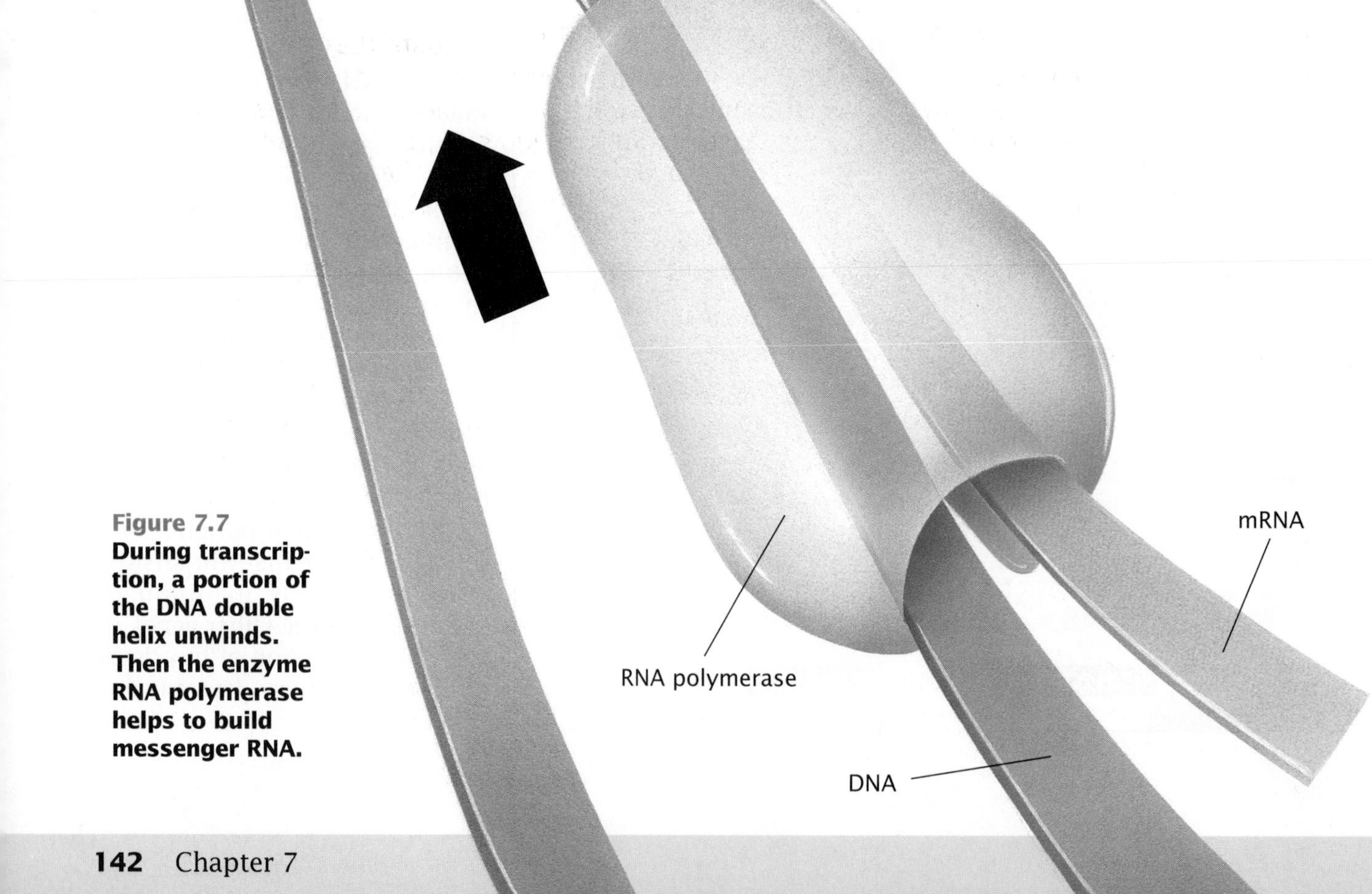

Figure 7.7
During transcription, a portion of the DNA double helix unwinds. Then the enzyme RNA polymerase helps to build messenger RNA.

RNA polymerase

mRNA

DNA

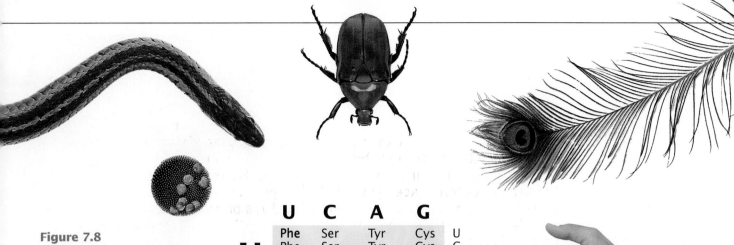

Figure 7.8

In the genetic code, each amino acid is coded for by three mRNA bases arranged in a specific sequence. The first base in a codon is found along the left side of the chart, the second base is at the top of the chart, and the last base is found along the right side of the chart. For example, the amino acid alanine (Ala) is coded for by the triplets GCU, GCC, GCA, and GCG.

	U	C	A	G	
U	Phe	Ser	Tyr	Cys	U
	Phe	Ser	Tyr	Cys	C
	Leu	Ser	stop	stop	A
	Leu	Ser	stop	Trp	G
C	Leu	Pro	His	Arg	U
	Leu	Pro	His	Arg	C
	Leu	Pro	Gln	Arg	A
	Leu	Pro	Gln	Arg	G
A	Ile	Thr	Asn	Ser	U
	Ile	Thr	Asn	Ser	C
	Ile	Thr	Lys	Arg	A
	Met	Thr	Lys	Arg	G
G	Val	Ala	Asp	Gly	U
	Val	Ala	Asp	Gly	C
	Val	Ala	Glu	Gly	A
	Val	Ala	Glu	Gly	G

The Genetic Code

Evolution

How is the DNA of all organisms similar?

How is mRNA translated into the sequence of amino acids that make up proteins? How can the four nucleotide bases found in mRNA carry instructions to build the thousands of proteins your body needs? Every three nucleotides in mRNA specify a particular amino acid. Each nucleotide triplet in mRNA is called a **codon** *(KOH dahn).* The order of bases in a codon determines which amino acid will be added to a growing protein chain. In turn, the order of amino acids will determine the structure and function of a protein.

To learn more about how mRNA directs amino acids to join in a specific order, biologists performed laboratory experiments using artificial mRNA to direct protein production. An mRNA that contained only the nucleotide uracil (U), for example, made a protein that consisted entirely of the amino acid phenylalanine (Phe). This information

told scientists that the codon "UUU" codes for the amino acid phenylalanine. These experiments ultimately revealed the genetic code, which is shown in **Figure 7.8**. The names of amino acids have been abbreviated in the chart. The **genetic code** is the correspondence between nucleotide triplets in DNA and the amino acids in proteins. Any of the four bases (U, C, A, G) found in mRNA can occur at any of the three positions of a codon. Thus, there are 64 different possible three-letter codons $(4 \times 4 \times 4 = 64)$ in the genetic code. Since there are 64 possible codons, but only 20 different amino acids occur in proteins, more than one codon may specify a single amino acid.

The genetic code is the same in nearly all organisms, so it is said to be universal. For example, the code for phenylalanine is the same in bacteria and humans.

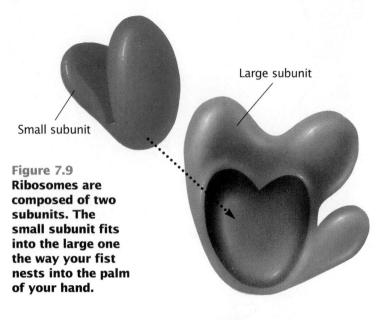

Small subunit

Large subunit

Figure 7.9
Ribosomes are composed of two subunits. The small subunit fits into the large one the way your fist nests into the palm of your hand.

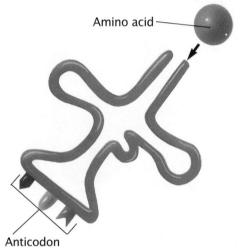

Amino acid

Anticodon

Figure 7.10
Transfer RNA molecules are chains about 80 nucleotides long, folded into a compact shape. The anticodon is a three-nucleotide sequence at one end of the tRNA. An amino acid is attached at the opposite end.

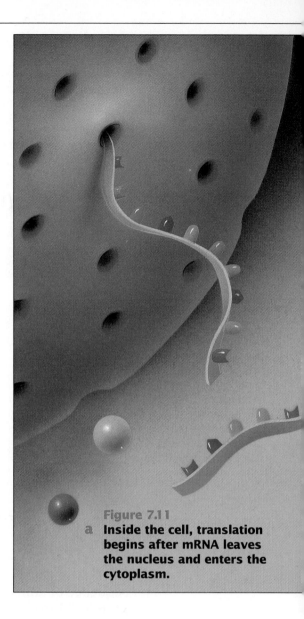

Figure 7.11
a **Inside the cell, translation begins after mRNA leaves the nucleus and enters the cytoplasm.**

How RNA Makes Proteins

After transcription in eukaryotes, the mRNA strand leaves the cell's nucleus and travels into the cytoplasm. During translation, the mRNA works with two other types of RNA to build proteins by joining amino acids. Translation occurs on ribosomes, complex organelles that contain a special kind of RNA called ribosomal RNA (rRNA). **Figure 7.9** shows that each ribosome consists of two subunits. The smaller ribosomal subunit contains a short rRNA sequence that is complementary to the mRNA codon that signals "start" in all genes. Translation begins when the mRNA "start" codon binds to the small ribosomal subunit. The large ribosomal subunit attaches to form a complete ribosome with a strand of mRNA running through it. Just as factories use

blueprints to direct the assembly of cars, so do ribosomes use mRNA to direct the assembly of proteins.

The pocket, or dent, in the small ribosomal subunit has just the right shape to bind a third kind of RNA molecule, transfer RNA (tRNA). Transfer RNA carries amino acids to the ribosome. The three-nucleotide sequence shown on one end of the transfer RNA molecule in **Figure 7.10** is called an **anticodon**. Anticodons are complementary to mRNA codons.

Like the address on an envelope, the anticodon ensures that an amino acid is delivered to its proper "address" on the mRNA as a protein is being assembled. And just as a mail carrier checks the address before delivering a letter, so the small ribosomal subunit checks

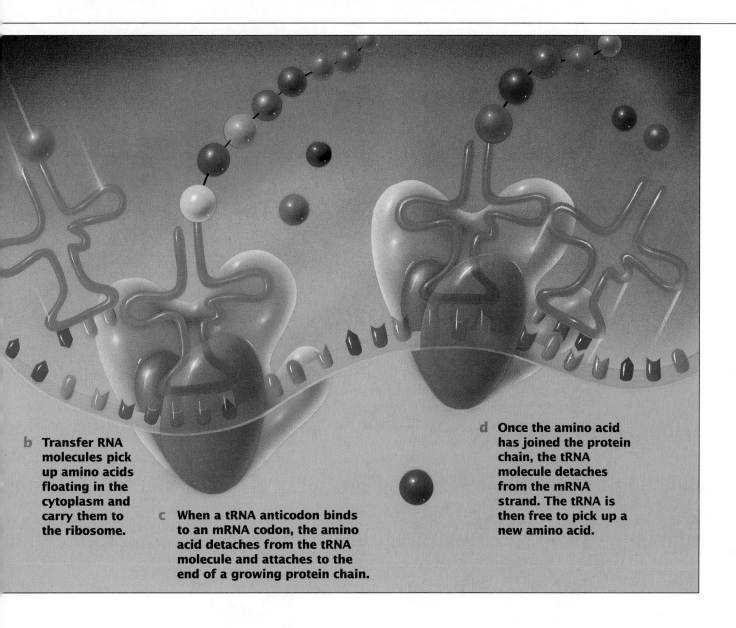

b Transfer RNA molecules pick up amino acids floating in the cytoplasm and carry them to the ribosome.

c When a tRNA anticodon binds to an mRNA codon, the amino acid detaches from the tRNA molecule and attaches to the end of a growing protein chain.

d Once the amino acid has joined the protein chain, the tRNA molecule detaches from the mRNA strand. The tRNA is then free to pick up a new amino acid.

the tRNA anticodon to see that it is complementary to the mRNA codon.

As the mRNA passes through the ribosome, one tRNA after another is selected to match the sequence of mRNA codons. Amino acids are added to the end of the growing protein chain until the end of the mRNA sequence is reached. At this point, a "stop" codon is encountered for which there is no anticodon on any tRNA molecule. With nothing to fit into the tRNA site, the ribosome complex falls apart and the newly assembled protein is released into the cell. **Figure 7.11** shows the steps of translation.

Section Review

❶ **Outline the path genetic information travels when proteins are made.**

❷ **Explain the process by which DNA is copied into RNA.**

❸ **What is the genetic code? Why is it said to be universal?**

❹ **Explain the process by which proteins are made from RNA.**

Diana Punales-Morejon:
Genetic Counselor

How I Became Interested in Genetic Counseling

"When I was five years old, my family moved to the United States from Cuba. We didn't speak any English, so I had to learn the language in school. It was very difficult at first, but soon I picked it up and began enjoying the time I spent at school, talking to my new friends. My favorite teacher was my high school biology teacher, a fantastic woman. She took the class on field trips to laboratories and science museums. We could sense that she really enjoyed biology and loved teaching; this made us all eager to learn.

"In college, I majored in biology and minored in psychology because I wanted to be a biology teacher. However, after graduation, I ended up working in a biomedical lab, doing research. I enjoyed the work, but most of all I missed the personal contact with people. I began looking for a career that could include both biology and psychology. That's when I came across genetic counseling. Genetics has always fascinated me and the counseling component satisfied my interest in psychology, so it sounded perfect.

"Today, as a genetic counselor, I deal with issues that can be difficult to understand and handle. It usually takes a lot of teaching to explain genetics to the people I'm counseling. But that's one of the things I like about my job. Most of the time they can't believe what technology is able to detect. It's mind-boggling for many people."

Diana enjoys hiking around town with her 10-month-old daughter, Amanda Isabel.

Name:	**Diana Punales-Morejon**
Home:	**Union City, New Jersey**
Employer:	**Beth Israel Medical Center, New York, NY**
Personal Traits:	• **Caring**
	• **Dedicated**
	• **Detail-oriented**
	• **Good communication skills**
	• **Sociable**

The Satisfaction of Genetic Counseling

Career Path

High School:
- Biology
- Chemistry
- Physics

College:
- Biology
- Microbiology
- Genetics
- Psychology
- Chemistry

Graduate School:
- Human and medical genetics
- Biochemistry

"I was hired in 1987 as a genetic counselor. Within a year, I was promoted to Coordinator of the Genetic Counseling Program. I teach medical, nursing, and graduate students. I also see patients for genetic counseling—women and couples that are at risk for having children with birth defects and genetic disease.

"It is the most incredible feeling to be able to tell someone that their child, who was at risk, will be normal. That feeling lasts forever. The worst part, the part that never gets easy for me, is when a couple has prenatal testing and the test shows the baby will be affected. That is the worst thing I will ever do in my life—

especially when people want and plan for the pregnancy.

"Fortunately, abnormalities occur only 5 percent of the time. The good far outweighs the bad, but it never gets any easier. A family having an abnormal pregnancy is faced with difficult and painful choices."

Research Focus

Diana coordinates all research projects for the genetic counseling department of her hospital. She is now studying social, psychological, and cultural issues by analyzing patients' perceptions and attitudes towards prenatal care and reproductive functions. She has developed a survey that is administered to patients who are referred for prenatal diagnosis and genetic counseling. The survey asks patients about their knowledge of genetics and prenatal tests as well as their feelings about terminating abnormal pregnancies.

Diana also looks at how demographic factors like race, religion, education, income, and gender impact patients' views about testing and reproductive options. She wants to see if patients of different races and cultures use testing differently. She is also interested in identifying patients' major concerns. Some individuals are more anxious about the pain involved; others are more worried about the length of time they must wait for results. It is important to give people the kind of information they're most anxious about first, before discussing their medical history.

THE TRANSLATION OF A GENE INTO A PROTEIN IS ONLY PART OF GENE EXPRESSION. EVERY CELL MUST ALSO BE ABLE TO REGULATE THE USE OF PARTICULAR GENES. JUST AS A CONDUCTOR CONTROLS HOW LOUD AND HOW FAST THE DIFFERENT INSTRUMENTS IN AN ORCHESTRA PLAY, A CELL DETERMINES WHEN PARTICULAR GENES ARE TRANSCRIBED BY CONTROLLING WHEN GENES ARE SWITCHED "ON" AND "OFF."

7.3 *Regulating Gene Expression*

Objectives

❶ Explain why cells must regulate gene expression.

❷ Summarize how a gene can be switched off and on.

❸ Distinguish between exons and introns.

❹ Define transposons and explain how they affect gene expression.

Switching Genes On and Off

Cells control the expression of their genes by determining when individual genes are to be transcribed. Each gene possesses special regulatory sites, which act as points of control. Specific regulatory proteins within the cell bind to these sites, switching transcription of the gene on or off.

The best understood regulatory mechanisms are those used by prokaryotes. Some genes in prokaryotes are expressed nearly all the time, while others are rarely used. Genes that are expressed only occasionally are said to be switched off. They are transcribed only when the proteins are needed. In these genes, transcription cannot occur because a large molecule called a **repressor protein** is bound to the DNA in front of the genes, as shown in **Figure 7.12a**. The repressor protein blocks transcription by preventing the RNA polymerase from moving along the gene. If someone placed a brick wall between your chair and desk, you could not begin your work until the wall was removed. In the same way, transcription cannot begin until the repressor protein is removed.

Stability

What mechanisms

provide a

eukaryotic

cell with the

proteins it needs

to function

properly?

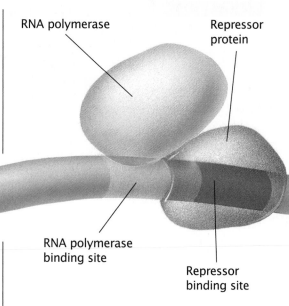

RNA polymerase

Repressor protein

RNA polymerase binding site

Repressor binding site

Figure 7.12

a **A repressor protein sits on the DNA and blocks RNA polymerase from attaching to its binding site.**

Transcription begins when an inducer is present

For transcription to begin, molecules called **inducers** must bind to the repressor protein. The binding causes the repressor protein to change its shape so that it no longer fits the DNA. As a result, the repressor protein falls off, removing the barrier to transcription. When this happens, the gene is switched on.

In bacteria, gene expression controls the digestion of lactose, the sugar found in milk. When a bacterium encounters lactose, the lactose acts as an inducer by binding to the repressor protein and altering its shape. The altered repressor protein then detaches from the DNA molecule, which allows the RNA polymerase to transcribe the genes needed to produce the enzyme responsible for digesting lactose.

Activators help DNA unwind

Because RNA polymerase binds to only one strand of the DNA double helix, it is necessary that the DNA molecule partially unwind to expose the bases at the promoter site. In many genes, this unwinding cannot take place without the help of a regulatory protein called an activator. An activator binds to the DNA in this region and helps it unwind.

Cells are able to switch genes "off" by binding signal molecules to the activator protein. This prevents the activator from binding to the DNA molecules. As a result, the double helix cannot unwind and the genes cannot be transcribed.

Activators enable a cell to carry out a second level of control. When a bacterium already has plenty of energy, the level of another signal molecule—a special "I-need-sugar" signal molecule —decreases. Without being prodded by this signal molecule, the activator protein cannot bind to the DNA molecule's unwinding site. As a result, the genes are not transcribed, even though the repressor protein does not block the RNA polymerase.

Enhancers expose binding sites in eukaryotic DNA

A third level of control is exercised by regulating access to the gene. For example, because human chromosomes are large and complex, it is not easy for RNA polymerase to find its way to the beginning of a particular gene. To make themselves more accessible to RNA polymerase when switched on, many genes in eukaryotes possess special sequences called enhancers. When activated by specific protein molecules, enhancers aid in exposing the RNA polymerase binding site. Unlike promoters and activator proteins, which are found at the beginning of a gene, enhancers are usually located far from the start of the gene, often as far as several thousand nucleotides away.

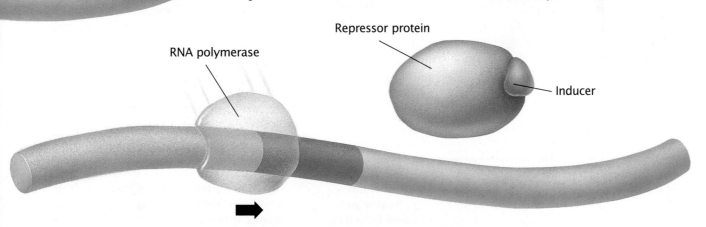

Genes for digesting lactose

Repressor protein

RNA polymerase

Inducer

b **When the inducer is present, it binds with the repressor protein and changes its shape. The altered repressor protein detaches from the DNA strand, enabling the polymerase to attach and begin transcribing the gene.**

Architecture of the Gene

While it is tempting to think of a gene as a single uninterrupted stretch of DNA that codes for proteins, this simple arrangement occurs only in bacteria. In eukaryotes, a gene contains a series of sequences called exons and introns. **Exons** are the portions of a gene that actually get translated into proteins. They are interrupted by noncoding portions of the DNA called **introns**. In most genes, the introns far outweigh the protein-encoding exons. In fact, in most cases, less than 10 percent of a human gene consists of exons. Like cars on a rural highway, exons are scattered here and there within genes. Introns are separated from exons during transcription.

Exons play a role in evolution

How could such a complicated system have survived the process of natural selection? The answer is that it adds evolutionary flexibility. Each exon encodes a different part of a protein. One exon may influence which molecules an enzyme is able to recognize, while another may determine whether a protein will respond to particular signal molecules. By possessing introns and exons, cells can shuffle exons between genes and create new combinations. Natural selection probably favored the intron-exon system of organization because cells are able to manufacture many different proteins by juggling exons between genes. The many thousands of proteins that occur in human cells appear to have arisen from only a few thousand exons!

Genes can jump to new locations

A few genes in chromosomes have the ability to move from one location in the chromosome to another. These genes are known as **transposons** (*tranz POH zahnz*). Once every few thousand cell divisions, a transposon jumps to a new location in the same chromosome or in a different chromosome. Transposons often inactivate the genes they jump into, creating mutations. Barbara McClintock, a geneticist working in the Cold Spring Harbor Laboratory on Long Island, New York, discovered transposons in the course of her studies of corn. The spotted and streaked patterns seen in the Indian corn shown in **Figure 7.13** result from the interactions of transposons that control its kernel pigments.

Figure 7.13
Barbara McClintock received a Nobel Prize in 1983 for her discovery of transposons in Indian corn.

Section Review

❶ **How do repressor proteins and inducer molecules affect transcription?**

❷ **How do activator proteins help RNA polymerase bind to a strand of DNA?**

❸ **How do the arrangements of genes differ in eukaryotes and prokaryotes?**

❹ **What effect do transposons have on other genes?**

Highlights

Watson and Crick built this model of DNA in 1953.

7.1 Understanding DNA

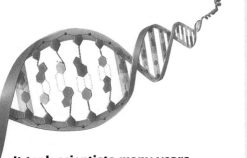

It took scientists many years to realize that DNA is the genetic material and to determine its structure.

Key Terms

transformation (p. 137)

nucleotide (p. 139)

purine (p. 139)

pyrimidine (p. 139)

double helix (p. 139)

replication (p. 140)

mutagen (p. 140)

Summary

- Griffith, Avery, and Hershey and Chase performed experiments that helped show that DNA is the hereditary material.

- DNA is composed of subunits called nucleotides. Each nucleotide contains a sugar, a phosphate group, and one of four bases.

- Watson and Crick showed that the DNA molecule is a double helix.

- Before cell division, DNA copies itself in a process called replication. The DNA separates into two strands, and new complementary bases attach to the exposed base.

7.2 How Proteins Are Made

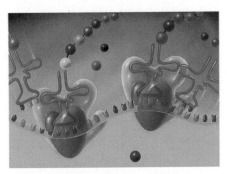

Proteins are made when amino acids are assembled by mRNA and tRNA molecules at the ribosome.

ribonucleic acid (p. 141)

gene expression (p. 141)

transcription (p. 141)

translation (p. 141)

codon (p. 143)

genetic code (p. 143)

anticodon (p. 144)

- During transcription, the genetic message from DNA is transferred to RNA.

- During translation, RNA directs the production of specific proteins encoded by genes.

- Each group of three nucleotides in mRNA is called a codon. Codons specify amino acids.

- Each sequence of three nucleotides in a tRNA molecule is called an anticodon. Anticodons complement the nucleotide sequence in mRNA codons.

7.3 Regulating Gene Expression

Cells can control gene activity. When an inducer binds to the repressor molecule, it falls off of DNA, allowing transcription to begin.

repressor protein (p. 148)

inducer (p. 149)

exon (p. 150)

intron (p. 150)

transposon (p. 150)

- In prokaryotes, transcription is regulated by repressor proteins. A repressor protein blocks RNA polymerase from transcribing a gene.

- In eukaryotes, genes are fragmented. Exons are the portions of a gene that are translated into proteins. Introns are noncoding regions of DNA.

- Transposons are genes that can jump to new locations on chromosomes.

Understanding Vocabulary

1. For each set of terms, explain the differences in their meanings.
   a. nucleotide, gene
   b. replication, transcription
   c. codon, anticodon
   d. exon, intron

Relating Concepts

2. Copy the unfinished concept map below onto a sheet of paper. Then complete the concept map by writing the correct word or phrase in each oval containing a question mark.

Understanding Concepts

Multiple Choice

3. Hershey and Chase's experiment with viruses that infect bacteria showed that
   a. protein gets into the bacterial cells.
   b. DNA is the genetic material.
   c. DNA contains radioactive sulfur.
   d. viruses undergo transformation.

4. The technology used by Franklin to investigate the structure of DNA is called
   a. translation.
   b. transcription.
   c. X-ray diffraction.
   d. photography.

5. Which represents the correct base pairing for DNA?
   a. G-T, A-C
   c. T-A, G-C
   b. A-G, A-T
   d. G-C, C-A

6. DNA and RNA are similar in that both have
   a. thymine as a nitrogen base.
   b. a single-stranded helix shape.
   c. nucleotides containing sugars, nitrogen bases, and phosphates.
   d. the same sequence of nucleotides for the amino acid phenylalanine.

7. If a segment of a DNA strand is CGTAGC, the complementary RNA strand is
   a. GCAUCG.
   c. ATGCAT.
   b. CGUAGC.
   d. AUGCAU.

8. The genetic code contains directions for
   a. copying DNA.
   b. constructing a double helix.
   c. ordering amino acids in proteins.
   d. removing repressor proteins.

9. The site where proteins are built from amino acids is the
   a. nucleus.
   b. endoplasmic reticulum.
   c. ribosome.
   d. gene.

10. To bring amino acids to the ribosome is the function of
    a. DNA.
    c. rRNA.
    b. mRNA.
    d. tRNA.

11. Which substance remains in the nucleus during translation?
    a. DNA
    c. tRNA
    b. mRNA
    d. rRNA

Completion

12. The first double helix model of DNA was constructed by _____ and _____ .

13. According to Chargaff's rule, in DNA the amount of _____ equals the amount of thymine, and the amount of _____ equals the amount of guanine.

14. The subunits that make up DNA are called _____ . They contain a sugar, a phosphate group, and a nitrogenous base.

15. The enzyme responsible for building a complementary strand of mRNA from DNA is called _____ .

16. Genes that move from one location on a chromosome to another and often inactivate other genes are called _____ , or jumping genes.

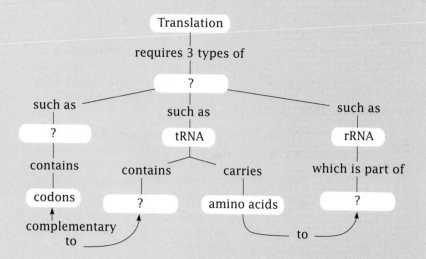

Short Answer

17. What was the importance of Griffith's work with strains of the bacterium *Streptococcus pneumoniae*?

18. Explain the process by which DNA is copied.

19. What is the genetic code? Why is it said to be universal?

20. What are the roles of the three types of RNA?

21. How do transposons inactivate genes?

Interpreting Graphics

22. The figure below shows the events of translation. Recall that translation follows transcription and is the process by which a protein molecule is assembled according to the mRNA code.

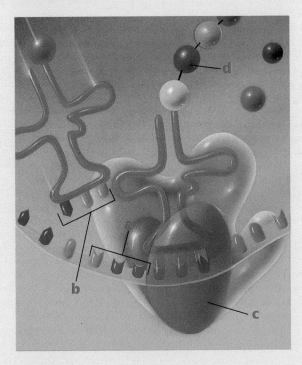

- What does **a** represent in the model?

- Part **c** represents the organelle inside the cell where translation occurs. What is the name of this organelle?

- What is the relationship between **a** and **b** during translation?

- What structure is represented by **d**?

Reviewing Themes

23. *Scale and Structure*
How is the design of the ribosome's structure appropriate for carrying out protein synthesis?

24. *Evolution*
What evidence is provided by the genetic code that all organisms have a common ancestry?

Thinking Critically

25. *Inferring Conclusions*
How would the operation of RNA polymerase be affected if the repressor protein were not bound to the proper site on a gene?

26. *Inferring Conclusions*
The codon "UUU" codes for the amino acid phenylalanine. What is the complementary DNA nucleotide to "UUU"?

27. *Building on What You Have Learned*
In Chapter 4 you learned about mitosis and cell division. Why is it important that DNA replication occur before mitosis and cell division?

Cross-Discipline Connection

28. *Biology and Language Arts*
As you learned in this chapter, proteins are compounds that consist of one or more chains of amino acids. Do library research to find the origin of the word "protein" and why it is used to describe amino acid chains.

Discovering Through Reading

29. Read pages 64–67 of the article "Life's Off-Switch" in *Discover*, July, 1991. How did Hwang's creative thought and persistence lead him to discover the protein that stops DNA replication? What was Kornberg's contribution to the discovery?

Investigation

How Does DNA Send Messages to the Cytoplasm?

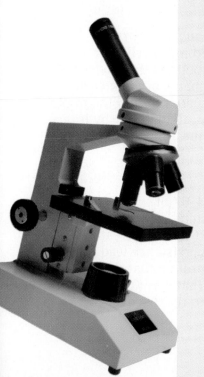

Objectives

In this investigation you will:

- *compare* and *contrast* the molecular structures of DNA and messenger RNA
- *design* models of DNA and messenger RNA to show how DNA sends messages to the cytoplasm

Materials

- soda straws (cut into 46 3-cm pieces)
- 13 red pushpins
- 9 blue pushpins
- 8 yellow pushpins
- 10 green pushpins
- 5 white pushpins
- 45 paper clips
- metric ruler
- a pair of scissors
- black felt-tip marker

Prelab Preparation

Review what you have learned about DNA by answering the following questions.

- What is the function of DNA?
- Where is DNA located in the eukaryotic cell?
- What is a nucleotide?
- Describe the basic structure of a DNA molecule.
- Where are proteins made in the cell?

Procedure: Making and Sending Messages From DNA

1. Form a cooperative team with another student to complete steps 2–11.

2. **CAUTION: Pointed objects can cause injury.** Cut soda straws into 3-cm segments to make 46 segments.

3. Insert a pushpin midway along the length of each segment of soda straw. Push a paper clip into one end of each 3-cm segment of straw until it touches the pushpin.

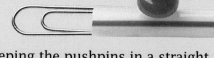

4. Keeping the pushpins in a straight line, insert the paper clip of a blue-pushpin segment into the open end of a red-pushpin segment. Add segments of straw to the red-pushpin segment end in the following order: blue, red, yellow, red, red, blue, green, red, yellow, yellow, yellow, blue, red, blue, green. Use a black felt-tip marker to label the blue segment on the end "top." This strand of segments is the first half of your DNA model.

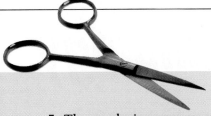

5. The pushpins represent nitrogen-containing bases, with red representing adenine (A), blue representing thymine (T), yellow representing guanine (G), and green representing cytosine (C). *What is the correct sequence of colors for a complementary strand for this model?* Construct this complementary strand using the remaining soda straw segments. *What do the pieces of soda straw represent?*

6. Place the first half of your DNA model (strand I) and its complementary strand (strand II) parallel to each other on the table. Make sure that the pushpins are in the correct order in both strands.

7. Make a sketch of your model and label the nucleotides A, T, G, or C.

8. Use a black marker to color a band around the remaining 15 pieces of soda straw. Each piece represents a phosphate group and the sugar ribose. These pieces are needed to make RNA nucleotides.

9. Separate strand I of the DNA molecule from strand II. Now use the ribose-containing nucleotides to form a strand of messenger RNA that pairs with strand I of the DNA molecule. Recall that RNA-DNA pairing is similar to DNA-DNA pairing except that uracil replaces thymine in the RNA molecule. Use white pushpins to represent uracil molecules. *What is the correct sequence of colors for an mRNA strand complementary to strand I of the DNA model?*

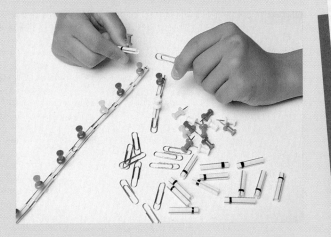

10. Make a sketch showing the pairing of the nitrogen-containing bases in the DNA nucleotides with those in the messenger RNA nucleotides. Label the bases A, T, G, C, or U.

11. Move the messenger RNA to the side and bring the two DNA strands back together.

Analysis

1. *Making Comparisons*
   How is messenger RNA similar to DNA? How is it different?

2. *Making Comparisons*
   Compare the sequence of nucleotides in strand I of the DNA molecule to the sequence of nucleotides in the messenger RNA.

3. *Making Comparisons*
   Compare the sequence of nucleotides in strand II of the DNA molecule with the sequence of nucleotides in the messenger RNA.

4. *Identifying Relationships*
   Why can messenger RNA be thought of as a "message" from DNA?

5. *Identifying Relationships*
   Where in the cell is messenger RNA made?

6. *Making Inferences*
   Where will the messenger RNA go after it is made?

Thinking Critically

1. What are the advantages of having DNA remain in the nucleus rather than allowing it to move about the cell?

2. According to your models, how might a mutation arise?

Gene Technology Today

Review

- **DNA structure (Section 7.1)**
- **replication (Section 7.1)**
- **protein synthesis (Section 7.2)**

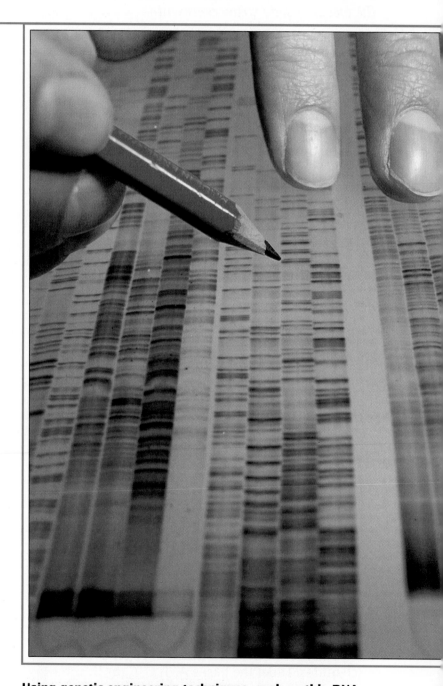

Using genetic engineering techniques, such as this DNA profile, scientists can isolate individual genes and transfer them from one species to another.

IN MYTHOLOGY, A CHIMERA IS A CREATURE WITH THE HEAD OF A LION, THE BODY OF A GOAT, AND THE TAIL OF A SERPENT. ALTHOUGH CHIMERAS DO NOT EXIST IN NATURE, SCIENTISTS CAN CREATE THEM IN LABORATORIES BY ALTERING THE GENES OF LIVING ORGANISMS. AS YOU WILL LEARN IN THIS CHAPTER, SCIENTISTS ARE USING THESE TECHNIQUES TO TRANSFORM AGRICULTURE AND REVOLUTIONIZE MEDICINE.

8.1　Genetic Engineering

Objectives

1 Outline the genetic engineering experiment performed by Cohen and Boyer.

2 Explain the four stages involved in a gene transfer experiment.

3 Describe how scientists use restriction enzymes in genetic engineering experiments.

4 Discuss some of the safety concerns associated with genetic engineering.

What Is Genetic Engineering?

The first chimera was created in 1973 by Stanley Cohen and Herbert Boyer, two geneticists at the University of California at San Francisco. Cohen and Boyer set out to insert a gene from an African clawed frog into a bacterium. They were hoping that the bacteria could then be used to make the protein encoded by the frog gene.

First, Cohen and Boyer isolated the gene that coded for frog ribosomal RNA. Then, they added that gene to bacterial DNA. Their experiment created the first living cells that had DNA from a foreign organism added to their own DNA. **Recombinant DNA** is a molecule formed when fragments of DNA from two or more different organisms are spliced together in a laboratory. Did the bacteria with the recombinant DNA use the frog gene? Yes! The genetically engineered bacteria cells produced frog rRNA, just as Cohen and Boyer had hoped. **Figure 8.1** outlines their experiment.

This experiment ushered in a new age in biology, one that explored the possibilities of moving genes from one organism to another. In approximately 20 years, the growing ability of researchers to transfer DNA from one organism into another has revolutionized biology. Moving genes from the chromosomes of one organism to those of another is called **genetic engineering**. Today specific genes from human chromosomes can be routinely transferred into bacteria.

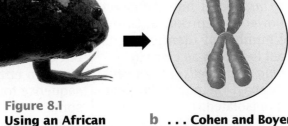

Figure 8.1

a **Using an African clawed frog as their experimental organism, . . .**

b **. . . Cohen and Boyer isolated an rRNA gene from one of its chromosomes . . .**

c **. . . and inserted the gene into a bacterium. This bacterium then produced frog rRNA.**

Figure 8.2

a **To isolate a gene from one of this frog's chromosomes, . . .**

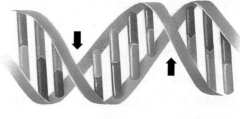

b **. . . Cohen and Boyer added the restriction enzyme *Eco*RI to cut the frog DNA at specific sites.**

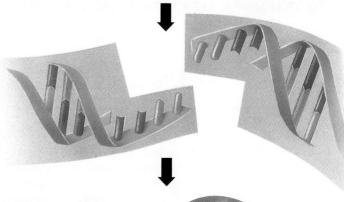

c **The result was many DNA fragments with "sticky" ends.**

How to Move a Gene From One Organism to Another

If you were a scientist working in a genetic engineering lab, how would you go about moving a gene? First you would need to find a source chromosome that contains the gene you wish to isolate. Next you would need to select a target cell into which the gene could be moved. And finally, you would need to figure out a way to transfer the isolated gene into the target cell. Every gene transfer experiment is performed in four distinct stages:

1. *Cleaving DNA* The source chromosome is cut into fragments of DNA.

2. *Producing recombinant DNA* The DNA fragments containing the desired gene are inserted into viral or bacterial DNA. The recombinant DNA is then allowed to infect target cells.

3. *Cloning target cells* Infected target cells are allowed to reproduce. Growing a large number of identical cells from one cell is known as **cloning**.

4. *Screening target cells* Target cells that have received the particular gene of interest are isolated.

Each of these stages will be explained in greater detail using the Cohen-Boyer experiment as an example.

Cleaving DNA is the first step of a gene transfer experiment

The first stage in any successful genetic engineering experiment involves cutting the source chromosome into fragments to obtain copies of the gene you wish to transfer. In their experiment, Cohen and Boyer used chromosomal DNA from *Xenopus laevis*, the African clawed frog.

In order to cut the frog DNA, Cohen and Boyer used special enzymes called **restriction enzymes**. Restriction enzymes recognize and bind to specific short sequences of DNA, and then cut the DNA at a specific site within that sequence. Cohen and Boyer used a restriction enzyme called *Eco*RI, which cuts DNA whenever it encounters the sequence CTTAAG. The sequence of the opposite strand is GAATTC, the same sequence written backward. **Figure 8.2b** shows the restriction enzyme *Eco*RI cutting a DNA sequence found within a frog chromosome.

Restriction enzymes do not make a straight cut through both strands of DNA. Instead, the cut is offset a few

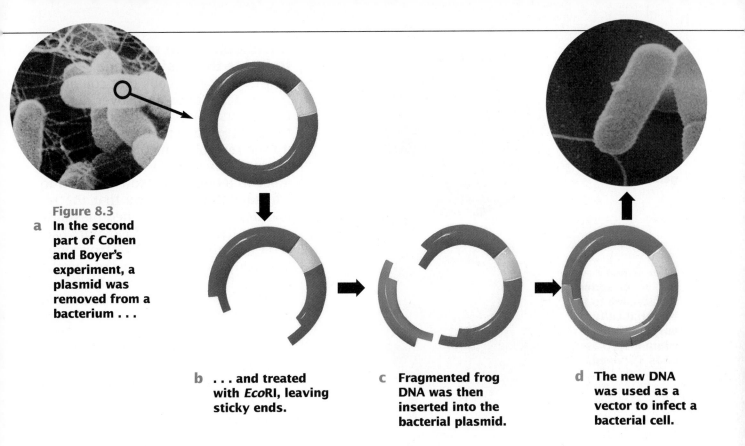

Figure 8.3

a In the second part of Cohen and Boyer's experiment, a plasmid was removed from a bacterium . . .

b . . . and treated with *Eco*RI, leaving sticky ends.

c Fragmented frog DNA was then inserted into the bacterial plasmid.

d The new DNA was used as a vector to infect a bacterial cell.

Evolution

What evidence

revealed by gene

technology

suggests that all

organisms are

related?

bases. For example, in the sequence CTTAAG, *Eco*RI cuts each strand between the A and G, a site four bases apart on both strands, creating DNA fragments with single-stranded "sticky ends."

Because the two single-stranded sticky ends have complementary sequences, they can pair back up and heal the break (which is why they are called sticky), *or they can pair with any other DNA fragment cut by the same enzyme.* DNA cut by the restriction enzyme *Eco*RI can be joined to DNA from any other organism that has also been cut by *Eco*RI, because they have the same sticky ends. One of the fragments Boyer and Cohen cleaved from frog DNA contained the rRNA gene they sought to transfer.

Restriction enzymes are among the basic tools of genetic engineering. More than 200 different restriction enzymes have been isolated and identified. Most restriction enzymes attack only one specific sequence. By trial and error, biologists can often find a restriction enzyme that cuts out the gene that they seek.

Restriction enzymes are used to produce recombinant DNA

Cohen and Boyer also used *Eco*RI to cut circular forms of bacterial DNA called **plasmids**. Because each plasmid contained only one *Eco*RI site, it was cut open in only one place. By mixing the frog DNA fragments with the plasmids cut open by *Eco*RI, Cohen and Boyer produced recombinant DNA.

It was easy for the ends of the opened plasmids to stick to the ends of the frog DNA fragments because they had complementary sticky ends. An enzyme that can join the ends was added to bond the plasmids and fragments together, resulting in new DNA. Once formed, the new DNA served as delivery agents, or **vectors**, which would carry the frog ribosomal RNA gene into the target bacterial cells.

Figure 8.3 shows how the frog DNA was inserted into a plasmid. The vector used by Cohen and Boyer contained two critical genes: one replicated the plasmid DNA, and the other made the cell resistant to an antibiotic called tetracycline. This second gene is important during the screening stage.

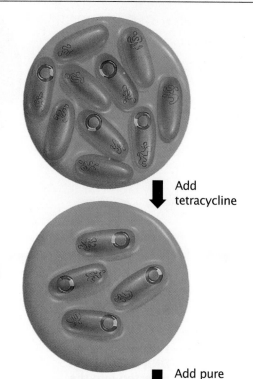

Figure 8.4

a Cohen and Boyer screened a population of bacteria to locate cells containing the frog rRNA gene.

Add tetracycline

b After Cohen and Boyer added tetracycline to the bacterial culture, only bacteria containing plasmids survived.

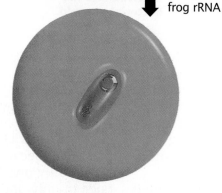

Add pure frog rRNA

c The pure frog rRNA that Cohen and Boyer added to the culture stuck to bacteria containing the one plasmid with the frog rRNA gene.

Target cells are cloned

To produce clones, Cohen and Boyer added recombinant DNA to a culture of bacteria. They placed the bacterial culture under conditions that encouraged the bacteria to take the recombinant DNA into their cells. Each bacterial cell then reproduced, forming clones. Some of the clones contained recombinant DNA, while others did not.

Scientists screen target cells to locate the desired gene

To find bacterial cells that contained frog genes, Cohen and Boyer did two things. First, they added tetracycline to the culture. Bacteria that did not take up the recombinant DNA were killed because they lacked the gene for resistance to tetracycline. Now they had to consider the possibility that the remaining bacteria contained plasmids without the frog gene. To identify bacteria containing the rRNA frog gene, Cohen and Boyer attempted to pair the bacterial DNA with pure frog rRNA. Only a cell carrying the frog rRNA gene would contain DNA that would stick to pure frog rRNA. Patient searching revealed a colony with cells containing DNA that stuck to the pure frog rRNA. These were the cells they sought—bacterial cells that carried the frog gene. These cells could be grown to produce large quantities of frog DNA and its rRNA gene product. **Figure 8.4** shows the steps involved in screening bacteria.

Precautions ensure that genetic engineering is safe

Because a genetic engineer can, in principle, move *any* gene from one organism to another, it is important to be careful not to introduce potentially dangerous genes into organisms that might escape from the laboratory. For this reason, particular care is taken when cancer cells or disease-causing organisms are used as gene sources in gene transfer experiments. Scientists select target cells that cannot survive outside the laboratory, and potentially dangerous experiments are forbidden. In more than a decade, no dangerous accident has ever occurred.

Section Review

❶ Describe the first experiment that successfully transferred a gene from a frog to a bacterium.

❷ Define the term "genetic engineering." List the four distinct stages involved in a gene transfer.

❸ How are restriction enzymes used in gene transfer experiments?

❹ How can scientists ensure that genetic engineering experiments are safe?

ONE OF THE GREATEST SUCCESSES OF GENETIC ENGINEERING HAS BEEN
THE MANIPULATION OF GENES IN CROP PLANTS AND LIVESTOCK. GENE
TRANSFERS HAVE MADE CROP PLANTS MORE RESISTANT TO DISEASE,
HERBICIDES, AND INSECTS. GENETIC ENGINEERING HAS ALSO BEEN USED
TO INCREASE THE MILK PRODUCTION AND GROWTH RATE OF LIVESTOCK.

8.2 Transforming Agriculture

Objectives

❶ **Explain the role of the Ti plasmid in agricultural research.**

❷ **Describe how herbicide-resistant genes in crop plants can benefit the environment.**

❸ **Explain how genetic engineering techniques have been used to improve crop yields.**

❹ **Describe how gene transfers are being used to make livestock more productive.**

The Ti Plasmid

Figure 8.5

a **The tumor-causing gene in the Ti plasmid . . .**

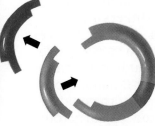

b **. . . is removed and replaced with the desired gene.**

c **The plasmid can then be inserted into bacteria, which can transfer the gene to plant cells.**

For years, genetic engineering in plants was difficult because scientists lacked an appropriate vector to deliver desirable genes. Unlike bacteria, plants normally contain few viruses or plasmids that can act as delivery agents. Recently, however, scientists discovered the Ti plasmid, an unusual bacterial plasmid shown in **Figure 8.5**. The Ti plasmid causes the development of large tumors in plants. **Figure 8.6** shows a tree that has been infected by a crown gall tumor, which is caused by the bacteria containing the Ti plasmid. When the Ti plasmid infects a plant cell, it inserts itself into the plant cell's chromosomes. To transform the Ti plasmid into an effective genetic engineering vehicle, scientists first remove the tumor-causing gene from the Ti plasmid. The vacant space in the now harmless plasmid can be filled by the desired gene. Unfortunately, bacteria carrying the Ti plasmid cannot be used to insert DNA into plants that produce cereal grains such as corn, rice, and wheat. Researchers are presently developing powerful new techniques for introducing useful genes into these plants, such as shooting the cells with "gene guns."

Figure 8.6

This tree has a crown gall tumor, an abnormal growth caused by the Ti plasmid.

Resistance to Plant-killing Chemicals

A recent improvement in agriculture has been the development of crop plants that are resistant to the chemical glyphosate, a powerful weed killer that also kills most actively growing plants. Glyphosate kills plants by destroying an enzyme that plants need to make certain amino acids. Genetic engineers found a kind of bacterium in which this enzyme is resistant to glyphosate. They then isolated the gene that codes for the enzyme. Shooting the gene in like a bullet, they successfully transferred the glyphosate-resistant gene into crop plants such as the wheat plant shown in **Figure 8.7**.

This advance is of great interest to farmers. The farmer simply treats a field with glyphosate, and all growing plants die except the crop, which is resistant to glyphosate. After it is applied, glyphosate is quickly broken down in the environment. Glyphosate is not harmful to humans because they do not make the amino acids it affects. These qualities make it a great improvement over most commercial weed-killing chemicals, which can be highly toxic. In addition, much of the tragic erosion of fertile topsoil could be prevented if cropland did not have to be intensively cultivated to remove weeds.

Figure 8.7
Scientists have made crop plants, like this wheat plant, resistant to plant-killing chemicals.

Nitrogen Fixation

Nitrogen is an element that plants must have in order to make proteins and DNA. The most abundant source of nitrogen in the environment is the atmospheric gas N_2. However, plants cannot obtain nitrogen from the air. All of the nitrogen that plants need must be obtained from the soil. Bacteria living in the roots of plants such as soybeans, peanuts, and clover provide plants with nitrogen by converting N_2 gas from the atmosphere into a form that plants can use. **Figure 8.8** shows these bacteria. The process of converting nitrogen into a form that plants can use is called **nitrogen fixation**.

Because crops use nitrogen rapidly, most farmers add high-nitrogen fertilizers to the soil. Worldwide, farmers applied more than 65 million metric tons of nitrogen fertilizers in 1990. Since high-nitrogen fertilizers are costly, using them adds a considerable expense to a farmer's budget. Farming would be much cheaper if major crops such as wheat, rice, and corn could be genetically engineered to carry out nitrogen fixation. The bacterial genes for nitrogen fixation have been successfully inserted into plants. However, these genes do not seem to function properly in their new hosts. Many experiments are being performed to find a way to overcome this difficulty.

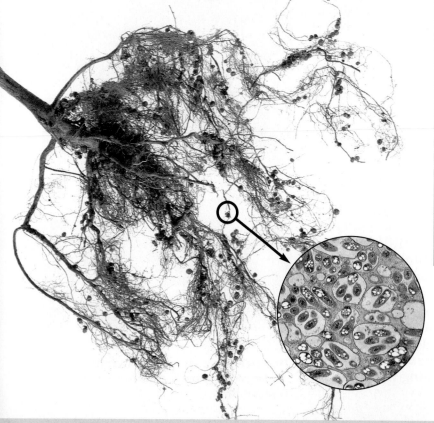

Figure 8.8
Nodules on the roots of this soybean plant contain populations of bacteria that convert nitrogen to a form the plant can use.

Resistance to Insects

Genetic engineering techniques can also be used to make crops resistant to destructive insects, such as the locust and other pests. Resistant crops would not need to be sprayed with insecticides, which are expensive and can be harmful to other organisms in the environment. Today more than 40 percent of all chemical insecticides are used to kill insects that eat cotton plants. Biologists are now trying to produce cotton plants that are resistant to these pests, so that insecticides would not be needed.

One successful approach to making plants resistant to insects uses bacteria that produce enzymes responsible for killing destructive caterpillars. When the genes coding for these enzymes are inserted into tomato plants, for example, the new enzymes make the plants highly toxic to insects called tomato hornworms. Hornworms that eat any of the tomato plants quickly die.

Many pests attack the roots of important plants. To combat these pests, genetic engineers are introducing the same insect-killing enzyme into a bacterium that colonizes the roots of crop plants. If an insect eats these roots, it will consume the bacteria and be killed by the enzyme. **Figure 8.9** shows how genetic engineering could improve a cotton plant.

Figure 8.9
Plants, such as this cotton plant, could be genetically engineered:

a to produce an enzyme that kills pests, so that highly toxic pesticides would not be needed,

b to resist weed-killing chemicals such as glyphosate, so that highly toxic herbicides would not be needed,

c and to carry out nitrogen fixation, so that farmers would not need to apply fertilizers.

Figure 8.10
When genetically–engineered bovine growth hormone is injected into dairy cattle such as these, each cow's milk production increases by about 10 percent. The United States government is presently reviewing this experimental procedure to determine if it is acceptable for widespread use by American farmers.

Genetic Engineering in Livestock

Patterns of Change

What are some

advantages of

genetically

engineered

proteins such as

human growth

hormone?

Genetic engineering techniques can also be used to produce bigger, more productive livestock. For example, injecting growth hormone into dairy cows, such as the ones shown in **Figure 8.10**, increases their milk production. The milk is no different from milk produced by other cows. Growth hormone is the natural signal that cows use to increase their milk production. In this case, the hormone is used in higher amounts than occur naturally.

Instead of extracting growth hormone from the pituitary glands of dead cows, the hormone is obtained in the laboratory using genetic engineering. The gene containing the instructions for producing growth hormone has been introduced into bacteria. These bacteria produce growth hormone so inexpensively that it can be added as a supplement to a cow's diet.

Biologists have introduced extra copies of the genes that code for growth hormones into the chromosomes of cows and hogs. These attempts are likely to create new, leaner, fast-growing cattle and hogs. Human growth hormone is now being tested in humans as a potential treatment for dwarfism, a disorder in which the pituitary gland fails to make enough growth hormone.

Section Review

1 How is the Ti plasmid used to insert genes into plant cells? What are its shortcomings as a genetic engineering tool?

2 Explain how genetic engineering techniques can make crop plants resistant to weed killers such as glyphosate.

3 How can genetic engineering reduce the amount of insecticides used in agriculture?

4 What effects can genetic engineering have on livestock?

MUCH OF THE EXCITEMENT SURROUNDING GENETIC ENGINEERING HAS FOCUSED ON ITS POTENTIAL TO AID IN PREVENTING AND CURING ILLNESS. MAJOR ADVANCES HAVE BEEN MADE IN THE PRODUCTION OF PROTEINS THAT CAN TREAT ILLNESS, IN THE DEVELOPMENT OF NEW VACCINES TO COMBAT DISEASE, AND IN THE REPLACEMENT OF DEFECTIVE GENES WITH HEALTHY ONES.

8.3 Advances in Medicine

Objectives

❶ Describe how transferring human genes into bacteria can benefit human health.

❷ Explain how gene transfers could be used to combat genetic disorders.

❸ Summarize the goals of the Human Genome Project and the ethical issues it raises.

❹ Describe three uses of DNA profiling.

Making Miracle Drugs

Figure 8.11

a **To make large amounts of insulin, scientists first obtain the gene that produces insulin.**

Many human illnesses occur because our bodies fail to make critical proteins. One example of such an illness is diabetes. In one form of diabetes a person cannot make a protein called insulin. As a result, they cannot regulate the levels of sugar in their blood. Diabetes can be treated if the body is supplied with insulin.

Until recently, it was not practical to use proteins as drugs because they were difficult to obtain. The body contains only small amounts of proteins like insulin, making them difficult and expensive to obtain. In most cases, proteins must be obtained from cow or pig tissue. Now genetic engineering can be used to provide large quantities of proteins in a short period of time. Genes that produce medically important proteins can be inserted into bacteria. Because bacteria can be easily grown in bulk, large amounts of the desired protein can be inexpensively prepared. **Figure 8.11** shows how genetic engineering is used to produce insulin.

Some proteins now being produced by genetically engineered bacteria dissolve blood clots, helping to prevent heart attacks and strokes. Genetically engineered bacteria are also used to produce proteins that prevent high blood pressure, and proteins that help regulate kidney function.

b **The insulin gene is then inserted into a bacterial plasmid.**

c **The bacteria containing the plasmid are cloned and insulin is produced.**

d **The insulin can be collected and injected into diabetics, such as this young man.**

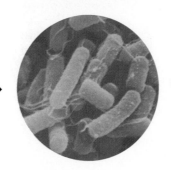

Making Vaccines

A **vaccine** is a harmless version of a disease-causing microbe (bacterium or virus) injected into animals or people so that their immune systems will develop defenses against the disease. The injected version serves as a model for the body, which responds by making defensive proteins called **antibodies**. If a vaccinated animal is exposed to the disease-causing microbe, it will immediately begin large-scale production of microbe-attacking antibodies. The antibodies will stop the growth of the microbe before the disease can develop.

Traditionally, vaccines have been prepared either by killing the microbe or rendering it unable to grow. This ensures that injecting a vaccine into your body will not make you sick. The problem with this approach is that any failure in the killing or weakening of the disease-causing microbe can result in introducing the disease into the very patients in need of protection. While the majority of such vaccines are extremely safe, a tiny percentage of vaccinated individuals contract the disease from the vaccine. This small, but real, danger is one reason why rabies vaccines are administered only when a person has actually been bitten by an animal suspected of carrying rabies.

Harmless microbes can be made into piggyback vaccines

Using genetic engineering techniques, genes from disease-causing microbes can be inserted into harmless bacteria or viruses. These harmless microbes can be used to stimulate your body to make disease-attacking antibodies. For example, a vaccine against the genital herpes virus, which produces small blisters on the genitals, has been made using genetic engineering techniques as shown in **Figure 8.12**. Genes that encode the surface proteins of the herpes virus are inserted into a harmless virus. This viral vector carries the genes "piggyback" into the human body, stimulating the person's immune system to make protective antibodies against the virus. Individuals who are injected with the vaccine are able to fight infections from the Herpesvirus II.

Genetically engineered vaccines also offer hope for other diseases. A vaccine against the hepatitis B virus, which causes a sometimes fatal inflammation of the liver, is also available. Scientists are also working to produce a vaccine that will protect people against malaria. Transmitted by mosquitoes, malaria affected more than 250 million people in 1992, and over 1 million of them died.

Figure 8.12

a **To produce a vaccine against genital herpes, the virus that causes genital herpes is obtained from the cells of a person with the disease.**

b **A DNA fragment containing the gene that codes for the herpes surface protein . . .**

c **. . . is inserted into a harmless cowpox virus. The fragment causes the virus to make herpes surface proteins.**

d **A person injected with the engineered virus containing the desired gene can make antibodies against the genital herpes virus.**

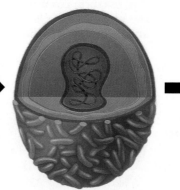
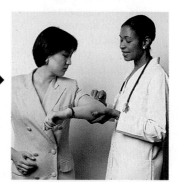

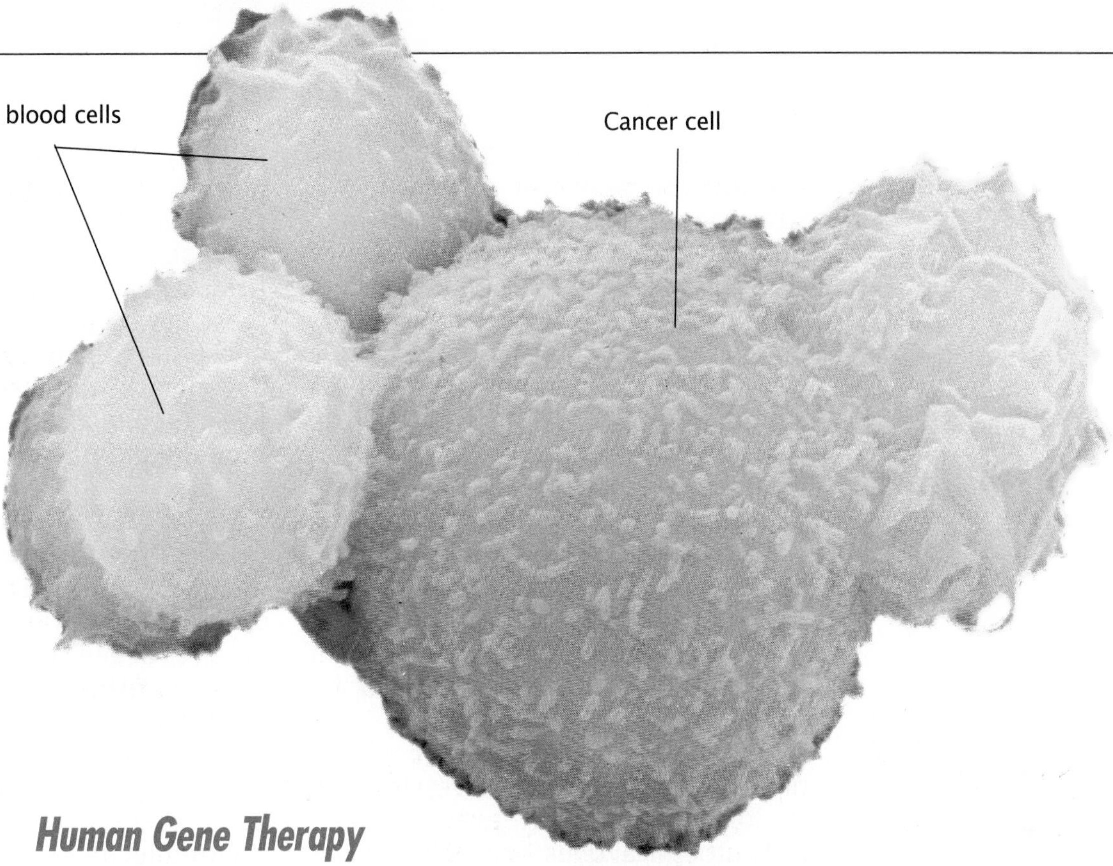

White blood cells

Cancer cell

Figure 8.13
White blood cells containing tumor necrosis factor are attacking this cancer cell.

Human Gene Therapy

Patterns of Change

If a woman with genetically altered bone marrow cells has children, would the genetic change be passed on to them?

Most human genetic diseases are due to an individual's lack of a normally functioning copy of a particular gene. The problem usually arises when both of the individual's parents contribute defective copies of a gene. One obvious way to cure such disorders is to give the person a working copy of the defective gene. Until recently, this approach was not practical for three reasons. First, the defective gene was difficult to identify and isolate. Second, it was hard to transfer a "healthy" copy of such a gene into the cells of body tissues that use it. Finally, it was necessary to find a way to keep the altered cells or their offspring alive in the body for a long time. With genetic engineering, it is now possible to overcome these difficulties. Gene transfers are being attempted as a way of combating a variety of genetic disorders, including cystic fibrosis and muscular dystrophy.

Human gene therapy has proven successful

Among the first successful attempts at human gene therapy was the transfer of an enzyme-encoding gene into a girl suffering from a rare blood disorder that was caused by the lack of this enzyme. Using genetic engineering, the gene was isolated, cloned, and inserted into cells taken from the girl's blood-cell-producing bone marrow. These cells were returned to the girl's bones where they began to produce the enzyme her body lacked. Because this kind of bone marrow cell actively divides within the bone marrow, researchers hope that offspring of the introduced cells will continue to secrete the enzyme into her blood for a long time.

Human gene transfers may also help in the battle against cancer

All humans possess white blood cells that secrete a protein called TNF (tumor necrosis factor) into the blood. TNF attacks and kills cancer cells. Unfortunately, TNF cannot work unless it encounters a cancer cell. This does not often happen. Recently, genetic engineers developed a method of adding the gene-encoding TNF protein to a kind of white blood cell that is very effective at locating cancer cells, but not very effective at harming them. Armed with this new TNF weapon, however, these white blood cells will become more like cruise missiles with a deadly payload homing in on cancer cells. **Figure 8.13** shows three white blood cells attacking a cancerous tumor.

Gene Sequencing

Scientists are mapping the human genome

Genetic engineers are presently attempting to catalog, locate, and sequence every gene on the 46 human chromosomes. Gene sequencing is the process of determining the order of nucleotide bases within a gene. Biologists call the entire collection of genes within human cells the **human genome**. The effort to determine the nucleotide base sequence of every human gene is called the **Human Genome Project**. Since the human genome contains approximately 3 billion nucleotides, the project is not a small one. This United States project was launched in 1988 and is expected to cost several billion dollars.

Our expanding knowledge of the human genome and our increasing ability to manipulate it present some difficult ethical questions. Who should have access to a person's genetic profile? Who decides what kinds of human gene transfer experiments should and should not be done? Which diseases should we attempt to cure first? There are no simple answers to these questions. These and many other questions will arise as gene technology becomes a more essential part of our lives. Each question will require careful thought and planning.

DNA profiling can be used to identify unknown DNA

In 1984, Alec Jeffreys, a British geneticist, devised a way to visually identify the DNA in genes. This technique, called **DNA profiling**, can identify the base sequences in a sample of DNA. The end result, shown in **Figure 8.14**, is a photograph with a pattern of dark bands that reflect the composition of a DNA molecule. Because these images can be used to establish the identity of a person, they are often called **DNA fingerprints**.

DNA profiling is based on the theory that it is extremely unlikely for two people to have identical DNA. Therefore, when the process is used to analyze DNA in blood, semen, bone, or hair, DNA profiling can help identify criminals. Criminal investigation is just one use for this new technology.

Scientists can isolate small fragments of DNA and locate genes that cause disorders such as neurofibromatosis, a painful condition causing benign tumors. In addition, doctors can use DNA profiling to reveal hereditary relationships and to measure the success of bone marrow transplants in leukemia patients.

Figure 8.14
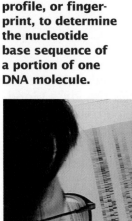
This scientist is examining a DNA profile, or fingerprint, to determine the nucleotide base sequence of a portion of one DNA molecule.

Section Review

❶ Explain how genetic engineering can be useful in the treatment of human illnesses such as diabetes.

❷ How could human genetic therapy cure genetic disorders such as cystic fibrosis?

❸ What are the goals of the Human Genome Project? What ethical issues does it raise?

❹ Describe how DNA profiling could be used to help determine a suspected criminal's innocence or guilt.

Gene therapy offers hope for patients with genetic disorders such as this child with cystic fibrosis.

	Key Terms	Summary
8.1 Genetic Engineering  **Bacterial plasmids make genetic engineering possible.**	recombinant DNA (p. 157) genetic engineering (p. 157) cloning (p. 158) restriction enzyme (p. 158) plasmid (p. 159) vector (p. 159)	• Over the last 20 years, genetic engineers have learned how to move genes from one organism to another. • Every gene transfer starts by cleaving DNA into small fragments using restriction enzymes. • Fragments containing the desired gene are then transferred to a target cell using a vector. • After removing uninfected cells, scientists search for cells that have taken up the desired gene.
8.2 Transforming Agriculture  **Cows injected with genetically engineered growth hormone will produce more milk.**	nitrogen fixation (p. 162)	• Genetic engineers have manipulated the genes of certain kinds of crop plants to make these plants resistant to weed killers and destructive pests. • Genetic engineers are looking for a way to transfer genes for nitrogen fixation into crop plants. • Adding genetically engineered growth hormone to the diet of livestock increases milk production in dairy cows and increases the weight of cattle and hogs.
8.3 Advances in Medicine  **Scientists hope that they will soon be able to use genetic engineering to treat diseases such as cancer.**	vaccine (p. 166) antibody (p. 166) human genome (p. 168) Human Genome Project (p. 168) DNA profiling (p. 168) DNA fingerprint (p. 168)	• Genetic engineering techniques can be used to manufacture proteins such as insulin and vaccines. • Some human genetic disorders are being treated and "corrected" by inserting copies of healthy genes into individuals lacking them. • The human genome is the entire collection of genes within human cells. The Human Genome Project seeks to locate, catalog, and read the base sequence of every human gene. • Scientists use DNA profiling to reveal DNA patterns that vary distinctively from one individual to the next.

Understanding Vocabulary

1. For each pair of terms, explain the differences in their meanings:
   a. recombinant DNA, restriction enzyme
   b. genetic engineering, DNA profiling
   c. vector, plasmid

Relating Concepts

2. Copy the unfinished concept map below onto a sheet of paper. Then complete the concept map by writing the correct word or phrase in each oval containing a question mark.

Understanding Concepts

Multiple Choice

3. The genetic engineering experiment of Cohen and Boyer produced
   a. frog clones.
   b. recombinant DNA.
   c. synthetic insulin.
   d. prokaryotic bacteria.

4. The screening stage of the Cohen and Boyer experiment involved
   a. locating cells that received the gene for frog rRNA.
   b. searching cells for plasmids not resistant to tetracycline.
   c. cutting frog DNA into fragments and inserting them into plasmids.
   d. marking DNA segments resistant to tetracycline.

5. Which is *not* one of the stages of gene transfer?
   a. DNA profiling
   b. producing recombinant DNA
   c. screening
   d. cleaving DNA

6. Scientists are able to transfer segments of DNA from one organism to another by using
   a. bacterial plasmids.
   b. frog chromosomes.
   c. cotton plants.
   d. piggyback vaccines.

7. Growing a large number of identical cells from one cell is known as
   a. nitrogen fixation.
   b genetic engineering.
   c. DNA profiling.
   d. cloning.

8. The role the Ti plasmid plays in agricultural research is that of a
   a. restriction enzyme.
   b. weed killer.
   c. nitrogen fixer.
   d. vector.

9. Genetically engineered plants that are able to fix nitrogen will
   a. make crop rotation a necessity.
   b. kill soil bacteria.
   c. reduce the use of fertilizers.
   d. be resistant to chemicals that kill insects.

10. An advantage of a genetically engineered human protein such as insulin is
    a. it can be produced in large amounts at low cost.
    b. less is needed to produce the same effect.
    c. it is safer than naturally produced insulin.
    d. it dissolves blood clots in addition to treating diabetes.

11. Cystic fibrosis, muscular dystrophy, and other genetic disorders
    a. are vectors used in genetic engineering.
    b. may be treated by replacing defective genes.
    c. can be cured in babies by genetic analysis.
    d. are too deadly to be considered by genetic engineers.

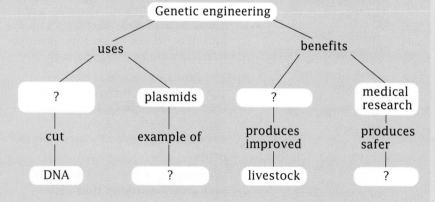

Completion

12. Restriction enzymes are used to cut DNA segments. The cut leaves _____ _____ .

13. To make "piggyback" vaccines, genetic engineers insert genes from a(n) _____ virus into the chromosomes of a(n) _____ virus, which is then put into the human body.

14. The complete collection of genes within human cells is called the _____ ; it contains about _____ nucleotides.

Short Answer

15. What makes the DNA fragments cut by restriction enzymes "sticky"?

16. In what ways can genetic engineering affect agriculture?

Interpreting Graphics

17. The figures below show four distinct stages of a gene transfer experiment.

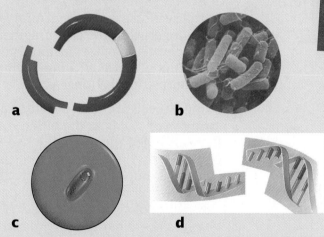

a

b

c

d

• How should the figures be ordered to show the proper sequence of events?

• What is occurring in **a**? In **c**?

Reviewing Themes

18. *Interacting Systems*
Why would farmers be interested in genetically engineered crop plants that are resistant to the weed-killing chemical glyphosate?

19. *Patterns of Change*
How has genetic engineering changed the way some vaccines are prepared?

Thinking Critically

20. *Inferring Conclusions*
Why are restriction enzymes said to be the basic tools of genetic engineering?

21. *Comparing and Contrasting*
How are organisms produced by genetic engineering different from organisms produced by sexual reproduction?

22. *Building on What You Have Learned*
In Chapter 7 you learned about the double helix model of DNA first constructed by Watson and Crick. How is Watson and Crick's model of DNA related to genetic engineering?

Cross-Discipline Connection

23. *Biology and Agriculture*
Several Texas farmers were recently asked to plant a genetically engineered variety of cotton that produces tan, green, and red cotton. What are the possible advantages of genetically engineering new varieties of different colored cotton?

Discovering Through Reading

24. Read the article "Barnyard Bioengineers" in *Newsweek*, September 9, 1991. How are animals being used to produce drugs to improve human health? Why are animal-rights activists interested in the work of "barnyard bioengineers"?

Investigation

How Are DNA Fingerprints Interpreted?

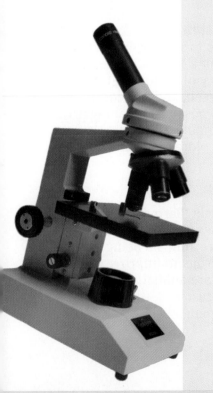

Objectives

In this investigation you will:
- *observe* and *interpret* similarities and differences among various representations of DNA fingerprints

Materials

- container
- colored beads (of various shapes and lengths)
- heavy-duty thread

Prelab Preparation

Review what you have learned about DNA by answering the following questions:
- Describe the structure of DNA.
- What are DNA fingerprints? What can they be used for?

Procedure: Interpreting DNA Fingerprints

1. The process of determining a DNA fingerprint involves three steps. First, DNA is extracted from cells and cut into small fragments by restriction enzymes. The containers on the desks in your classroom contain beads that represent such fragments of DNA. Each container represents a different source. Remove the beads from the container at your desk and sort them into groups based on size and shape. No two containers hold the exact same assortment of beads. However, one container holds an assortment of beads similar to those in the container at your teacher's desk. These beads represent fragments of DNA taken from cells found at the scene of a crime.

2. In the second step of determining a DNA fingerprint, the DNA fragments are separated by size. Arrange your collection of beads in a line from the shortest to the longest.

3. Using a piece of heavy-duty thread, string your beads in the order you have arranged them. Tie the two ends together in a knot so that the beads will not fall off. Compare the numbers and sizes of your beads with those of your classmates. *How does the combination of beads on your string compare to other combinations of beads in the class?*

4. In the third step of determining a DNA fingerprint, the DNA fragments are treated with a radioactive substance so they can be seen. The substance sticks to specific parts of the fragments and produces a pattern of stripes when exposed to X-ray film. The resulting pattern is unique for every individual. Study

the figure below, which shows how the patterns of stripes might appear in a photograph. These illustrations represent DNA fingerprints from four separate sources.

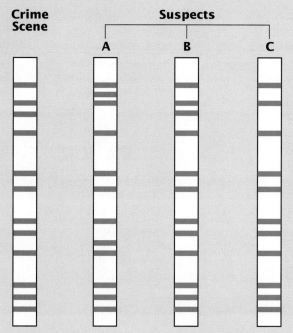

1. *Summarizing Observations*
   Describe the similarities and differences between the DNA fingerprint from the crime scene sample and the DNA fingerprints of the suspects shown in the figure to the left.

2. *Analyzing Data*
   What evidence is there that the cells left at the crime scene belong to one of the suspects?

3. *Drawing Conclusions*
   Why is it unlikely that two individuals would have identical DNA?

4. *Evaluating Methods*
   Why is the technique of labeling and photographing DNA fragments called DNA fingerprinting?

5. *Inferring Relationships*
   Why do similarities exist among all the samples in the figure to the left?

5. Imagine that a crime has been committed and that a sample of cells belonging to the criminal is collected at the crime scene. The DNA fingerprint of these cells is represented by the illustration labeled "Crime Scene." Study the pattern of stripes in this DNA fingerprint.

6. Compare the DNA fingerprint of the cells obtained at the crime scene with the DNA fingerprints of the samples collected from the three suspects labeled A, B, and C. *Which suspect's DNA fingerprint most closely matches the DNA fingerprint of the cells obtained at the crime scene?*

7. Under direction from your teacher, compare the arrangement of beads on your string with the arrangement of beads on the string in the container in the front of the room. *How do your beads compare with the evidence?*

Thinking Critically

1. Imagine you are on a jury and DNA fingerprinting evidence is introduced. Explain how you would regard such evidence.

2. Explain how the DNA fingerprinting procedure could be used by scientists who study fossils.

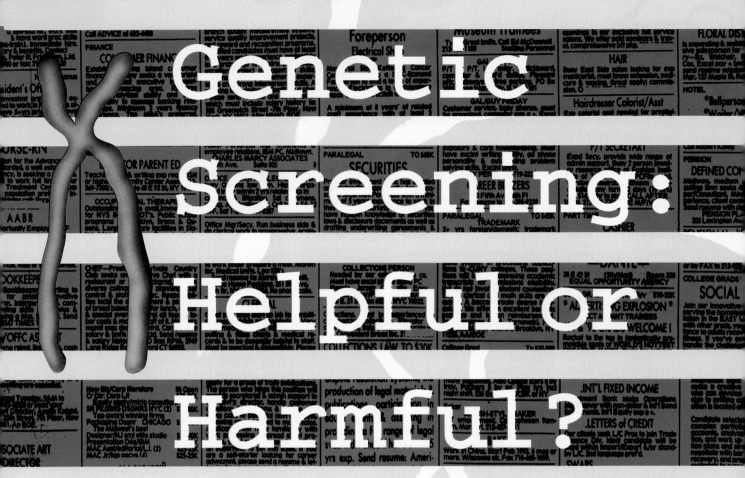

Genetic Screening: Helpful or Harmful?

Have you ever wondered about your future? Do you wonder if you will be rich or famous? Do you wonder if you will be healthy? The time is rapidly approaching when scientists will help foretell the future. The Human Genome Project could provide the tools geneticists need for diagnosing and treating many disorders. But many people are against this project. Read on to find out why.

It is the year 2020 and Bill Islet has applied for a job with a company in his area. During the interview, Bill is told that as a condition of employment he will have to undergo some DNA testing. This testing will determine if Bill is a carrier of any genetic diseases. Because some genetic diseases are very serious and require extensive medical treatment, the company cannot obtain health insurance coverage if it temploys people likely to have these diseases.

"But I am not sick," says Bill. "Why would you test me for a disease that I don't have?"

Bill's situation could very well occur as a spinoff from the Human Genome Project. The Human Genome Project is an effort by many scientists to map human chromosomes. These maps will tell scientists what genes you have and where they are located.

The technology to screen for many genetic diseases or conditions could be available from the gene mapping project. Specific policies for how this information will be used are yet to be developed. There is the very real possibility that this technology could be misused. Because of these concerns about misuse of genetic screening information, there are many opponents of the Human Genome Project.

Some people would like to know if they are carrying a gene that could result in a disease, such as Huntington's disease. Symptoms of Huntington's disease do not appear until middle age. The disease begins with lapses of memory and with irritability. As it progresses, the victim loses muscle control, experiencing spasms and extreme mental illness. The disease eventually causes death. The unfortunate aspect of Huntington's disease is that by the time symptoms appear, genes for the disease have already been passed on to offspring.

Those against testing contend that because the disease is so devastating and is incurable, no one would want to know if they had it. Forcing someone to be tested could become difficult, because such tests are seen as an invasion of privacy. If a person has the test, the next issue becomes who should be given the results of such testing? Consider Bill's case. Should the insurance company or his prospective employer have access to his genetic profile? What about Bill's family? If Bill is married, his wife might want to know his genetic profile. Bill's offspring might inherit a genetic disease from him. Should he be legally bound to share this information with his wife?

Under most circumstances, courts have ruled that medical records are private and cannot be accessed without permission. A patient must sign a release if he or she wants an employer, insurance company, or relatives to see medical test results. However, insurance companies can require certain tests before agreeing to provide insurance. Employers can also make the passing of a medical

exam a condition of employment. Such practices place people in situations where their records become known. Insurance companies that provide health care and life insurance want healthy clients. Insurance companies do not want to pay for long, expensive treatments for those who have the potential to become seriously ill. For example, if Bill was shown to carry a cystic fibrosis gene (a recessive trait), he would not be affected by the disease. But if his wife also carried the gene, their child might be affected by the disease. Health care for a child with cystic fibrosis (CF) is quite expensive. Therefore, if Bill were a carrier of CF, he could have trouble getting health insurance for himself, his wife, and their child.

Some people see the value in mapping the human genome, but feel that the project should stop there. Spinoff technologies, like genetic screening tests, should not be pursued. Those supporting the project, however, see reasons to explore the positive uses of this technology. For example, gene therapy could be used to cure fatal or painful diseases.

Should the development of a technology be limited when we know it can help some, yet harm others?

Thinking Critically

❶ If you were in Bill's situation, would you agree to be tested?

❷ The cost of the Human Genome Project will be about $88 million per year for 15 years. Should the government spend money on this project? Why or why not?

❸ Some people say that knowing about our own personal genome could change the human population physiologically. How do you think the human population could change? What benefits could result from these changes?

❹ List three disadvantages of creating an extensive bank of criminal genomes.

Acting on the Issue

❶ Find out what diseases, if any, are screened for by law in your state.

❷ Find a genetic counselor at your local hospital or clinic. Ask the counselor to come and speak to your class about his or her role in

advising prospective parents. How does the counselor expect his or her role to change in the future?

❸ Propose possible legislation that could result from the Human Genome Project.

❹ Write for information on the Human Genome Project:
Human Genome Project
Cold Spring Harbor, NY 10098

Evolution and Natural Selection

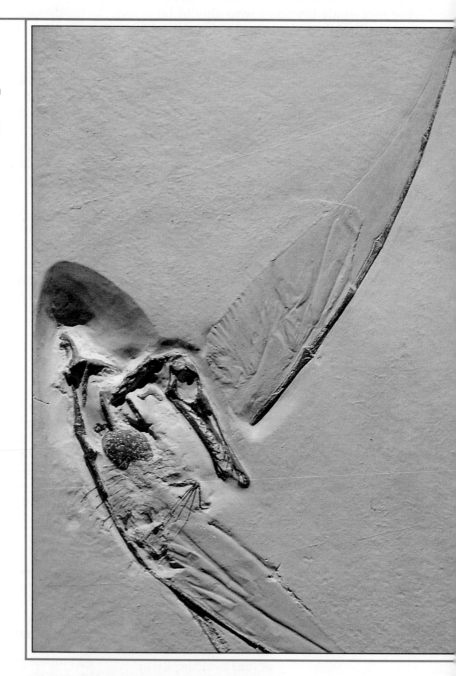

Fossils, such as this 150 million-year-old pterosaur, a flying reptile, are one line of evidence for the occurrence of evolution.

OUR EARTH IS RICH IN LIFE. ABOUT 1.4 MILLION SPECIES HAVE BEEN NAMED AND MILLIONS MORE ARE THOUGHT TO EXIST. WHY ARE SOME SPECIES MORE ALIKE THAN OTHERS? WERE THEY ALL CREATED AT ONCE, OR HAVE SPECIES EVOLVED? THESE QUESTIONS HAVE BEEN ASKED FOR THOUSANDS OF YEARS, BUT IT WAS NOT UNTIL 1859 THAT THE WORK OF CHARLES DARWIN OFFERED ANSWERS.

9.1 Charles Darwin

Objectives

❶ **Summarize Darwin's beliefs about the origin of species before he sailed around the world.**

❷ **Identify two observations from Darwin's voyage that led him to question his beliefs.**

❸ **Describe the two major ideas Darwin put forth in *The Origin of Species*.**

Voyage of the Beagle

Figure 9.1
Below you can see the course of the *Beagle*, the ship in which Darwin sailed around the world. On this voyage, Darwin collected thousands of specimens of plants, animals, and fossils.

On December 27, 1831, H.M.S. *Beagle* sailed from England to survey the coast of South America. On board as the ship's unpaid naturalist was Charles Darwin, a 22-year-old who had graduated from Cambridge University. Darwin's observations during his five years at sea would eventually change the way we think of ourselves and our world.

The son of a wealthy doctor, Darwin was not an attentive student, spending more time outdoors than in school. As a medical student in Edinburgh, he was horrified by operations, which were performed without anesthetic. For two years Darwin skipped lectures to spend time collecting biological specimens. In desperation, his father sent him to Cambridge University to train to be a minister. After graduating from Cambridge in 1831, Darwin was recommended by one of his professors for the position on the *Beagle*. Follow the voyage of the *Beagle* in **Figure 9.1**.

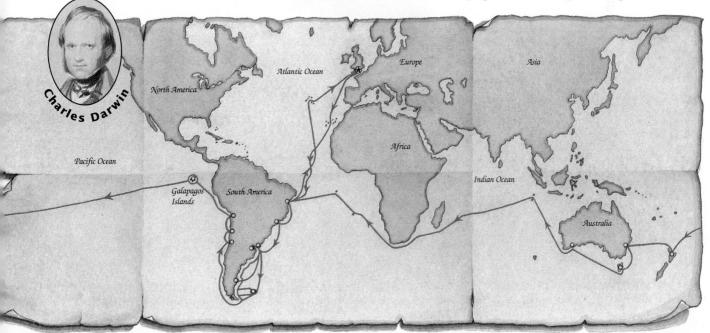

When the *Beagle* sailed from England, Darwin, like most people of his time, believed in creationism. Creationism is the idea that God was responsible for the creation of new species, or kinds, of organisms. According to creationists, God designed each kind of animal and plant to match its particular habitat. Thus, places with similar environments have similar kinds of plants and animals. Moreover, creationists believe that species are unchanging.

During his journey Darwin often left the ship to collect specimens of animals, plants, and fossils. His observations led him to doubt creationism and its assumption of unchanging species.

Darwin's Finches

Darwin repeatedly saw patterns in how kinds of animals and plants differed, patterns suggesting that species changed over time and gave rise to new species. On the Galapagos Islands, 1,000 km (600 mi.) from the coast of Ecuador, Darwin collected several species of finches. All these species were similar, but each was specialized to catch food in a different way, as shown by the different shapes of the birds' bills in **Figure 9.2**. Some species had heavy bills for cracking open tough seeds. Others had slender bills for catching insects. One finch species even used a twig to probe for insects.

All the species of finches closely resembled one species of South American finch. In fact, all the plants and animals of the Galapagos Islands were very similar to those of the nearby coast of South America. If each one of these plants and animals had been created to match the habitat on the Galapagos Islands, why did they not resemble the plants and animals of islands with similar environments that lie off the coast of Africa? Why did they instead resemble those of the adjacent South American continent? Darwin felt that the simplest explanation was that a few organisms from South America must have migrated to the Galapagos Islands in the past. These few kinds of animals and plants then changed during the years they lived in their new home, giving rise to many new species. Change in species over time is known as evolution.

Figure 9.2
The blue-black grassquit (inset), native to the Pacific coast from Mexico to Chile, is thought to be the ancestor of the Galapagos finches (below). Darwin attributed the differences in bill size and feeding habits among these finches to evolution that occurred after the birds migrated to the Galapagos Islands.

a **The woodpecker finch captures insects with its grasping bill.**

b **The crushing bill of the large ground finch enables it to feed on seeds.**

c **The cactus finch uses its probing bill to feed on cactuses.**

Figure 9.3
In 1859 Darwin published his famous book, *The Origin of Species*. He accomplished much of his work in his study at Down House in Kent, England (above, right). Darwin is shown at age 73 (above).

Alfred Russel Wallace

Darwin's Mechanism for Evolution

Darwin returned to England in 1836 and for 20 years gathered evidence supporting his ideas about evolution—but he did not publish them. He accomplished much of his work in Down House in Kent, England, shown in **Figure 9.3**. Then in 1858 another biologist, Alfred Russel Wallace, sent Darwin an essay putting forth these same ideas. This drove Darwin to finally publish his work.

When Darwin's book, *On the Origin of Species by Means of Natural Selection*, appeared in November of 1859, it stirred up great controversy. Darwin's conclusion that species changed over time and gave rise to new species contradicted the prevailing beliefs that God created all living things and that living things did not change. Furthermore, the implication that apes were close relatives of humans was unacceptable to many people.

In *The Origin of Species*, as the book is commonly known, Darwin not only presented much evidence that evolution occurred, but he also proposed that natural selection was its mechanism. **Natural selection** is a process by which organisms with traits well suited to an environment survive and reproduce at a greater rate than organisms less suited to that environment.

Because Darwin presented a mechanism as well as evidence for evolution, his arguments were compelling. His views were soon accepted by biologists around the world. Since the rediscovery of Mendel's ideas about genetics in the early 1900s, genetic principles have been added to Darwin's ideas, forming the modern theory of evolution.

Section Review

❶ **Identify two major beliefs of creationism.**

❷ **Describe two observations Darwin made on his voyage that led him to doubt creationism.**

❸ **Explain the two major ideas Darwin presented in *The Origin of Species*.**

MUCH OF WHAT YOU HAVE READ SO FAR IN THIS BOOK WAS UNKNOWN TO DARWIN WHEN HE WROTE *THE ORIGIN OF SPECIES*. THE NUCLEIC ACIDS DNA AND RNA HAD NOT BEEN DISCOVERED. THE STRUCTURE OF PROTEINS WAS UNKNOWN. MENDEL'S GENETIC EXPERIMENTS WERE UNPUBLISHED. THESE AND OTHER DISCOVERIES HAVE SINCE CONTRIBUTED ADDITIONAL EVIDENCE FOR EVOLUTION.

9.2 The Evidence for Evolution

Objectives

❶ Describe the conditions necessary for fossils to form.

❷ List one example in the fossil record indicating that evolution has occurred.

❸ Explain how comparisons of organisms can reveal evidence of evolution.

❹ Describe the important evidence for evolution found in proteins and DNA.

Understanding the Fossil Record

More than a century has passed since Darwin's death in 1882. During this period, a great deal of new evidence has accumulated supporting the theory of evolution, much of it far stronger than that available to Darwin and his contemporaries. This evidence has come from a variety of sources, including studies of fossils, comparisons of the structures of organisms, and the rapidly expanding knowledge about DNA and proteins.

Fossils are any traces of dead organisms

What are fossils? Most people think of fossils as shells or old bones. Actually, fossils are any traces of dead organisms. Tracks of dinosaurs, footprints of human ancestors, insects trapped in sticky tree sap, impressions of leaves or skin, and animals buried in tar are fossils. A photograph of a fossil is shown in **Figure 9.4**. For fossils to form, very special conditions are necessary. If a skeleton or shell is to fossilize, the dead animal must be buried by sediment. Burial usually only occurs on the ocean floor, in swamps, in mud, or in tar pits. Calcium in the bone or in the shell is slowly replaced by other harder minerals. Unless the sediment is very fine and no oxygen is present to promote decay, soft tissues such as those found in skin or muscle do not fossilize.

Figure 9.4
This photograph shows the fossil of a fish that formed in limestone in Brazil 110 million years ago.

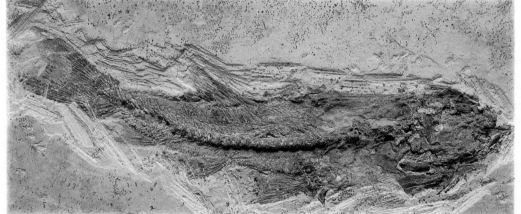

How Fossils Are Dated

Since the late 1940s scientists have been able to determine the ages of rocks and fossils by measuring the amount of radioactive decay, or breakdown, of radioactive atoms in the rock. A radioactive atom contains an unstable combination of protons and neutrons. Since it is unstable, a radioactive atom will eventually change into a more stable atom of another element. For example, carbon-14, a rare form of carbon found in tiny amounts in living things, decays into nitrogen. The term **half-life** describes how long it takes for one-half of the radioactive atoms in a sample to decay. For example, the half-life of carbon-14 is 5,730 years. Thus, a sample that initially contained 12 g of carbon-14 will have 6 g of carbon-14 left after 5,730 years and 3 g of carbon-14 left after 11,460 years. Since carbon-14 decays relatively rapidly, other isotopes with longer half-lives are more often used to date fossils.

Because the rate of decay of a radioactive element is constant, scientists can use the amount of radioactive element remaining in a rock or fossil to determine its age. This technique is called **radioactive dating**.

Evolution is a very slow process; the transformation of one species into another requires at least thousands of years. Using radioactive dating, scientists have determined that the Earth is about 4.5 billion years old, ancient enough for all species to have been formed through evolution.

Transitional forms link new species to old

Because new species form from existing species, Darwin predicted that transitional forms, intermediate stages between older and newer species, would be found in the fossil record. When *The Origin of Species* was published, no intermediates had been found. Darwin recognized that this was a weakness in his theory. But there are now many good examples of evolutionary transitions. For instance, modern horses are descendants of dog-sized animals with four toes on each front foot and three toes on each back foot—modern horses have only one toe per foot. As you can see in **Figure 9.5**, fossil intermediates between this 60 million-year-old ancestor of the horse and the modern horse reveal a history of slow transformation. Over time, the length and size of the limbs increased, and the number of toes on each foot decreased until only one toe was left. A detailed picture of horse evolution would be somewhat more complicated than that shown in **Figure 9.5**, but there is no doubt that a general trend toward the modern one-toed horse did occur.

Figure 9.5
The evolution of the modern horse began with *Hyracotherium* (below) about 60 million years ago. Notice the change from four toes to one toe on each front foot.

a *Hyracotherium*
60 million years ago

b *Mesohippus*
About 30 million years ago

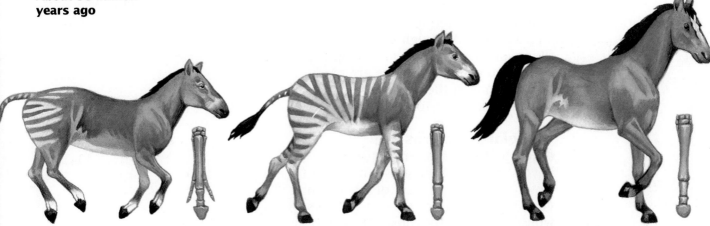

c *Merychippus*
About 20 million years ago

d *Pliohippus*
14 million to 7 million years ago

e *Equus* (modern horse)
10,000 years ago

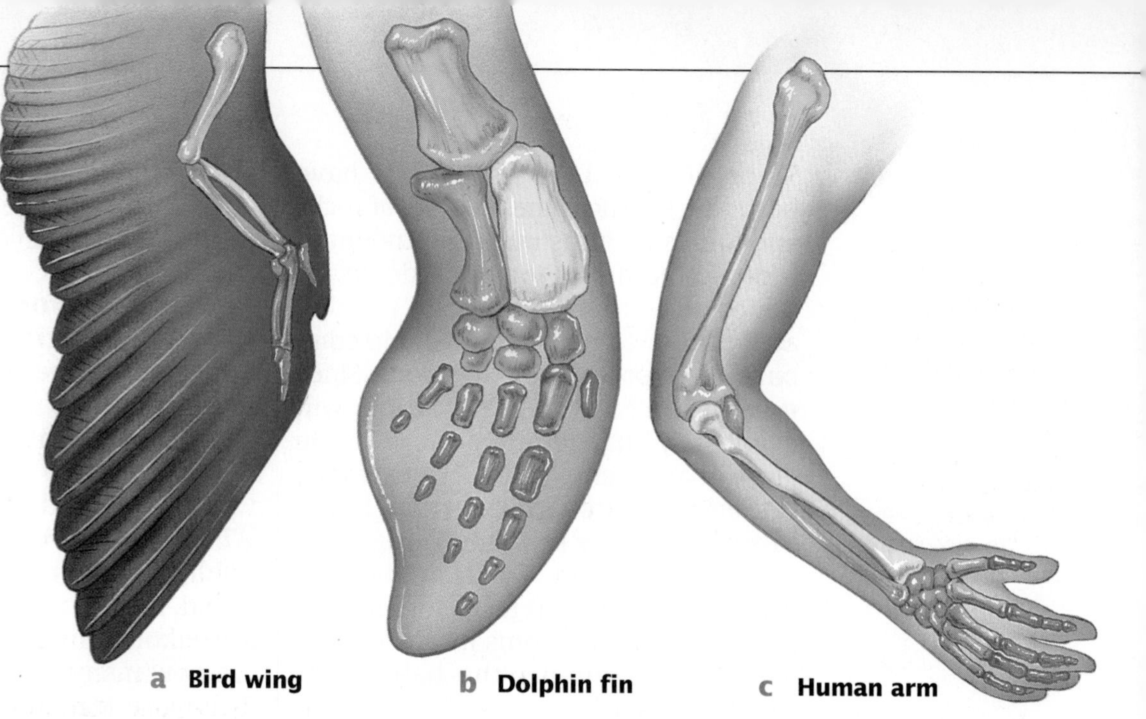

Figure 9.6
The bones in the front limbs of the bird, the dolphin, and the human are homologous structures. Homologous bones are shown in the same color on each diagram.

a **Bird wing**   **b** **Dolphin fin**   **c** **Human arm**

Comparing Organisms

Comparing the way organisms are put together provides important evidence for evolution. Your arm appears quite different from the wing of a bird or the front fin of a dolphin. Yet if you examine **Figure 9.6**, you can see that the placement and order of bones in these limbs are very similar. Biologists say that these three limbs are homologous. **Homologous structures** are structures that share a common ancestry. Homologous structures are similar because they are modified versions of structures that occurred in a common ancestor. Although suited for flying, swimming, and grasping, the limbs of the animals above are modified versions of the front fins of their common fish ancestor.

Vestigial structures are clues to evolutionary origins
If you were designing a submarine, would you include a set of wheels in your design? Of course not. Wheels would serve no purpose on a submarine. However, structures without function are found in living things. A whale propels itself with its powerful tail and has no need for hind limbs or

Scale and Structure

Name two

structures in your

cells that are

homologous to

structures found

in bacterial cells.

the pelvis to which they attach. Nevertheless, whales still have a reduced pelvis that serves no apparent purpose, as shown in **Figure 9.7**. Structures with no purpose are known as **vestigial structures**. Vestigial structures are remnants of an organism's evolutionary past. The whale's pelvis is evidence of its evolution from four-legged, land-dwelling mammals.

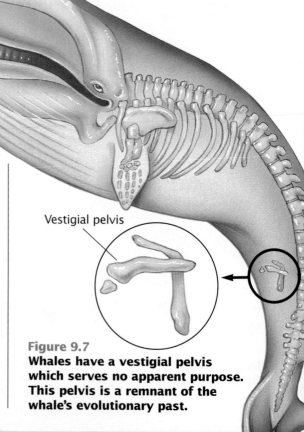

Vestigial pelvis

Figure 9.7
Whales have a vestigial pelvis which serves no apparent purpose. This pelvis is a remnant of the whale's evolutionary past.

Developmental patterns show evolutionary relationships

Much of our evolutionary history can be seen in the way human embryos develop. Early in development, human embryos and embryos of all other vertebrates are strikingly similar, as shown in **Figure 9.8**. In later stages of development, a human embryo develops a coat of fine fur. The similarity of these early developmental forms strongly suggests that the process of development has evolved. New instructions on how to grow have been added to old instructions inherited from ancestors.

Figure 9.8
The five-week-old human embryo (a) and the four-day-old chicken embryo (b) each have a bony tail and gill pouches similar to those of fishes.

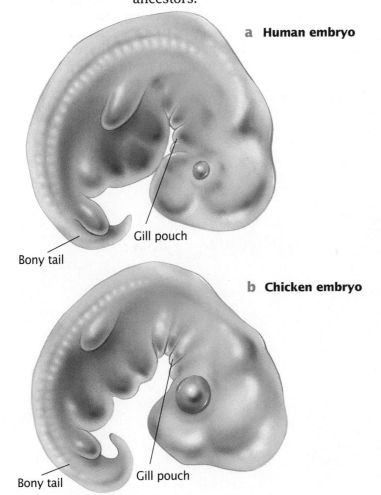

a **Human embryo**

Gill pouch

Bony tail

b **Chicken embryo**

Gill pouch

Bony tail

DNA and proteins contain evidence of evolution

Although complete fossil histories for living organisms are rare, an organism's history is written in the sequence of nucleotides making up its DNA. If species have changed over time, their genes also should have changed. The theory of evolution predicts that genes will accumulate more alterations in their nucleotide sequences over time. Thus, if we compare the genes of several species, closely related species will show more similarities in nucleotide sequences than will distantly related species. Closely related species also will show more similarities in the amino acid sequences in their proteins. This is because the amino acid sequence in a protein reflects the nucleotide sequence of the gene coding for that protein.

For example, to see how closely related chimpanzees, dogs, and rattlesnakes are to humans, scientists examined the sequence of amino acids in the protein cytochrome *c*, an essential participant in cellular respiration. They found that human cytochrome *c* and chimp cytochrome *c* are identical in all 104 amino acids. This high degree of similarity indicates our very close kinship to chimpanzees. A dog's cytochrome *c* differs from human cytochrome *c* in 13 amino acids, indicating that dogs are fairly distant relatives. But dogs are more closely related to us than are rattlesnakes, whose cytochrome *c* differs from ours in 20 amino acids. In most cases, the evolutionary relationships indicated by DNA or protein sequences confirm those suggested by comparative anatomy and by developmental patterns.

Section Review

❶ Why is it unlikely that you will be fossilized?

❷ Explain why transitional species, such as the sequence of ancestral horses, are crucial evidence for evolution.

❸ How does the whale's vestigial pelvis provide evidence in support of evolution?

❹ Explain how sequences of amino acids in proteins can be used to reveal relationships among organisms.

Sexual Selection

The Mating Game

Why does the male peacock (below, right) have brilliant plumage, while the female (below) is brown? Darwin proposed that male peacocks evolved beautiful feathers because males with long tails and bright plumage attracted more mates than males with dull plumage. This evolutionary mechanism is called sexual selection.

The plumage of a male peacock is brilliant green and blue. The peacock's brightly colored, elaborate tail can measure five times its body length. In contrast, the female, or peahen, is drab brown and has a small tail. How have these differences between the sexes evolved? A long tail and bright plumage cannot be essential for survival, since the peahen survives without them. In fact, the male's showiness hampers his survival by making him more visible to predators. It seems that the gaudy plumage of

the peacock should never have been permitted by natural selection.

Charles Darwin proposed the mechanism of sexual selection to account for the evolution of traits like the peacock's tail. In sexual selection, traits that enhance the ability of individuals to acquire mates increase in frequency. For example, the long tails and brilliant plumage of peacocks probably evolved because males with these traits were able to attract more females than were males with short tails and dull plumage.

Sexual Selection at Work

What kinds of traits are advantageous for acquiring mates? Darwin noticed that in most species males compete for opportunities to mate with females. This competition may involve direct interactions between males, such as threatening displays or combat. Therefore, traits that make males more intimidating or better at combat will be favored by sexual selection. Examples of these traits include large body size, antlers of deer, manes of lions, horns of bighorn sheep, and the oversized jaws of the beetle shown at the left.

Male–male competition can also take subtler forms. A male can gain a reproductive advantage over other males by interfering with their reproduction. In some species of worms, butterflies, and snakes, for instance, the male seals the female's reproductive tract after mating so that other males cannot mate with her.

Darwin also recognized that in many species females choose their mates. Consequently, males with characteristics preferred by females are chosen more often and so have more offspring than males with less attractive traits. Therefore, traits attractive to females become more common.

The antlers of the male red deer (top) and the jaws of the hercules beetle above are favored by sexual selection because they are useful in combat against other males.

Female Choice in Selecting a Mate

Darwin's suggestion that female choice could influence evolution was initially rejected by most scientists. However, recent research has supported Darwin's proposal. For instance, zoologist Malte Andersson of the University of Gothenburg, in Sweden, studied female preference in the long-tailed widowbird of Kenya. Like the peacock, the male widowbird has an extremely long tail, which is about 0.5 m (19 in.) in length. The female's tail, in contrast, is only about 7 cm (less than 3 in.) long. Andersson shortened the tails of one group of males by clipping their tail feathers. He glued the tail pieces removed from these males onto the tails of another group of males. After releasing the birds, Andersson found that four times as many females settled in the territories of the males with lengthened tails. Since females usually mate with the male that controls the territory they settle in, this experiment indirectly reveals a preference for long tails among female widowbirds.

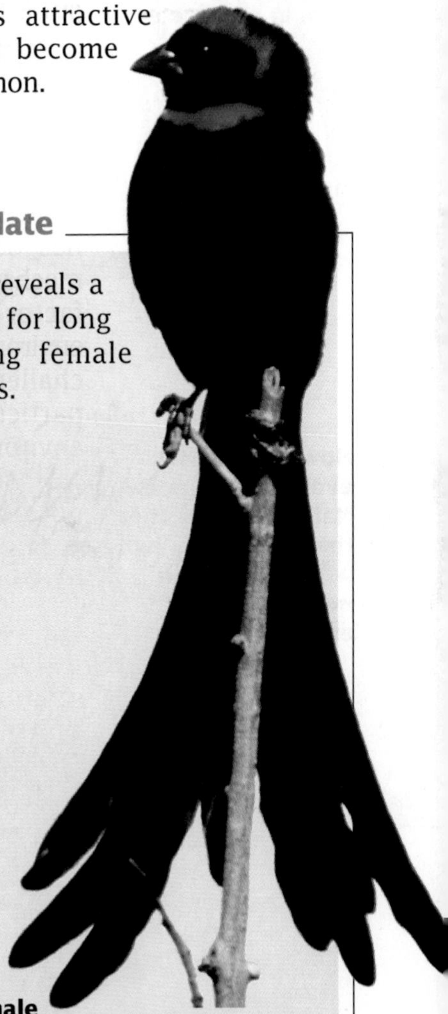

Like male peacocks, male widowbirds have very long tails. One study indirectly shows that female widowbirds prefer males with long tails. Why females prefer these males is unknown.

How Species Form

Because natural selection favors changes that increase an organism's chances of surviving and reproducing, it will continuously shape a species to improve the fit between the species and its environment. Recall from Chapter 1 that a species is a group of individuals that can interbreed and produce fertile offspring, but that cannot breed with any other such group. When populations of a species are found in several different kinds of environments, selection will act to make each population suit its particular environment.

Populations in different places thus become increasingly different, as each becomes better suited to the particular challenges of living where it does.

Separate populations of a species can eventually become quite distinct, if their environments differ enough. These populations form what biologists call ecological races, as shown by the example of the sparrows in **Figure 9.12**. **Ecological races** are populations of the same species that differ genetically because they have adapted to different living conditions.

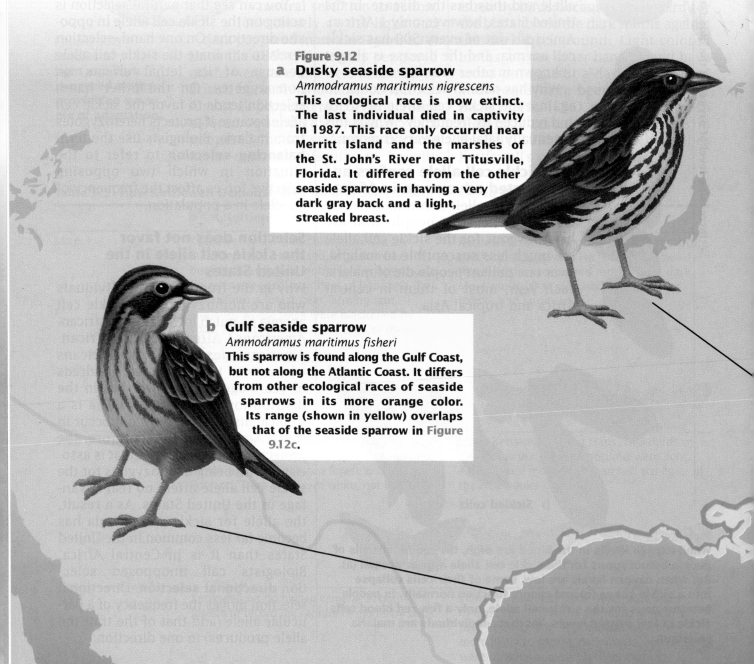

Figure 9.12

a Dusky seaside sparrow
Ammodramus maritimus nigrescens
This ecological race is now extinct. The last individual died in captivity in 1987. This race only occurred near Merritt Island and the marshes of the St. John's River near Titusville, Florida. It differed from the other seaside sparrows in having a very dark gray back and a light, streaked breast.

b Gulf seaside sparrow
Ammodramus maritimus fisheri
This sparrow is found along the Gulf Coast, but not along the Atlantic Coast. It differs from other ecological races of seaside sparrows in its more orange color. Its range (shown in yellow) overlaps that of the seaside sparrow in Figure 9.12c.

Members of different ecological races are not yet different enough to belong to different species, but they have taken the first step. The differences among human races, while they may seem large to some of us, are actually very small in an evolutionary sense. Notice that each sparrow in **Figure 9.12** has a three-word scientific name, which indicates a subspecies, a distinct group within a species.

Ecological races form new species

Ecological races often become increasingly different. The accumulation of differences between species or populations is called **divergence**. Divergence occurs because natural selection favors different survival strategies in different environments. Eventually, races can accumulate so many differences that biologists consider them separate species.

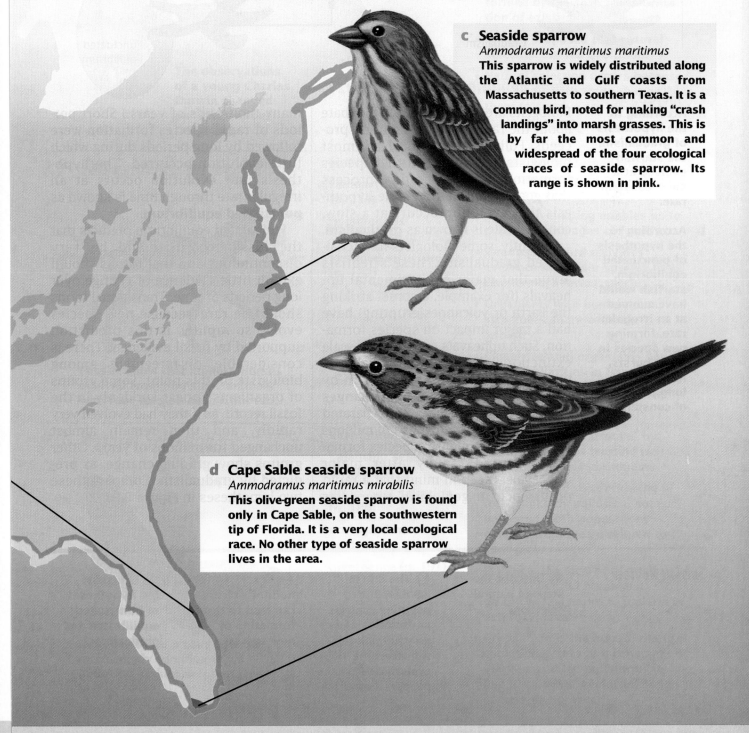

c Seaside sparrow
Ammodramus maritimus maritimus
This sparrow is widely distributed along the Atlantic and Gulf coasts from Massachusetts to southern Texas. It is a common bird, noted for making "crash landings" into marsh grasses. This is by far the most common and widespread of the four ecological races of seaside sparrow. Its range is shown in pink.

d Cape Sable seaside sparrow
Ammodramus maritimus mirabilis
This olive-green seaside sparrow is found only in Cape Sable, on the southwestern tip of Florida. It is a very local ecological race. No other type of seaside sparrow lives in the area.

Understanding Vocabulary

1. For each pair of terms, explain the differences in their meanings.
   a. adaptation, natural selection
   b. homologous structures, vestigial structures
   c. balancing selection, directional selection
   d. gradualism, punctuated equilibrium

Relating Concepts

2. Copy the unfinished concept map below onto a sheet of paper. Then complete the concept map by writing the correct word or phrase in each oval containing a question mark.

Understanding Concepts

Multiple Choice

3. Darwin was prompted to publish his ideas by
   a. the arrival of Wallace's essay.
   b. his declining health.
   c. permission from the Queen.
   d. encouragement from his family.

4. Darwin concluded that new species arose from
   a. divine creation.
   b. nonliving matter.
   c. existing species.
   d. extinct species.

5. Which of the following is not an example of directional selection?
   a. sickle cell anemia in Africa
   b. color change in peppered moths
   c. sickle cell anemia in the United States
   d. evolution of the modern horse

6. Evolutionary relationships among species can be deduced from
   a. nucleotide sequences in DNA.
   b. scientists using small clocks.
   c. radioactive atoms.
   d. ecological races.

7. Which species' cytochrome *c* is most similar to our own?
   a. chimpanzee   b. dog
   c. rattlesnake   d. poppy

8. The mechanism proposed by Darwin to explain how evolution occurs is
   a. speciation.   c. adaptation.
   b. natural selection.   d. punctuated equilibrium.

9. Using radioactive dating, scientists have determined that
   a. 90 percent of the earth was once covered by water.
   b. the earth is about 10 billion years old.
   c. carbon-14 has a half-life of 500 years.
   d. the Earth is old enough for evolution to have produced all species.

10. Which of the following is *not* a factor in natural selection?
    a. genetic variation
    b. evolution of the individual
    c. survival
    d. struggle for existence

11. Studies of the peppered moth in England illustrate the operation of
    a. adaptation.   c. speciation.
    b. gradualism.   d. punctuated equilibrium.

Completion

12. Darwin proposed the mechanism of _____ _____ to explain evolution. This mechanism is described in his book entitled _____ .

13. On the _____ Islands, Darwin witnessed an example of evolution in the beaks of small birds called _____ .

14. Before his voyage on the *Beagle*, Darwin believed that _____ best explained the origin of life on Earth.

15. Naturally preserved remains or traces of dead organisms are called _____ .

16. The time required for one-half of the radioactive atoms in a sample to decay is its _____ _____ .

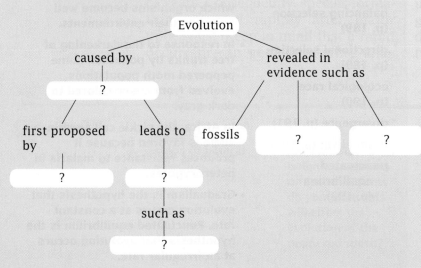

Evolution

caused by — ?

first proposed by — ?

leads to — ?

revealed in evidence such as — fossils, ?, ?

such as — ?

17. Human structures with no apparent purpose include the tail bone and wisdom teeth. These structures are called _____ _____ .

18. Populations of the same species that differ genetically as a result of their different living conditions are called _____ _____ .

Short Answer

19. What conditions are necessary for an animal fossil to form?

20. What does the whale's vestigial pelvis reveal about the evolution of whales?

21. Explain how the hypotheses of gradualism and punctuated equilibrium differ.

Interpreting Graphics

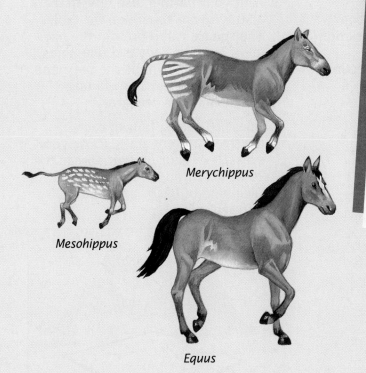

Merychippus

Mesohippus

Equus

22. These sketches show three stages in the evolution of the horse.

- List the three stages in their correct evolutionary order.

- Suppose that radioactive dating showed that *Equus* was older than *Mesohippus*. How would this discovery weaken the evidence for evolution provided by this example?

Reviewing Themes

23. *Scale and Structure*
Darwin noted that the Galapagos finches had specialized bills. What relationship did Darwin observe between the birds' bill types and the ways in which they caught food?

24. *Scale and Structure*
The bones in the human arm are very similar to the bones in a bird's wing and a dolphin's flipper. What does this similarity in structure tell about the ancestry of these animals?

25. *Patterns of Change*
What evidence supports the idea of evolution as an ongoing process?

Thinking Critically

26. *Inferring Conclusions*
Punctuated equilibrium supports the notion that transitional forms of life are rare in the fossil record because evolution occurred so rapidly. What else could explain the absence of transitional forms?

27. *Building on What You Have Learned*
In Chapter 7 you learned that DNA is a molecule made up of nucleotides that code for amino acids. What can biologists learn about two species by comparing their DNA sequences?

Cross-Discipline Connection

28. *Biology and Art*
Darwin observed how different finches adapted to a specific environment. Considering Darwin's observations, identify an environment and then draw a picture showing the human adaptations needed to match that environment.

Discovering Through Reading

29. Read "In the Beginning," in *Discover*, October, 1990, pages 98–102. What conclusions about the origin of life on Earth did Dr. J. William Schopf draw from his examination of the Precambrian rocks?

Investigation

How Does Natural Selection Work?

Objectives

In this investigation you will:
- *model* the process of natural selection

Materials

- beads of five different sizes and shapes
- masking tape
- meter stick

Prelab Preparation

Review what you have learned about natural selection by answering the following questions:
- What is natural selection?
- What is the role of variation in natural selection?
- How does natural selection lead to adaptation?

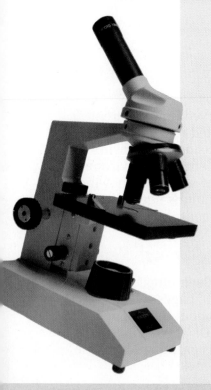

Procedure: Modeling Natural Selection

1. Make a data table like the one shown below.

	Survivors of each generation			
Prey Type	First	Second	Third	Fourth
1				
2				
3				
4				
5				

2. Form a cooperative group of four students. Work with one member of your group to complete steps 3–13.

3. With your partner, use tape to mark a 50 × 50 cm square on the table. The square will represent the habitat of one predator and five kinds of prey. Beads will be the prey, and you and your partner will take turns being the predator.

4. Select 20 beads, four of each of the five different sizes. Close your eyes. Have your partner spread the 20 beads throughout the habitat. Each kind of bead represents a different prey type.

5. To hunt, have your partner position your hand so that it is approximately 5–10 mm above the table surface at one edge of the square. Slowly move your hand across the square, trying to keep it at the same height above the surface. *Why is it important to keep your hand at a constant height?*

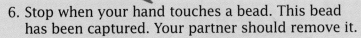

6. Stop when your hand touches a bead. This bead has been captured. Your partner should remove it.

7. Continue moving your hand across the square, stopping to capture each bead you touch. If you reach the edge of the square, have your partner reposition your hand at a new spot along the edge of the square, then start hunting again.

8. Continue hunting until 50 percent of the prey have been captured.

9. Count the survivors of each kind of prey. Record these data in the column labeled "First" in your data table.

10. The prey now reproduce. Assume each survivor has one offspring. Add the appropriate number of beads to the habitat.

11. Conduct another round of hunting, again stopping after 50 percent of the prey have been "eaten." Count the survivors and record their numbers in your data table in the column labeled "Second."

12. Now allow the prey to reproduce. Again, each survivor produces one offspring. Distribute the offspring in the habitat.

13. Exchange roles with your partner and repeat steps 4–12, recording the number of survivors in the third and fourth generations.

14. Pool the data for both teams in your group. Plot the number of survivors of each kind of prey over the four generations in a graph.

Analysis

1. *Summarizing Data*
   Describe what happened to the number of each type of prey over the four generations.

2. *Analyzing Data*
   What evidence suggests that certain types of prey are better adapted to the habitat than others?

3. *Making Inferences*
   What adaptation of the prey contributes most to their chance of survival in this habitat?

4. *Making Inferences*
   Describe an example in nature in which larger size might be a disadvantage.

5. *Making Inferences*
   Why might the shapes of similarly sized beads affect your ability to capture them?

Thinking Critically

1. Suppose that you slid your hand along the surface of the table, instead of moving it slightly above the surface. How might this different way of hunting change your results?

2. In what way does this experiment differ from events that might occur in a natural habitat?

History of Life on Earth

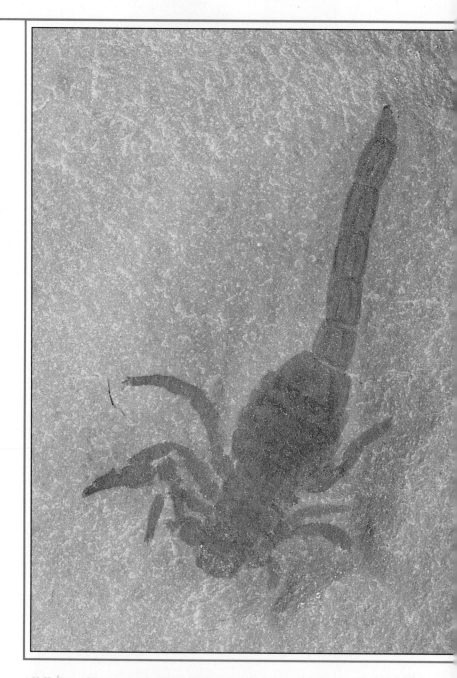

All living things, including this scorpion that lived nearly 410 million years ago, evolved from the same simple life forms that were found on the young planet Earth.

WITH LIFE ALL AROUND US, IT IS DIFFICULT TO IMAGINE A TIME WHEN THERE WAS NO LIFE ON EARTH, WHEN NO GRASSES GREW AND NO FISHES SWAM IN THE SEA. THE EARTH IS MUCH OLDER THAN LIFE, HOWEVER. THE STUDY OF RADIOACTIVE DECAY IN ROCKS REVEALS THAT THE EARTH IS SOME 4.5 BILLION YEARS OLD, 1 BILLION YEARS OLDER THAN THE OLDEST FOSSILS. WHERE DID LIFE COME FROM?

10.1   *Origin of Life*

Objectives

❶ Contrast the three explanations for the origin of life.

❷ Describe the importance of the Miller-Urey experiment.

❸ Summarize the reasons scientists think RNA, not DNA, was the first genetic material.

❹ Describe how the first cells might have evolved.

How Did Life Begin?

There were no witnesses to the origin of life, and you cannot go back in time to see for yourself. But if you could, what would the possibilities be? In principle, there are at least three ways life could have begun:

1. *Extraterrestrial origin* Some scientists hypothesize that life originated on another planet outside our solar system. Life was then carried here on a meteorite or an asteroid and colonized the Earth. How life arose on other planets, if it did, is a question we cannot hope to answer soon.

2. *Creation* Many people believe that life was put on Earth by divine forces. In this view, common to many of the world's religions, the forces leading to life cannot be explained by science.

3. *Origin from nonliving matter* Most scientists think that life arose on Earth from inanimate matter, after the newly formed Earth had cooled, as shown in **Figure 10.1**. First, random events produced stable molecules that could reproduce themselves. Then, natural selection favored changes in these molecules that increased their rate of reproduction, leading eventually to the first cell.

This chapter will examine the third option. The first two possibilities are not considered because they are not testable and thus fall outside the realm of science. However, the demonstration that life could have originated from nonliving matter does not rule out the other two possibilities. Since these possibilities cannot be tested, either can be accepted as a matter of faith.

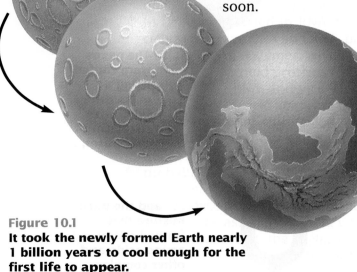

Figure 10.1
It took the newly formed Earth nearly 1 billion years to cool enough for the first life to appear.

Figure 11.2
The large eyes of this loris reveal that it is a nighttime hunter. Native to India, Sri Lanka, and Southeast Asia, lorises creep slowly through the trees, hunting insects and small animals. They grow to about 25 cm (10 in.) in length.

Another adaptation to living in trees is the position of the eyes in the skull. The eyes of rats and squirrels are in the sides of their heads. The fields of vision of their eyes do not overlap. The eyes of primates, in contrast, are located in the front of the face, like the eyes of the loris in Figure 11.2. Each eye of a primate sees a slightly different view of the same scene. The brain merges the two views in perceiving the distances to objects. This type of vision is called **binocular vision**. The ability to judge distance is advantageous when jumping from one branch to another or when stalking prey.

Other mammals, such as cats, have binocular vision, but only primates have both binocular vision and grasping hands. These features require an enlarged brain to process information from the eyes and to coordinate hand movements. Large brains and increased intelligence thus became hallmarks of the primates.

Prosimians were the first primates

The first primates were **prosimians** (meaning "before monkeys"). Fossils 30 million to 40 million years old show that prosimians were common in North America, Europe, Asia, and Africa. Only a few species of prosimians, such as the loris shown in Figure 11.2, survive today. Many of these surviving species are nighttime hunters. You can tell by the disproportionately large size of their eyes. Large eyes are necessary to capture what little light is available at night or in dark forests. All 24 surviving species of lemurs, cat-sized prosimians with long tails for balancing, live on Madagascar, an island about 180 km (300 mi.) off the east coast of Africa. Today the native vegetation of Madagascar is being rapidly destroyed by an expanding human population. As the lemur's forest home disappears, some of our oldest living relatives are becoming extinct in the wild.

Scale and Structure

House cats lack

color vision. From

this information,

what can you

conclude about

the time of day

during which

house cats

probably are

most active?

About 35 million to 40 million years ago, a revolutionary change occurred in how primates lived: they became active during daytime. How do we know this change occurred? Fossil skulls of primates that lived at this time have much smaller eye sockets than do prosimian skulls, suggesting that they were active during the day. The new day-active primates were **anthropoids.** Monkeys, apes, and humans are the existing anthropoids.

Since daytime activity places different demands on the eye, many changes in eye design probably evolved at this time. One of these changes was the development of color vision. Of course, we can't actually examine 35 million-year-old anthropoid eyes to see if they could see color. Soft tissues such as those of the eye are rarely preserved as fossils. Instead, we infer that early anthropoids had color vision because all living anthropoids see color. Therefore, color vision probably arose early in anthropoid history.

The brains of anthropoids are larger than prosimian brains. Larger brain size seems to be associated with the more complicated behavior patterns of anthropoids. Anthropoids replaced prosimians rather rapidly. In the fossil record, prosimian fossils become rare as anthropoid fossils become common.

Figure 11.3
Monkeys, such as the adult baboon (below, left), take care of their young for a longer time than do most other mammals.

Figure 11.4
This spider monkey, like many other New World monkeys, has a long tail that functions as a limb for grasping branches.

Scientists don't know exactly why prosimians lost the evolutionary competition, but they suspect that the larger brains and color vision of anthropoids better adapted them to living in trees.

Monkeys have complex social interactions

Monkeys are anthropoids with tails. Two groups of monkeys occur today. Old World monkeys, such as the baboons in **Figure 11.3**, live in Asia and Africa. New World monkeys, such as the spider monkey in **Figure 11.4**, inhabit Central America and South America. Monkeys feed mainly on fruits and leaves rather than on insects. They live in groups in which complex social interactions occur. Monkeys tend to care for their young for a longer time than most other mammals, except for humans and apes. This long period of dependency seems to be necessary for the development of the large brains of monkeys, apes, and humans.

IDENTIFYING THE GENETIC MOLECULE

1850

1871 Friedrich Miescher, a German scientist, isolates the nucleic acids DNA and RNA from cell nuclei.

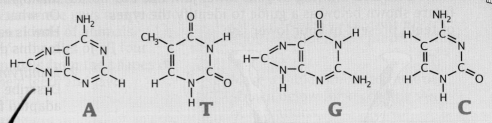

A   T   G   C

1903 American biochemist **Phoebus A. Levene** shows that DNA contains four nitrogenous bases: adenine (A), thymine (T), guanine (G), and cytosine (C), which are shown in the diagram above.

Lubber grasshopper

1910–1915 Thomas Hunt Morgan, an American geneticist, confirms that the genes of fruit flies occur in linear sequences on chromosomes.

Thomas Hunt Morgan

Chromosomes stained with fuchsin

1902 Walter S. Sutton, an American cytologist studying grasshopper genetics, proposes that genes are found on chromosomes. Based on the work of Gregor Mendel, scientists know that genes control an organism's traits.

1914 Robert Fuelgen, a German scientist, discovers that nucleic acids can be stained with the red dye fuchsin, helping confirm that nucleic acids are found in chromosomes.

1950

X-ray diffraction photograph of DNA

1952 American scientists **Alfred Hershey** and **Martha Chase** purify DNA and show that DNA, rather than protein, is the genetic material. The race to discover the structure of DNA quickens.

1961 Francis Crick confirms the hypothesis that three sequential nucleotides in DNA code for one amino acid.

1953 American biologist **James Watson** and British biophysicist **Francis Crick** determine that the structure of DNA is indeed a double helix.

1961 American scientists **Marshall Nirenberg**, **J.H. Matthei**, and **Severo Ochoa** demonstrate the three-base genetic code for DNA and mRNA.

1954 The use of Chargaff's ratios brings scientists to the realization that the sequence of DNA bases represents a code that carries hereditary information.

1951–1952 British scientists **Maurice Wilkins** and **Rosalind Franklin** use X-ray diffraction to produce images of DNA. In 1952, Franklin produces a DNA image suggesting that DNA is a spiral molecule called a helix.

Rosalind Franklin

1971 Biochemist **Marie Maynard Daly**, an African American, researches the atomic bonding characteristics of nucleic acids.

1941 American scientists **George Beadle** and **Edward Tatum** show that each gene mutation in bread mold leads to a change in one of the mold's enzymes. This links changes in genes to changes in proteins.

George Beadle

Edward Tatum

1950

1947 Erwin Chargaff shows that for every adenine there is a thymine, and for every guanine there is a cytosine. No one can explain why these proportions of A to T and G to C are found in DNA.

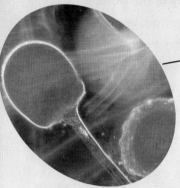

Reproductive structures of black bread mold

1944 American scientists **Oswald Avery**, **Maclyn McCarty**, and **Colin MacLeod** provide evidence that DNA is the genetic material, an idea greeted with skepticism. Others believe that proteins are the genetic material.

1992

1972–1973 American molecular biologists **Paul Berg**, **Stanley Cohen**, and **Herbert Boyer** cut genes into pieces using enzymes. These pieces are then inserted into another organism's genes. This technique, similar to combining two strips of film, is called gene splicing.

1975 Scientist **F. Agnes Stroud-Lee**, a Santa Clara Pueblo (Tewa) Native American, studies abnormalities in chromosomes resulting from birth defects, chemicals, and radiation.

F. Agnes Stroud-Lee

1992 American geneticist **Mark Dubnick**, who is visually challenged, writes scientific computer applications for genetic engineering, such as DNA sequencing, cloning, and mapping human genomes.

1988 James Watson becomes the director of the Human Genome Project (HUGO). Scientists working on this project plan to map the sequence of the 3 billion nucleotides that make up the human genome.

Susumu Tonegawa

Illustration of a genetically engineered circular DNA molecule

1987 Geneticist **Susumu Tonegawa**, a Japanese-American, is awarded the Nobel Prize for discovering how genes change to create antibodies. His findings open the door to new experimental ideas regarding cells of the immune system.

One of the primary concerns of modern biology is the threat to the environment posed by today's high-tech society. Global warming, acid rain, ozone holes, disappearing rain forests—our world is under great stress, and great care will be needed to prevent further damage. In this unit you will discover how living communities like this pine forest function, how human activities are damaging many of the world's natural communities, and what is being done today to protect the fragile Earth. Few areas of biology are as important to your future, and to that of your children.

The Environment

Ecosystems

Review

- **photosynthesis**
  (Section 5.3)

- **cellular respiration**
  (Section 5.4)

- **metabolism**
  (Section 5.4)

This slug, mushroom, and spider are just three of the many organisms found in a deciduous forest ecosystem.

YOU CANNOT PICK UP A NEWSPAPER TODAY WITHOUT SEEING NEWS ABOUT THE ENVIRONMENT. ENVIRONMENTAL ISSUES ARE IMPORTANT TO EVERYONE BECAUSE WE ALL HAVE TO LIVE IN A WORLD WE SEEM TO BE DESTROYING. WE NEED DETAILED KNOWLEDGE OF HOW THE WORLD WORKS SO THAT WE CAN PREVENT FURTHER ABUSE TO OUR PLANET AND PERHAPS BEGIN TO REPAIR THE DAMAGE WE ALREADY HAVE DONE.

12.1   What Is an Ecosystem?

Objectives

❶ **Identify the components of an ecosystem.**

❷ **Discuss the flow of energy through ecosystems.**

❸ **Identify the different trophic levels in an ecosystem.**

❹ **Explain why ecosystems can only contain a few trophic levels.**

State of Our World

More than 5 billion humans live on Earth. Scientists estimate that between 10 million and 80 million other species share the world with us. Yet we seem to be rapidly destroying our planet's ability to support us and its other inhabitants. For example, here are a few changes humans have made to the Earth *in just the last year.* About 17 million hectares (40 million acres) of forest have been burned down or cut, like the forest shown in **Figure 12.1**. That is an area almost as large as the state of Washington. As many as 50,000 species of animals and plants have become extinct, disappearing forever. The world's human population has grown by 92 million people, mostly in the world's poorer nations. These 92 million people were born into a world in which more than 1 billion people are not adequately nourished.

For centuries people have believed that the environment was there for them to use as they saw fit. But, today's environmental problems teach us that the environment is not a passive stage on which we can act as we please. Rather, we share the environment with other organisms, resulting in a complex network of interactions on which we all depend. Changes made to the environment can have serious consequences, not all of which are predictable.

Figure 12.1
This section of forest in the Quinleute Reservation in Washington has been logged. Although the trees will provide many useful products, cutting the forest has a devastating effect on the organisms living there.

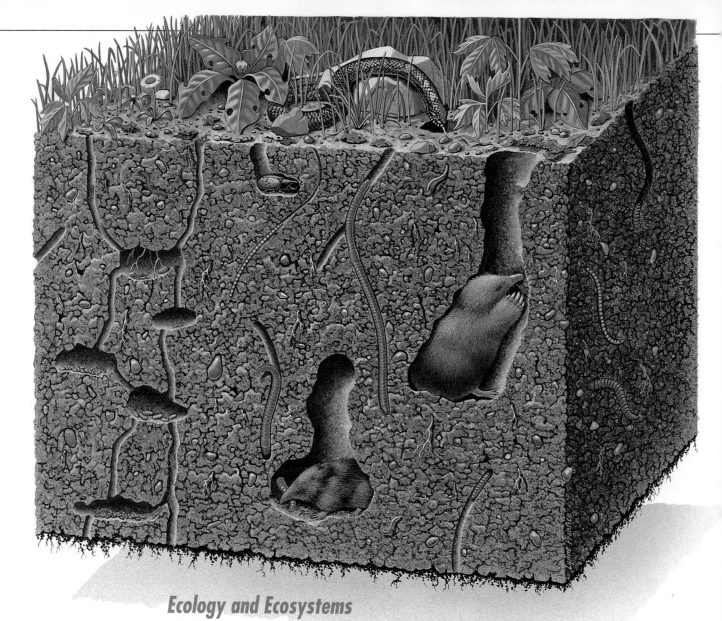

Ecology and Ecosystems

Figure 12.2
The inhabitants of this soil ecosystem include earthworms, insects, snakes, moles, bacteria, and a variety of plants and fungi.

In 1866, the German biologist Ernst Haeckel gave a name to the study of how organisms fit into their environment. He called it **ecology**, from the Greek words *oikos* (house, place where one lives) and *logos* (study of). Ecology, then, is the study of the "house" in which we live. Most of our environmental problems could be avoided if we treated the world in which we live the same way we treat our own homes.

Ecology is the study of the interactions of organisms with one another and with their physical environment. The organisms that live in a particular place, such as a forest, are known as a **community**. Ecologists, the scientists who specialize in ecology, call the physical location of a community its **habitat**. You can think of a habitat as a neighborhood and of a community as the residents of the neighborhood. The sum of the community and habitat is called an ecological system or ecosystem. An **ecosystem** is a self-sustaining collection of organisms and their physical environment.

Imagine that you could collect every organism living in an ecosystem. **Figure 12.2** shows just some of the inhabitants of a soil ecosystem. If you visited a tropical rain forest, you could collect many more species than occur in this soil ecosystem. As many as 100 species of trees can be found in 1 hectare (2.5 acres) of South American rain forest. The **diversity** of an ecosystem is a measure of the number of species living there and how common each species is. Tropical rain forests are the most diverse terrestrial ecosystems.

Why Study Ecology?

You study ecology because you need to know about the place in which you live. If you are going to prevent pollution, conserve resources, and save the world for your children to live in, then you need to know how your world works—just as you need to study how any complex machine works in order to keep it running properly. Do you think a car would run for long if its owner had no idea of the need for water, oil, and gasoline? Remember that there is a fundamental difference between a car and an ecosystem, however. A car that receives no attention from its owner will eventually break down. Ecosystems, on the other hand, have been functioning for billions of years without human tending. Only now, when we have caused them great damage, do some ecosystems require our help to continue.

Ecosystems are very complex

It is very difficult to understand how an ecosystem works, because it can contain hundreds or even thousands of interacting species. Nevertheless, you can gain a basic understanding of how an ecosystem works by asking two questions. From where does the energy needed by particular animals and plants come? How do organisms in ecosystems maintain adequate amounts of the minerals and other inorganic substances they need?

Answering these questions will give you a pretty good idea of how an ecosystem normally works. You will then be in a position to ask how an ecosystem might be expected to respond to a disturbance. To make such predictions, ecologists build a model, a simplified version of the ecosystem. An ecosystem model consists of a series of hypotheses that describe how the ecosystem functions: how energy moves through the ecosystem, how species interact, and so on. In some cases, ecologists express their models as mathematical equations and then solve these equations for various situations. **Figure 12.3** is a schematic representation of an ecological model.

Because ecosystems are so complex, no ecosystem model can consider all the factors affecting that ecosystem. Nevertheless, seeing what happens to the ecosystem model when a variable is changed helps ecologists predict what might happen if some component of the real ecosystem is altered. If, for example, one species in an ecosystem becomes extinct, or the amount of available energy declines, an ecological model can help scientists predict the possible consequences. Ecological models allow you to look into the future, but your vision can be only as accurate as the hypotheses used to build the model.

Figure 12.3
Ecologists use models to predict the possible outcomes of a disturbance in an ecosystem such as the soil ecosystem shown below.

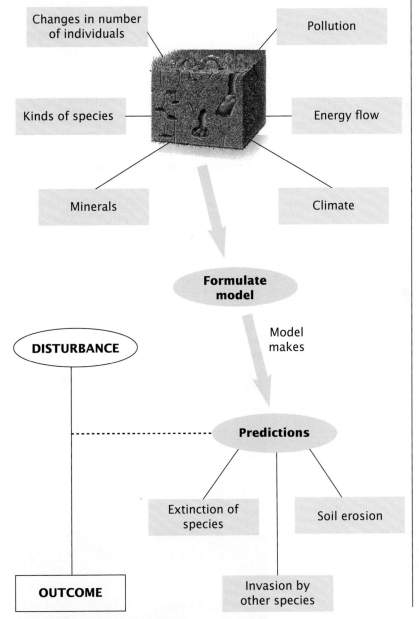

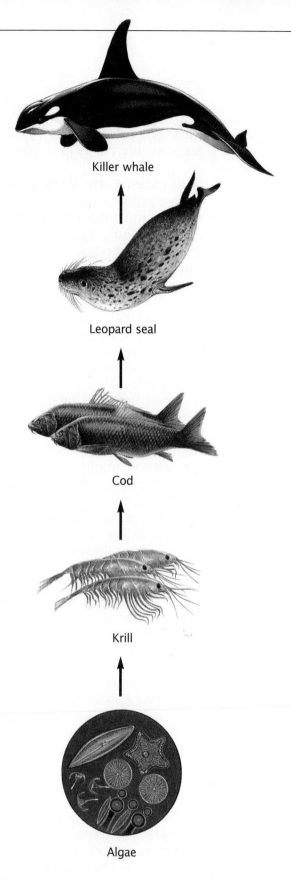

Energy in Ecosystems

An ecosystem uses energy because its living members use energy. A flower blooming, a squirrel running along a tree branch, a worm burrowing through the soil—every action of even the tiniest creatures requires energy. The most important factor determining how many and what kinds of organisms live in an ecosystem is the amount of energy available.

From where do the organisms in an ecosystem obtain energy? Life exists because organisms called **producers** take in energy from their surroundings and store it in complex molecules. Except for a few kinds of bacteria, producers capture energy from the sun through photosynthesis. Plants, some bacteria, and algae are producers. All other organisms are called **consumers**. They obtain their energy by consuming other organisms. Animals, most protists and bacteria, and all fungi are consumers. The food chain in **Figure 12.4** shows how energy is transferred from producers to consumers in an ocean ecosystem.

Each ecosystem contains consumers called **decomposers**. Decomposers obtain energy by consuming organic wastes (feces, urine, fallen leaves) and dead bodies. Fungi, such as the mushrooms shown in **Figure 12.5**, and some species of bacteria, are decomposers.

Figure 12.4
This food chain shows the path of energy transfer in an ocean ecosystem. Algae, which are photosynthetic, are the producers in this ecosystem.

Figure 12.5
These scarlet waxy caps are decomposers.

If pollution drastically reduced the number of algae in the food web shown below, would killer whales be affected? Explain.

Energy flows from producers to consumers

To follow the movement of energy through an ecosystem, ecologists assign each organism to a **trophic** (feeding) **level**. A trophic level is a group of organisms whose energy sources are the same number of steps away from the sun. Producers such as plants are in the first trophic level. Plants and other organisms that make their own food are **autotrophs**. Animals that eat plants are in the second trophic level. Animals that feed on plant-eaters are in the third trophic level.

Creatures in the second trophic level are **herbivores** (plant eaters). Cows, caterpillars, elephants, and ducks are herbivores. All organisms at the third trophic level or above are **carnivores** (flesh eaters). Carnivores at the third level feed on herbivores, while carnivores above the third trophic level feed on other carnivores. Tigers, hawks, weasels, pelicans, and killer whales are carnivores. Some animals cannot be classified as either carnivores or herbivores. **Omnivores** such as bears and humans eat both plants and animals. Because they cannot make their own foods, organisms living in trophic levels above the first trophic level are **heterotrophs**. Most animals feed at more than one trophic level and eat several different species at each trophic level. As shown in **Figure 12.6**, energy moves through an ecosystem in a complex network of feeding relationships called a food web. Notice that the food chain in **Figure 12.4** is just one part of this food web.

Figure 12.6
This food web shows how energy flows through an ocean ecosystem as one organism is eaten by another.

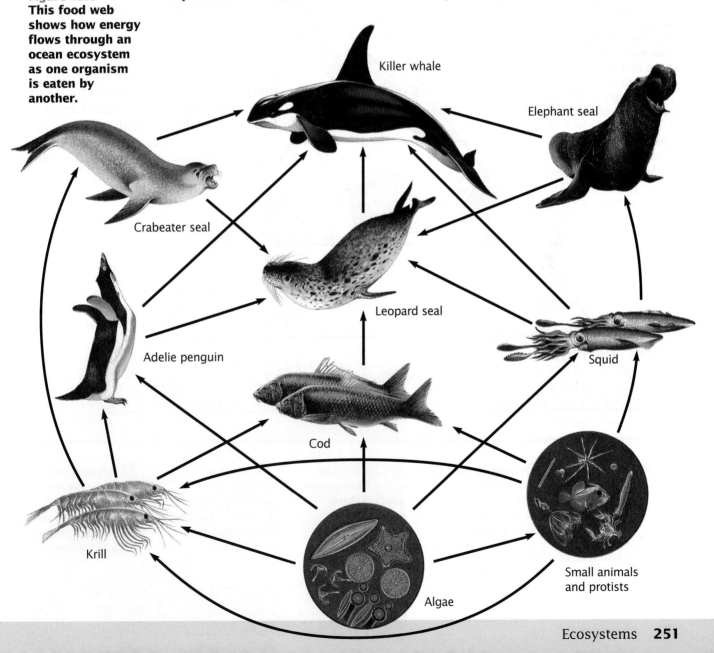

Killer whale

Elephant seal

Crabeater seal

Leopard seal

Adelie penguin

Squid

Cod

Krill

Algae

Small animals and protists

Figure 12.9
The white-tailed deer and speckled alder play important roles in the nitrogen cycle in an ecosystem found in the northeastern United States.

Bacteria play a key role in the nitrogen cycle

Organisms must have nitrogen to make proteins and nucleic acids. **Figure 12.9** shows how nitrogen cycles through an ecosystem in the northeastern United States. As you learned in Chapter 8, most living things cannot use the nitrogen gas in the air. The two nitrogen atoms that make a molecule of nitrogen gas are held together by a strong chemical bond that is difficult to break. The variety of life found on earth is possible only because a few kinds of bacteria have enzymes that can break this strong bond. Nitrogen atoms are then free to bond with hydrogen atoms to form ammonia molecules. Conversion of nitrogen gas to ammonia is called **nitrogen fixation**. Ammonia is a form of nitrogen that plants can absorb and use to make proteins. Since animals cannot absorb nitrogen from the soil, they must obtain nitrogen by eating plants or other animals.

Nitrogen-fixing bacteria live in the soil or within the roots of plants such as peas, clover, alfalfa, beans, and alder trees. The growth of plants in ecosystems is often severely limited by the availability of nitrogen in the soil. When an organism dies, the nitrogen in its body is released by decomposers. Animal wastes, such as dung and urine, as well as plant materials like leaves and bark, also contain nitrogen. These materials are also broken down by decomposers. Thus, decomposers play a vital role in ecosystems by returning nitrogen to the soil.

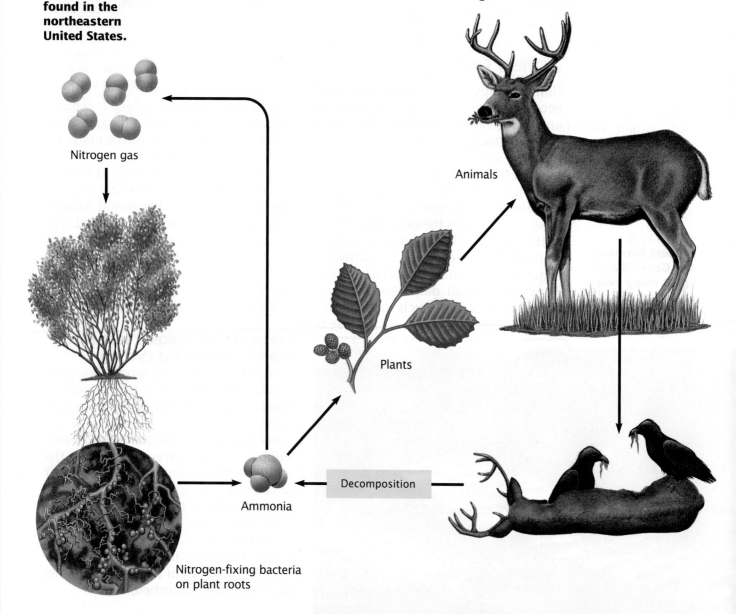

Nitrogen gas

Animals

Plants

Decomposition

Ammonia

Nitrogen-fixing bacteria on plant roots

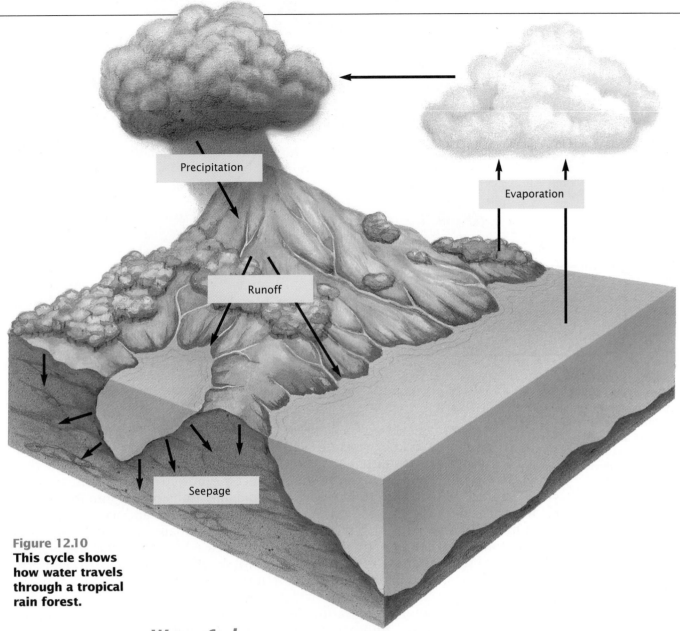

Precipitation

Runoff

Evaporation

Seepage

Figure 12.10
This cycle shows how water travels through a tropical rain forest.

Water Cycle

Water is perhaps the most important nonliving component of an ecosystem. To a large degree, availability of water determines the diversity of organisms in an ecosystem. Few species live in the desert where there is little water, while many species live in the tropical rain forest where water is plentiful. Water is constantly moving within ecosystems in the cycle illustrated in **Figure 12.10**.

Plants play an important role in the water cycle

In tropical rain forests, where there are dense concentrations of trees and other plants, more than 90 percent of the moisture that enters the ecosystem passes through plants and evaporates from their leaves. In a very real sense, these plants create their own rain.

When forests are cut down, the water cycle is broken. Moisture cannot be returned to the atmosphere by plants. Instead, water drains into streams and rivers and eventually flows into the ocean. Moreover, without protection by the roots of trees and other plants, the soil is easily carried away by runoff. As a result, nutrient cycles also are broken. Because both water and nutrients can no longer cycle in a forest ecosystem after the trees are cut down, extensive cutting can convert lush forests into deserts. Tragically, such a transformation is presently occurring in many tropical rain forests.

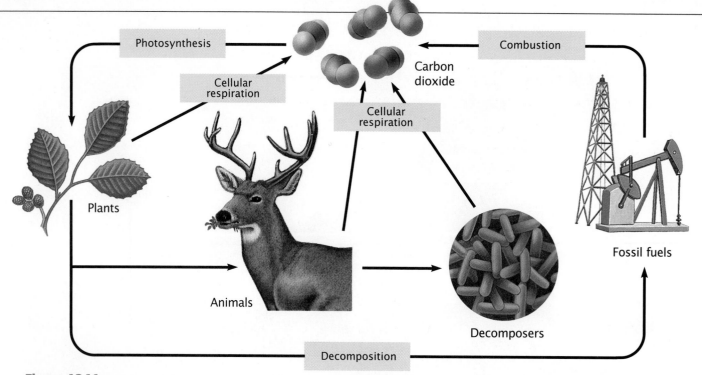

Figure 12.11
This cycle shows how carbon moves within an ecosystem in an industrialized nation such as the United States.

Carbon Cycle

Like water, carbon also cycles between the environment and organisms. The Earth's atmosphere contains carbon in the form of carbon dioxide. Plants use carbon dioxide to build organic molecules during photosynthesis. Consumers obtain energy-rich molecules that contain carbon by eating plants or other animals. As these molecules are broken down, carbon dioxide is produced and released into the Earth's atmosphere. Cellular respiration by decomposers and photosynthetic organisms also returns carbon dioxide to the atmosphere. **Figure 12.11** shows how carbon cycles within an ecosystem.

Humans are overloading the carbon cycle

Large amounts of carbon are tied up in wood and may stay trapped there for hundreds of years, only returning to the atmosphere when the wood is burned. Over millions of years, plants that become buried in sediment may be gradually transformed into fossil fuels such as coal, oil, and natural gas. The carbon originally trapped by plants is not released back into the atmosphere until fossil fuels are burned. By burning large amounts of fossil fuels, humans are increasing the concentration of carbon dioxide in the atmosphere. Carbon dioxide traps heat from the sun within the atmosphere, much like glass panes trap the sun's heat in a greenhouse. The ability of gases such as carbon dioxide to retain the sun's heat, and in so doing to warm the atmosphere, is called the **greenhouse effect**. You will learn more about the greenhouse effect in Chapter 14.

Section Review

❶ Explain the significance of Bormann and Likens's experiments at Hubbard Brook.

❷ Explain the role of nitrogen-fixing bacteria in the nitrogen cycle.

❸ How do you think deforestation affects the water cycle?

❹ Describe how human interference in the carbon cycle may be causing an increase in global temperatures.

ALL ECOSYSTEMS ARE CONNECTED. THE DESTRUCTION OF THE RAIN FORESTS IN BRAZIL, FOR EXAMPLE, WILL PERSONALLY AFFECT YOU AND EVERYONE ELSE IN THE WORLD. IF HUMANS ARE GOING TO TAKE ACTION TO PRESERVE THE WORLD FOR FUTURE GENERATIONS, IT IS IMPORTANT TO UNDERSTAND WHAT THE MAJOR ECOSYSTEMS OF THE WORLD ARE LIKE AND HOW THEY FUNCTION.

12.3 Kinds of Ecosystems

Objectives

❶ **Identify the importance of plankton in freshwater ecosystems.**

❷ **Identify the factors that determine the ecosystem type found in a particular area.**

❸ **Contrast the seven major terrestrial ecosystems.**

❹ **Identify the major ocean ecosystems.**

Freshwater Ecosystems

Freshwater ecosystems include lakes, ponds, and rivers. These ecosystems are very limited in area. Inland lakes cover 1.8 percent of the Earth's surface, and rivers and streams cover about 0.3 percent. Although small in total area, freshwater ecosystems support a rich array of life, including fishes, amphibians, insects, turtles, crocodiles, and many plants. A diverse biological community of microscopic organisms called **plankton** lives near the surface of lakes and ponds. Plankton contain photosynthetic organisms that are the base of aquatic food webs. All freshwater habitats are strongly connected to land ecosystems. Nutrients flow from terrestrial ecosystems into freshwater ecosystems. In addition, many land animals come to the water to feed or reproduce.

Figure 12.12 shows a freshwater ecosystem. Ponds and lakes usually have three zones in which organisms occur: a shallow "edge" zone, an open-water surface zone, and, in deep lakes and ponds, a deep-water zone to which little light can penetrate.

Figure 12.12
Freshwater ecosystems, such as this pond, support a wide variety of plants and animals.

Terrestrial Ecosystems

Major ecosystems that occur over wide areas on land are called **biomes**. The seven biomes are: (1) tropical rain forests; (2) savannas; (3) deserts; (4) temperate grasslands; (5) deciduous forests; (6) coniferous forests; and (7) tundra. These biomes differ remarkably from one another because they evolved in different geographic locations. The kinds of animals and plants that live in an ecosystem depend on the physical nature of the habitat: the soils, the terrain, and the climate. **Table 12.1** compares the Earth's seven biomes.

Table 12.1 Biomes

Biome	Yearly Precipitation	Characteristics
a **Tropical rain forests**	250 cm (100 in.)	Little temperature variation; abundant moisture
b **Savannas**	90–150 cm (36–60 in.)	Open; widely spaced trees; seasonal rainfall
c **Deserts**	20 cm (8 in.)	Dry; sparse vegetation; scattered grasses
d **Temperate grasslands**	10–60 cm (4–24 in.)	Rich soil; tall, dense grasses
e **Deciduous forests**	75–250 cm (30–100 in.)	Warm summers; cool winters
f **Coniferous forests**	20–60 cm (8–24 in.)	Short growing season; cold winters
g **Tundra**	25 cm (10 in.)	Open; wind-swept; dry; ground always frozen

a Tropical rain forest, Australia

b Savanna, Kenya

c Desert, Australia

d **Temperate grassland, Oklahoma**

e **Deciduous forest, New Hampshire**

Inhabitants	Location	Comments
More species than any other biome	Tropical regions of South America, Central America, Asia, Africa, Australia	Forests and their species being destroyed at an alarming rate
Lions, rhinoceroses, elephants, giraffes, gazelles	Parts of Africa, South America, Australia	Conversion to agriculture threatens inhabitants
Kangaroo rats, camels, cactuses	Parts of Africa, Asia, Australia, North America	Inhabitants must conserve water
Buffalo, prairie grasses	Central North America, central Asia	American prairies were once the home of huge numbers of bison
Raccoons, deer, maples, oaks, hickories	Europe, north-eastern United States, eastern Canada	Deciduous trees lose their leaves every year
Elk, moose, needle-leaved evergreens	Northern Asia, northern North America	Primary source of the world's lumber
Caribou, lemmings, wolves, mosses, lichens	Far northern Asia, northern North America	Covers one-fifth of the Earth's land surface

g **Tundra, Alaska**

f **Coniferous forest, Montana**

Ocean Ecosystems

Nearly three-quarters of the Earth's surface is covered by ocean. Three types of ocean ecosystems are shown in **Figure 12.13**. Shallow ocean waters are small in area but contain most of the ocean's diversity. Many fishes swim in the open ocean surface, feeding on plankton. Photosynthetic plankton account for about 40 percent of all photosynthesis on earth. There is increasing evidence that pollution is harming photosynthetic plankton. If significant numbers of plankton are destroyed, the oxygen you breathe will be slowly depleted from the Earth's atmosphere. The deep ocean waters are cold and dark. Among the few residents of the deep ocean are some of the most bizarre organisms found on earth. Many organisms in the deep ocean have light-producing body parts that they use to attract mates or lure prey. Because photosynthesis cannot occur in the deep ocean, most of these organisms prey on other deep sea residents or scavenge the dead bodies of organisms that have fallen from above.

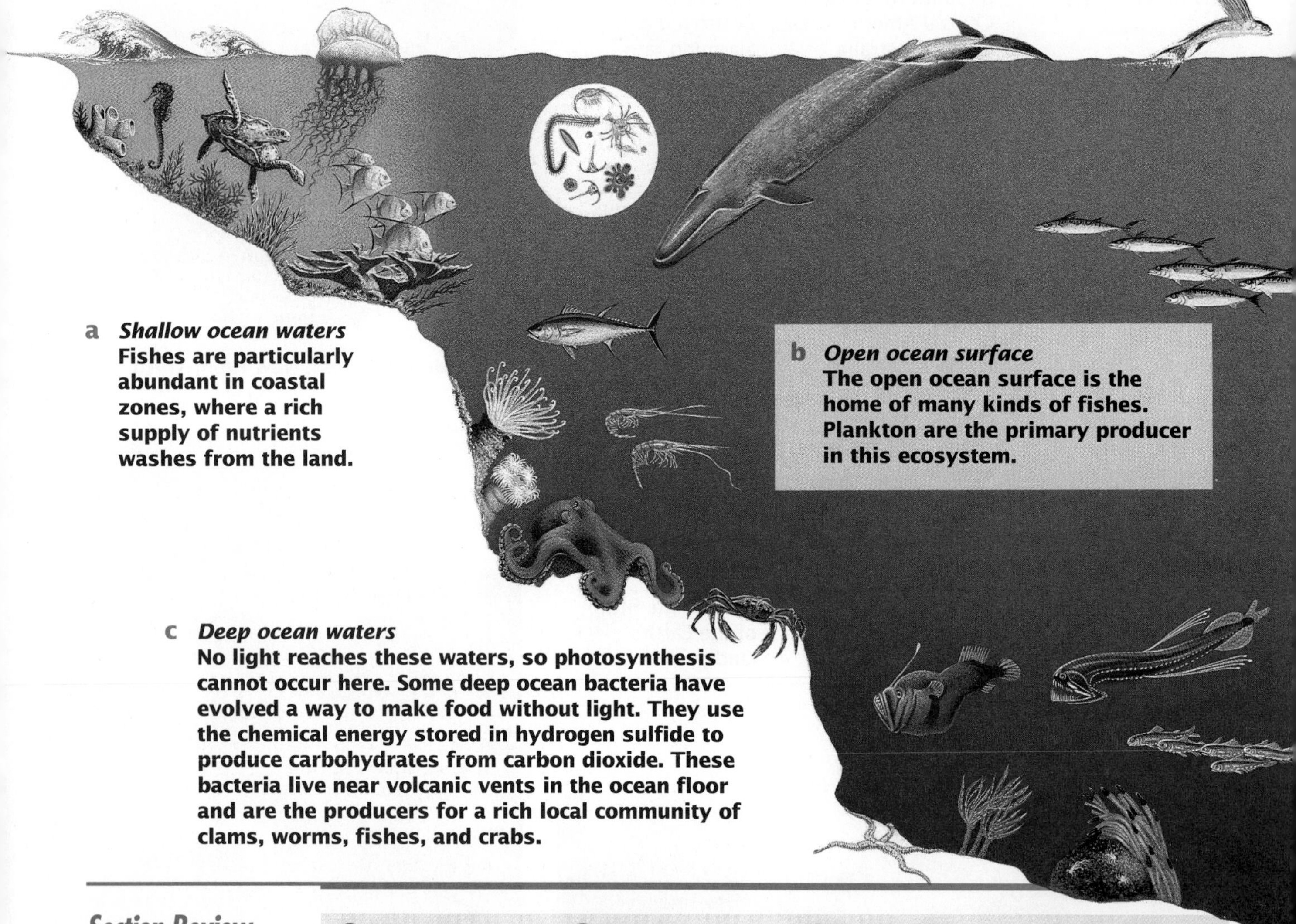

Figure 12.13
Each type of ocean ecosystem supports different kinds of living organisms. This representation is not drawn to scale.

a *Shallow ocean waters*
Fishes are particularly abundant in coastal zones, where a rich supply of nutrients washes from the land.

b *Open ocean surface*
The open ocean surface is the home of many kinds of fishes. Plankton are the primary producer in this ecosystem.

c *Deep ocean waters*
No light reaches these waters, so photosynthesis cannot occur here. Some deep ocean bacteria have evolved a way to make food without light. They use the chemical energy stored in hydrogen sulfide to produce carbohydrates from carbon dioxide. These bacteria live near volcanic vents in the ocean floor and are the producers for a rich local community of clams, worms, fishes, and crabs.

Section Review

1 What role do plankton play in a freshwater ecosystem?

2 List two reasons why tropical rain forests do not occur in the United States.

3 Name two biomes that occur where there is very little precipitation.

4 Describe how organisms living deep in the ocean obtain their food.

If the tropical rain forests disappear, so will this butterfly, from Trinidad.

Key Terms	Summary

12.1 What Is an Ecosystem?

Decomposers, such as these mushrooms, are an important part of every ecosystem.

Key Terms

ecology (p. 248)

community (p. 248)

habitat (p. 248)

ecosystem (p. 248)

diversity (p. 248)

producer (p. 250)

consumer (p. 250)

decomposer (p. 250)

trophic level (p. 251)

autotroph (p. 251)

herbivore (p. 251)

carnivore (p. 251)

omnivore (p. 251)

heterotroph (p. 251)

Summary

• Ecology is the study of how living things fit into their environment.

• An ecosystem is a group of interacting organisms and their physical environment.

• Producers capture energy and store it in complex molecules. Consumers obtain energy by feeding on producers or other consumers. Decomposers eat dead organisms, animal wastes, fallen leaves, twigs, and other debris.

• Autotrophs are organisms such as plants that make their own food. Heterotrophs cannot make their own food and must obtain it from other organisms.

12.2 Cycles Within Ecosystems

Bormann and Likens studied how nutrients cycle within ecosystems at this site in New Hampshire.

Key Terms

nitrogen fixation (p. 254)

greenhouse effect (p. 256)

Summary

• Materials such as water, nitrogen, and carbon move through ecosystems in cycles.

• Some bacteria absorb nitrogen gas and convert it to ammonia. The process of transforming nitrogen gas into ammonia is known as nitrogen fixation.

• The burning of fossil fuels releases large amounts of carbon dioxide into the atmosphere. Carbon dioxide in the atmosphere retains heat from the sun, a phenomenon known as the greenhouse effect.

12.3 Kinds of Ecosystems

The tundra is one of the seven biomes.

Key Terms

plankton (p. 257)

biome (p. 258)

Summary

• Photosynthetic plankton are the basis of the food web in aquatic ecosystems.

• On land, there are seven major types of ecosystems, which are called biomes.

• There are three major types of ecosystems found in the ocean.

Understanding Vocabulary

1. For each set of terms, identify the term that does not fit and explain why.
   a. producers, decomposers, herbivores, omnivores
   b. water, energy, carbon, nitrogen
   c. photosynthesis, cellular respiration, decomposition, combustion

Relating Concepts

2. Copy the unfinished concept map below onto a sheet of paper. Then complete the concept map by writing the correct word or phrase in each oval containing a question mark.

Understanding Concepts

Multiple Choice

3. Which of the following organisms would *not* be part of the same trophic level?
   a. photosynthetic bacteria
   b. fungi
   c. grass
   d. trees

4. Bormann and Likens found that cutting trees from part of the Hubbard Brook ecosystem
   a. reduced calcium loss.
   b. decreased rainfall.
   c. increased calcium loss.
   d. increased rainfall.

5. The form of nitrogen usable by plants is
   a. ammonia.
   b. nitrous oxide.
   c. nitrogen gas.
   d. nucleic acids.

6. Which does *not* add carbon dioxide to the atmosphere?
   a. cellular respiration
   b. gasoline-burning cars
   c. photosynthesis
   d. forest fires

7. The biome with the greatest diversity is
   a. grassland.
   b. temperate deciduous forest.
   c. coniferous forest.
   d. tropical rain forest.

8. Tundra and desert are both
   a. cold.
   b. very diverse.
   c. dry.
   d. near the equator.

9. Which biome is *not* found in the United States?
   a. desert
   b. tropical rain forest
   c. deciduous forest
   d. temperate grasslands

10. The first trophic level in the open ocean contains
    a. plankton.
    b. herbivorous clams.
    c. carnivorous sharks.
    d. bacteria and fungi.

11. Which organisms would *not* be in a deep ocean ecosystem?
    a. photosynthetic bacteria
    b. clams
    c. crabs
    d. carnivorous fishes

Completion

12. Ernst Haeckel coined the term _____ to describe the study of how organisms interact with each other and with their environment.

13. Nitrogen gas is converted to ammonia during the process of _____ .

14. Because they eat only plants, cows are called _____ . Humans who eat plants and animals are called _____ .

15. The African savanna is the _____ inhabited by lions, giraffes, and cheetahs. Lions, giraffes, cheetahs, and other organisms that live on the African savanna make up a(n) _____ .

Short Answer

16. What is the ecological role of decomposers such as bacteria and fungi?

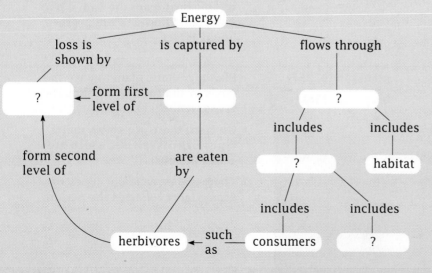

17. What are the benefits of nitrogen fixation to plants?

18. What is an ecological model? How are ecological models used by ecologists?

19. How is diversity of an ecosystem measured?

Interpreting Graphics

20. Look at the food chain of a marine ecosystem shown below.

   - What is the producer in this ecosystem?

   - Which organisms are carnivores?

   - Which organisms are herbivores?

   - Which organisms should be least common? Explain.

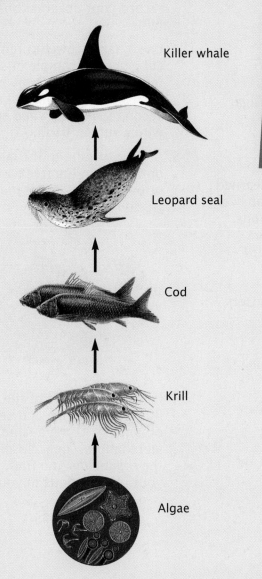

Killer whale

Leopard seal

Cod

Krill

Algae

Reviewing Themes

21. *Interacting Systems*
    If destruction of tropical rain forests continues at its present rate, most of the world's rain forests will be gone within 40 years. What are two likely effects of the destruction of tropical rain forests?

Thinking Critically

22. *Inferring Conclusions*
    Clear-cutting—cutting all trees in a certain area—was a practice used extensively by loggers in the early part of this century. It had a devastating effect on the water cycle. What can be done to restore the water cycle in clear-cut areas?

23. *Comparing and Contrasting*
    Describe the characteristics of the biome in which you live. What plants and animals live there? Which of the biomes described in this chapter is most like the one in which you live?

24. *Building on What You Have Learned*
    Based on what you learned about photosynthesis and cellular respiration in Chapter 5, why can plants act as producers and consumers, while animals can only act as consumers?

Cross-Discipline Connection

25. *Biology and History*
    By watching Native Americans, European settlers learned that placing a dead fish and corn seeds in the same hole produced a greater yield. How did this affect soil nutrients? How is this practice continued today?

Discovering Through Reading

26. Read the article "Why American Songbirds Are Vanishing," in *Scientific American*, May 1992, pages 98–104. What types of ecological damage explain the vanishing of many bird species? What can be done to increase the number of songbirds in North America?

Chapter 12 Investigation

Exploring a Soil Community

Objectives

In this investigation you will:
- *observe* a soil community
- *measure* the total mass present at each trophic level in the community

Materials

- metric ruler
- plastic bags
- rubber bands
- hand trowel
- freezer
- pan
- forceps
- hand lens or stereomicroscope
- triple-beam balance
- test tube
- distilled water
- test tube rack
- sterile swab
- petri dish of nutrient agar
- tape

Prelab Preparation

Review what you have learned about communities by answering the following questions:
- How does energy move through a community?
- What is an ecological pyramid?
- What role do decomposers play in a community?

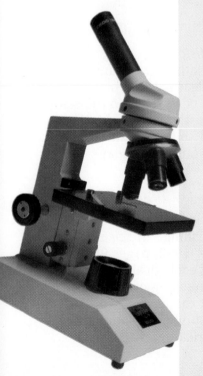

Procedure: Exploring a Soil Community

1. Form a cooperative group of four students. Work with one member of your group to complete steps 2–13. Make a table like the one shown on the opposite page. Steps 2–4 should be carried out in a nearby field or forest. The less disturbed your chosen area is, the better your results will be.

2. Collect the leaf litter from a 25 x 25 cm (10 x 10 in.) area and place it in a plastic bag. Use a rubber band to seal the bag.

3. Observe the surface of the soil and record your observations in your table.

4. Use a hand trowel to remove a cube-shaped sample of soil that measures about 15 cm (6 in.) on each side. Place the soil sample in a second plastic bag and seal the bag with a rubber band.

5. After returning to the lab, place both bags in the freezer for 5 hours or overnight. *What is the purpose of freezing the bags?*

6. Pour the contents of the bag of leaf litter into the pan and search for organisms. Using a hand lens or stereomicroscope, closely observe the organisms you find. Record your observations in your table.

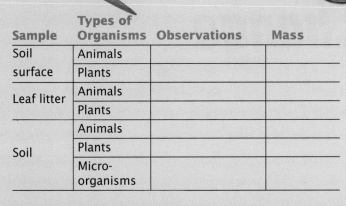

Sample	Types of Organisms	Observations	Mass
Soil surface	Animals		
	Plants		
Leaf litter	Animals		
	Plants		
Soil	Animals		
	Plants		
	Micro-organisms		

7. Using forceps, separate the animal and plant material into two piles. Use the balance to find the mass of each pile. Return the leaf litter to the plastic bag.

8. Open the plastic bag containing your soil sample. Place a sample of soil about the size of a pea in a test tube. Add enough distilled water to the test tube so that the soil sample is covered in water. Set the tube aside.

9. Now perform steps 6 and 7 for the remainder of the soil sample.

10. Remove a sterile swab from its package. Dip the swab into the test tube containing the soil-water mixture. Remove excess moisture by rotating the swab against the side of the tube. Open the cover of the petri dish and lightly rub the moistened swab over the surface of the agar in a zigzag pattern. Close the lid of the dish and seal it with tape, as your teacher directs.

11. Label the petri dish and incubate it at room temperature for one to two days. Then record your observations. **Do not open the petri dish**. After making your observations, dispose of the dish as directed by your teacher.
What kinds of organisms are growing on the petri dish?

12. Share your observations with the other members of your group.

13. Clean up your materials and wash your hands before leaving the lab.

Analysis

1. *Summarizing Observations*
Compare the leaf litter to the soil in terms of the mass and variety of animals.

2. *Analyzing Observations*
What evidence suggests that an ecological pyramid exists in this community?

3. *Analyzing Observations*
What evidence suggests that different trophic levels exist in this community?

4. *Making References*
On what do the soil microorganisms feed in this community?

5. *Comparing Observations*
How does the community you investigated compare to the community studied by the other team in your group?

Thinking Critically

1. Is there evidence that predators occur in this community? Explain your answer.

2. Where does a soil ecosystem start and end? What are the boundaries of an ecosystem?

3. Suppose that a fire destroyed the leaf litter of this community. How would this disturbance affect the community?

How Ecosystems Change

Ecosystems are constantly changing. Here in Yellowstone National Park, naturally occurring forest fires enable new trees to grow.

Every ecosystem on Earth, whether it is frozen tundra or tropical rain forest, is a complex network of interacting species. To preserve the Earth's fast-disappearing ecosystems, it is essential to understand the nature of these interactions and how they are shaped by natural selection and the physical environment.

13.1 Interactions Within Ecosystems

Objectives

1 Recognize the role of coevolution in shaping the structure of ecosystems.

2 Relate the characteristics of flowers to their coevolution with insects.

3 Describe how plants and their herbivores have coevolved.

4 Contrast parasitism, mutualism, and commensalism.

Evolution and Ecosystems

Species evolve in response to the challenges posed by their environments. As a result, animals, plants, and other creatures in an ecosystem possess characteristics that fine-tune them for living where they do. For example, many plants in desert ecosystems have waxy coatings on their leaves that help them retain water. For the same reason, desert animals often hide underground during the hottest part of the day; an animal's behavior is just as much an adaptation to its environment as are its physical characteristics.

An organism's survival and reproduction also depend on interactions with other living members of its ecosystem, as seen in the photograph in **Figure 13.1**. These interactions influence the evolution of a species just as do its interactions with the physical environment. For instance, many plant species have evolved tough leaves that protect against being eaten by herbivores. Of course herbivores evolve too, and many have evolved flatter, larger teeth that are better suited to grinding the very tough leaves they eat. Cows and horses have teeth such as these. **Coevolution** occurs when two or more species evolve in response to each other.

Figure 13.1
In this photograph, you can see that the protective coloration of this spider conceals it from both predators and prey.

Coevolution Shapes Species Interactions

Coevolution shapes the many ways animals and plants interact within ecosystems, creating a complex web of interactions among the organisms that live there. For example, producers, herbivores, and carnivores in a mature forest ecosystem have adjusted to one another during millions of years of evolution.

One of the most dramatic examples of coevolution is the side-by-side evolution of flowering plants with insects. Plants are immobile and cannot actively search for mates. Instead of moving to seek a mate, many species of flowering plants rely upon animals to transport their male gametes within grains of pollen. Insects are attracted to a plant's flowers by bright colors, pleasant odors, and food, which is usually sweet nectar. As the insect feeds, sticky pollen attaches to its body. After feeding at the flowers of one individual, the pollen-covered insect flies to flowers of another individual. At this next feeding stop, some of the insect's load of pollen is rubbed onto the female reproductive structures of the flowers. Fertilization is now able to take place. Animals that carry pollen from flower to flower are known as pollinators.

Within each species of flowering plant, those individuals that are better at attracting pollinators will leave more offspring. Thus, the attractive features of each species of flowering plant have evolved in concert with the preferences of its pollinators. For instance, bees cannot see the color red, and plant species pollinated by bees rarely have red flowers. Bee-pollinated flowers are usually yellow or blue. In comparison, the *Rafflesia* flower in **Figure 13.2** is pollinated by flies that feed on dead organisms. To attract its pollinators, this flower releases a powerful, nauseous odor like the stench of rotting flesh.

Pollinators, in turn, have evolved traits that enable certain species to specialize on particular species of flowers when seeking nectar. Scarlet lobelia flowers, for example, produce nectar at the bottom of a very long, tube-shaped flower. The hawk moth is the only pollinator with a tongue long enough to reach the bottom of the flower.

Figure 13.2
The pungent aroma of rotting meat attracts flies to this giant *Rafflesia* flower. The flies will pollinate it, and then transfer pollen from this flower to another *Rafflesia* flower.

Avoiding Being Eaten: Plants and Herbivores

Being eaten is not beneficial for plants. Herbivores can kill plants by feeding on them, just as carnivores kill their prey. Therefore, characteristics that enable plants to protect themselves from herbivores are favored by natural selection. The biological structures of many of the Earth's ecosystems have been determined largely by the ways plants avoid being eaten and the ways by which herbivores succeed in eating them.

Plants defend themselves from herbivores

Touch the stem of a rose bush and you might experience a familiar plant defense against herbivores. Many species of plants, such as roses and the cactus in **Figure 13.3**, employ physical defenses such as thorns, prickles, sticky hairs, and tough leaves. The most crucial plant defenses, however, are chemical. Virtually all plant species produce chemicals that protect them against herbivores. Some of these defensive chemicals are poisons that kill the animal that eats the plant, while other chemicals simply make the plant taste bad. Most herbivores learn to avoid plants that possess defensive chemicals. Anyone who has broken out in a rash from contact with poison ivy, shown in **Figure 13.4**, has been the victim of plant chemical defenses. Poison ivy produces a gummy oil called urushiol that causes severe blistering in many people.

As a rule, each group of closely related plant species has a unique battery of chemical defenses. Poison ivy, poison oak, and poison sumac produce urushiol. The plants of the mustard family produce a group of defensive chemicals called mustard oils. Mustard oils are the source of the pungent aromas and tastes characteristic of such plants as mustard, cabbage, radish, capers, and horseradish. The same tastes that we enjoy signal the presence of chemicals that are toxic to many groups of insects.

Many herbivores overcome plant defenses

Over time, herbivores have evolved ways to overcome the chemical defenses of plants. Because different plant species produce different chemicals, coevolution has resulted in a very specialized pattern of feeding; certain kinds of herbivores feed exclusively on particular kinds of plants.

Cabbage butterflies provide a good example of how counteractive measures have evolved in response to the chemical defenses of plants. Although most insects avoid plants of the mustard family, the caterpillars of cabbage butterflies eat these plants voraciously. These caterpillars are able to eat plants of the mustard family because they have evolved the ability to break down mustard oils into harmless chemicals. As a result of this evolutionary breakthrough, cabbage butterflies have been able to use a new food resource — plants of the mustard family — without competition from other insect herbivores.

Figure 13.3
The spines of this golden barrel cactus protect it from herbivores.

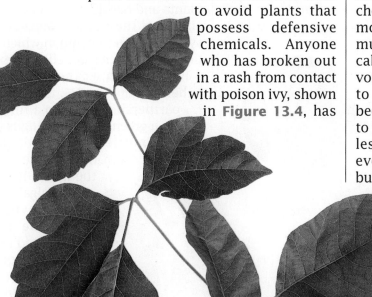

Figure 13.4
Touch the leaves of poison ivy and the chances are good that you'll get a rash from the toxins on their surfaces.

Three Types of Close Species Interactions

Evolution

Variation is

essential for

natural selection.

What role did

variation play in

the long-term

survival of rabbits

in Australia?

Three types of species interactions involve particularly close relationships among the participants. **Symbiosis** is a close, long-term association between two or more species. The three types of symbiotic relationships are parasitism, mutualism, and commensalism.

Parasites and their hosts coevolve

Worldwide, between 200 million and 300 million people suffer from malaria. People who have malaria play host to a single-celled parasite that was injected into their blood by the bite of a mosquito. Parasites obtain nutrition by feeding on their host. How is a parasite different from a predator? A parasite usually does not kill its host, and it is usually smaller than the organism on which it feeds. The relationship between a parasite and its host is called **parasitism**.

Both host and parasite coevolve in response to each other. In 1859, 12 rabbits were introduced to Australia. No rabbits naturally occur on the Australian continent. By the 1940s there were millions of rabbits swarming the countryside. To control the rabbit population, the Australian government introduced the viral disease myxomatosis (*mihk suh muh TOH suhs*) from South America. At first, myxomatosis was very deadly to the rabbits. More than 99 percent of the Australian rabbit population was killed by the initial introduction of the virus. The few surviving rabbits continued to breed, so the virus was reintroduced. This time, 90 percent of the rabbits died. By the third introduction of the virus, only 50 percent of the rabbits were killed.

Why did a smaller percentage of the rabbit population die each time the virus was introduced? Tests on the rabbits and the virus showed that both were evolving. Because rabbits resistant to myxomatosis were more likely to survive and reproduce, the rabbit population contained a greater proportion of resistant individuals after each introduction. Also, instead of becoming more deadly to overcome the rabbit's resistance, the virus had actually become less virulent, or deadly. If a virus kills its host rabbit too quickly, that rabbit cannot spread the virus to other rabbits. A virus that allows its hosts to live longer is able to infect more hosts.

All parties benefit in mutualism

Not all coevolution involves antagonistic relationships such as those between plants and herbivores or between parasites and their hosts. **Mutualism** is a symbiotic relationship in which all participating species benefit. For example, a lichen, such as the British soldier lichen in **Figure 13.5**, is a mutualistic partnership between a fungus and a green alga. The fungus absorbs nutrients for both partners from the surface on which the lichen is growing. The alga carries out photosynthesis to provide food for itself and its fungal partner. Mycorrhizae, which you read about in Chapter 10, are mutualistic associations between plants and fungi. In Chapter 12, you

Figure 13.5
The British soldier lichen is made of a mutualistic relationship between a fungus and an alga.

Alga

Fungus

Figure 13.6
Stinging anemones live on the claws of this female boxing crab. She guards herself and the red eggs she is carrying under her abdomen by using the anemones like boxing gloves, jabbing them at attacking predators. This is an example of a commensal relationship.

Commensalism is taking without harming

Commensalism is an ecological relationship in which one species benefits and the other is not obviously affected. A very intriguing example of commensalism is the boxing crab, which is described in **Figure 13.6**. The crab benefits from the protection anemones provide, and the anemones apparently are not harmed or helped.

The crusty growths seen on the back of the gray whale in **Figure 13.7** are actually small animals called barnacles. Barnacles hitch a ride on the whale. In doing so, they gain protection from predators and transportation to new sources of food (tiny animals they filter from the water). Apparently, the whale neither benefits nor is harmed by the presence of barnacles.

learned about the relationship between nitrogen-fixing bacteria and the roots of plants such as peas and beans. These bacteria provide the plants with a source of nitrogen. In exchange, the nitrogen-fixing bacteria receive a place to live (swellings on the plant roots) and sugars produced by the plant.

Mutualism also has played an important role in the coevolution of humans and the microbes that live in our intestines. Within the human large intestine live immense colonies of the bacterium *Escherichia coli*. These bacteria have ready access to food while providing us with vitamin K, which is necessary for blood to clot. Animals that lack these bacteria, such as birds, must consume food that contains the necessary amounts of vitamin K.

Figure 13.7
Barnacles on this gray whale have found a safe way to obtain food by riding the whale's back. The whale does not seem to be bothered by the presence of barnacles.

Section Review

❶ **Describe one example of coevolution.**

❷ **Do you think that the drab flowers of some grasses and trees are pollinated by insects? Explain.**

❸ **Describe some changes that have occurred in caterpillars and their food plants as a result of coevolution.**

❹ **What kind of ecological relationship do your cells have with their mitochondria? Explain.**

Competition and Ecosystem Development

Competition plays an important role in how ecosystems develop. The role of competition in the development of ecosystems is most easily seen when a serious disruption creates new habitat. New habitat is formed, for example, when a volcano forms a new island, or a glacier recedes and exposes bare soil, or a fire burns all the vegetation in an area. In every case scientists have been able to study, the empty habitat is quickly occupied. **Figure 13.10** shows some of the events that followed the eruption of Mount St. Helens in Washington. The first organisms to move into new habitat are small, fast-growing plants—you would probably call them "weeds." These early "settlers" are specialized for life under harsh conditions, such as on bare rock, and are able to eke out a living where few others could.

Figure 13.10

a A coniferous forest covered the slopes of Mount St. Helens before its eruption on May 18, 1980.

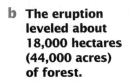

b The eruption leveled about 18,000 hectares (44,000 acres) of forest.

c Now, more than 10 years later, a new ecosystem has become established in the devastated areas.

Competition drives change in a developing ecosystem

The initial weedy colonists do not remain in the ecosystem very long, however, because their pioneering efforts soon make the ground more hospitable. As a result of this improvement, later plant arrivals soon out-compete and replace the original inhabitants. This second wave of immigrants is replaced in turn by still other species better able to compete in the new environment. As the ecosystem matures, niches become more and more finely subdivided, and species become more and more interdependent. Hence, the diversity of the ecosystem usually increases as succession proceeds.

The regular progression of species replacement in a developing ecosystem is called **succession**. When succession takes place on land where nothing has ever grown before, it is called **primary succession**. When it occurs in areas where there has been previous growth, as in abandoned fields or forest clearings, it is called **secondary succession**. Succession does not continue indefinitely. Eventually, if the ecosystem is undisturbed for a long time, a community that is resistant to change results. Although no two episodes of succession are exactly alike, the progression of species replacement tends to result in similar communities in similar physical conditions. That is why biomes such as tropical rain forests are so similar wherever they occur.

Ecosystem Stability

Why does succession stop? The last stage in a series of successional stages is able to absorb disruption without major change better than could its predecessors. Early successional stages, for instance, are easily changed by the invasion of new competing species. In contrast, the final stage of succession is able to resist invasion by potential competitors.

The ability of an ecosystem to resist change in the face of disturbance is known as **stability**. During succession, early stages show low stability, later stages are more stable, and the final community is the most stable.

What factors promote stability?

What makes some ecosystems more stable than others? Most ecologists now agree that more diverse ecosystems will be more stable than less diverse ecosystems. A more diverse ecosystem contains a more complex web of interactions among species than does a less diverse ecosystem. Alternate links in the web of species interactions are more likely to be available to compensate for the disruptions such as the loss of a species.

Even ecosystems that are very diverse contain points of vulnerability, however. A species whose niche affects many others in the ecosystem and that cannot be readily replaced if lost is called a **keystone species**. Because keystone species are the focus of many biological interactions, these species represent points where the web of species interactions can come unraveled.

In the 1960s, ecologist Robert Paine of the University of Washington discovered an excellent example of a keystone species. Paine worked on a 15-species ecosystem along the Washington coast. As shown in **Figure 13.11**, when he removed all the sea stars from this ecosystem, one species of mussel that the sea star ate began to thrive and out-competed many other species in the ecosystem. The number of species in this ecosystem fell from 15 to 8. The sea star in Paine's ecosystem was a keystone species.

To preserve natural ecosystems, it is essential to promote their biological diversity. It is very important to realize that diverse ecosystems can be damaged if key species are lost.

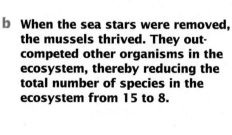

Figure 13.11

a The sea star in the above photograph is prying open a mussel. It is the primary predator in a 15-species ecosystem that was studied by Robert Paine. His experiments showed the importance of a keystone species.

b When the sea stars were removed, the mussels thrived. They out-competed other organisms in the ecosystem, thereby reducing the total number of species in the ecosystem from 15 to 8.

History of an Ecosystem

As species invade a new habitat, such as when a glacier recedes, they alter the habitat, making it possible for other species to enter and replace them. This process is called succession.

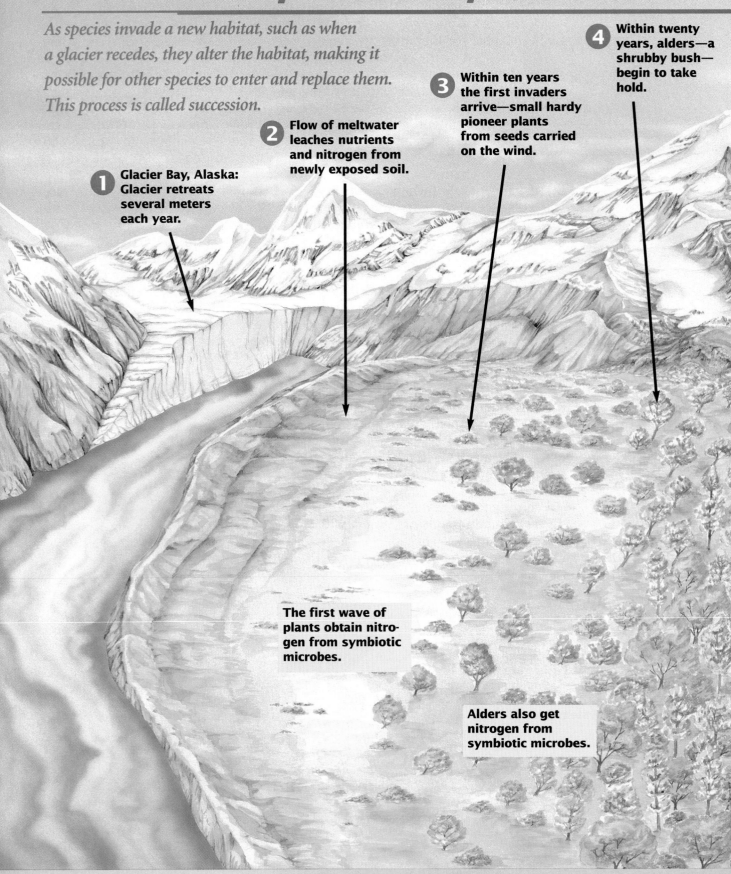

2 Flow of meltwater leaches nutrients and nitrogen from newly exposed soil.

1 Glacier Bay, Alaska: Glacier retreats several meters each year.

3 Within ten years the first invaders arrive—small hardy pioneer plants from seeds carried on the wind.

4 Within twenty years, alders—a shrubby bush—begin to take hold.

The first wave of plants obtain nitrogen from symbiotic microbes.

Alders also get nitrogen from symbiotic microbes.

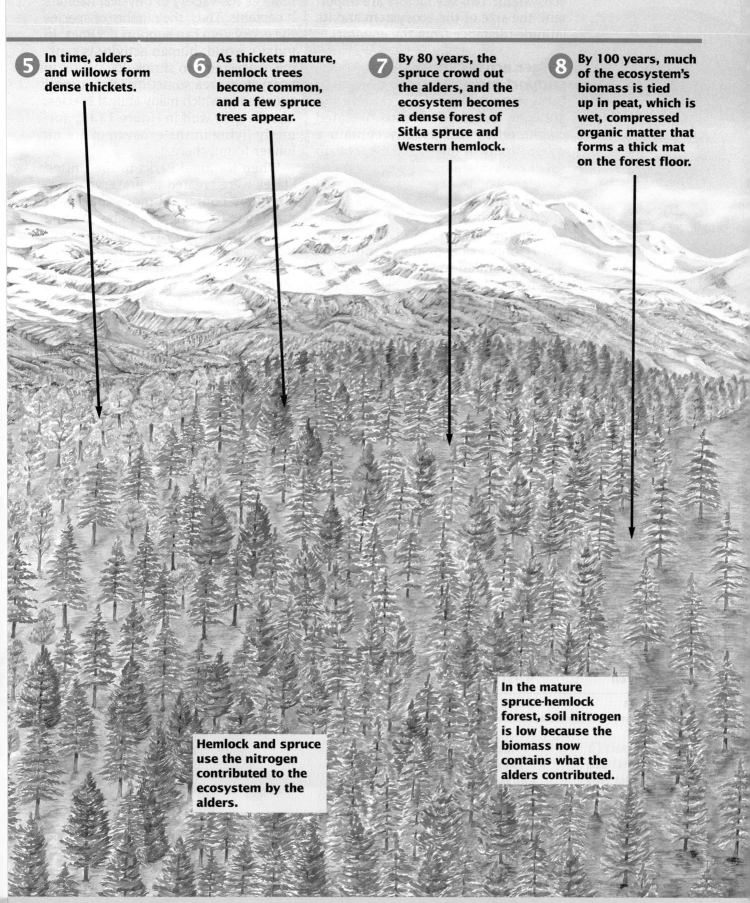

5 In time, alders and willows form dense thickets.

6 As thickets mature, hemlock trees become common, and a few spruce trees appear.

7 By 80 years, the spruce crowd out the alders, and the ecosystem becomes a dense forest of Sitka spruce and Western hemlock.

8 By 100 years, much of the ecosystem's biomass is tied up in peat, which is wet, compressed organic matter that forms a thick mat on the forest floor.

Hemlock and spruce use the nitrogen contributed to the ecosystem by the alders.

In the mature spruce-hemlock forest, soil nitrogen is low because the biomass now contains what the alders contributed.

How Ecosystems Change **277**

Carmen R. Cid: Ecologist

How I Became Interested in Ecology

"Ever since I was a little girl, my father, a well-educated man, fostered my interest in biology. When I was 12, my family moved from Havana, Cuba, to Brooklyn, New York. In high school, my curiosity about biology deepened thanks to my biology teacher. She was a great role model who shared her love of biology while demanding respect from a tough audience of inner-city teenagers.

"I entered college with the intention of becoming a biology teacher. However, in my junior year, I took a plant ecology course and fell in love with the outdoors. The professor, an enthusiastic naturalist, would take us out of New York City into the nearby woods and wetlands. I had always lived in big cities and this was my first opportunity to visit the lovely, unspoiled environment of a forest. Until then, I had thought of trees and other plants only as decorations along sidewalks or in parks. I never thought of plants as being very much alive. When you grow up with no connection to plants, and then someone gives names to things you thought were lifeless, it's like making a whole bunch of new friends. My professor taught me a new way to look at nature, and I soon saw plants as living, growing organisms, interacting with each other."

Carmen loves to spend her spare time traveling to places with waterfalls and lush, green mountains.

Name:	**Carmen R. Cid**
Home:	**Willimantic, Connecticut**
Employer:	**Eastern Connecticut State University**
Personal Traits:	• **Curiosity**
	• **Ambition**
	• **Patience**
	• **Energy**
	• **Attention to detail**

Research Focus

Dr. Cid is currently examining the effect of ground leaf litter (the layer of dead leaves) that accumulates in the fall on seedling growth and competition in the spring. She is also investigating the factors that affect diversity and regeneration of plant species in wetlands. The results of her research can be used to restore and manage previously disturbed forest and wetland communities.

Career Path:

High School:
• Biology
College:
• Biology
• Ecology
Graduate School:
• Botany
• Ecology

The Excitement of Ecology

"Ecology is a challenging science that seeks to discover how a community of living organisms, such as a forest, is organized.

"This calls for exciting detective work. As an ecologist, I design experiments that enable me to piece together the puzzle of forest species interactions. One of my projects researched short-lived annual plants living in forests where mostly long-lived species such as trees are found. Why weren't these annuals found in surrounding fields where more light and nutrients were available? Why did these annual plants grow in patches throughout the forest? Once I answered these and many other questions, I had the information to solve the puzzle of species interactions in the forest. Ultimately, the data I collect from all my experiments can show me how to manage, restore, and maintain species diversity in forests and wetlands. My work gives me the satisfaction of knowing that I am helping to preserve a place of great natural beauty for future generations.

"As an associate professor of biology, I enjoy introducing my students to ecology by taking them to meadows, forests, and wetlands, just as my college professor did. My goal is to make science accessible and interesting to all students. I am also striving to increase participation of women and minorities in all areas of biology."

HUMAN ACTIVITIES ARE DISRUPTING THE DELICATE BALANCE OF THE WORLD'S ECOSYSTEMS. THE FATE OF EVERY ECOSYSTEM ON EARTH DURING THE NEXT CENTURY MAY BE INFLUENCED MORE BY ITS ABILITY TO SURVIVE HUMAN DISRUPTION THAN BY ANY OTHER FACTOR. WE MUST UNDERSTAND HOW WE ARE DISRUPTING ECOSYSTEMS IN ORDER TO REDUCE THE MOST DAMAGING ASPECTS OF THAT INTERFERENCE.

13.3 How Humans Disrupt Ecosystems

Objectives

❶ Explain how changes in natural habitats can have a drastic impact on ecosystems.

❷ Recognize the devastating effects that exotic species have on native organisms.

❸ List three ways to reduce human impact on ecosystems.

Human Impact on Ecosystems

Climate, coevolution, competition—these driving forces of evolutionary change have molded ecosystems over long periods of time to create the world seen today. The most important single influence on natural ecosystems today, however, is human activity. To get some sense of how ecosystems are responding to the massive impact of our species, it is important to realize that disruption and disturbance are a natural part of the life of any ecosystem. For example, fires ignited by lightning normally burn forest ecosystems, and usually the forest quickly returns. Today, however, humans are disturbing ecosystems on a greater scale and to a greater degree than ever before. By making large-scale changes, like the development of ocean-front property shown in **Figure 13.13**, humans tear apart the webs of species interactions that stabilize ecosystems and enable them to bounce back from disturbances.

Disrupting physical habitat

Ever since humans learned to grow crops about 10,000 years ago, natural habitats have been altered and replaced with habitats of human construction. Forests are felled, swamps are drained and filled, and rivers are diverted, all to make room for buildings, parking lots, roads, and farms. Now humans are changing not only local habitats but the whole globe. Burning of fossil fuels may be changing the world's climate, and industrial chemicals are destroying the Earth's ozone shield. These changes to natural habitats reduce their ability to support living things.

Figure 13.13
Construction of this housing development in the Florida Keys greatly altered the habitat of the island.

Stability

How can the introduction of an exotic species reduce the stability of an ecosystem?

Decreasing species diversity

Altering habitats in drastic ways often exterminates the native organisms that have evolved to fit natural habitats, not the artificial habitats humans create. For instance, conversion of forest to farmland or pasture reduces the number of species from hundreds or thousands to only a few. Similarly, massive logging of virgin forests followed by the planting of a single species of tree as a future lumber "crop" diminishes diversity of the forest ecosystem.

Destroying interactions among species

Intentionally or unintentionally, the kinds of species interactions that promote diversity often are eliminated. Removal of predators, whether wolves or insects, often reduces diversity, as you saw in the earlier keystone species example. The intentional or accidental introduction of exotic species from other parts of the world, such as the kudzu plant shown in **Figure 13.14**, can disrupt competitive balances that have coevolved among native species. Freed from the controls imposed by their natural predators and diseases, exotic species often easily out-compete native species, displacing many or all of them. The result is a simplified ecosystem in which far less competition occurs. By removing or introducing organisms without regard to their effects on native species, humans often wreak havoc on the complex web of biological relationships within ecosystems.

Figure 13.14
The Japanese vine kudzu was originally introduced to North America for its decorative qualities. With no natural predators or diseases, its growth was unchecked. It is now a devastating weed in the southeastern United States.

Modifying the Environment

Figure 13.15

a **These cypress trees are located in the Okefenokee swamp in Georgia. The Okefenokee is a protected wetland.**

It is unrealistic to expect that human-kind will be able to avoid disrupting the environment in the future. The growth of the human population is simply too great, and the appetite of industrial society is too voracious. But the environment must be modified in such a way that the fabric of the Earth's ecosystems is disturbed as little as possible. **Figure 13.15** shows some examples of what should be done:

1. *Disrupt the physical habitat as little as possible* A great deal of the damage done today is unnecessary. Tidal wetlands could be protected from conversion to home sites and garbage dumps, without impairing our standard of living.

2. *Avoid decreasing species diversity* Replacement of forests need not be with stands of only one species of tree, as often occurs when logged forests are replanted. Single-species stands are created simply because they make logging in the area easier.

3. *Avoid disrupting species interactions* Humans should not intrude into natural ecosystems unless compelled to do so to preserve them or by economic necessity. Introductions of exotic species should be avoided.

c **Curious nature enthusiasts can disrupt the serenity of bird nesting sites.**

b **This black rhino in the Nairobi National Park in Kenya is protected by law. Because the horns of the rhino are so valuable, poachers have killed off large numbers of these animals.**

Section Review

❶ **How can a change in a natural habitat reduce the number of species occurring there?**

❷ **How are exotic species such as kudzu able to spread rapidly when introduced to a new environment?**

❸ **List three steps that can be taken to minimize human impact on the environment.**

Naturally occurring forest fires are part of the process of succession.

	Key Terms	**Summary**

13.1 Interactions Within Ecosystems

This boxing crab has a commensal relationship with the anemones she is wearing.

Key Terms

coevolution (p. 267)

symbiosis (p. 270)

parasitism (p. 270)

mutualism (p. 270)

commensalism (p. 271)

Summary

- Species evolve in response to other living members of their ecosystems. This process, called coevolution, shapes the species interactions in an ecosystem.

- Insects and flowers coevolve as do plants and the herbivores that attempt to eat them.

- A symbiosis is a close, long-term relationship between species.

- In parasitism, one species (the parasite) lives on or in another species (the host). Mutualism is a symbiotic relationship in which all parties benefit. Commensalism is a relationship in which one species benefits and the other is neither helped nor harmed.

13.2 Ecosystem Development and Change

The sea star in the 15-species ecosystem studied by Robert Paine is an example of a key-stone predator.

Key Terms

niche (p. 272)

fundamental niche (p. 272)

competition (p. 273)

realized niche (p.273)

competitive exclusion (p. 273)

succession (p. 274)

primary succession (p. 274)

secondary succession (p. 274)

stability (p. 275)

keystone species (p. 275)

Summary

- An organism's niche is the sum of all its interactions in its environment, including interactions with other organisms.

- Competition occurs when organisms attempt to use the same resource.

- Succession, the regular progression of species replacements in a developing ecosystem, is driven by competition.

- Stability is the ability of an ecosystem to resist change. More diverse ecosystems are usually more stable than less diverse ecosystems.

- An ecosystem's diversity is partly determined by its latitude and its size.

13.3 How Humans Disrupt Ecosystems

The introduction of exotic species, such as kudzu, can damage ecosystems.

- Human activities disrupt ecosystems in three main ways: by altering natural habitats, by reducing species diversity, and by destroying interactions among species.

Deforestation: Are We Losing Only the Trees?

This logging truck is removing trees that were once part of the tropical rain forest.

How many products made of paper or wood do you use in a day? We are highly dependent on trees as a natural resource, yet the amount of forested land is dwindling worldwide.

After trees of the tropical rain forest are cut down, the land often erodes.

Worldwide, the rate of deforestation—or forest loss—has reached record levels. Each year, about 17 million hectares (40 million acres) are deforested—usually by cutting or burning. Deforestation is most severe in the tropical regions of Africa, Asia, Central America, and South America. More than 50 percent of the rain forests have already been cut for firewood and timber or burned to open up land for agriculture or livestock. About 2 percent of the remaining area of rain forest is destroyed each year. Two percent may sound like a small amount, but it means that an area the size of Florida is cleared each year. At this rate, no rain forest will remain by the middle of the next century.

The countries with large stands of tropical rain forest, such as Brazil, Indonesia, and Zaire, are poor and have rapidly increasing populations. Timber from the rain forest provides these countries with essential income. Clearing the forest makes land available for the expanding populations.

Deforestation has disastrous consequences on both a local and global scale. Cropland or pasture created by clearing rain forest is productive for only a short time—sometimes as little as two years. Although it supports luxuriant growth, tropical soil is usually poor in nutrients. Most of the nutrients are held within the tissues of plants. Burning or cutting the vegetation breaks the nutrient cycles that sustain the forest. Unless supplemented by artificial fertilizers, the soil is quickly exhausted by agriculture or ranching. Without the protection that vegetation gives, topsoil on cleared patches is easily carried away by wind and water. Trees also affect the local climate. Water that evaporates from their leaves falls back on the forest as rain. Deforested areas often rapidly turn into virtual deserts.

Tropical rain forest also affects the levels of carbon dioxide in the atmosphere, thereby influencing global climate. In the tropics, the growing season is year-round. Photosynthesis by the lush vegetation withdraws large amounts of carbon dioxide from the atmosphere.

Reduced forest area means lessened absorption of carbon dioxide. Even worse, burning of rain forests releases carbon dioxide; about 20 percent of the carbon dioxide added to the atmosphere each year comes from burning rain forest. Thus, deforestation is a major contributor to the rising carbon dioxide levels that are projected to escalate global warming.

Another serious consequence of deforestation is extinction of species. Tropical rain forests are the most diverse terrestrial ecosystems, home to more than one-half of the world's species. The diversity in the rain forests is so great that most of the species are unknown to science. As the rain forests are destroyed, species are becoming extinct much faster

than they can be identified and studied. Many of these species may be beneficial. For instance, in 1978 a species of grass closely related to corn was discovered in Mexico. The newly discovered species is much more disease-resistant than corn and is a perennial (it lives longer than one growing season, unlike corn). If genes from this grass could be transferred to corn, a strain of disease-resistant, perennial corn could be created to increase food production. But this potentially important new species was discovered in the nick of time—just one week before its habitat was scheduled to be burned and cleared.

Valuable species are threatened by forest clearing throughout the world. One of these is the Pacific yew tree, which grows only in old growth forests of the Pacific Northwest. Old-growth forests are ancient ecosystems: some of their trees may be more than 1,000 years old. More than 90 percent of the old-growth forest in the lower 48 states has been logged and replanted. Yew trees were not replanted because they were considered useless by the lumber industry. The bark of the "useless" Pacific yew yields taxol, an effective treatment for a variety of cancers. No one knows how many other medicines, foods, and other products are disappearing with the world's forests.

Pacific yew tree

Thinking Critically

❶ Think of three ways that the rain forest might be more valuable when left standing than when cut.

❷ List two ways that wood could still be harvested without destroying the forest.

❸ Give three reasons why scientists (not environmentalists) would want to preserve the rain forest.

Acting on the Issue

❶ Look for rain forest products in your supermarket. How does buying these products benefit the rain forest?

❷ Write a letter to a logging company in the Northwest to find out how the company manages its forest land.

❸ Research 25 species endangered by the destruction of the rain forest. What efforts are being made to save these species?

WHEN EUROPEAN SETTLERS ARRIVED IN NORTH AMERICA, 60 MILLION
TO 125 MILLION BISON LIVED HERE. BY 1889, ONLY 85 BISON REMAINED.
CONSERVATION EFFORTS HAVE ENABLED THE BISON POPULATION TO
REBOUND TO ABOUT 65,000. SAVING THE BISON FROM EXTINCTION IS
ONE EXAMPLE OF AN ENVIRONMENTAL PROBLEM IN WHICH ACTION LED
TO A SOLUTION.

14.3 Solving Environmental Problems

Objectives

1 List some examples of successful solutions to environmental problems.

2 List the five basic elements necessary to solve any environmental problem.

3 Recognize your role in solving environmental problems.

Environmental Problems Can Be Solved

The most important fact to remember about the environmental crisis is that each of its many problems is solvable. A polluted lake can be cleaned up, a dirty smokestack altered to remove toxic gases, and the waste of key resources stopped. Each requires a clear understanding of the problem and a commitment to doing something about it. The extent to which American families recycle aluminum cans and newspapers shows that people want to become part of the solution, rather than remaining part of the problem. Progress is being made. Newspapers carry success stories daily. In the United States since 1970 eight national parks have been established. As a result, 80 million acres of land have been protected as wilderness and 34 million

Figure 14.12
Peregrine falcons are found throughout the continental United States. They can often be seen roosting on buildings or bridges in small towns as well as large cities.

acres of farmland particularly vulnerable to soil erosion have been withdrawn from production. Many previously endangered species are better off than they were in 1970, including the pronghorn antelope, the wild turkey, the bald eagle, and the peregrine falcon, which is shown in **Figure 14.12**.

Pollution control efforts have been particularly successful. Emissions of sulfur dioxide, carbon monoxide, and soot, which were 200 million tons in 1970, have been reduced by more than 30 percent. The release of toxic chemicals into the environment (notably the insecticide DDT, and the carcinogens asbestos and dioxin) has been banned outright. The Environmental Protection Agency estimates that private firms and public agencies are spending about $100 billion per year on pollution control, double the figure of 10 years ago and five times the figure of 1970. In the same period the population of the United States increased by about 40 million people and the number of cars grew by more than 60 million vehicles. Had this progress not been made, environmental quality in the United States would almost certainly have declined dramatically.

Steps Toward Saving the Environment

There are five steps to solving an environmental problem: assessment, risk analysis, public education, political action, and follow-through. **Table 14.1** outlines these steps using as an example a lake damaged by chemicals.

Table 14.1 Steps Toward Solving an Environmental Problem

Steps	Plan of action	What you can do
Assessment	Data must be collected and experiments performed in order for scientists to construct an ecological model of an ecosystem. The model predicts how the environment will respond to changes.	You can volunteer to test the water of a nearby lake to determine the degree of damage done by chemical pollution. You also can record changes in the numbers of animals and plants found in and around the lake.
Risk Analysis	Using the ecological model, scientists can predict the effects of environmental intervention. Scientists will evaluate the potential for solving the environmental problem as well as the potential for any adverse effects of the proposed solution.	A local college or university can test the water samples that you collect. Scientists there can suggest ways to decrease the amounts of chemicals entering the lake.
Public Education	When a clear choice can be made, the public must be informed. This involves explaining the problem in terms people can understand. Costs and expected results of each alternative should be presented and explained.	The people of your town or city will have to decide whether they are willing to pay for the ditches necessary to divert agricultural runoff from the lake. You could go to a city council or town meeting and emphasize the importance of saving the lake for recreation.
Political Action	The public, through its elected officials, selects a course of action and implements the plan. Individuals can have a major impact by exercising their right to vote. Many voters do not understand the magnitude of what they can achieve by writing letters and supporting special interest groups.	It is important to start learning about environmental issues now, so that you can make good decisions when you are old enough to vote. Voters in your town or city will need to ask themselves if the cost of preventing water pollution is worth the potential benefits.
Follow-through	The results of any action taken should be monitored carefully to see if the environmental problem is being solved and to evaluate and improve the initial assessment and modeling of the problem. We learn by doing.	If the town does agree to channel runoff water away from the lake, you can volunteer to continue collecting water samples. From these samples, scientists can determine how well the lake is recovering.

Chapter 14 · Investigation

How Does Pollution Affect Organisms?

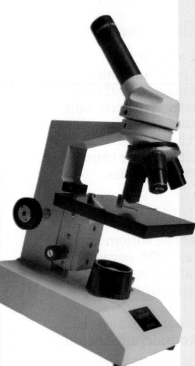

Objectives

In this investigation you will:

- *observe* the effects of two forms of pollution on microorganisms
- *relate* environmental conditions to an organism's ability to survive

Materials

- cold pack or ice in self-sealing plastic bag
- hot pack or hot water in self-sealing plastic bag
- thermometer
- distilled water
- *Daphnia* culture
- 50-mL beaker
- 5-mL pipette
- modeling clay
- paper towels
- *Amoeba* culture
- medicine droppers
- depression slide
- coverslip
- compound light microscope
- 10-percent table salt (NaCl) solution
- stereomicroscope

Prelab Preparation

1. Review what you have learned about pollution by answering the following questions:
   - What is pollution?
   - What are some examples of pollutants found in the environment?
2. Review the procedures for proper use of the microscope in the Appendix.
3. Review the procedures for making a wet-mount slide in the Appendix.

Procedure: Investigating the Effects of Pollution

1. Form a cooperative group with another student to complete steps 2–10.
2. Obtain a hot pack or fill a plastic bag with hot tap water having a temperature of at least 60° C. Obtain a cold pack or fill a plastic bag with ice cubes. Record the actual temperature of each pack or bag.
3. Mix approximately 3 mL of distilled water and approximately 3 mL of *Daphnia* culture in a 50-mL beaker. Swirl the mixture gently. Fill the pipette with the mixture from the 50-mL beaker. Use modeling clay to close the narrow tip of the pipette. Use modeling clay to close the top of the pipette.
4. Place a paper towel on the surface of your work area. Lay the pipette on the paper towel. Cover one end of the pipette with a hot pack or bag of hot water. Cover the other end of the pipette with a cold pack or bag of ice. Note the time and set the apparatus aside.

5. Make a wet-mount slide with a drop of *Amoeba* culture. Observe the slide under low power. Make a

312   Chapter 14

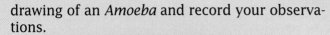

drawing of an *Amoeba* and record your observations.

6. Use a clean medicine dropper to place a drop of 10-percent NaCl solution at the edge of the coverslip. Draw the solution under the coverslip by placing a small piece of paper towel against the opposite edge of the coverslip.

7. Examine the slide under high power. *How does an* Amoeba *react to the salt solution?* Record your observations.

8. Pick up the pipette after 15 minutes. Feel the pipette's surface at both ends and in the middle. Record your observations.

9. Use a stereomicroscope to examine the culture in each part of the pipette. *Where have most of the* Daphnia *gathered?* Record your observations.

10. Dispose of the *Daphnia* mixture as directed by your teacher. Clean up your materials and wash your hands before leaving the lab.

Analysis

1. *Comparing Observations* Compare the response of *Daphnia* to heat with the response of *Amoeba* to salt.

2. *Making Inferences* Heat and salt are necessary for life. Explain why they could be considered pollutants.

3. *Making Inferences* How might excessive heat affect an organism that cannot move?

4. *Making Predictions* How might heat pollution affect the fish populations of a river?

5. *Identifying Relationships* What would happen to an *Amoeba* if it could not move away from the salt solution?

6. *Evaluating Methods* Suppose you neglected to completely fill the pipette before corking it with modeling clay. How would this error affect your results?

Thinking Critically

1. Why do you think it is advantageous for an organism to avoid temperature extremes?

2. Near volcanic vents on the ocean floor, bacteria live in water with a temperature of 250°C (480°F). What adaptations would you expect these bacteria to have that enable them to live at such high temperatures?

AGRICULTURE AND TECHNOLOGY: FEEDING

8000 B.C.

8000 B.C. The **Natufians of Palestine** are considered the first farmers.

6000 B.C. Domesticated cattle, cultivated crops, massive pit silos, and granaries are found in parts of **ancient Greece**. Nitrogen-fixation and crop rotation maintain soil fertility. Irrigation is used in drier areas.

Egyptian farmers

1750 B.C. Tradition gives Chinese **Emperor Shen Nung** the title "inventor of agriculture." It is said that he carved a piece of wood into a plowshare, bent another piece to make a handle, and taught the world the advantages of plowing and weeding.

Emperor Shen Nung

1300 B.C. Ancient Egyptians prosper from agricultural development and an integrated social system. Wider uses of plants advance the field of medicine. Farming techniques like plowing, raking, and manuring are practiced in Egypt. Animal breeding for special purposes is used.

1896

1896 George Washington Carver, an African-American agricultural chemist, revolutionizes Southern agriculture. He discovers 325 different products that can be made from peanuts, 118 products from the sweet potato, 75 from the pecan, and hundreds of products from cotton and corn stalks.

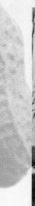

George Washington Carver

1962 Naturalist author **Rachel Carson** warns of the severe environmental impact of DDT and the danger it poses to human food sources. DDT sprayed on crops destroys insect pests and endangers other animal life.

1975 Richard E. French, a Yakima Native American and forester, protects cultural foods in forest areas. He monitors fire management, fire and timber trespass activities, disease control, and insect control for the Bureau of Indian Affairs in Portland, Oregon.

Rachel Carson

THE WORLD'S POPULATION

Pliny the Elder

A.D. 43 Roman historian **Pliny the Elder** writes about revolutionary changes in Roman agriculture, including the introduction of complex plows fitted with wheeled forecarriages pulled by oxen or donkeys.

1450 Native Americans develop advanced systems of agriculture. Native American farmers, in various parts of the Americas, are producing crops such as corn, peanuts, peppers, cocoa beans, squash, rubber trees, tobacco, and tomatoes. Europeans first learn of these crops from Native Americans.

1793

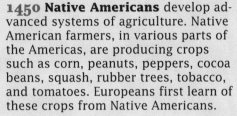

Native American farmers

1793 Catherine Green develops plans for the cotton gin. Fearful of public scorn, she directs her employee **Eli Whitney** to build the machine.

Cotton gin

1991

Deforestation of rain forest by gold prospectors

1980's Ecological consequences of **tropical deforestation** begin to receive widespread notice—land is stripped of nutrients; extensive erosion occurs; watersheds are destroyed.

1989 Discoveries of **Dr. J.E. Henry** lead to the commercialization of protozoans as bacterial insecticides. Dr. Henry is a Chippewa Native American and a professor at Montana State University.

1991 Ben Villalón is a Hispanic-American virologist and geneticist at Texas A&M University's Agricultural Experiment Station in Weslaco, Texas. He breeds disease- and insect-resistant varieties of chilies. He and his group have developed 15 new varieties, including the world's first mild jalapeño.

1990 University of Georgia agronomist **Dr. David E. Radcliffe**, physically challenged, studies the physical behavior of soil when subjected to a variety of disturbances. He provides valuable agricultural information, determining which types of soil are more subject to runoff, or erosion, and their tendency towards crusting or compaction.

David Radcliffe

Chili peppers

Diversity
of Life

T he greatest hallmark of life
on Earth is its incredible
diversity. Living things occupy
every available nook and cranny,
from the boiling waters of hot
springs to the tangled growth of
rain forests. Even the perpetual
darkness at the bottom of the
oceans, too deep for light to
penetrate, is teeming with life,
bizarre forms unlike any that live
near the sea's surface. In this unit
you will discover some of the
richness of life's tapestry,
exploring the many living things
that are NOT animals like you.

Chapter 15

Classifying Living Things

Review

- characteristics of living things (Section 2.1)

- natural selection (Section 9.1)

- five kingdoms (Section 10.2)

Earth is teeming with life. To make order out of the chaos, scientists use a five-kingdom classification system.

DO THE WORDS "CANE," "CHIEN," AND "HUND" MEAN ANYTHING TO YOU? THESE ARE ITALIAN, FRENCH, AND GERMAN WORDS FOR "DOG." COULD YOU CONDUCT A CONVERSATION ABOUT DOGS WITH SOMEONE WHO SPOKE ONLY ONE OF THESE LANGUAGES? SCIENTISTS—REGARDLESS OF THEIR NATIVE LANGUAGES—CAN COMMUNICATE ABOUT DOGS BECAUSE THEY HAVE THE SAME NAME FOR THE DOG: *CANIS FAMILIARIS*.

15.1 The Need for Naming

Objectives

❶ Explain why scientists use scientific names instead of common names.

❷ Describe the scientific system of naming organisms.

❸ Explain why scientific names are in Latin.

❹ Recognize the role of Linnaeus in creating the modern system of naming organisms.

The Importance of Scientific Names

What do you and your classmates call the "bug" shown in **Figure 15.1**? You will probably have many common names for this animal, including sow bug, pill bug, wood louse, roly-poly, and potato bug. If some of your classmates are from other countries, you could collect an even longer list of names.

Yet if you asked a biologist to name this creature, you might receive only one answer: *Porcellio scaber*. Each kind of organism on Earth is assigned a unique two-word **scientific name**.

Porcellio scaber is the scientific name of the animal shown below. *Homo sapiens*, which means "wise man," is our species' scientific name. All biologists, regardless of their native languages, use scientific names when speaking or writing about organisms.

Most organisms also have common names. Why don't scientists use common names? Although adequate for everyday use, common names are too ambiguous for scientific communication. For one thing, as you have seen, an organism can have more than one common name. In addition, science is an international endeavor, and an organism rarely has the same name in different languages. Finally, one common name often refers to more than one kind of organism. The plant we know as corn in North America is called maize in Great Britain. To a resident of Britain, corn is the plant we call wheat. When a biologist writes a scientific paper on *Zea mays*, however, other scientists know the subject of the paper is the American "corn" plant. The use of scientific names enables all scientists to exchange information about an organism and to be certain that they are referring to the same organism.

Figure 15.1
Below are some of the common names for this animal (240X). Scientists have assigned it a single scientific name: *Porcellio scaber*.

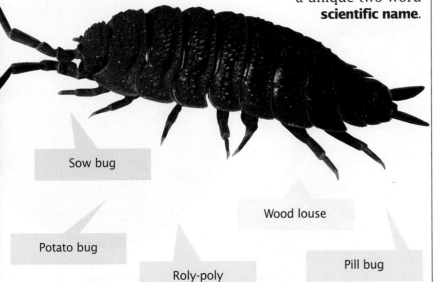

Sow bug

Wood louse

Potato bug

Roly-poly

Pill bug

What's in a Scientific Name?

The first word in a scientific name describes the organism in a general way. The second word identifies the exact kind of living thing. We use a similar naming system in our everyday speech. For instance, your last name identifies your family, while your first name specifies exactly who you are.

The first word of a scientific name is the name of the **genus** (*JEE nuhs*) to which the organism belongs. (The plural of genus is "genera.") A genus is a group of organisms that share major characteristics. For example, all oak trees produce acorns. Therefore, all oak trees are assigned to the genus *Quercus*, which means "oak" in Latin.

There are dozens of different kinds of oak trees. One kind is tall, while another is low and spreading. Leaves and acorns of different oak trees can vary in size and shape, as shown in **Table 15.1**. The second word in a scientific name identifies one particular kind of organism within the genus. For example, *Quercus rubra* is the red oak. *Quercus phellos* is the willow oak. Scientists call each different kind of organism a **species** (*SPEE sheez*). (The plural of species is "species.") The correct name for an organism must include *both* parts of its scientific name. The red oak is properly called *Quercus rubra*, not just *rubra*.

Table 15.1 Comparison of Red Oak and Willow Oak

	Red oak	Willow oak
Genus name	*Quercus*	*Quercus*
Scientific name	*Quercus rubra*	*Quercus phellos*
Traits	Acorns about 25 mm (1 in.) long	Acorns about 15 mm (0.5 in.) long
	Common in open Northeastern forests; tolerant of city soot and cold temperatures	Popular shade tree found in the South; grows well in rich, moist soil
	Lobed leaves	Unlobed, narrow leaves

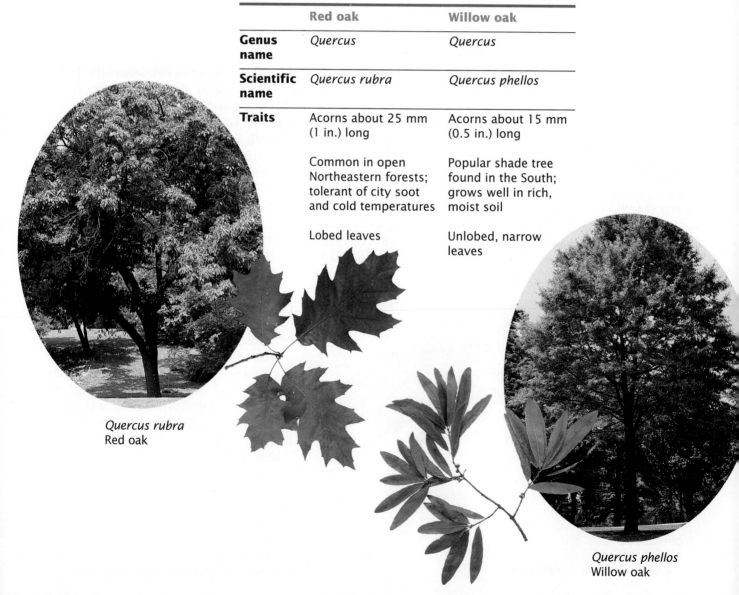

Quercus rubra
Red oak

Quercus phellos
Willow oak

Figure 15.2

a The second word of a scientific name can come from various sources. The green anole lizard *Anolis carolinensis* and the chickadee *Parus carolinensis* are both found in North Carolina and South Carolina.

b *Tyrannosaurus rex*, which means "tyrant-lizard king," was named for its enormous teeth and tremendous size. This dinosaur measured about 15 m (49 ft.) in length.

Scientific names must conform to a set of rules

The name given to a newly discovered species must conform to rigorous rules formulated by an international commission of scientists. All scientific names must be made of Latin words or of terms constructed according to the rules of Latin grammar. Two different organisms cannot be assigned the same scientific name. Since all members of a genus will share their genus name, the second word in the name of each member of that genus must be different. Only one member of the genus *Homo* can be given the name *sapiens*. Organisms in different genera cannot have the same genus name but can share the second word of their scientific names.

For example, the green anole lizard *Anolis carolinensis* and the chickadee *Parus carolinensis*, shown in **Figure 15.2a**, share the name *carolinensis* because they both occur in North Carolina and South Carolina.

When choosing a name for a species, biologists often pick a name that describes the appearance or lifestyle of an organism. One of the most vivid names belongs to the fierce-looking dinosaur *Tyrannosaurus rex*, shown in **Figure 15.2b**. *Tyrannosaurus rex* means "tyrant-lizard king," a fitting name for a carnivore with teeth 15 cm (6 in.) long. Sometimes scientific names are a tribute to the discoverer of a species or to an admired colleague or teacher of the discoverer. The frog *Rhinoderma darwinii* shown in **Figure 15.2c**, the lizard *Liolaemus darwinii*, and the bird *Rhea darwinii* are three of the many species named for Charles Darwin.

c The frog *Rhinoderma darwinii* was named to honor Charles Darwin.

Why are scientific names in Latin?

Why are scientific names written in a language that is no longer spoken? Why aren't they in English, or Russian, or Chinese? In the Middle Ages, when scientists began to name organisms, Latin was used in academic circles. Hence, scientists and other scholars found it easier to communicate with each other in Latin. They wrote books in Latin, wrote letters to each other in Latin, spoke Latin when they met, and named organisms in Latin.

Although scientists no longer communicate with each other in Latin, it is easier and more logical to retain the Latin names for living things than to rename all 1.5 million living organisms in a more modern language. **Figure 15.3** shows how useful a Latin name can be when scientists from around the world communicate with each other.

Figure 15.3
If you asked scientists from around the world what they might call the animal in Figure 15.1, they would all agree on one name: *Porcellio scaber*.

"טחבית (*tah hah VEET*) or *Porcellio scaber*"
—Yehoshua Anikster, Botanist
Israel

"Cloporte (*klo PORT*) or *Porcellio scaber*"
—Paul Melançon, Chemist and biochemist
Canada

"Mangy sowbug or *Porcellio scaber*"
—Maria Alma Solis, Entomologist
United States

"Cucaracha (*koo kah RAH chah*) or *Porcellio scaber*"
—Ernest H. Williams, Jr., Marine biologist

—Lucy Bunkley-Williams, Microbiologist
Puerto Rico

Linnaeus devised the two-name system

The modern system of naming organisms was the brainchild of Swedish botanist Carl von Linné (1707–1778). As was the fashion in his time, Von Linné gave himself a Latin name: Carolus Linnaeus. In Linnaeus' day, organisms were given very long Latin names (sometimes more than 15 words), which were often changed according to the whims of particular authors. Linnaeus assigned a standard, two-word Latin name to each organism known in his time.

Writing a scientific name is simple

When you write a scientific name, always capitalize the genus name. Begin the second word with a lowercase letter. Both parts of a scientific name are underlined or written in italics. The scientific name for humans can be written either Homo sapiens or *Homo sapiens*. After the first use of the full scientific name, the genus name can be abbreviated as a single letter if the meaning is clear. For example, since *Homo sapiens* has just been mentioned, *H. sapiens* is acceptable.

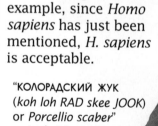

"КОЛОРАДСКИЙ ЖУК (*koh loh RAD skee JOOK*) or *Porcellio scaber*"
—Andrei Lapenis, Earth systems scientist
Russia

Section Review

❶ List the problems associated with the use of common names to identify living things.

❷ Why is it necessary to use both words of a scientific name to correctly identify an organism?

❸ Explain the advantages of using Latin for scientific names.

❹ Evaluate the accuracy of this statement: "Linnaeus was the first scientist to give species Latin names."

BECAUSE THE PRODUCTS IN A STORE ARE ORGANIZED, YOU CAN GO STRAIGHT TO THE AISLE WHERE SCHOOL AND OFFICE SUPPLIES ARE KEPT TO FIND A PEN, INSTEAD OF SEARCHING THE ENTIRE STORE. LIVING THINGS HAVE BEEN SIMILARLY ORGANIZED BY SCIENTISTS, BUT NOT ONLY FOR CONVENIENCE. LIVING THINGS ARE ORGANIZED INTO GROUPS TO REFLECT EVOLUTIONARY RELATIONSHIPS.

15.2 Classification: Organizing Life

Objectives

❶ Describe the system scientists use to classify organisms.

❷ Describe how the classification of living things reflects their evolutionary history.

❸ Summarize the methods of classification used by taxonomists.

❹ Define the term "species."

Classification of Living Things

Figure 15.4
This page from a manual called a herbal was published in Turkey in the tenth century. The Greek text classifies the plant illustrated below according to its medicinal uses.

Humans have been classifying organisms for thousands of years. The Greek philosopher Aristotle (384–322 B.C.) grouped animals according to their physical similarities. Aristotle's classification included some of the groups recognized today, such as the mammals. In the Middle Ages, herbalists used plants to treat disease and needed to know how to identify which plants were poisonous and which had healing powers. Herbalists produced manuals of plant types, known as "herbals," like the one shown in **Figure 15.4**. In herbals, plants were organized by their medicinal uses.

Today biologists classify organisms not by their usefulness but by their physical, chemical, and behavioral similarities. These similarities reveal evolutionary relationships. The science of classifying living things is called **taxonomy**. Taxonomists are scientists who practice taxonomy.

Like the system for naming organisms, the system of classification was derived by Linnaeus. Linnaeus wrote a huge encyclopedia of life, the *Systema Naturae* ("system of nature"). This work described all organisms then known. It classified living things into a hierarchy in which individuals are assigned to groups, groups are collected into larger groups, and these larger groups are part of still larger groups. A similar system is used in the U.S. Army. Each soldier belongs to a squad containing about nine soldiers. Four squads are organized into a platoon. Each platoon belongs to a company of about 150 people, and so on.

Chapter 16

Bacteria and Viruses

Review

- **prokaryotes (Section 3.3)**
- **cell division (Section 4.3)**
- **five kingdoms (Section 10.2)**

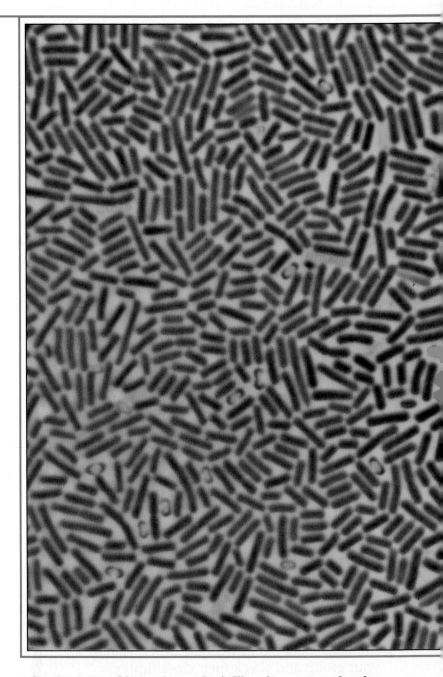

A large group of bacteria can look like abstract art, but it is really life at its least complex level.

ALTHOUGH YOU DO NOT SEE THEM, BACTERIA AFFECT YOUR LIFE IN NUMEROUS WAYS. IN YOUR LARGE INTESTINE, BACTERIA SYNTHESIZE VITAMIN K, WHICH IS ESSENTIAL FOR YOUR NUTRITION. BACTERIA THRIVE IN YOUR MOUTH AND MAY CAUSE CAVITIES IN YOUR TEETH. BACTERIA ALSO CAUSE MANY DISEASES, INCLUDING CHOLERA, TYPHUS, PNEUMONIA, AND TUBERCULOSIS.

16.1 Bacteria

Objectives

❶ **Describe the structure of a bacterial cell.**

❷ **Contrast a bacterial cell with a eukaryotic cell.**

❸ **Explain how bacteria reproduce.**

❹ **Recognize the diverse ways bacteria obtain nutrition.**

Bacteria Are Small, Simple, and Successful

Figure 16.1
A human red blood cell (250X) is approximately 100 times larger than the bacteria shown below and to the right.

In many ways, bacteria are the most successful organisms on earth. For one thing, bacteria are the oldest group of organisms. The earliest known fossils are 3.5 billion-year-old bacteria. Today, bacteria can be found living almost everywhere on the globe, even in some very hostile habitats. Certain bacteria live beneath more than 400 m (1,200 ft.) of ice in Antarctica, and others live near deep sea volcanic vents where temperatures reach 360° C. Bacteria also occur in great abundance. One gram ($\frac{1}{28}$th of an ounce) of rich soil can contain around 2.5 billion bacteria.

It is obvious that bacteria are very small,

otherwise you could see them without a microscope. To give you an idea of how small bacteria are, look at **Figure 16.1**. A human red blood cell dwarfs the *Escherichia coli* bacterium that is normally found in the human intestine. If you made a chain of *E. coli* bacteria laid end to end, the chain would have to be more than 250 bacteria long just to be visible to the unaided eye.

Most species of bacteria are one of three different shapes: spherical, spiral, or rod-shaped. **Figure 16.1** clearly illustrates these shapes. *Streptococcus* bacteria that cause strep throat are spherical, for example. Spherical bacteria often link to form long chains of cells.

a These rodshaped bacteria, *Bacillus subtilis,* produce antibiotics.

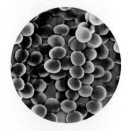

b These spherical bacteria, *Staphylococcus aureus,* cause skin infections.

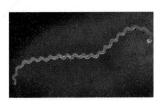

c This spiral bacterium, *Leptospira,* sometimes causes liver and kidney damage.

Bacteria reproduce rapidly

Bacteria, unlike most other organisms, do not undergo mitosis or meiosis. Instead, a bacterium first duplicates its DNA so that there is enough DNA for two cells. Then the bacterium splits into two identical cells. Each cell receives one molecule of DNA and some cytoplasm, a process you saw in Chapter 4. Some kinds of bacteria are able to divide as much as five times in an hour.

Bacterial reproduction does not include meiosis, so it does not allow for the recombination that crossing over and sexual reproduction provide.

Reproduction occurs very rapidly in bacteria. If you were to place a single bacterium into a culture dish containing an abundant supply of food, you could find more than 600,000 bacteria in the dish after only four hours. After six hours, the bacterial population of the dish could reach 476 million.

How Bacteria Obtain Nutrition

Energy and Life

Bacteria use many

different energy

sources. How does

this enable them to

live in a wide

variety of places?

A glass of milk left out of the refrigerator provides a wealth of food for bacteria. Within hours, bacteria colonize the milk and break down its supply of sugar, causing the milk to curdle. Other species of bacteria feed on organic material in sewage. Still other species consume industrial products such as nylon, and pesticides. Some bacteria are able to metabolize petroleum and may be used to clean oil spills. A major reason for the success of bacteria is the wide variety of foods they can use.

Some bacteria are autotrophs

Autotrophic organisms make their own food by using simple molecules. All autotrophic eukaryotes are photosynthetic: they capture solar energy and use it to make food. Many autotrophic bacteria, such as the freshwater cyanobacteria shown in **Figure 16.3**, are also photosynthetic. As you learned in Chapter 10, cyanobacteria were probably the first photosynthetic organisms to produce oxygen, starting about 3 billion years ago.

Figure 16.3

a **Microscopic examination of the water in freshwater ponds such as this one in Kansas reveals . . .**

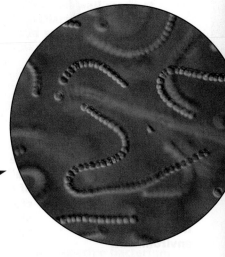

b **. . . that photosynthetic cyanobacteria such as *Nostoc* are a major component of the pond's ecosystem.**

Not all autotrophic bacteria are photosynthetic, however. In Chapter 12 you studied the communities that surround volcanic vents on the ocean floor. The organisms in these very deep ocean communities cannot perform photosynthesis because no light reaches these depths. Instead, bacteria such as those shown in **Figure 16.4** are able to use the energy stored in the inorganic compound hydrogen sulfide. Similarly, some kinds of soil-dwelling bacteria obtain energy from ammonia. Other bacteria that live in swamps use methane. These bacteria use a process called **chemosynthesis** to make complex organic molecules from the energy in inorganic molecules. All chemosynthetic organisms are prokaryotes.

Heterotrophic bacteria are consumers

Most bacteria cannot make their own food and are therefore heterotrophs. Many feed on dead animals and animal wastes; dead plants; and fallen leaves, branches, and fruit. Other types of heterotrophic bacteria are parasites. Parasitic bacteria cause many diseases, as you will see later in the chapter.

Figure 16.4

a The canals in Venice, Italy are one of the world's great tourist attractions, but an examination of the water reveals . . .

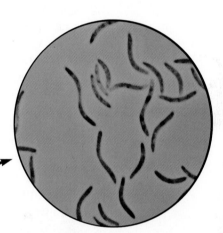

b . . . the presence of large numbers of sulfur bacteria much like *Desulfovibrio gigas*. These bacteria metabolize sulfur, releasing an odor that is similar to rotten eggs. The smell detracts from the beauty of the canals.

Section Review

❶ **How does Gram-staining help a doctor prescribe treatment for a bacterial infection?**

❷ **List three structures found in a eukaryotic cell but not in a bacterial cell.**

❸ **How do bacteria reproduce?**

❹ **Explain the two ways in which autotrophic bacteria obtain energy.**

Lyme Disease

Biology and the Latest in Outdoor Fashion

"Tuck your pant legs into your socks. Roll down your shirt sleeves. Those clothes are too dark." These are not the comments of a friend who doesn't like the fashion statement you make. They are advice from the Centers for Disease Control suggesting ways you can reduce your chances of contracting an ailment called Lyme disease.

Lyme disease is caused by a spiral-shaped bacterium called *Borrelia burgdorferi*. This bacterium lives inside several species of ticks, the most common being members of the genus *Ixodes*. When a tick carrying these bacteria bites a human to feed on blood, the bacteria can be transferred to a human, who may develop the disease.

Lyme disease was first identified in the town of Lyme, Connecticut, in 1975. The disease, however, is found in almost every state. It is common among pets, especially dogs. Ticks carrying the Lyme disease bacteria are spread by deer, rodents, other mammals, and birds.

The first symptom of Lyme disease is usually a red rash that begins at the site of the tick bite and spreads outward. Other early symptoms include fever and chills, nausea, fatigue, and headache. If untreated, later symptoms include severe pain, arthritis, meningitis, and internal organ damage, including brain damage.

Lyme disease is treated with various antibiotics. With early treatment, a patient can be cured before any severe damage occurs. However, diagnosis is difficult to confirm because no reliable

Lyme disease is caused by this bacterium, *Borrelia burgdorferi*. This bacterium is carried mainly by ticks of the genus *Ixodes*. When young, these ticks are as small as poppy seeds. Adults that have recently fed can be as large as jelly beans.

The center of the characteristic "bulls-eye" rash of Lyme disease is the tick bite.

White-tailed deer carry *Borrelia*, but are not affected by it.

test for the presence of this bacterium exists. The best "cure" is prevention. Cover up when you journey into wooded, grassy, or bushy areas. Long sleeves and pants, tucked-in pant legs, and hats help prevent ticks from coming into contact with your skin. Wear light-colored clothing so ticks on your clothes can be seen easily. You can also use tick repellents sparingly on your clothing and skin.

After you have been in tick-infested areas, inspect your body for ticks. Ticks often move to warm, hair-covered areas of the body. Get help inspecting your scalp. Carefully remove attached ticks with tweezers, wrap them in tissue, and flush them down the toilet. Do not smash or burn them as you may release the bacteria they contain. Do not attempt to remove ticks using nail polish, ointments, or a hot match head, as doing so may force the tick's gut contents—which may include the disease-causing bacteria—into your body.

Inspect pets that have been in wooded or grassy areas. Comb out loose ticks and remove any attached ticks with tweezers. Tick repellents for pets also can be used. Have dogs vaccinated against Lyme disease.

By taking a few simple steps and using some common sense, you can enjoy the outdoors, free of big worries about tiny pests. If you think you may have Lyme disease, see your doctor immediately. Treatment is more successful when initiated early.

A Tick's Life

The deer tick, *Ixodes dammini,* has three life stages that differ from each other in size: nymph (about the size of a poppy seed), larva, and adult (when swollen with blood, it can be the size of a jelly bean). During the adult stage, deer ticks lay large numbers of eggs.

Deer ticks usually feed on field mice and white-tailed deer and only need three blood meals during their two-year life cycle. *Borrelia*, one of several kinds of internal parasites carried by ticks, can be ingested during one of these blood meals and transferred to a host during a subsequent blood meal.

Since ticks are unable to fly and can only crawl slowly, they hide in trees or tall grass, waiting for a possible host to wander into their jumping range. They have poor vision and rely on their keen sense of smell. They are especially sensitive to carbon dioxide and butyric acid, a rancid-smelling chemical occurring on the skin of many animals, including humans. The scents immediately activate the hungry tick, sending it leaping in the direction of the odor.

Deer ticks have eight legs and a flat body. Like lobsters, crabs, and insects, they have a hard outer covering that must be periodically shed as the animal grows.

THE DISEASE TUBERCULOSIS IS CAUSED BY THE BACTERIUM *MYCOBACTERIUM TUBERCULOSIS.* TUBERCULOSIS PATIENTS ARE OFTEN TREATED WITH STREPTOMYCIN, AN ANTIBACTERIAL DRUG THAT IS PRODUCED BY BACTERIA OF THE GENUS *STREPTOMYCES.* THIS EXAMPLE SHOWS HOW BACTERIA ARE BOTH HARMFUL AND BENEFICIAL TO HUMANS.

16.2 How Bacteria Affect Humans

Objectives

① **Describe three beneficial effects of bacteria.**

② **List five human diseases caused by bacteria.**

③ **Summarize three ways to prevent bacterial diseases.**

④ **Recognize the importance of antibiotics in fighting bacterial diseases.**

Beneficial Bacteria

Although you probably know that some bacteria can cause harm—such as when they destroy food or make you ill—you might not be fully aware of the tremendous benefits bacteria provide. For instance, bacteria maintain crucial links in nutrient cycles that make essential elements such as nitrogen and sulfur available to plants and, indirectly, to humans. Bacteria are also very important in the manufacture of food and life-saving drugs. **Figure 16.5** summarizes the ways bacteria affect humans in their everyday lives.

Decomposers are nutrient recyclers

Recall from Chapter 12 that decomposers are organisms that return nutrients to the environment by breaking down organic matter. When a plant or animal dies, bacteria from the air and soil settle on the dead organism. The bacteria begin to grow, releasing carbon dioxide, water, nitrogen, phosphorus, and sulfur—nutrients that plants need to grow. Without decomposers, most of these nutrients would be locked away in the bodies of dead organisms.

Figure 16.5
Bacteria in a soybean field may have beneficial effects, whereas other bacteria can have damaging effects on human beings.

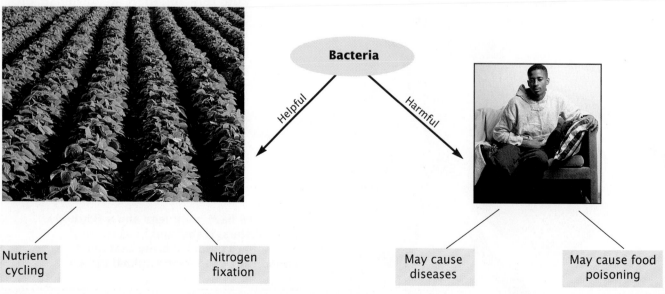

Bacteria

Helpful   Harmful

Nutrient cycling   Nitrogen fixation   May cause diseases   May cause food poisoning

How would an

ecosystem be

affected if its

nitrogen-fixing

bacteria were

destroyed?

Nitrogen-fixing bacteria enrich the soil

Because nitrogen is a component of proteins, plants cannot photosynthesize, grow, or reproduce without it. Much of the nitrogen available for plants is produced by nitrogen-fixing bacteria. Nitrogen-fixing bacteria transform atmospheric nitrogen, which cannot be absorbed by plants, into ammonia, a nitrogen compound that plants can absorb. No other organisms have this ability.

Nitrogen-fixing bacteria are found in the soil, in aquatic ecosystems, and within the roots of some plant species. The legumes, plants including the peas and beans, contain nitrogen-fixing bacteria in swellings on their roots. These bacteria enable legumes to grow in nitrogen-poor soils where other plants cannot.

Figure 16.6
The young woman below is eating yogurt, while the young man on the right is eating olives. Yogurt and olives are made when the bacteria indicated are added during the food manufacturing process.

Bacteria are used to manufacture food and drugs

Like the young woman and man in **Figure 16.6**, when you eat yogurt or olives, you are eating foods that are the product of bacterial decomposition. Humans have learned that decomposition is occasionally beneficial because it adds flavor to food. For example, bacteria convert cabbage and cucumbers into tangy sauerkraut and pickles.

Modern technology is taking advantage of the enormous genetic diversity among the bacteria. Using genetic engineering technology, biologists can now "reprogram" bacteria to manufacture any protein for which a gene has been isolated. For example, bacteria now produce most of the insulin needed by diabetics in the United States. Before genetic engineering, insulin had to be isolated from the pancreases of animals killed in slaughterhouses. Other drugs and products produced by genetically engineered bacteria are described in Chapter 8.

Streptococcus thermophilus and *Lactobacillus bulgaricus*

Leuconostoc mesenteroide and *Lactobacillus plantarum*

Yogurt

Bacteria and Disease

Occasionally your body is a temporary home for parasitic bacteria with serious effects—they cause disease or infection. A disease-causing agent is called a **pathogen**. Pathogenic bacteria are harmful because they damage their host's tissues. This damage results from either direct attacks on the host's cells or from poisonous substances called toxins that many bacteria release.

How are bacterial diseases actually transmitted? Each species of pathogenic bacteria has a characteristic way of being carried to new hosts: in water, in the air, in food, by insects, or by direct human contact. **Table 16.1** lists some bacterial diseases and their modes of transmission.

Water can carry pathogens

Cholera is a serious disease transmitted through polluted water. Cholera bacteria produce a strong toxin that causes acute diarrhea and vomiting, which can lead to rapid dehydration. If untreated, this severe loss of water can be fatal within 24 hours. Cholera bacteria are spread in drinking water that has been contaminated by the feces of infected individuals. Any fish that are caught in contaminated water can also carry cholera bacteria. In 1991 a large cholera epidemic struck Peru and rapidly spread to the rest of South America and Central America. At least 1,700 people died in the first year of this devastating outbreak.

Table 16.1 Common Bacterial Diseases

Disease	Mode of Transmission	Symptoms
Tuberculosis	Airborne water droplets	Fatigue, persistent cough, bleeding in lungs; can be fatal
Diphtheria	Airborne water droplets	Fever, sore throat, fatigue
Scarlet fever	Airborne water droplets	Rash, fever, sore throat
Bubonic plague	Fleas	Swollen glands, bleeding under skin; often fatal
Typhus	Lice	Rash, chills, fever; often fatal
Tetanus	Dirty wounds	Severe, prolonged muscle spasms
Cholera	Contaminated water	Severe diarrhea, vomiting; often fatal
Typhoid	Contaminated water and food	Headaches, fever, diarrhea, rash; often fatal
Leprosy	Personal contact	Nerve damage, skin lesions, tissue degeneration
Lyme disease	Ticks	Rash, pain, swelling in joints

Figure 16.7
Just in case the chicken you are preparing is contaminated with *Salmonella*, you should use hot, soapy water to wash the surfaces that have been touched by the raw chicken.

Figure 16.8
a A sneeze sends thousands of bacteria into the air. Some of these bacteria may be pathogens.

Food can be contaminated by bacteria

Although certain bacteria can produce yogurt and cheese, most bacteria in food are not helpful. Pathogenic bacteria can contaminate foods and cause food poisoning.

One of the most dangerous kinds of food poisoning is botulism, which is caused by a toxin released by the bacterium *Clostridium botulinum*. Consumption of less than one-millionth of a gram of this toxin causes paralysis and death. *C. botulinum* normally lives in the soil, but it can grow in canned foods that have not been properly sterilized. Because oxygen kills botulism bacteria, they cannot grow in fresh and frozen foods, which contain oxygen.

Another type of food poisoning is caused by *Salmonella* bacteria found in pork, eggs, poultry, and other foods. The symptoms of *Salmonella* food poisoning are diarrhea, vomiting, and abdominal cramps. Although the symptoms usually pass quickly, a serious infection can cause dehydration, a drastic loss of water from the body. In the very young or in the elderly, severe dehydration can be fatal. Hot, soapy water will destroy *Salmonella* bacteria encountered during food preparation, as shown in **Figure 16.7**.

Bacteria travel through the air

A sneeze or cough sprays out a shower of tiny droplets, as you can see in **Figure 16.8**. Visible in the circular photograph are some of the 10,000 to 100,000 bacteria that are carried in the water droplets produced by a single sneeze. The bacteria that cause diphtheria, scarlet fever, whooping cough, and tuberculosis drift through the air in droplets such as these. Most airborne diseases affect the respiratory systems of their victims. Tuberculosis bacteria, for example, invade the lungs. The scars they leave are often visible as very large shadows on X rays of the lungs.

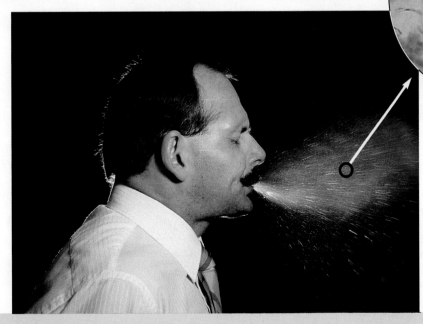

b This view shows the bacteria that can be found in a single water droplet of a sneeze. *Mycobacterium tuberculosis*, the bacterium that causes the disease tuberculosis, can be transmitted in airborne water droplets such as these.

Controlling Bacterial Diseases

Today most Americans have little experience with serious bacterial diseases. Advances in medicine and sanitation have brought freedom from bacterial disease to the industrialized countries of the world only in the last 100 years. Much of the world's population is still subject to these life-threatening but preventable bacterial diseases.

Sanitation and hygiene prevent bacterial diseases

In the industrialized countries, drinking water is filtered and then purified with chlorine, a chemical that kills bacteria. Similarly, sewage is collected and treated to remove pathogens before it is discharged into rivers or the ocean. Thus, cholera, typhoid, and other diseases of contaminated water are almost unknown in the industrialized countries.

The clean conditions found in many of the industrialized countries are not characteristic of some of the less-developed countries. Because of poverty, many of these countries cannot provide clean drinking water for all of their citizens. In the poorest areas, as shown in **Figure 16.9**, the same river that provides water for drinking and bathing might also serve as a sewer. Each year about 25 million people die from typhoid, cholera, and other diseases of contaminated water in the less-developed countries .

Figure 16.9

a **Water sanitation is of critical importance in controlling bacterial diseases. In poor, rural areas, such as this area of Peru, some of the poorest people are forced to cook, drink, and bathe in the same water.**

b **Modern methods of water purification, such as those used in these treatment plants, can almost eliminate the possibility of bacterial contamination of water.**

Energy and Life

How do viruses

reproduce if they

cannot use energy?

Heat and cold protect food from bacterial contamination

Bacterial contamination of food can be prevented with either heat or cold. **Pasteurization** involves heating food so that the threat of bacterial contamination is removed. Pasteurization is doubly effective because it kills most bacteria and destroys their toxins. Pasteurization eliminates the possibility of contracting diseases such as botulism from canned food, or brucellosis or tuberculosis from milk.

Cooling food to a few degrees above freezing also prevents bacterial contamination. Although bacteria are not killed by the cold, their rate of growth is greatly slowed, and their population size remains small. Most salmonella infections are caused by food that has been left out of the refrigerator too long, allowing bacteria to grow.

Vaccination stimulates the body's defenses

Preventing the contamination of food and drinking water are only two ways to protect yourself from infection. Vaccinating against disease is another way. Before you started school, you were probably vaccinated against bacterial diseases such as diphtheria, whooping cough, and tetanus. A **vaccine** is a solution containing pathogens or their toxins that have been made harmless, usually by treatment with heat, chemicals, or by genetic engineering. How does a vaccine protect you against pathogenic bacteria? As shown in **Figure 16.10**, a vaccine against whooping cough, for example, is a solution of whooping cough bacteria that have been killed with heat or chemicals. These dead bacteria cannot cause disease, but once inside your body they stimulate your immune system to make defenses against them. Once you have been vaccinated against whooping cough, your immune system has defenses ready to destroy any live whooping cough bacteria before they have a chance to make you ill.

Figure 16.10
The preparation of vaccines involves inactivation of the disease-causing organism and its subsequent introduction into the body.

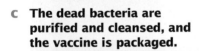

a **Whooping cough vaccine is prepared by first culturing the bacteria responsible for whooping cough.**

b **The bacteria, *Bordetella pertussis*, are then killed using heat or chemicals.**

c **The dead bacteria are purified and cleansed, and the vaccine is packaged.**

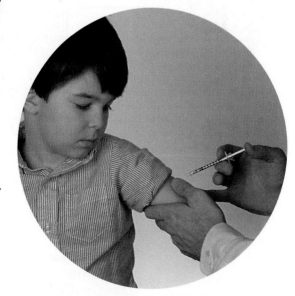

d **The vaccine containing the dead bacteria is injected. This child's body will now produce antibodies that fight *Bordetella pertussis*.**

Figure 16.11
Alexander Fleming's discovery of penicillin is one of the most important medical milestones of the century. Millions of lives have been saved because of his observations. The fungus that produces penicillin is shown below.

Antibiotics are used to treat bacterial diseases

Before the 1940s, doctors had few treatments for bacterial diseases. Whether a patient recovered or died often depended more on the type of disease and the strength of the patient than it did on the efforts of the doctor. This grim situation changed because of a chance event in 1928. In that year, Alexander Fleming, the British physician shown in **Figure 16.11**, found a blue-green mold growing on one of his bacterial samples. At first Fleming was angry because his experiment had been contaminated. Before throwing out the sample, however, he noticed that bacteria did not grow near the mold. The mold was apparently releasing a chemical that was poisonous to bacteria. Fleming isolated this substance and named it **penicillin**, after the *Penicillium* mold that produced it.

In the early 1940s, scientists following up on Fleming's discovery showed that penicillin was a very effective treatment for many bacterial diseases. Antibacterial drugs such as penicillin are called **antibiotics**. There have been many other antibiotics discovered since penicillin; tetracycline and streptomycin are two common examples. Some antibiotics prevent bacteria from making new cell walls.

b The organism is microscopic and reproduces by forming spores on the ends of fungal branches. It is easy to culture, and modern manufacturing methods can produce large quantities of penicillin.

c Penicillin is still one of the most common medicines used today to combat bacterial infections.

a The genus *Penicillium* is a common fungus. It is the common blue mold often seen on an old orange.

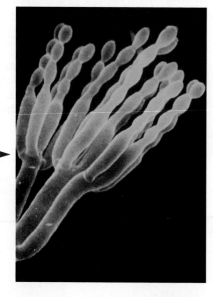

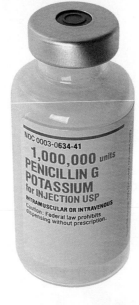

Section Review

❶ Describe one way in which bacteria are used to treat a disease.

❷ Name four diseases caused by bacteria.

❸ Explain why refrigeration and pasteurization are effective in preventing bacterial contamination of food.

❹ Suggest a hypothesis to explain why antibiotics will kill bacteria without destroying cells in your body.

THE WORLD HEALTH ORGANIZATION ESTIMATES THAT MORE THAN 10 MILLION PEOPLE THROUGHOUT THE WORLD ARE INFECTED BY **HIV**, THE VIRUS THAT CAUSES **AIDS**. THE NUMBER OF INFECTED PEOPLE GROWS RAPIDLY EACH YEAR. TO FIND A CURE FOR **AIDS** OR A VACCINE AGAINST **HIV** WILL REQUIRE A CLEAR UNDERSTANDING OF WHAT VIRUSES ARE, HOW THEY REPRODUCE, AND HOW THEY AFFECT THEIR HOSTS.

16.3 Viruses

Objectives

❶ Describe the structure of a virus.

❷ Explain why viruses are not living organisms.

❸ Describe how a virus reproduces.

❹ List four diseases that are caused by viruses.

What Is a Virus?

Think back to the last time you had a cold or the flu. Did you ask the doctor for antibiotics to kill the flu or cold "bacteria" that were making you so miserable? If you did, the doctor would have explained that colds and flu are not caused by bacteria, so antibiotics would have no effect. Colds and flu are caused by **viruses**. Viruses are microscopic particles that invade the cells of plants, animals, fungi, and bacteria. Viruses often destroy the cells they invade.

Figure 16.12
The computer-generated colors on this image of the polio virus represent the different proteins present on the surface of the virus.

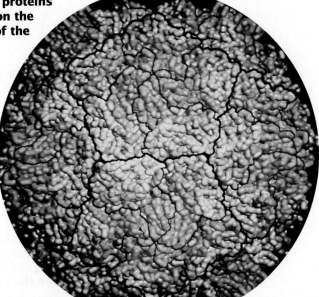

Viruses are small, simple particles

If you could open a virus, what would you find inside? Would you find cytoplasm, ribosomes, and mitochondria, like you would find in one of your own cells? No, because a virus is not a cell. A typical virus, such as the polio virus shown in **Figure 16.12**, is composed of a core of genetic material surrounded by a protein "coat." The protein coat protects the genetic material and enables the virus to invade its host cell. The *Tour of a Virus* on the next page shows the structure of the Human Immunodeficiency Virus (**HIV**). This virus causes Acquired Immune Deficiency Syndrome (**AIDS**).

In many viruses, DNA is the genetic material. A few viruses have RNA instead and are known as RNA viruses. The viruses that cause AIDS, polio, and the flu are RNA viruses. Viruses are parasitic and can only reproduce inside the cells of their hosts. Because viruses are so small, viral genetic material has room for only a few genes, usually only genes coding for the protein coat and for enzymes that enable the virus to take over its host cell.

Tour of a Virus

The disease AIDS is caused by a human virus called HIV, human immunodeficiency virus.

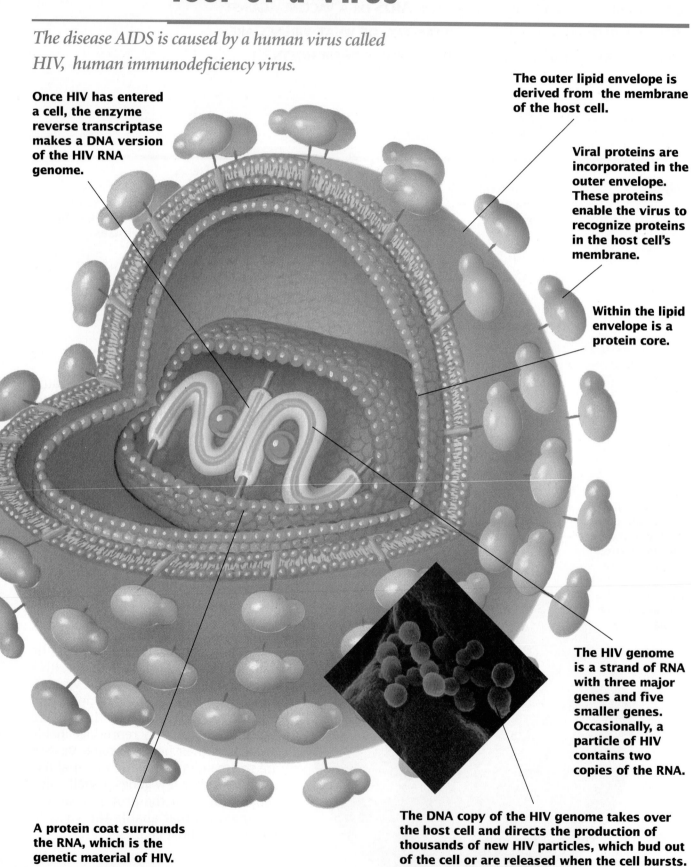

Once HIV has entered a cell, the enzyme reverse transcriptase makes a DNA version of the HIV RNA genome.

The outer lipid envelope is derived from the membrane of the host cell.

Viral proteins are incorporated in the outer envelope. These proteins enable the virus to recognize proteins in the host cell's membrane.

Within the lipid envelope is a protein core.

The HIV genome is a strand of RNA with three major genes and five smaller genes. Occasionally, a particle of HIV contains two copies of the RNA.

A protein coat surrounds the RNA, which is the genetic material of HIV.

The DNA copy of the HIV genome takes over the host cell and directs the production of thousands of new HIV particles, which bud out of the cell or are released when the cell bursts.

Are viruses alive?

Viruses do have some characteristics of living things. They have genetic material that is transmitted to future generations and that can change over time. Therefore, viruses are able to evolve. On the other hand, viruses lack three very critical features of living things: they are not made of cells, they cannot make proteins, and they cannot use energy. Even though viruses can reproduce, they are able to do so only when inside living cells. Because viruses lack these essential features, biologists consider viruses nonliving. Nevertheless, since viruses are active inside living cells, the study of viruses is part of biology.

How Viruses Reproduce

To reproduce, a virus must insert its genetic material, which contains the instructions for making new viruses, into a host cell. This viral genetic material seizes control of its host cell and transforms it into a virus factory. All viruses, whether they attack your cells or bacteria, reproduce by taking over the reproductive machinery of a cell. Follow how HIV reproduces in the example below.

Figure 16.13
HIV enters a human helper T cell and then takes over the cell's machinery. The cell is directed to make HIV proteins and RNA. Newly made viruses then leave the cell and continue the cycle of infection.

HIV seizes control of host cells

HIV cannot infect a cell it cannot enter. HIV is able to enter only those cells that have a particular receptor protein in their cell membranes. HIV recognizes these cells because the virus contains a protein that will bind to the receptor protein in the cell membrane. You can think of the relationship between the viral protein and the cell membrane protein as being like the relationship between a key and lock. If the key (the viral protein) fits the lock (the cell protein), then the cell is opened to invasion. Otherwise, the virus is locked out.

The main host cell for HIV is a type of white blood cell known as the helper T cell. Helper T cells occur in the blood and in the lymphatic system. They play a very crucial role in the body's ability to fight infection. Follow the events that occur when HIV infects a helper T cell in **Figure 16.13**.

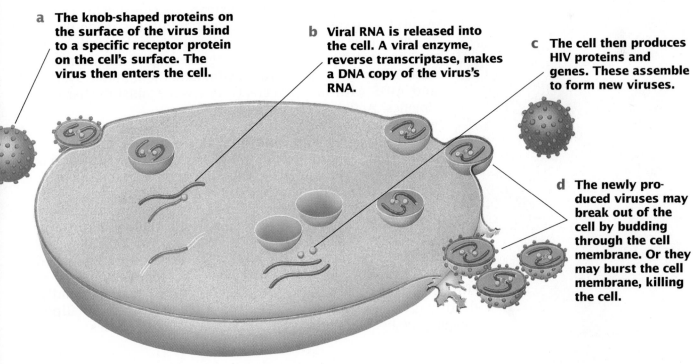

a The knob-shaped proteins on the surface of the virus bind to a specific receptor protein on the cell's surface. The virus then enters the cell.

b Viral RNA is released into the cell. A viral enzyme, reverse transcriptase, makes a DNA copy of the virus's RNA.

c The cell then produces HIV proteins and genes. These assemble to form new viruses.

d The newly produced viruses may break out of the cell by budding through the cell membrane. Or they may burst the cell membrane, killing the cell.

AIDS is fatal

HIV causes AIDS by killing helper T cells, which coordinate the body's attack on pathogens. However, HIV may not begin to destroy large numbers of helper T cells for a period lasting from a few months to over 10 years after infection. During this period, an infected person may experience only mild or flu-like symptoms, or no symptoms at all. An infected person can transmit HIV to others during this period. Scientists now think that practically everyone infected with HIV will develop AIDS. Eventually, the virus begins to destroy a large percentage of an infected person's helper T cells. When the number of helper T cells has fallen to very low levels, the person is said to have AIDS. AIDS patients usually die of cancer or diseases a healthy immune system would defeat. You will read more about HIV transmission and AIDS prevention in Chapters 32 and 34.

Diseases Caused by Viruses

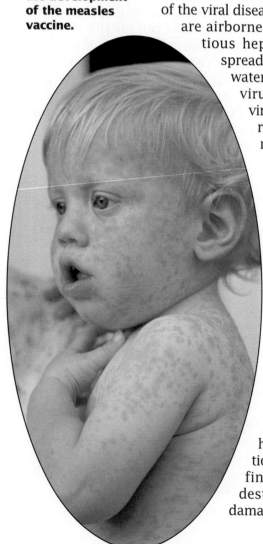

Figure 16.14
The tell-tale blotchy rash of measles has become relatively uncommon since the development of the measles vaccine.

Viruses cause many serious human diseases in addition to AIDS, such as measles, shown in **Figure 16.14**, and smallpox. Like pathogenic bacteria, pathogenic viruses are transmitted from host to host in characteristic ways. Most of the viral diseases listed in **Table 16.2** are airborne. A few, such as infectious hepatitis and polio, can spread through contaminated water. Insects also transport viruses. The yellow fever virus, common in tropical regions, is carried by mosquitoes.

There are defenses against viruses

Why don't physicians treat viral diseases such as colds, flu, and AIDS with antibiotics? Antibiotics work by interfering with cellular processes such as protein production or cell-wall synthesis, which do not occur in viruses. Moreover, since a virus uses its host cell for reproduction, it is very difficult to find any drugs that will destroy the virus without damaging the host.

A drug called azidothymidine (AZT) blocks an enzyme essential for DNA replication. Many AIDS patients are now being treated with AZT, which can prolong the lives of many of these patients. Unfortunately AZT cannot cure AIDS; it only slows the course of the disease. Furthermore, AZT is very toxic and has many side effects.

Vaccination also protects against viral diseases

Vaccination is the only effective defense against most viral diseases. Recall that vaccines against bacterial disease are composed of dead bacteria. But viruses are not alive and cannot be killed. Instead, vaccines against viral diseases contain viruses made harmless by treatment with chemicals or by genetic engineering. These harmless viruses stimulate the immune system to create defenses against the harmful form of the virus.

Smallpox is no longer a killer

Eliminating smallpox from the world is vaccination's greatest triumph. Smallpox virus produced tiny pustules or sores (small "pox") on its victim's skin. These pustules developed scabs and often turned into permanent, disfiguring scars. Far worse, smallpox killed half of those who contracted it.

Because smallpox was so deadly, cures and ways to prevent smallpox had been sought for centuries.

In 1798, an English doctor named Edward Jenner, shown in **Figure 16.15**, discovered a way to immunize people against smallpox. He noticed that milkmaids, who were continually exposed to cattle, did not develop smallpox. He formed the hypothesis that milkmaids developed immunity to smallpox because they had been exposed to cowpox, a disease like smallpox that infects cattle but produces only very mild symptoms in humans. Jenner tested his hypothesis by injecting seepage from cowpox sores into a boy. The boy did not get smallpox, even when Jenner deliberately exposed him to smallpox virus.

Vaccination against smallpox became commonplace in the industrialized countries. As a result, smallpox rapidly disappeared from these countries. The last case in the United States occurred in 1949. Because the disease persisted in some of the poorest nations, the World Health Organization launched a worldwide vaccination campaign against smallpox. In 1967, the year this campaign was launched, 10 million to 15 million cases of the disease occurred. Just 11 years later, smallpox was eliminated. The last known person to contract smallpox is shown in **Figure 16.16**.

Why does injection with cowpox virus provoke immunity to smallpox? Jenner couldn't explain how his vaccine worked, but scientists now can. The cowpox virus and the smallpox virus have very similar protein coats. After vaccination, the immune system creates defenses that recognize the shape of the cowpox protein coat. These defenses cannot differentiate a cowpox virus from a smallpox virus, so they destroy either virus whenever it is encountered.

Figure 16.15
Edward Jenner was able to decrease the threat of smallpox by vaccinating people with the cowpox virus.

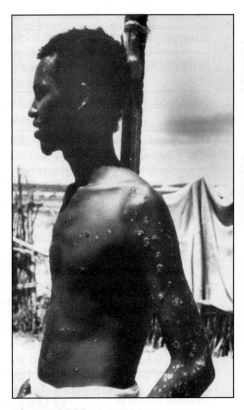

Figure 16.16
Ali Maow Maalin of Merka, Somalia, contracted smallpox in 1977 at the age of 23. His was the last known case of smallpox in the world.

LOOK FOR CARRAGEENAN IN THE LIST OF INGREDIENTS OF YOGURT, ICE CREAM, AND OTHER DAIRY PRODUCTS. IT IS ADDED TO THESE FOODS AND TO PAINTS AND COSMETICS TO CREATE A SMOOTH TEXTURE. CARRAGEENAN, A JELLYLIKE MATERIAL EXTRACTED FROM THE CELL WALLS OF A RED ALGA KNOWN AS IRISH MOSS, IS ONE OF MANY PRODUCTS WE OBTAIN FROM PROTISTS.

17.2 Protist Diversity

Objectives

❶ Recognize the variation in shape, size, and method of obtaining energy among protists.

❷ Explain the role autotrophic protists play in ecosystems.

❸ Recognize the evolutionary connection between green algae and plants.

❹ Identify three kinds of heterotrophic protists.

Classification of the Protists

Figure 17.6
Although these two organisms look very different, they are both members of the kingdom Protista.

Look at the two dissimilar organisms in **Figure 17.6**. **Figures 17.6a** and **17.6b** show a brown alga, or kelp, which is a type of seaweed. Kelp is multicellular and autotrophic. Some kelps can grow to 100 m (328 ft.) in length. The creature in **Figure 17.6c** is *Didinium*. It is single-celled and also heterotrophic. *Didinium* is so small that it cannot be seen without a microscope. Both of these organisms are classified as protists. The members of the kingdom Protista exhibit a greater range of sizes and a greater variety of structure than do the members of any other kingdom.

Great diversity is also found in protist metabolism. Among the eukaryotic kingdoms, the kingdom Protista is the only one that includes both heterotrophs and autotrophs. Autotrophic protists, including both unicellular and multicellular forms, are photosynthetic, as are plants. Heterotrophic protists, such as *Paramecium* and *Amoeba*, obtain their energy by consuming other organisms or by parasitism. You can see the structure of a heterotrophic protist in the *Tour of a Protist* on the next page.

Separating the protists into autotrophs and heterotrophs is an artificial classification that does not show the complexity of their evolutionary relationships. Because of its diversity, biologists disagree on how many phyla the kingdom Protista should contain.

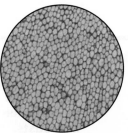

a A kelp, such as this *Fucus*, is a protist. It is a large organism that is autotrophic . . .

b . . . and is composed of millions of cells.

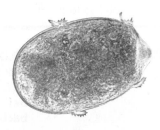

c *Didinium* (150X), also a protist, is unicellular and heterotrophic.

Tour of a Protist

Paramecium is a kind of protist called a ciliate. It is covered with fields of flagella called cilia that move it through the water.

Paramecium spontaneously absorbs water. Excess water is squeezed back out by contractile vacuoles.

Paramecium has two nuclei. The macronucleus contains fragmented chromosomes used in routine cellular functions, and divides by pinching in two.

The micronucleus contains the cell's chromosomes, and divides by mitosis.

Paramecium ejects wastes by exocytosis.

Food is ingested through a cilia-lined gullet and taken into the cell by endocytosis as food vacuoles.

A *Paramecium* recognizes members of the opposite sex by special protein molecules that ring the gullet like a necklace.

Paramecium is found in streams and ponds.

Paramecium is unicellular but so complex that some biologists prefer to call them "organisms without cell boundaries" rather than single cells.

Autotrophic Protists

Do you think anything you did today involved autotrophic protists? Before you answer "no" without hesitation, consider these activities. Did you drive or ride in a car? Did you brush your teeth? Did you eat ice cream, cheese, or yogurt? Did you eat fish or other seafood? Autotrophic protists or their products are involved in all these activities. Autotrophic protists are often called algae (*AL jee*). Despite being grouped under one name, not all kinds of algae are closely related.

Although they can be found in moist soil, on damp rocks, and on the shady sides of trees, most species of algae live in fresh water or salt water, as shown in **Figure 17.7**. Algae serve the same ecological function in aquatic ecosystems as plants serve in terrestrial ecosystems. They form the base of the aquatic food web, directly or indirectly providing food for all aquatic consumers. Thus, fishes, whales, and humans who eat seafood depend on algae. Algae also produce large amounts of oxygen. About one-third of the oxygen in the atmosphere is produced by algae.

Figure 17.7
Autotrophic protists are found in the open ocean, along the shore, and in freshwater habitats. They range from unicellular to multicellular, yet all are photosynthetic.

Most kinds of algae are unicellular

Unicellular algae are one part of a community of aquatic organisms called plankton. The organisms found in plankton float near the surface in fresh water or salt water. Heterotrophic bacteria, fish larvae, crustaceans, and heterotrophic protists are also considered to be part of the plankton. Photosynthetic plankton are called **phytoplankton** (*fy tuh PLANK tuhn*), which means "plant plankton," because they are photosynthetic.

Phytoplankton require light for photosynthesis, and so they usually are found near the surface. A wide variety of adaptations help them stay afloat. Some beat their flagella and "tread water." Others have fins and spines that act as water wings. Others store extra food as oil, which buoys them up near the surface. When algae full of oil die and sink to the bottom, they can be buried under mud and sand. Over millions of years, heat and pressure within the earth transform the oil from the algae into crude-oil deposits. The

a Microscopic algae like this diatom serve as food for many marine animals.

b Red algae such as this *Porphyra* often are attached to rocks at the edge of the sea.

c Brown algae such as *Fucus* can be many meters in length.

gasoline used today was formed partly from the remains of algae that died millions of years ago.

Look at the object in **Figure 17.7a**. These objects may look like hubcaps or jewelry, but they are the remains of aquatic algae called diatoms. Diatoms form intricately patterned shells made of silica, a glasslike substance. When diatoms die, their shells settle to the ocean floor. Humans harvest fossilized deposits of these shells to make abrasives for toothpaste and silver polish.

If you visit the beach, you might experience a colorful but dangerous effect of marine algae. Occasionally, the ocean turns reddish and thousands of dead fishes, seabirds, turtles, and dolphins wash up onto the beach. These "red tides" are caused by phytoplankton known as dinoflagellates (*deye noh FLAJ uh lihtz*). Population explosions of certain species of dinoflagellates occur irregularly in coastal areas. The dinoflagellates release toxins that are poisonous to vertebrates. Because fishes and shellfish concentrate large amounts of these toxic secretions in their internal organs, health authorities often do not allow fishing during red tides.

Many green, brown, and red algae are multicellular

There are three groups of multicellular algae—the green, brown, and red algae. The brown algae and the red algae are commonly called "seaweeds."

You might have seen algae like the ones in **Figure 17.7b** and **17.7c** attached to rocks at low tide or washed up on a beach.

Green algae, like the pond scum in **Figure 17.7d**, are distinguished from other algae by their kinds of photosynthetic pigments, by their use of starch to store food, and by their unique form of cell division. Plants share these characteristics with the green algae. Therefore, scientists think that green algae are the ancestors of the plant kingdom.

Kelp, which you saw in **Figure 17.6a**, is a brown alga that often grows in dense stands or thick floating mats. Kelp "forests" occur off the coast of California. The Sargasso Sea, an area of the Atlantic Ocean east of Florida, is so thick with floating kelp that early European explorers feared their sailing ships could be trapped there (an unreasonable fear, it turns out).

Red algae tend to be smaller and less spectacular than the brown algae. Distinctive photosynthetic pigments give red algae their color. These same pigments enable some species to photosynthesize in waters more than 200 m (650 ft.) deep, where there is little light.

Red and green algae are harvested for many uses. *Porphyra*, one of the red algae, is particularly popular in Far Eastern countries such as Japan and China, where it is known as *nori*. Nori is cultivated as a crop on coastal algae "farms."

d Many kinds of green algae live in freshwater habitats. These filamentous green algae, called *Hydrodictyon*, are microscopic. But large numbers of green algae often are visible as pond scum.

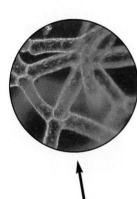

Heterotrophic Protists

The word "carnivore" brings to mind fierce, sharp-toothed animals like lions, tigers, wolves, and sharks. The word "herbivore," on the other hand, calls forth images of dull, plodding creatures like cows and sheep. You don't normally think of tiny one-celled protists as carnivores and herbivores, but most heterotrophic protists play one of these two ecological roles. In the plankton, for example, some heterotrophic protists "graze" on phytoplankton while others prey on these grazers. Other important heterotrophic protists are parasites or decomposers.

Slime molds and water molds superficially resemble fungi

If you looked carefully through moist, decaying leaves on a forest floor, you might find a slime mold. A slime mold spends much of its life as a single-celled mass called an amoeba that is found in damp environments and feeds on bacteria and decaying matter. If food or even moisture become scarce, the individual amoebas come together, forming a mass resembling the slime mold shown in **Figure 17.8a**. The mass produces reproductive structures, also shown in **Figure 17.8a** (inset).

If your ancestors arrived in the United States from Ireland in the 1840s, they were probably fleeing the effects of a heterotrophic protist. In Ireland, from 1845 to 1847, a famine killed more than 1 million people and forced another 3 million to emigrate, mainly to the United States. This famine occurred because the potato crop (the staple of the Irish diet) was almost wiped out by an outbreak of late blight. Late blight is a disease caused by *Phytophthora infestans*, a protist that is a water mold, shown in **Figure 17.8b**.

Despite some similarities, slime molds and water molds are not close relatives of fungi. For one thing, fungi have cell walls made of chitin, a complex polysaccharide. Slime molds and water molds either have cellulose cell walls or have no cell walls at all.

Figure 17.8
Heterotrophic protists are extremely diverse. Two very different species are shown below.

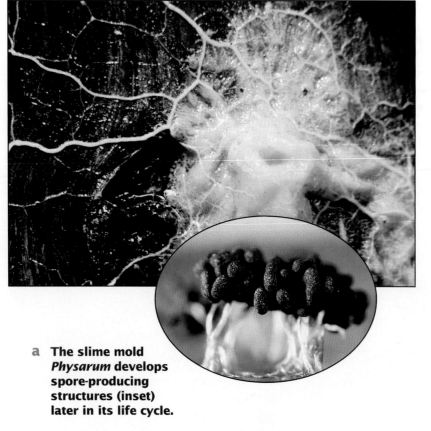

a The slime mold *Physarum* develops spore-producing structures (inset) later in its life cycle.

b Water molds are microscopic organisms. The water mold *Phytophthora infestans,* which causes late blight of potatoes, has profoundly influenced history.

Potato plant

Protozoa include the ancestors of animals

About 300 years ago, scientists using the first microscopes saw what appeared to be tiny animals darting about in drops of water. They called these organisms protozoa, which means "first animals." Today scientists do not classify protozoa into the animal kingdom, but animals are thought to have descended from extinct protozoa.

Paramecium, Didinium, and the *Stentor* in **Figure 17.9** are members of a group of protozoa called ciliates. Ciliates are covered with numerous cilia, which function in locomotion and feeding.

Not all groups of protozoa are closely related. Indeed, based on recent molecular analysis of DNA, some biologists argue that ciliates should be placed in a separate kingdom of their own.

The ancestor of animals probably belonged to a group of protozoa known as the zoomastigotes (*ZOH mast ih gohtz*). Zoomastigotes are one-celled, heterotrophic protists that have at least one flagellum. Some zoomastigotes, such as the *Trichonympha* illustrated in **Figure 17.10**, are covered with flagella. If your house has ever been attacked by termites, you can blame *Trichonympha*. Without these protists living in their intestines, termites could not digest wood.

Figure 17.9
Stentor **is a ciliate that lives anchored to a solid surface and gathers food by the whirling action of its cilia.**

The tiny, ornate shells shown in **Figure 17.11** are made by marine protozoa called foraminiferans (*fawr uh MIHN ih fur ihnz*). When foraminiferans die, their calcium shells sink to the ocean floor. Over millions of years, these shells have accumulated, forming huge limestone deposits like the White Cliffs of Dover in southeastern England.

Figure 17.10
Trichonympha **is a zoomastigote that lives in a termite's intestines.**

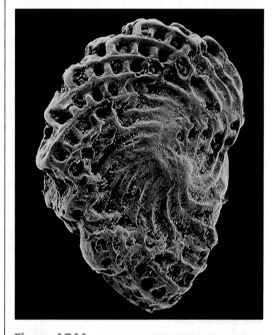

Figure 17.11
Foraminiferans secrete tiny, ornate shells that accumulate on the bottom of the sea.

Section Review

1 **Explain why the statement, "All protists are microscopic" is untrue.**

2 **Summarize the importance of algae in aquatic ecosystems.**

3 **Describe two pieces of evidence indicating that plants evolved from green algae.**

4 **Name two kinds of heterotrophic protists.**

What Are Slime Molds Like?

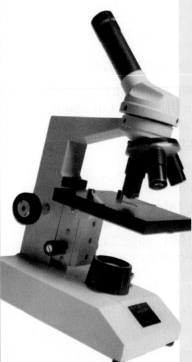

Objectives

In this investigation you will:
- *observe* slime molds
- *contrast* slime molds with animals and plants

Materials

- slime mold culture
- compound microscope
- wax pencil
- oatmeal flakes
- sewing needle
- single-edged razor blade
- glass rod
- beaker containing a small amount of vinegar

Prelab Preparation

1. Review what you have learned about slime molds by answering the following questions:
   - To which kingdom do slime molds belong?
   - Slime molds were once classified in the kingdom Fungi. Why are they no longer considered to be members of this kingdom?

2. Review the procedures for using the compound microscope in the Appendix.

Procedure: Observing Slime Mold

1. Form a cooperative group with another student. Work with your partner to complete steps 2–10 below. Obtain a petri dish containing a slime mold culture.

2. Remove the cover from the petri dish and observe the slime mold. Record your observations.

3. Make a sketch of the petri dish and its contents.

4. With the lid of the petri dish closed, turn the dish upside down and place it on the stage of the microscope. Observe one branch of the slime mold under low power for two minutes. Record your observations. Make a sketch of the slime mold as it appears under low power.

5. Remove the dish from the microscope stage. Keep the dish upside down. With a wax pencil, trace the outline of one branch of the slime mold on the bottom surface of the dish. Turn the dish right side up and open the lid. Place a flake of oatmeal 1–2 mm from the branch you traced. Do not disturb this branch for a few minutes as you conduct the next three procedures.

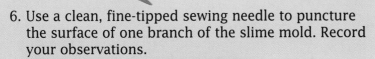

6. Use a clean, fine-tipped sewing needle to puncture the surface of one branch of the slime mold. Record your observations.

7. **CAUTION: Razor blades are sharp and can cause injury.**
Use a razor blade to cut a small piece of slime mold from one branch. Place the piece next to the body of the slime mold. Observe the slime mold for a few minutes. *What happened to the separated piece?* Record your observations.

8. Dip a clean glass rod into a beaker of vinegar. While observing a branch of the slime mold, touch it with the glass rod. *How does the slime mold respond to vinegar?* Record your observations.

9. Now observe the flake of oatmeal. *How did the slime mold react to the oatmeal? How did tracing the outline of the slime mold help you to observe its behavior?*

10. Your teacher has prepared a petri dish containing a slime mold that has been left in the light for 24 hours. Examine the slime mold in this dish and make a drawing of your observations.

11. Clean up your materials and wash your hands before leaving the lab.

Analysis

1. *Analyzing Observations*
What evidence suggests that the slime mold is alive?

2. *Making Comparisons*
Describe two differences between a plant and a slime mold.

3. *Making Comparisons*
In what ways do slime molds resemble animals?

Thinking Critically

Slime molds live among decaying material on the forest floor. Why is it advantageous for the slime mold to be able to form fruiting bodies when exposed to light?

Chapter 18

Fungi and Plants

Review

- **ratio of surface area to volume (Section 3.1)**
- **invasions of the land (Section 10.3)**
- **five kingdoms (Section 15.3)**

Mushrooms, club mosses, and seed plants can all be found growing in the leaf litter that carpets the forest floor.

FUNGI AND PLANTS WERE THE FIRST EUKARYOTIC ORGANISMS TO INVADE LAND FROM THE SEA. TOGETHER WITH BACTERIA, FUNGI DECOMPOSE ORGANIC MATTER IN THE ENVIRONMENT. THEIR ACTIVITIES ARE AS NECESSARY TO THE CONTINUED EXISTENCE OF THE WORLD AS ARE THOSE OF PLANTS. AS YOU WILL SEE, FUNGI ALSO HAVE A DIRECT IMPACT ON HUMAN LIFE, PROVIDING FOOD AND MEDICINE.

18.1 Fungi

Objectives

❶ **List the features shared by most fungi.**

❷ **Explain how fungi obtain nutrients.**

❸ **Describe the main differences among the four divisions of fungi.**

❹ **Summarize the ecologic and economic roles of fungi.**

The Kingdom Fungi

Fungi (*FUHN jeye*) are a group of eukaryotic organisms, most of which are multicellular. Although little is known about their origins, fungi are at least 400 million years old and probably older. Like plants, fungi are terrestrial. Fungi make up a diverse kingdom that includes the mold and mushrooms shown in **Figure 18.1**. The body of a fungus is made up of many slender filaments called **hyphae** (*HY fee*), which are barely visible to the naked eye. Each hypha of a fungus has cell walls that contain chitin (*KYT uhn*). Chitin is a polysaccharide that is also found in the outer skeletons of insects.

Fungi digest organic matter

You may think of fungi as mold growing on a loaf of bread or on an orange. Just as you use these items for food, fungi can also use them for nutrients. Unlike plants, fungi lack chloroplasts and cannot carry out photosynthesis. Instead, like animals, they are heterotrophs and obtain energy and nutrients from other organisms. Whereas animals digest their food with enzymes in their stomachs, fungi secrete enzymes that digest food outside their bodies. They then absorb the nutrients. Thus, mold growing on bread is actually digesting it a little at a time, and absorbing the nutrients.

Figure 18.1
Fungi come in many forms, . . .

a . . . from mold growing on a melon, . . .

b . . . to this poisonous fly agaric mushroom . . .

c . . . and this edible shiitake mushroom.

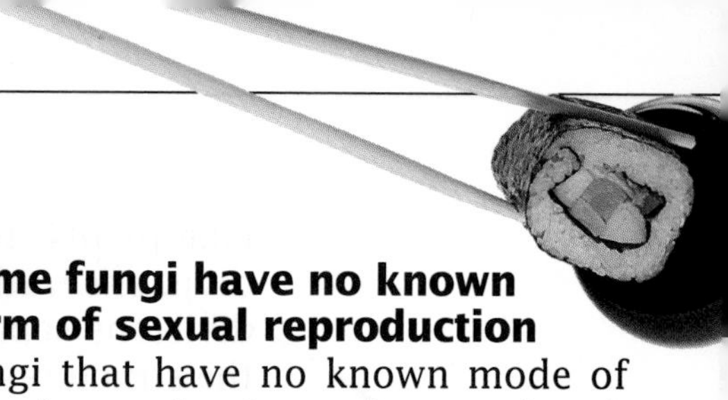

Many fungi form clublike structures

Most of the organisms you call mushrooms belong to a third division of fungi, the basidiomycetes (*buh sid ee oh MY seets*). This division also includes toadstools, puffballs, rusts, and smuts. These fungi form reproductive structures shaped like wooden clubs, and thus are called club fungi (the prefix "basidio" means "little club"). The clublike structures are found under the caps of mushrooms and toadstools or under the "shelves" of bracket fungi. Spores form and mature on the tips of the clubs and are released.

Some fungi have no known form of sexual reproduction

Fungi that have no known mode of sexual reproduction make up a fourth division called the deuteromycetes (*doot uh roh MEYE seets*), or the Fungi Imperfecti. Some of these fungi have particular economic importance. The flavors of gourmet cheeses such as Roquefort and Camembert are produced by particular strains of molds in this group. Another species is used to ferment soy sauce, shown in **Figure 18.3**. In addition, most of the fungi that cause human skin diseases, such as athlete's foot, are also deuteromycetes.

Figure 18.3
Soy sauce, used to flavor sushi (right) and other foods, is made by the fermentation of soybeans by certain fungi.

Fungi in Nature

Figure 18.4
Lichens are resistant to extremes in temperature and moisture. The lichen *Caloplica elegans* is shown here growing on a rock overlooking the St. Lawrence River in upstate New York.

There are very few places on land that at least one species of fungi does not call home. Fungal spores are found in almost any environment. That is why many molds and mushrooms seem to spring up in any location with the right amount of moisture and food. As you read in Chapter 10, many fungi grow in or on the roots of certain plants, in symbiotic associations called mycorrhizae. When mycorrhizal fungi are present, they help transfer nutrients from the soil to the roots and enable plants to thrive. The symbiosis of fungi and plants in mycorrhizae played a critical role in the successful invasion of land by life from the sea. Many of the earliest known plants had mycorrhizae. Fungi provided mineral nutrients for the plants, and plants provided photosynthetically produced energy for the fungi. Recently, researchers have found that the destruction of mycorrhizae by acid rain is playing a key role in the deaths of many forests. Without the mycorrhizae, forest trees are unable to absorb minerals from soil.

If you have ever been in the woods, you may have noticed rocks or logs spotted with orange or green patches called lichens, as shown in **Figure 18.4**. A **lichen** (*LY kuhn*) is an organism that consists of a fungus and an alga living in a symbiotic relationship. The fungus absorbs minerals and other nutrients from the rock and retains water the alga needs for photosynthesis. In turn, the alga produces carbohydrates that the fungus absorbs as food. Lichens can live in very harsh environments. For instance, they are found high on the slopes of Mt. Everest in the Himalaya Mountains of Asia.

Because they absorb water and nutrients directly from the air, lichens are extremely susceptible to environmental pollution. For example, lichens are generally absent in and around cities because automobile traffic and industrial activity produce pollutants that destroy chlorophyll. As a result, the alga cannot make food for itself or for the fungus. Because of this sensitivity, scientists can use lichens as indicators of air quality.

Fungi and Human Life

Figure 18.5

In 1928, Alexander Fleming discovered that *Penicillium notatum* killed bacteria. Ten years later, the drug penicillin was purified. Penicillin is effective in curing several bacterial diseases including pneumonia, scarlet fever, diphtheria, and many others. Today, penicillin is the world's most widely used antibiotic.

When most of us think of fungi, we think of decay—mold spoiling bread, mildew speckling a shower curtain, or athlete's foot between our toes. However, there are many ways in which fungi work for our benefit.

Take the single-celled ascomycete yeast, for example. How would you be able to enjoy freshly baked bread without *Saccharomyces cerevisiae*, better known as common baker's yeast? *Saccharomyces* is a fungus that is able to turn sugar into carbon dioxide or alcohol. These byproducts enable us to make bread and alcoholic beverages.

Fungi provide medically valuable compounds

If you forget about that orange in the kitchen, it may become covered with a bluish-white fungus called *Penicillium*. In 1928 Alexander Fleming noticed that bacteria he was growing in petri dishes had been killed by the fungus *Penicillium notatum*.

Fleming's accidental discovery, shown in **Figure 18.5**, led to the development of the world's most widely used antibiotic, penicillin.

Other fungi have also been used to revolutionize the field of medicine. In 1972 the Swiss immunologist Jean Borel found a type of fungus that produces a substance capable of suppressing the immune system's response to transplanted organs. This substance, called cyclosporine, opened up a new frontier in medicine. Before it became available in 1979, fewer than half of all kidney transplants were successful. Now, with the widespread use of cyclosporine, the survival rate has risen to 90 percent.

Yeast is now a hot topic in many scientific laboratories. Because yeast cells are eukaryotes, they make better genetic engineering subjects for proteins than do bacteria. Research with yeast may lead to innovative treatments for diseases such as cancer and AIDS.

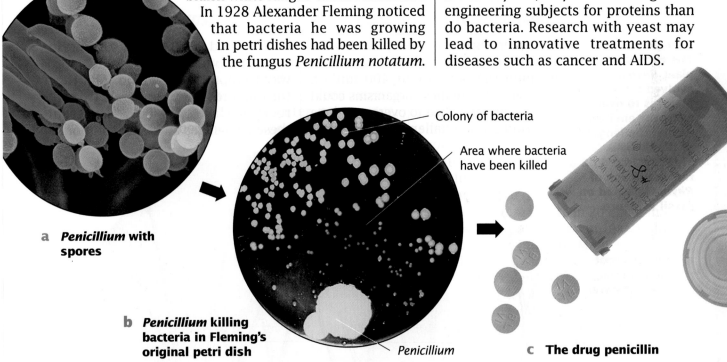

a **Penicillium* with spores*

b **Penicillium* killing bacteria in Fleming's original petri dish*

Colony of bacteria

Area where bacteria have been killed

Penicillium

c **The drug penicillin**

Section Review

❶ **Describe some reasons why biologists have classified fungi into a separate kingdom.**

❷ **How do fungi obtain nutrients?**

❸ **Summarize the methods of reproduction for the four divisions of the kingdom Fungi.**

❹ **How have fungi proven valuable to human life?**

Does Soil Type Affect the Germination of Seeds?

Objectives

In this investigation you will:
- *design* an experiment to test a hypothesis
- *compare* the germination of seeds to the conditions of different soil types

Materials

- lima bean seeds
- paper cups (or small flowerpots)
- potting soil
- sand
- gravel
- powdered clay
- water

Prelab Preparation

1. Review what you have learned about scientific methods by answering the following questions:
   - What is a hypothesis?
   - What is a control experiment?

2. A gardener wonders how the conditions of the soil in the backyard might affect the germination of lima bean seeds. What properties of the soil might affect the rate of germination of the seeds?

Procedure: Determining If Soil Type Affects Germination

1. Form a cooperative group of four students. Work with a member of your team to complete steps 2–11.

2. You and your partner are to design an experiment that demonstrates the effects of soil type on seed germination. With your partner, discuss how you might complete your task, using the materials on your desk.

3. State a hypothesis that addresses the following question: *Does the type of soil affect the germination of a seed?*

4. Using the materials on your table, design a control experiment that tests your hypothesis. Keep in mind that lima bean seeds will usually germinate when they are planted about 2 cm (1 in.) below the surface of the soil and are adequately watered. *What steps will you use in your experiment?*

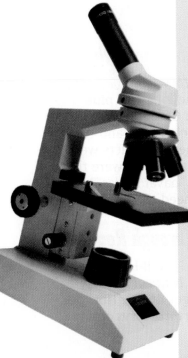

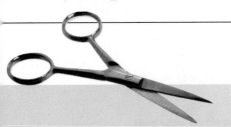

5. For each step specify the materials you will use, the amounts that will be used, how measurements will be made, and the conditions under which the procedures will take place.

6. Working with your partner, proceed with your experiment only after having your experimental design approved by your teacher. *What makes your experiment a control experiment?*

7. Determine a period of time to observe your experiment. Make a table similar to the one shown below for recording your data.

Observation of Seeds Grown in Different Types of Soils

Soil type	Day 1	Day 2	Day 3	Day 4	Day 5
Potting soil					
Sand					
Gravel					
Clay					

8. Record your observations in your table. After the period of observation, combine your data with that of the other team in your group.

9. Make a bar graph showing the number of seeds that germinated in each type of soil.

10. Explain how the information does or does not support your hypothesis. If your data show the need to change your hypothesis, state a new one. Explain why you changed your original hypothesis.

11. Clean up your materials and wash your hands before leaving the lab.

Analysis

1. *Summarizing Data*
   Summarize the data collected throughout your experiment.

2. *Analyzing Information*
   Do the results of your observations support your hypothesis? Explain.

3. *Analyzing Methods*
   What is the importance of keeping careful records when doing an experiment?

4. *Communicating Information*
   Of what value is sharing observations with other scientists?

5. *Making Inferences*
   What aspect of soil is most important to the germination of seeds?

6. *Applying Concepts*
   How might a gardener apply the principles illustrated in this investigation?

Thinking Critically

1. What relationship do you observe between rates of growth of the lima bean seeds and the type of soil?

2. Is a soil best suited for germinating seeds likely to be the best soil for growing plants? Explain your reasoning.

Chapter 19

Plant Form and Function

Review

- ratio of surface area to volume (section 3.1)
- plant cells (Section 3.3)
- photosynthesis (Section 5.3)
- the terms *osmosis* and *cotyledon* (Glossary)

In addition to their beauty, plants provide us with oxygen to breathe and food to eat. In this chapter you will learn how the structure of a plant helps it carry out its daily activities.

THE STRUCTURE OF A PLANT ENABLES IT TO CAPTURE ENERGY AND BUILD MOLECULES DURING PHOTOSYNTHESIS. BELOW THE GROUND, ROOTS ANCHOR THE PLANT AND ABSORB WATER AND MINERALS. ABOVE THE GROUND, SHOOTS HOLD THE LEAVES HIGH TO ABSORB SUNLIGHT AND TAKE IN CARBON DIOXIDE FOR PHOTOSYNTHESIS. NO ANIMAL MAKES ITS OWN FOOD IN THIS WAY.

19.1　The Plant Body

Objectives

❶ Describe the functions of roots and shoots.

❷ Compare and contrast vascular tissues, ground tissue, and dermal tissue.

❸ Describe the role of meristems in plant growth.

❹ Distinguish between monocots and dicots.

Roots

Although orchids, cactuses, and pine trees do not look alike, these and all plants are similar in the way they grow. Throughout their lives, vascular plants develop by adding new cells at the ends of roots and shoots. The roots are the part of the plant below the ground. They anchor the plant in the soil and absorb water and minerals from the soil. **Figure 19.1** shows the roots of a radish. Roots make up about one-third of the total dry weight of a plant. The roots of most plants do not usually extend into the earth beyond a depth of 3 m to 5 m (9–15 ft.). However, the roots of mesquite trees can grow to be more than 50 m (about 164 ft.) long. The roots of some plants have specialized functions such as food storage or water storage. For example, when you eat a carrot or sweet potato, you are eating a root that stores large amounts of carbohydrates.

Root hairs increase the surface area of a root

When a seed germinates, the first part to start growing is the root of the new plant. The end of the root is covered by the root cap, a thimble-like cluster of cells that protects the tip of the root. The root elongates in the region just above its root cap. As the root grows longer, the cells on the outside of the root cap are sloughed off and replaced by the new cells underneath. Above this region are slender projections called root hairs. Each individual root hair is a single cell that penetrates the space between soil particles. Virtually all absorption of water takes place through root hairs, which greatly increase the surface area of the root.

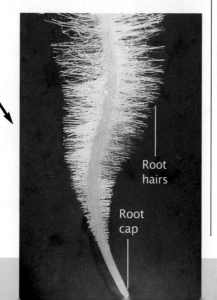

Figure 19.1
The roots of plants absorb water and minerals necessary for growth. Absorption occurs through the root hairs. Cell division occurs inside the root cap.

Root hairs

Root cap

Shoots

Figure 19.2
To survive in different environments, plants have evolved different types of leaves and stems.

The shoot is the portion of the plant that consists of stems and leaves and in many cases, flowers and fruits. The stem supports the leaves and enables them to receive sunlight. Leaves are the major site of photosynthesis. Compare the shoots of the rose plant, cactus, and strawberry plant shown in **Figure 19.2**.

Stems connect roots to leaves

Stems support leaves, flowers, and fruits, usually holding them off the ground. They vary greatly in shape and size from one plant species to another, as is evident when comparing a redwood trunk with a grass stalk. Stems contain vascular tissues that transport substances between roots and leaves. Some stems are modified for storage. For example, cactuses are able to store a considerable amount of water inside their stems. White potatoes are underground stems swollen with stored food, usually in the form of starch.

Leaves are the main sites of photosynthesis

In Chapter 5 you learned that photosynthesis enables plants to capture sunlight energy and use it to make carbohydrates. Leaves are specialized to carry out photosynthesis. They have many cells packed with chloroplasts, the chlorophyll-containing organelles essential to photosynthesis.

Regardless of their size, most leaves are relatively thin and flat. A thin, flat shape maximizes the ratio of surface area to volume. This helps a plant efficiently capture the sunlight and carbon dioxide needed for photosynthesis. A waxy outer layer called a cuticle prevents the leaf from losing too much water and drying out. Carbon dioxide enters the leaf, and water and oxygen exit through tiny pores called **stomata** (*stoh MAH tuh*). Inside the leaf are layers of photosynthetic cells containing chloroplasts. Bundles of vascular tissue run through the leaf, moving water with its dissolved minerals to photosynthetic cells and carrying the products of photosynthesis away from them. These bundles are the veins you see when you hold a leaf up to the light. You can read more about leaves in the *Tour of a Leaf* on page 405.

a The stem of this rose plant has prickles, sharp outgrowths that serve as a type of protection.

b This cactus has modified leaves called spines that protect it from predators and extreme temperatures.

c The horizontal stems of this strawberry plant spread over the ground and ultimately may root to establish a new plant.

Tour of a Leaf

A leaf is not as simple as it might first appear. It has an intricate architecture well suited to its function—photosynthesis.

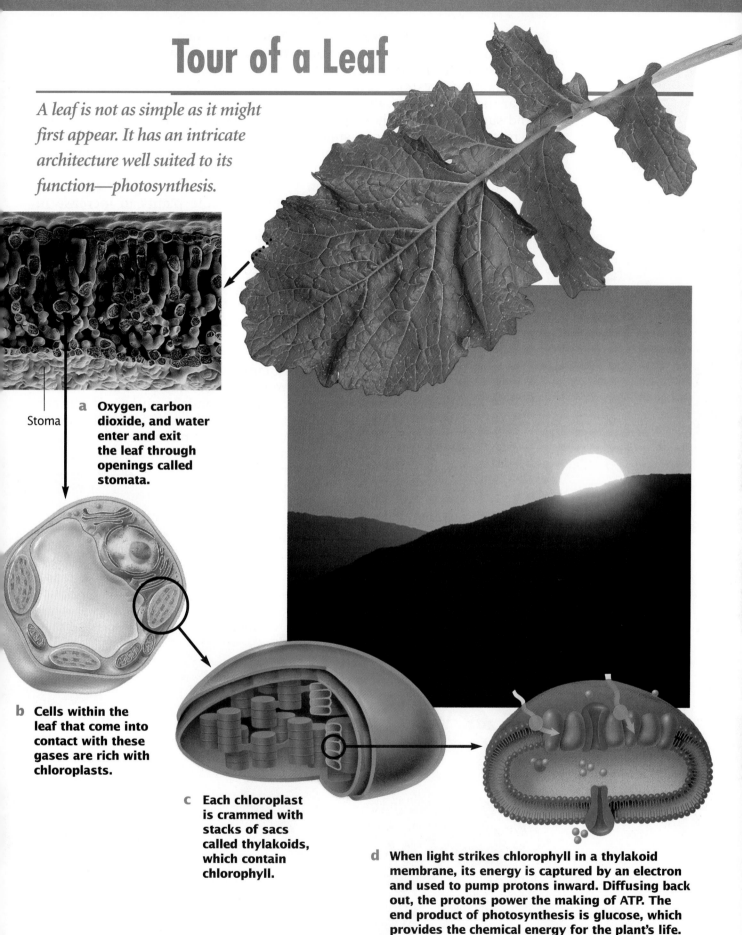

Stoma

a **Oxygen, carbon dioxide, and water enter and exit the leaf through openings called stomata.**

b **Cells within the leaf that come into contact with these gases are rich with chloroplasts.**

c **Each chloroplast is crammed with stacks of sacs called thylakoids, which contain chlorophyll.**

d **When light strikes chlorophyll in a thylakoid membrane, its energy is captured by an electron and used to pump protons inward. Diffusing back out, the protons power the making of ATP. The end product of photosynthesis is glucose, which provides the chemical energy for the plant's life.**

Plant Tissues

The leaves, roots, and stems of plants are made of different tissues. As you learned in Chapter 18, vascular tissue is the "plumbing system" of a plant. Vascular tissue enables water, minerals, and sugars made by photosynthesis to move through the roots, stems, and leaves of a plant. Water and minerals are transported through vascular tissue called **xylem** (*ZY luhm*), shown in **Figure 19.3a**. Sugars are transported through another vascular tissue called **phloem** (*FLOH ehm*).

The bulk of the plant body is made of **ground tissue**, which supports the vascular tissue. Some cell types found in ground tissue are designed for storage, while other cell types have thickened walls that lend support to the plant. A layer of tightly packed, flattened cells makes up the plant's **epidermis**. These cells secrete a waxy substance that protects the plant from water loss.

Meristems are regions of active cell division

Plants grow in regions of active cell division called **meristems** (*MEHR uh stemz*), shown in **Figure 19.3b**. Every time a cell in the meristem divides, one cell remains in the meristem and the other cell becomes more specialized as it matures.

Meristems at the tips of roots and shoots enable plants to increase in length. The lengthening of roots and shoots is called primary growth. Annuals—plants that die after one season of growth—may have only primary growth. Woody plants—trees and shrubs—show secondary growth at meristems that run like cylinders through stems. Secondary growth causes plant bodies to thicken by producing new xylem and new phloem. Wood consists mainly of accumulated xylem. Secondary phloem forms the inner part of bark. If a ring of bark is stripped from the trunk a tree's roots will not receive the sugars produced during photosynthesis, and the tree will die.

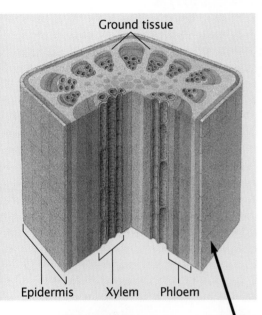

Ground tissue

Epidermis   Xylem   Phloem

Figure 19.3

a **The xylem of this *Coleus* plant is made of elongated cells that connect end to end and transport water throughout the plant. The phloem contains cells that transport sugars. The epidermis covers and protects the plant body. The ground tissue supports the vascular tissue.**

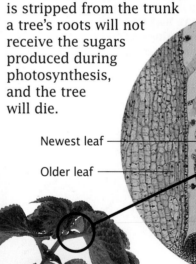

Newest leaf

Older leaf

Bud

b **This vertical cross section through the shoot tip of a *Coleus* plant shows the meristem that produces new plant cells.**

Coleus plant

Flowering plants are either monocots or dicots

As you learned in Chapter 18, cotyledons are leaflike structures that store or absorb food in a seed. The number of cotyledons is used to classify flowering plants into two large groups: the monocots (about 65,000 species) and the dicots (about 170,000 species). Monocots include grasses, palms, lilies, and orchids. The crop plants wheat, corn, rice, rye, and barley are also monocots. Dicots include many of the familiar flowering plants, shrubs, trees, and cactuses. Monocots and dicots are similar in structure and function, but they differ in distinctive ways. Evidence suggests that monocots may have evolved from primitive dicots. **Table 19.1** shows some of the characteristic differences between most monocots and dicots.

Table 19.1 Summary of Differences Between Monocots and Dicots

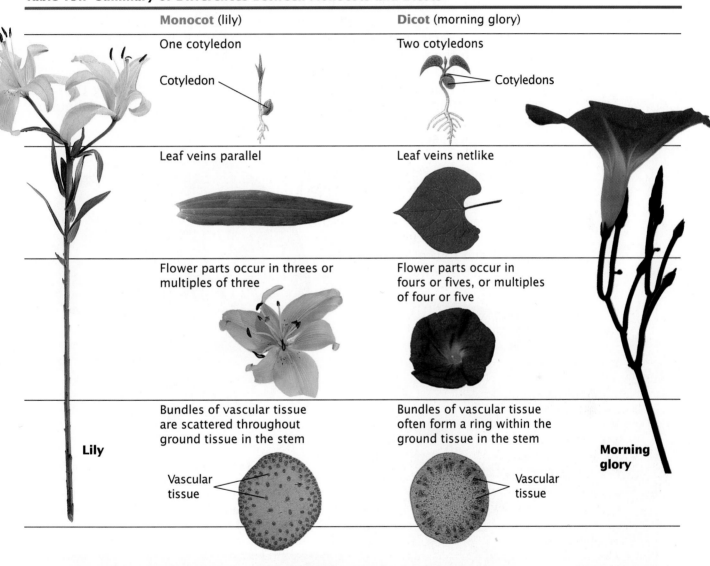

Monocot (lily)	Dicot (morning glory)
One cotyledon	Two cotyledons
Leaf veins parallel	Leaf veins netlike
Flower parts occur in threes or multiples of three	Flower parts occur in fours or fives, or multiples of four or five
Bundles of vascular tissue are scattered throughout ground tissue in the stem	Bundles of vascular tissue often form a ring within the ground tissue in the stem

Lily

Morning glory

Section Review

❶ What are the functions of roots and shoots?

❷ Explain the functions of vascular tissue, ground tissue, and epidermis.

❸ How do meristems enable a plant to grow?

❹ What do the similarities between monocots and dicots suggest?

Eloy Rodriguez: Biochemist

How I Became Interested in Biology

"I grew up in the county with the lowest average income per person in the United States—Hidalgo, Texas. I lived close to an open field where my grandfather would take me on nature walks. He would stop and point out different plants and insects, telling me all he knew about them. We would often gather seeds and bring them home to germinate and cultivate. The plants we grew brought life to our otherwise dreary-looking house. Since then, I have always been fascinated by the life around me.

"My family was very close. I was one of 67 cousins living within a five-block radius. The entire family stressed the importance of education. As a result, 64 of us earned undergraduate degrees, and four of us eventually received Ph.D.s. It was a good thing that I paid no attention to a high school counselor who advised me to go to a technical school, which was standard advice for many minority students at the time.

As an undergraduate, I was hired to clean a lab. I met a researcher who disliked laboratory work. He accepted my offer of help and I learned to extract compounds. I loved doing those experiments. My work in the laboratory changed my life."

In addition to his research, Dr. Rodriguez also runs several programs that help minority and female students to become engaged in science and the process of critical thinking.

Name:	**Eloy Rodriguez**
Home:	**Irvine, California**
Employer:	**University of California–Irvine**
Personal Traits:	• **Creative**
	• **Hard-working**
	• **Fair**
	• **Demanding**
	• **Compassionate**

Research Focus

Dr. Rodriguez and his colleague Dr. Richard Wrangham of Harvard have developed a new scientific discipline, which they call zoopharmacognosy—the chemistry of plants used medicinally by animals. It combines anthropology, behavioral ecology, organic chemistry, and botany. Dr. Rodriguez's research has led to the discovery of new drugs that are useful against cancer and viruses.

In his research, Dr. Rodriguez isolates natural drugs, determines their structure, and tries to understand how they work. How do they stop cancer cells? How do they kill parasites? By combining medicine, chemistry, the study of animal behavior, and botany, Dr. Rodriguez and his colleagues have come up with a multidisciplinary approach to answering these questions.

One plant Dr. Rodriguez has studied is used by wild chimpanzees to remove parasites. By learning about the plant's structure, he discovered a drug that is effective against human intestinal parasites, fungi that cause skin infections, and possibly cancer.

The Wild World of Biochemistry

Career Path

High School:
- Chemistry
- Math

College:
- Zoology
- Chemistry
- Taxonomy
- Genetics
- Field Ecology

Graduate School:
- Biology
- Plant Chemistry

To determine how a plant can cure a disease or ailment, Dr. Rodriguez studies chemicals extracted from plant cells, such as this one.

"I have a passion for what I do. I search the world for natural medicines. In the past 10 years, I've visited the rain forests deep in South America and Africa. The indigenous people taught me about their medicines and the plants from which they are derived. To find out more about these natural medicines, I brought many plants back to my laboratory. By extracting chemicals out of the plant cells and running various experiments on them, I've been able to study how these chemicals affect human cells that are infected by bacteria or other damaging organisms. Essentially, I try to figure out how these plant compounds act as medicines.

"Having investigated the biochemistry of many plants from all over the world, I believe that thousands of species of plants possess medicinal properties and other undiscovered uses. Many are potential sources of oils, food, pesticides, and other products. However, these applications may go undiscovered unless we start conserving our limited resources, many of which are already endangered. I hope that many other people will become interested in researching plants that may be useful to humans. Low-cost, natural drugs have a bright future in developing nations where more sophisticated and expensive synthetic medicines are beyond the reach of the average pocketbook."

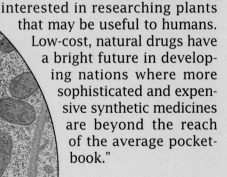

ALTHOUGH A PLANT MAY NOT APPEAR AS COMPLEX AS AN ANIMAL, ITS INTERNAL STRUCTURE IS MORE COMPLEX THAN YOU MIGHT THINK. ITS VASCULAR SYSTEM SENDS WATER, CARBOHYDRATES, AND MINERALS FROM ONE PLANT PART TO ANOTHER. LIKE YOUR BODY, A PLANT REGULATES ITS GROWTH WITH HORMONES, CHEMICALS THAT ACT AS MESSENGERS TO COORDINATE ACTIVITIES.

19.2 How Plants Function

Objectives

❶ Summarize how water is transported through plants.

❷ Explain the manner in which carbohydrates are moved through plants.

❸ State the functions of three kinds of plant hormones.

❹ Explain how plant growth responds to changes in seasons.

How Water Moves Through Plants

Scale and Structure

Explain how the structures of the vascular tissues make them an efficient means of transportation for water, minerals, and sugars.

In northern California, giant redwood trees can be up to 117 m (348 ft.) tall. How do trees move water from roots deep in the soil up to branches and leaves at such heights? Water enters the vascular system by osmosis. But it takes more than just osmosis to move water up to the top of a tall tree!

Several forces cause the movement of water in plants. Water enters root hairs and moves into the xylem tissue. Xylem cells form long narrow tubes. The water in these tubes is attracted to the xylem cell walls. This attraction tends to pull water up the tubes. The force of attraction that causes water to move up narrow tubes such as those in the xylem is called **capillary action**. **Figure 19.4a** shows the xylem of a zucchini plant's stem.

The major force behind the movement of water in plants, however, is the fact that when water evaporates from leaves, the upward movement of one molecule tugs on the molecules below. Thus the loss of water from the leaves is responsible for water flow through the plant. More than 90

Figure 19.4

a Capillary action is the name for one of the mechanisms that moves water from the roots to the leaves.

Xylem   Phloem

Zucchini plant

b Transpiration is the process by which water is lost to the environment through the above-ground portion of the plant.

c Stomata, most frequently found on the underside of a leaf, open and close to regulate water loss.

Phloem

Xylem

d Translocation is the process by which carbohydrates made during photosynthesis are carried to the rest of the plant via phloem.

percent of the water taken up by the roots of a plant is lost to the atmosphere as water vapor. Water passes primarily through the stomata in the leaves, as shown in **Figure 19.4b**. The loss of water by the leaves and the stem of the plant is called **transpiration**. Water that evaporates from leaves is continually replaced with water entering the roots.

Stomata regulate water loss

The only way that plants can control water loss is to close their stomata. Stomata are numerous—an average-sized sunflower leaf has about 2 million stomata. Each stoma is formed by two pickle-shaped cells called guard cells. The walls of guard cells are flexible and thicker on the inner surface than on the outer surface. When guard cells are swollen with water, they curve outward and the stoma opens. When guard cells lose water, they collapse and the stoma closes. The stomata of the zucchini plant, shown in **Figure 19.4c**, are open during the day and closed at night. Therefore, very little water is lost from leaves at night.

How carbohydrates are transported

Most of the carbohydrates made in the photosynthetic green parts of plants are moved through the phloem tissue to other parts of the plant. **Figure 19.4d** shows the phloem tissue of a zucchini plant. The transport of sugars made by photosynthesis from the leaves to the rest of the plant is known as **translocation**. Translocation is responsible for moving carbohydrate building blocks to actively growing regions of the plant. In some plants, such as potatoes or sugar beets, carbohydrates are concentrated in storage structures. When the plants need these carbohydrates, they can be broken down into smaller molecules and moved through the phloem. **Figure 19.4** shows the processes involved in the transport of water, minerals, and carbohydrates through the tissues of a zucchini plant.

Regulating Plant Growth: Plant Hormones

Plants continue to grow throughout their lives. With plenty of water, air, and sunlight, leaves and branches will form and develop, flowers will blossom, and fruit will ripen and drop. Where and how these events occur is controlled by chemical messengers called hormones. As you will see, hormones control many aspects of plant life.

Auxins stimulate the elongation of plant cells

Have you ever seen a houseplant bending toward the light? This growth pattern is caused by plant hormones called auxins (*AWK sihns*). Auxins cause cell walls to become more flexible. As a result, a cell grows longer. Experiments have shown that auxins migrate to dark sides of plants in response to light, causing the cells there to elongate.

The presence of auxins is what makes a plant bend toward the light as it grows. As shown in **Figure 19.5,** when auxins are present, cells in that portion of the plant elongate.

Gardeners often use synthetic auxin in a commercial powdered product to dust the cut ends of leaf cuttings. The powder contains auxins that speed the formation of roots for the new plant. Some types of auxins have been used as herbicides that cause weeds to literally grow themselves to death.

Auxins can also inhibit plant growth. For example, the presence of auxins at the tip of the shoot blocks the active growth of the branches along the sides of a plant. Gardeners often make plants grow more side branches by pinching off the shoot tip at the top of the plant where auxins are produced.

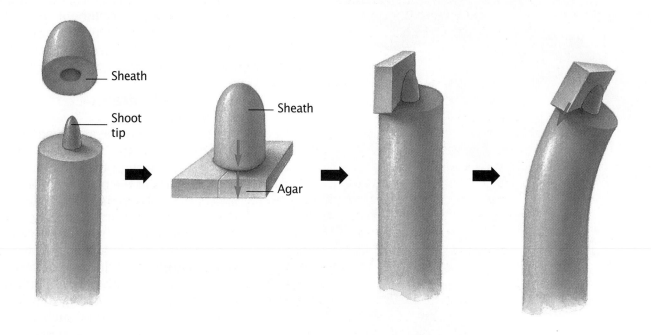

Figure 19.5

a **To better understand how auxins work, scientists cut off the protective sheath surrounding the tip of a young oat seedling . . .**

b **. . . and placed it on a block of a gelatinlike substance called agar. Auxins (red arrow) diffused into the agar.**

c **When a piece of the block was put on one side of another seedling's tip, . . .**

d **. . . the plant began to grow in the opposite direction. Cells in contact with auxins in the agar elongated.**

Figure 19.6

a This California poppy was grown under normal conditions.

b This California poppy was treated with gibberellin so that it would grow more quickly, and be larger than a normal poppy.

Gibberellin stimulates rapid growth

In 1926 Japanese scientists reported that rice seedlings infected with the fungus *Gibberella* had stems that grew abnormally long. They found that extracts of the fungus brought about the same effects in uninfected plants. Nine years later, the substance was identified and called gibberellin, after the fungus the scientists studied. Gibberellins not only dramatically increase stem growth, but also cause some kinds of plants to germinate and to flower. Their most dramatic effects can be seen in dwarf or miniature strains of plants. These strains are often genetic mutants that do not produce gibberellins. When treated with gibberellins, they grow as tall as normal varieties. Gibberellins are used to increase the size of plants, such as the California poppy shown in **Figure 19.6b**. They are also used to delay the ripening of citrus fruits and to speed the flowering of strawberries.

Ethylene controls the ripening of some fruits

In ancient China, farmers ripened fruits in rooms that contained burning incense. More recently, citrus growers ripened their fruits in rooms that contained stoves burning kerosene. In both instances, ethylene gas brought about the results. In 1934 scientists discovered that ethylene is produced naturally by fruits, such as the bananas shown in **Figure 19.7**, as well as by flowers, seeds, leaves, and even roots. Ethylene also appears to be the main factor in the formation of specialized cells that form before leaves drop off plants.

Today ethylene is used commercially to ripen bananas, honeydew melons, and mangoes. These fruits are harvested when still green. Some growers still use natural ethylene to ripen pears and peaches by wrapping each fruit individually in tissue paper. The paper keeps ethylene from escaping from the fruit and hastens ripening.

Figure 19.7
Ethylene is the hormone that determines when bananas and other fruits ripen.

Other Factors Affecting Plant Growth

The growth of many plants varies with the seasons. Have you ever wondered why? The growth of plants such as strawberry plants and apple trees is affected by the length of daylight. These plants contain a pigment that is sensitive to the amount of daylight. As light gives way to darkness, the pigment initiates changes that alter growth.

Some plants respond to changes in day length by producing flowers. Long-day plants, such as the irises shown in **Figure 19.8**, produce flowers during the longer days of summer, when the nights are shorter. Short-day plants, such as goldenrods, begin to produce flowers during the short days of spring or fall. The response of plants to periods of light and dark is called **photoperiodism**.

Plants have other mechanisms that enable them to grow in response to changes in their environment. Occasionally, some parts of plants grow faster than other parts, causing the plant to bend. You have already read that uneven light can cause auxins to stimulate growth of cells on the shady sides of stems. Uneven growth causes the stem to bend toward the light. Roots respond differently; they grow downward, whereas stems grow upward. These movements are clearly beneficial. Stems that grow upward can receive more light than those that do not. Roots that grow downward will reach water and minerals. Differences in concentrations of auxins are responsible for this growth with or against gravity.

Length of Darkness Determines When Plants Bloom

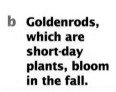

Figure 19.8

a Irises are long-day plants. They bloom in the summer when the days are longer.

b Goldenrods, which are short-day plants, bloom in the fall.

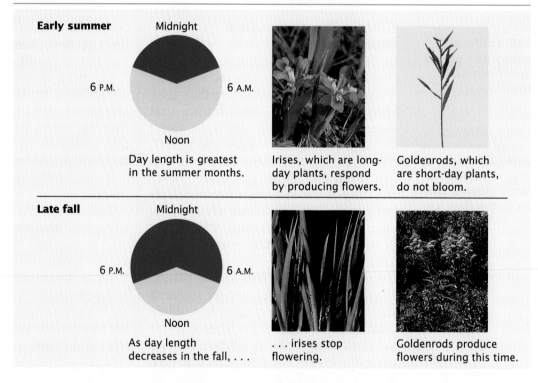

Early summer

Day length is greatest in the summer months.

Irises, which are long-day plants, respond by producing flowers.

Goldenrods, which are short-day plants, do not bloom.

Late fall

As day length decreases in the fall, . . .

. . . irises stop flowering.

Goldenrods produce flowers during this time.

Section Review

❶ Trace the path of water through a plant.

❷ How are carbohydrates moved through a plant?

❸ List three plant hormones and explain their effects on plant development.

❹ How can a change in season affect plant growth?

FLOWERING PLANTS ARE THE DOMINANT FORM OF PLANT LIFE FOUND ON EARTH TODAY. WHILE YOU MAY APPRECIATE FLOWERS FOR THEIR DECORATIVE ROLE, THEY ARE ACTUALLY THE VISIBLE EVIDENCE OF A FLOWERING PLANT'S ABILITY TO REPRODUCE. EACH FLOWER HELPS ENSURE THAT A PARTICULAR PLANT SPECIES WILL PRODUCE MORE OF ITS OWN KIND.

19.3 *Reproduction in Flowering Plants*

Objectives

❶ **Describe the structure of a flower.**

❷ **Summarize the processes of pollination and fertilization in typical flowering plants.**

❸ **Explain the survival value of double fertilization.**

❹ **Describe some adaptations of flowers that help ensure pollination.**

Architecture of a Flower

Figure 19.9 shows the sexual reproductive structure of angiosperms, the flower. Most flowers share certain basic features. They consist of four whorls, or circles, of parts. The two outer whorls of a flower, the sepals and petals, protect the flower and attract insects and animal pollinators. The inner whorls of a flower contain the male and female gametophytes. One inner whorl of the flower consists of **stamens** (*STAY mehnz*), the male parts of the flower that produce pollen grains. Pollen is produced in the tip of the stamen, an area called the **anther**. The innermost whorl of the flower consists of the pistil. The base of the **pistil** is the ovary.

Inside an ovary, egg cells are produced within **ovules**. The tip of the pistil is the stigma, the site of pollination.

Petal
Sepal

Figure 19.9
If you slice the flower of a lily in half, you can see all of its reproductive structures.

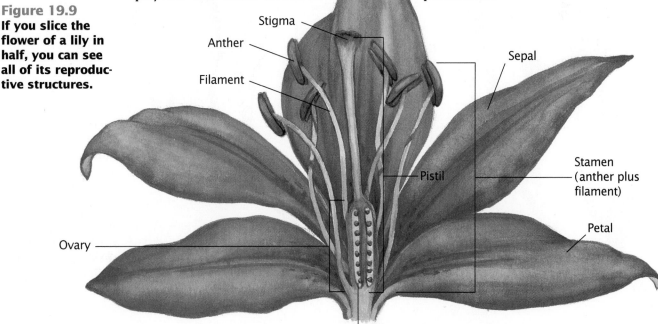

Stigma
Anther
Filament
Sepal
Pistil
Stamen (anther plus filament)
Ovary
Petal
Ovule

Pollination and Fertilization

For sexual reproduction to occur, pollen containing the male gametes must reach the stigma. Since flowering plants cannot move about to seek their mates, they must rely on other methods to move pollen. Pollination is the transfer of pollen grains from the anther to the stigma, the top of the pistil. In some plants, pollen falls from the anthers onto the stigma of the same flower, which may result in **self-pollination**. Wind can transfer pollen from the flowers of one plant to the flowers of another. However, most flowering plants, from magnolia trees to orchids to the morning glory in **Figure 19.10a**, depend on animals to transport their pollen. The flowers of the plants must attract pollinators. In Chapter 13 you learned how flowers and their pollinators coevolved, or evolved in response to changes in each other. Some flowers, such as the morning glory shown in **Figure 19.10b**, have attention-catching "advertisements" such as brightly colored petals or scents. Others secrete a sugary liquid called nectar, which pollinators use as food. As an animal explores these flowers, pollen sticks on its body. When the animal wanders away, it transfers the pollen to the stigma of another flower. By attracting animals, flowering plants can ensure **cross-pollination**. Cross-pollination is the transfer of pollen to another plant of the same species. This ensures genetic recombination, and thus produces wider genetic variety than does self-pollination.

Fertilization in flowers occurs in two stages

When the pollen grain of a flowering plant reaches the stigma, a pollen tube may begin growing through the tissue of the pistil and into the ovule, which contains egg cells. Each pollen grain contains two sperm, which travel down

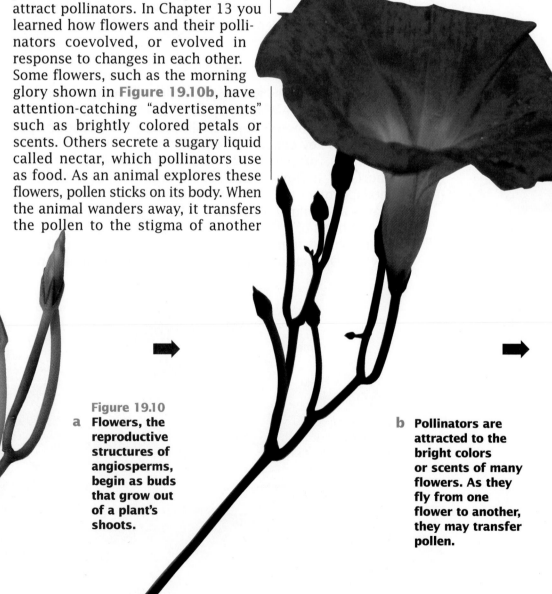

Figure 19.10

a **Flowers, the reproductive structures of angiosperms, begin as buds that grow out of a plant's shoots.**

b **Pollinators are attracted to the bright colors or scents of many flowers. As they fly from one flower to another, they may transfer pollen.**

the pollen tube. One sperm fuses with the egg to form a zygote, as shown in **Figure 19.10c**. The other sperm fuses with two nuclei inside the ovule, forming a tissue that will become a source of nutrition for the plant developing inside the seed. This event, in which one sperm fuses with an egg and a second sperm fuses with two nuclei, is called **double fertilization**. Double fertilization has great survival value because each new generation carries its own initial source of nutrition.

Remember that pollination is not the equivalent of fertilization. Fertilization involves the union of egg and sperm and may not occur until weeks or months after pollination has taken place. Sometimes it may not follow pollination at all.

Seeds and fruits form after fertilization

Fertilization causes rapid changes to occur in the flower. The ovule develops into a seed, often with a tough coat protecting the developing plant and its food supply. As you read in Chapter 18,

development of the seed was a major factor in the success of flowering plants on land. Seeds enable plant species to survive in unfavorable environments. As the seed develops, the ovary grows larger and develops into a **fruit**. A fruit is an enlarged ovary of a flowering plant that contains seeds. You can see the fruit of a morning glory plant in **Figure 19.10d**. Many "vegetables," including tomatoes, string beans, cucumbers, and squash, are technically fruits. Whether a flower's ovary will turn into a fruit depends on whether it is fertilized. If the egg cells go unfertilized, the flower will normally wither and drop.

Some fruits lack seeds

Have you ever wondered why navel oranges and pineapples have no seeds? Scientists describe the ability of plants to develop fruits without the fertilization of eggs as parthenocarpy (*pahr thuh noh KAHR pee*). Seedless varieties of grapes and citrus fruits occur naturally because they contain high levels of the plant hormone auxin. Researchers can induce parthenocarpy in fruits such as tomatoes and watermelons by applying auxin to plants at certain stages in their development.

Pollen tube

Pistil

Nuclei

Pollen tube

Ovule

Egg cell

Sperm

c One sperm in a pollen grain fertilizes an egg. A second sperm fuses with two nuclei inside the ovule to form nutritive tissue. The petals, anthers, and sepals wither as the fruit begins to develop.

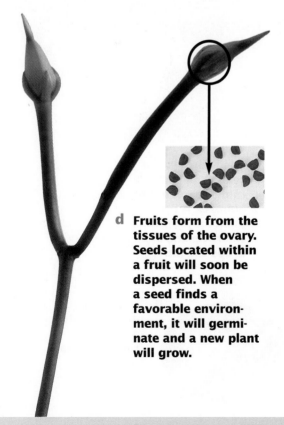

d Fruits form from the tissues of the ovary. Seeds located within a fruit will soon be dispersed. When a seed finds a favorable environment, it will germinate and a new plant will grow.

How Seeds Are Dispersed

Figure 19.11
Plant seeds are dispersed by a variety of mechanisms. Several examples are shown below.

Once mature, seeds are ready to be dispersed. **Figure 19.11** shows how various types of seeds are dispersed. Some fruits, such as the plump, fleshy peaches, tomatoes, and watermelons you buy at the grocery store, are eaten by animals. When fleshy fruits ripen, their sugar content increases and the fruit becomes soft and juicy. Their colors often change from green to bright red, yellow, or orange. These changes, signs that the seeds are ripe and ready for dispersal, aid in catching the attention of hungry animals. When fruits are eaten by birds or mammals, the seeds are spread by passing unharmed through the digestive tract. The success of these sweet, colorful fruits as a method of seed dispersal is important in the coevolution of animals and flowering plants. Plants such as maple trees, tumbleweeds, and dandelions have extremely light fruits and seeds that can be carried by the wind. The fruits and seeds of many plants growing in or near water, like coconuts, are adapted for floating. Some fruits or seeds, such as burdocks, have hooks that catch the feathers, fur, or clothing of passing animals, giving the seed a free ride to a new home.

d Humans can disperse seeds too. Have you ever spit watermelon seeds on the ground? One of those seeds may have germinated and grown into a new watermelon plant.

c The succulent fruit of a tomato plant attracts animals. The seeds are deposited elsewhere after passing through an animal's digestive system.

a The sharp, tough coat of a peach pit cannot be eaten by animals. A new plant may germinate wherever the animal drops the pit.

b Squirrels and other animals carry seeds from one area to another.

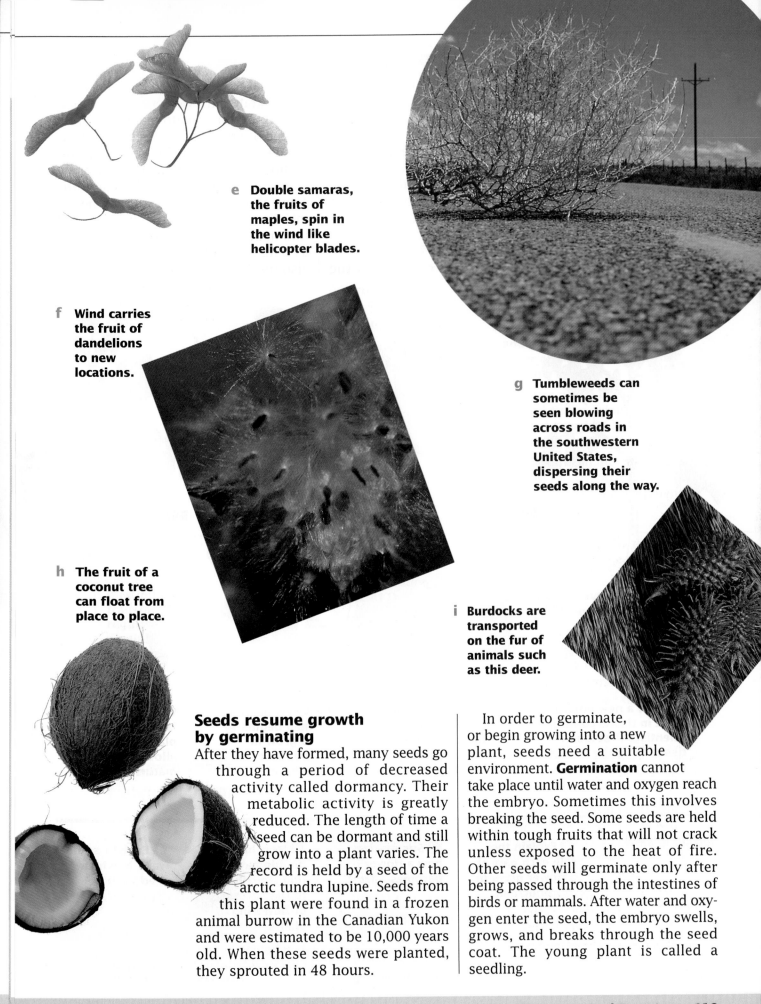

e Double samaras, the fruits of maples, spin in the wind like helicopter blades.

f Wind carries the fruit of dandelions to new locations.

g Tumbleweeds can sometimes be seen blowing across roads in the southwestern United States, dispersing their seeds along the way.

h The fruit of a coconut tree can float from place to place.

i Burdocks are transported on the fur of animals such as this deer.

Seeds resume growth by germinating

After they have formed, many seeds go through a period of decreased activity called dormancy. Their metabolic activity is greatly reduced. The length of time a seed can be dormant and still grow into a plant varies. The record is held by a seed of the arctic tundra lupine. Seeds from this plant were found in a frozen animal burrow in the Canadian Yukon and were estimated to be 10,000 years old. When these seeds were planted, they sprouted in 48 hours.

In order to germinate, or begin growing into a new plant, seeds need a suitable environment. **Germination** cannot take place until water and oxygen reach the embryo. Sometimes this involves breaking the seed. Some seeds are held within tough fruits that will not crack unless exposed to the heat of fire. Other seeds will germinate only after being passed through the intestines of birds or mammals. After water and oxygen enter the seed, the embryo swells, grows, and breaks through the seed coat. The young plant is called a seedling.

Fermentation

Fuel From Plants

Imagine pulling up to a gas station and filling up your car—not with gasoline, but with corn fuel. Corn, as well as wheat, wood chips, grass clippings, and other plants, can be fermented to form so-called biomass fuels such as ethanol and methanol. Some biomass fuels are already being used on a limited scale. Ethanol, for example, can be added to gasoline to improve combustion. Methanol might someday become the fuel of choice in California, which is struggling to cut down on air pollution from millions of motor vehicles.

Concerns about global warming and decreasing supplies of petroleum have served as an incentive to increase research on alternative fuels. Biomass fuels, say advocates, have a lot of attractive characteristics. They are renewable, since the starting materials are plants or plant wastes such as corn plants left in the field after the corn is harvested. They burn cleaner than petroleum-derived fuels. And they can help reduce global warming, since the carbon dioxide gas released during combustion is recycled by the plants grown to make the fuel.

At first glance, fermentation is a relatively simple process. Certain bacteria and yeasts can break down sugars for energy even when oxygen is limited or completely absent. The simple compounds left at the end of the process—alcohols and fatty acids—still contain energy. These chemicals can be collected and burned as fuel.

Fermentation on an industrial scale, however, is not quite that simple. Different organisms metabolize particular sugars under specific conditions. For instance, some organisms can ferment only five-carbon sugars such as

Fermentation vats

Since ancient times, people have used fermentation to convert sugar to alcohol to make beverages such as wine.

The process of winemaking is not very different from the process that produces ethanol. After the grapes are crushed, yeast is added in an oxygen-free environment so that fermentation can occur. One-way gas valves enable the carbon dioxide that is produced to escape without allowing oxygen to enter. In this environment, alcohol is produced.

arabinose and xylose. Other microbes can handle only six-carbon sugars such as glucose and mannose. Given the mix of sugars in various plant materials, coming up with the right combination of microbes can be quite a challenge.

One method of biomass fuel production that is currently under study is a technique called simultaneous saccharification-fermentation (SSF). In SSF, plant materials like corncobs are treated with dilute sulfuric acid. This releases cellulose and lignin, polysaccharides that give plants their rigid cell structure. Then the mixture is put into a container with the enzyme cellulase and yeast. The enzyme breaks the polysaccharides down into glucose, which the yeast rapidly ferments to ethanol.

Scientists working on fermentation technology are also looking at genetically engineered organisms to perform fermentation. For example, one researcher has used the bacterium *Zymomonas mobilis* as a source of genes for two enzymes that produce ethanol during fermentation. Splicing the *Zymomonas* genes into another bacterium, *Escherichia coli*, has resulted in a strain of microbe that ferments all the sugars found in different plants. The genetically engineered microbe has shown promise in laboratory experiments.

Ethanol From Wood

Fermentation of Wood Chips

1. *Wood is treated with dilute sulfuric acid:* cellulose and xylan are released.
2. *Two rounds of enzyme treatments are applied:* large compounds are broken down into five- and six-carbon sugars.
3. *Bacteria and yeast ferment the sugars to ethanol.*

Not everyone believes it is necessary to rely on genetically engineered organisms to produce ethanol by fermentation. Some scientists think that the right combination of naturally occurring microbes and enzymes can do the trick. One system, still in the experimental stage, uses a

process to make fuel from hardwood wastes left over by the forestry industry. Wood chips are treated with dilute sulfuric acid to free cellulose and xylan, another polysaccharide. Additional enzyme treatments break these large compounds down into five- and six-carbon sugars. Finally, an assortment of bacteria and yeasts are added to vats to ferment the sugars to ethanol.

As researchers find microbes that perform fermentation more quickly and that efficiently handle the types of plant matter available, the use of biomass fuels will increase. As much as 50 percent of the nation's liquid fuel needs could eventually be met with fuels generated from biomass.

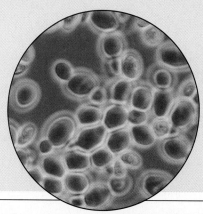

Wood chips (above) will be treated with dilute sulfuric acid and enzymes. In fermentation vats (left), yeast (right) ferments the sugars in the wood chips to ethanol.

PLANTS HAVE MANY VALUABLE USES IN ADDITION TO PROVIDING FOOD FOR ANIMALS AND PEOPLE. TREES PROVIDE LUMBER AND COOKING FUEL FOR HEAT, AND RAW MATERIALS FOR MANUFACTURED PRODUCTS SUCH AS PAPER, PLASTICS, AND RAYON FABRIC. IN ADDITION TO TREES, OTHER PLANTS ARE USED TO MAKE FIBERS, MEDICINES, AND FOOD SEASONINGS.

20.2 Other Uses for Plants

Objectives

① **List three important wood products.**

② **Recognize the role of plants as sources for drugs.**

③ **Describe how turpentine and rubber are made.**

④ **Identify the source of two kinds of natural fibers.**

What Is Wood?

After food and oxygen, the most valuable resource that plants produce for people is wood. In Chapter 19 you learned that plants grow when the cells in their meristems divide. When plants that live more than one growing season get thicker, their meristems are producing secondary xylem and secondary phloem. **Wood,** which is shown in **Figure 20.7a,** is secondary xylem. Thousands of products are made from wood, including paper goods of all kinds.

Trees are harvested for lumber

Wood from trees that have been cut down and sawed into boards and planks is called lumber. Nearly three-quarters of the lumber in the United States is used for construction. The rest goes to factories that make products such as boxes, crates, toys, railroad cars, boats, and items shown in **Figure 20.7b–d.** Wood chips and sawdust can be treated with chemicals to make wood pulp. Manufacturers use wood pulp to make paper, rayon, and other products.

Bark
Sapwood
Heartwood

Figure 20.7

a This cross section of a cedar tree shows the nonliving heartwood, the water-conducting sapwood, and the bark. Many products are made from wood. One of the most important is paper, which forms the graphic backdrop of these two pages.

Some wood is used as fuel

Wood has been used as a fuel since prehistoric times. It is the main fuel for most people in developing countries and for more than half the people in the world. A family that uses wood for heating and cooking burns about a ton of wood each year. As a result, wood is being cut for fuel faster than trees can regrow. In some African countries, wood has become so scarce that the average family spends one-third of its income on wood.

Paper is one of our most important plant products

Papermaking begins when wood chips are ground and chemically treated to produce wood pulp. Wood pulp contains cellulose fibers, which are found in all plant cell walls. When a mixture of cellulose fiber and water is filtered through a fine screen, the fibers tangle in a mass. This mass is pressed between huge rollers and allowed to dry, forming paper. Papermaking fibers come from many different plants, including bamboo, cotton, sugar cane, wheat, and rice.

Paper can be recycled and the fibers used to make new paper. But because it is expensive to remove inks and dyes, the cost of recycling paper often exceeds the cost of making paper from wood.

c Many musical instruments, like this violin, are made from wood.

b Wood has been used in art for centuries. This sculpture was created by a Native American who lived in the Florida Keys before Columbus arrived in the New World.

d Furniture is one of the many things that are made from wood. This chair is called an Eames chair and it is made of laminated plywood.

Plants in Our Lives   **437**

Figure 20.8
The Native American doctor (left) is teaching the Western doctor (right) how to use plants native to the Amazonian rain forest to treat some illnesses.

Drugs From Plants

People have always used plants to relieve pain and cure ailments or disease. Some people still use plants to cure ailments, and many doctors, such as the one in **Figure 20.8**, are studying the cures these plants provide. Today plants produce many substances used in medicinal drugs. For instance, sweet potatoes provide an extract used in producing steroid hormones for birth control pills and cortisone. **Figure 20.9** shows foxglove, a plant that produces digitalis, used to treat heart problems. High blood pressure is controlled by reserpine (*REHS ur pihn*), a drug obtained from the shrub *Rauwolfia*. The May apple was used by Cherokees to kill parasitic worms. Research on the May apple has led to drugs that kill viruses and treat cancer of the testis, and to a spray that protects crops from beetles.

Quinine, aspirin, and ephedrine were first derived from plants

Throughout recorded history, no disease has ever caused more human deaths than malaria. The first medicine used to treat malaria successfully came from the cinchona (*sihn KOH nuh*) tree in South America. In the middle of the seventeenth century, Jesuit missionaries found that Native Americans had a remedy for malaria made by boiling cinchona bark in water. The Native

Americans called the tree bark "quina." The medicine isolated from its bark was called quinine, (*KWEYE neyen*). Today quinine is made synthetically. It is still widely used to prevent malaria.

Solutions made by soaking leaves of the white willow were often placed on aching areas of the body. The ingredient in willow that reduced the pain was isolated in 1827 and called salicin. A derivative called acetylsalicylic acid can be swallowed to relieve all types of pain. The original makers of this drug called it aspirin. Aspirin is the most widely used drug in the world today.

Ephedrine (*eh FEH drihn*) can be obtained by soaking the dried stems of the gymnosperm *Ephedra sinica*. These stems have been prescribed in China for centuries as a stimulant, and for the treatment of high blood pressure, hay fever, and asthma. Ephedrine is used today as an ingredient in decongestants.

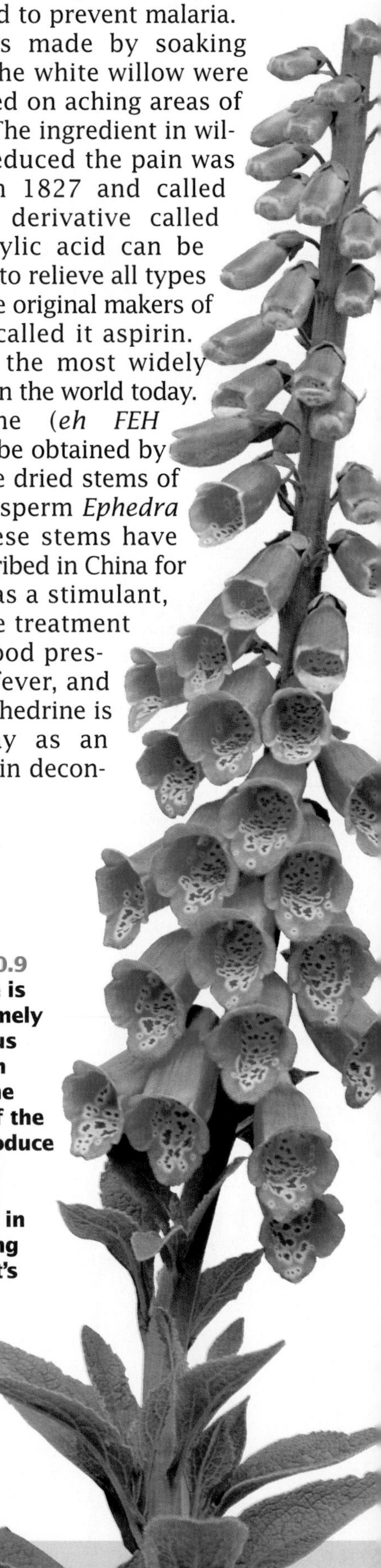

Figure 20.9
a **Foxglove is an extremely poisonous European plant. The leaves of the plant produce digitalis, which is effective in stabilizing the heart's action.**

Figure 20.10
The rosy periwinkle, found in Madagascar, is a plant that produces the chemical vincristine, which is used to treat childhood leukemia.

A chemical from yew trees may help combat cancer

Scientists have recently discovered that a chemical isolated from the yew tree, *Taxus*, appears to have cancer-fighting properties. The chemical, called taxol, is found in the bark of yew trees. Clinical tests have shown that taxol reduces the sizes of cancerous tumors in some patients. An obstacle to further testing, however, is the fact that only very small amounts of taxol are made by a single yew tree. Approximately 8,000 pounds of yew bark are needed to produce one pound of taxol. Unfortunately, scientists have not yet found a way to make taxol in the laboratory, and yew trees remain the only source. Most of the taxol used for clinical testing comes from the bark of the Pacific yew, shown in **Figure 20.11**. It is a tree that has been widely eliminated by clearcutting. Biologists are now searching for yew varieties that contain high levels of taxol. If these varieties could be grown in cultivation, the availablity of taxol would greatly increase.

Like the rosy periwinkle and the yew, many other plant species might also have valuable medicinal properties. The cancer-fighting ability of taxol is yet another example of why species preservation may be important for our own survival.

The rosy periwinkle produces a leukemia-fighting drug

One of the most recently discovered drug-producing plants is the rosy periwinkle, *Catharanthus roseus*, shown in **Figure 20.10**. It originally came from Madagascar, an island near the southeastern coast of Africa, and is often planted in flower beds. The plant contains a chemical called vincristine (*vihn KRIHS teen*), which is used to treat certain forms of leukemia. This disease once killed nearly every child who had it, but thanks to vincristine, the lives of many patients now are saved.

Figure 20.11
An ethical question has arisen concerning the yew tree. Should the yew trees be destroyed to benefit those few individuals who need cancer treatment? Or should the people go untreated to preserve the yew tree?

HUMANS AND THEIR VIRUSES

900

900 Rhazes, a Persian physician, writes the first accurate descriptions of infectious diseases: plague, measles, consumption, tuberculosis, smallpox, and rabies.

Thomas Thatcher

1678 British physician **Thomas Thatcher** writes the first medical treatise published in America on smallpox and measles.

1721 Onesimus, an African slave, describes to his American owner the process of inoculation used in Africa for the treatment of smallpox.

1790s English physician **Edward Jenner** observes that milkmaids who have had cowpox rarely contract smallpox. He develops a vaccination against smallpox.

1930

Wendell Stanley

1930s American biochemist **Wendell Stanley** proves that viruses are not living organisms but are chemical matter.

1937 Max Theiler, a South African microbiologist, develops the vaccine for yellow fever.

Max Theiler

1943 Infantile paralysis epidemic kills almost 1,200 in U.S. and cripples thousands more.

1949 African-American physician **Dr. Jane C. Wright** continues the work of her father, **Dr. Louis Tompkins Wright**. They believe cancer stems from viral agents, and they study the effect of various drugs for the treatment of cancer.

1954 American physician **Dr. Jonas Salk** develops poliomyelitis vaccine.

Polio virus

Yellow fever epidemic, Memphis.

1930

Yellow fever virus

1878 Memphis, Tennessee, reels from an outbreak of **yellow fever**. Fifty-two hundred of the 19,600 residents die. Memphis loses its city charter due to the decrease in population.

1881 Cuban physician **Carlos Finlay** suggests that yellow fever is transmitted by the bite of the common household mosquito.

1900 U.S. Army physician **Walter Reed** establishes that the bite of certain mosquitoes transmits yellow fever. The experiments also show how the fever might be controlled.

1901 Cuban physician **Juan Guiteras** verifies the cause of yellow fever independently of Walter Reed.

1927 The **yellow fever virus** is isolated.

Walter Reed

1966 American **Francis Peyton Rous** receives the Nobel Prize in medicine and physiology for discovery of a cancer virus.

PRESENT

1969 Americans **M. Delbrück, A. D. Hershey**, and **S. E. Luria** win Nobel Prizes for discovering the genetic structure of viruses.

1978 Smallpox virus is now extinct in the wild. Scientists debate whether the only smallpox virus in existence (in two high-security laboratories) should be eliminated. The world's last known case of naturally occurring smallpox was reported in Ali Maow Maalin of Merka, Somalia, in 1977.

HIV

1981 The Centers for Disease Control in Atlanta, Georgia, discovers in its study of Kaposi's sarcoma more than 500 cases of a mysterious disease that knocks out the immune system.

1983 HIV, the virus that causes **AIDS**, is identified.

Ali Maow Maalin

1980s - present Antiviral drug AZT slows the reproduction of HIV.

Animal Kingdom

You are an animal, and share a common heritage with earthworms and dinosaurs, butterflies and sea stars. It is no accident that the fingers of your hand have bones like those in a bird's wing. In this unit you will discover how the animal body has been shaped by its long evolutionary journey, from the simplest sponge to worms, insects, and vertebrates. Evolution has molded animals to suit many ways of living, often altering their design to take advantage of new opportunities in the environment.

The Animal Body

Review

- **natural selection (Section 9.3)**
- **animal evolution (Sections 10.3 and 10.4)**
- **classifying living things (Section 15.2)**

These arctic fox cubs are vertebrates. Their body plan is the result of more than 3 billion years of evolution.

FROM MICROSCOPIC WORMS TO BLUE WHALES, ANIMALS OCCUR IN A GREAT VARIETY OF SIZES AND SHAPES. THIS CHAPTER DESCRIBES THE EVOLUTIONARY JOURNEY LEADING TO TODAY'S GREAT DIVERSITY OF ANIMALS, IN TERMS OF A SERIES OF KEY ADAPTATIONS IN BODY ARCHITECTURE. THESE ADAPTATIONS REVEAL BOTH THE PROGRESSIVE NATURE OF EVOLUTION AND THE IMPORTANCE OF BODY DESIGN.

21.1 The Advent of Tissues

Objectives

❶ List three characteristics of animals.

❷ Recognize the advantage of multicellularity.

❸ Identify the difference between tissues and specialized cells.

❹ List three features found in cnidarians but not in sponges.

Animal Body Plans

**Figure 21.1
These animals show just a part of the diversity of the kingdom Animalia.**

What do the animals in **Figure 21.1** have in common? These animals, and all the other members of the kingdom Animalia, share three characteristics. First, all animals are heterotrophs that ingest their food, digesting it within the body. Second, all animals are multicellular. Third, as you learned in Chapter 3, animal cells lack cell walls.

Despite these similarities, it is obvious that the animals below differ in shape and structure. The **body plan** of an animal is its overall structure, the way its parts fit together. You learned in Chapter 15 that animals evolved from heterotrophic protists. The first animals probably evolved from colonies of protist-like cells. How could

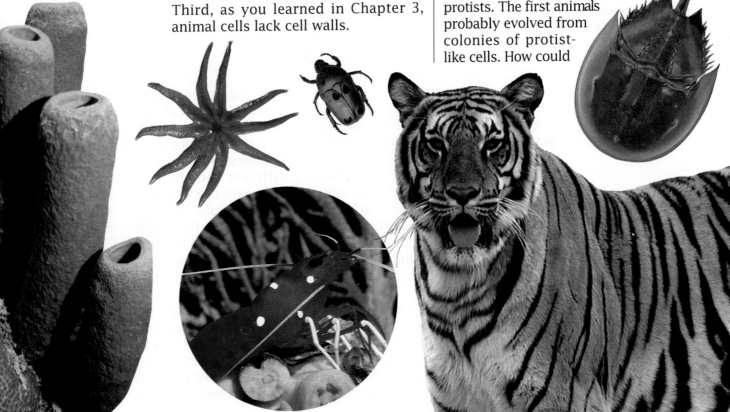

Alan Shipley: Research Associate

How I Became Interested in Science

"I left school at age 15, then traveled around the country and joined the Navy when I was 17. That was in the '60s—the Vietnam War era. Both my parents were in the Navy so I felt a strong sense of responsibility and patriotism and wanted to serve my country, and to explore life. I had to experience the real world. I got a GED and a high school diploma through the Navy at a high school in Hawaii.

"In the Navy, I was a sonar technician. That's where I learned electronics, physics, and oceanography. After six years of active service, I left the Navy in 1976.

"As a kid I was interested in science. I've always had an inherent curiosity. Though I left school prematurely, I've never stopped learning. I continually read and learn from others by doing."

In his free time, Alan Shipley can occasionally be found on a boat, fishing in the waters off the coast of Massachusetts with his good friend Patrick O'Malley, a fellow engineer at the Marine Biological Laboratory. He also has a side business in electronics that he does on weekends.

Name:	**Alan Shipley**
Home:	**Sandwich, Massachusetts**
Employer:	**Marine Biological Laboratory, Woods Hole, MA**
Personal Traits:	• **Honest**
	• **Resourceful**
	• **Friendly**
	• **Helpful**
	• **Cooperative**

Research Focus

Mr. Shipley works for Dr. Lionel Jaffe at the National Vibrating Probe Facility located at the Marine Biological Laboratory in Woods Hole, Massachusetts. They are funded by the National Institute of Health, Division of Research Resources, to design, develop, build, and teach visitors the use of new and innovative techniques in biological research. Thus, the laboratory provides an invaluable research environment for visiting investigators from around the world. The lab's main focus is the investigation of ionic currents and gradients as they relate to living organisms. Dr. Jaffe has perfected a technique of measuring electrical fields around living organisms. It is a non-invasive technique and is applicable to many different disciplines of scientific research. Mr. Shipley is also involved in trying to introduce an inexpensive way to study environmental pollution at minute levels in order to avoid damage before it becomes apparent by other means.

The Satisfaction of Working With Scientists

Career Path
....
U.S. Navy:
- Electronics
- Physics
- Oceanography

Junior College:
- Business
- Electronics
- Labor Relations

"A jack-of-all-trades like me was ideal to fit into this environment. I had to be able to do a wide range of things in order to effectively assist the visiting investigators. Working for Dr. Jaffe for ten years, I've learned by doing. I've worked on a few hundred projects that have involved different aspects of biology and physical science. We're on the cutting edge of technology. Everything we do is brand new. The education I receive here every day is irreplaceable.

"Recently I got a call from a student who was trying to find a good Ph.D. thesis. I suggested that he work here. After less than a week, he had accurate data with an interesting story. I have had two other Ph.D. students in the past do their work in our lab. I've been proud to be able to work with and assist them. It's been very satisfying for me. As an NIH facility, the Marine Biological Laboratory provides a marvelous opportunity for scientists to do a good project, publish and try to get some more funding."

Stage 3: Internal Organs, Bilateral Symmetry, and Cephalization

The evolution of the mesoderm allowed the formation of organs, which first appeared in flatworms such as the liver fluke. Flatworms are bilaterally symmetrical and have a distinct head. The body plan of the liver fluke is composed of solid layers of tissues surrounding a central gut.

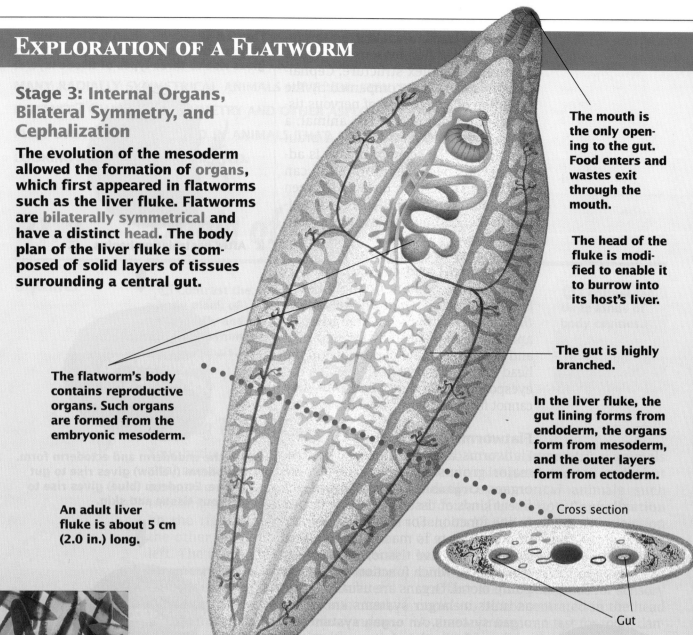

The mouth is the only opening to the gut. Food enters and wastes exit through the mouth.

The head of the fluke is modified to enable it to burrow into its host's liver.

The gut is highly branched.

In the liver fluke, the gut lining forms from endoderm, the organs form from mesoderm, and the outer layers form from ectoderm.

Cross section

Gut

The flatworm's body contains reproductive organs. Such organs are formed from the embryonic mesoderm.

An adult liver fluke is about 5 cm (2.0 in.) long.

Figure 21.8
Liver flukes are parasitic flatworms. Their acoelomate body plans require that they be thin to allow substances to pass easily to all organs.

Flatworms are solid worms

If you were to cut a flatworm in half across its body, as shown in the **Exploration** above, you would see that the gut is completely surrounded by tissue and organs. This solid body construction is termed **acoelomate** (*ay SEEL oh mayt*), meaning without a body cavity.

Flatworms are thin because of their acoelomate body construction. Dissolved substances such as carbon dioxide, oxygen, and nutrients cannot diffuse rapidly through the solid bodies of flatworms, such as the liver fluke in **Figure 21.8**. Flatworms are small or thin (or both), which shortens the distance these substances must move. In addition, the gut of a flatworm is highly branched so that it runs close to most of the tissues.

The guts of all flatworms have only one opening, the mouth. Because these animals consume food and eliminate wastes through the same opening, two-way movement of material occurs within the gut. The two-way gut is less efficient at extracting nutrients than a one-way gut such as yours. If animals with an acoelomate body plan eat when food is already in the gut, newly consumed food can mix with partially digested food and wastes.

A One-Way Gut and a Body Cavity: Roundworms

Figure 21.9
Roundworms, such as this nematode, have a body cavity, the pseudocoelom. The presence of the pseudocoelom means that roundworms are not packed with solid tissues, as are flatworms. The pseudocoelom allows for more efficient diffusion of nutrients to body organs.

Roundworms (phylum Nematoda), or nematodes, such as *Caenorhabditis elegans* in **Figure 21.9**, have a cavity within the body and a one-way gut with two openings. Food is taken in through the mouth, and wastes are eliminated through an opening at the other end of the gut, the **anus**. This arrangement allows a one-way movement of food through the gut. Also, different regions of the digestive tube can be specialized for different digestive activities. The front part of the gut is adapted for ingesting food. The middle region breaks down food and absorbs nutrients. The last region expels waste products.

Specialization of different regions of the gut brings with it a potential problem, however. Since the absorption of nutrients occurs in only one part of the digestive tract, there must be some effective way of distributing those absorbed nutrients to other parts of the body. Roundworms have a fluid-filled cavity between the gut and the body wall, as illustrated in the **Exploration**. This body cavity is called a **pseudocoelom** (*SOO doh see luhm*), which means "false body cavity." This term is somewhat misleading. The cavity is real, but it differs from the body cavities of most other animals.

The pseudocoelom permits rapid diffusion of nutrients and other dissolved substances over the short distances between tissues and organs. Diffusion is enhanced by body movements, which cause the fluid within the body cavity to move. Nevertheless, diffusion is a slow process. Pseudocoelomate animals must either be very small—most are less than 2.5 mm (0.1 in.) in length—or have body shapes that maintain short distances between organs and the body surface. For this reason, nematodes are usually thin and threadlike.

EXPLORATION OF A ROUNDWORM

Stage 4: Body Cavity

The major body plan innovation in roundworms, such as this nematode, is the presence of a body cavity between the gut and the body wall. This cavity is the pseudocoelom. After its evolution, animals were not constrained by a solid body, as were flatworms. Nematode organs could now form away from the gut because nutrients could diffuse through the new body cavity.

Nematode adults consist of very few cells. The nematode *C. elegans* has only 1,000 cells and is the only animal whose complete cellular anatomy is known.

Anus

Muscles extend along the length of the worm's body, rather than encircling it.

A cross section of a roundworm shows how the tissues are arranged around the pseudocoelom. The body cavity of a nematode separates the endoderm-lined gut from the rest of the body.

This roundworm, like all terrestrial nematodes, must be small because organs have to be close to the pseudocoelom to receive diffused nutrients. Notice the presence of a mouth and an anus, indicating a one-way digestive tract.

Mouth

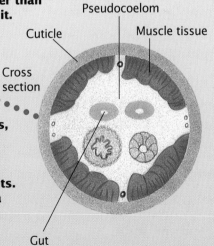

Pseudocoelom

Cuticle

Muscle tissue

Cross section

Gut

A Better Body Cavity: Mollusks

Most animals with a body cavity have a coelom, or true body cavity. A **coelom** (*SEE luhm*) is a fluid-filled body cavity that lies completely within the mesoderm. The coelom separates the muscles of the body wall from the muscles that surround the gut.

A major advantage of the coelom is that it allows interactions between mesoderm and endoderm to occur during development. This interaction is necessary in order for local regions of the digestive tract to become highly specialized. For example, your stomach is a locally specialized portion of the gut that developed from both endoderm and mesoderm. In a pseudocoelomate, by contrast, mesoderm and endoderm are separated by the fluid-filled pseudocoelom. Interactions between these two layers are limited, and a high degree of digestive specialization is not possible. The fluid-filled coelom also provides a body cavity in which organs can develop and against which muscles can operate. The body constructions of acoelomates, pseudocoelomates, and coelomates are compared in **Figure 21.10**.

In coelomates, as in flatworms, the gut tube is surrounded by solid tissue that is a barrier to rapid diffusion. But most coelomates have a **circulatory system**, a network of blood-carrying vessels. The circulatory system brings nutrients and oxygen to the tissues and removes wastes and carbon dioxide. Blood is usually propelled through the circulatory system by contractions of one or more muscular hearts.

Figure 21.10
These cross sections show the differences between acoelomate, pseudocoelomate, and coelomate body constructions.

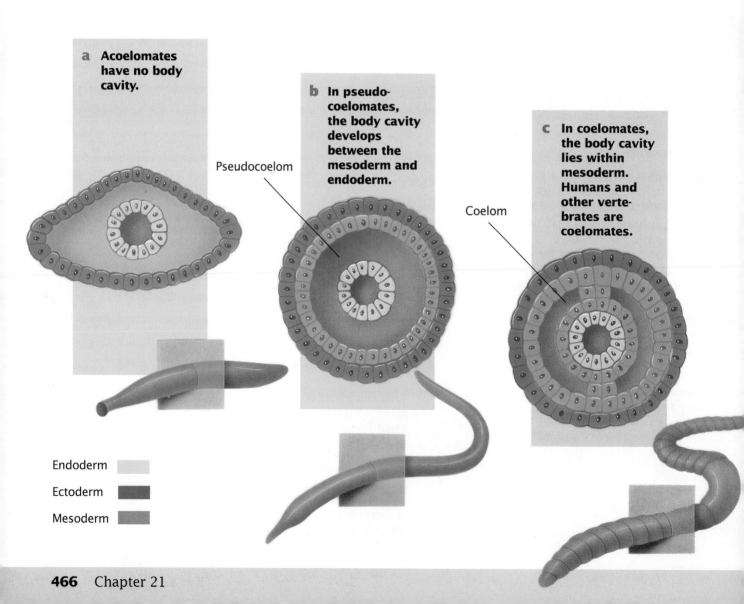

a Acoelomates have no body cavity.

b In pseudocoelomates, the body cavity develops between the mesoderm and endoderm.

Pseudocoelom

c In coelomates, the body cavity lies within mesoderm. Humans and other vertebrates are coelomates.

Coelom

Endoderm
Ectoderm
Mesoderm

EXPLORATION OF A MOLLUSK

Stage 5: Coelom

This snail is a mollusk. Mollusks have a coelom and a circulatory system. The presence of a coelom allows interaction between the mesoderm and the endoderm. This interaction enables development of highly specialized organs such as a stomach.

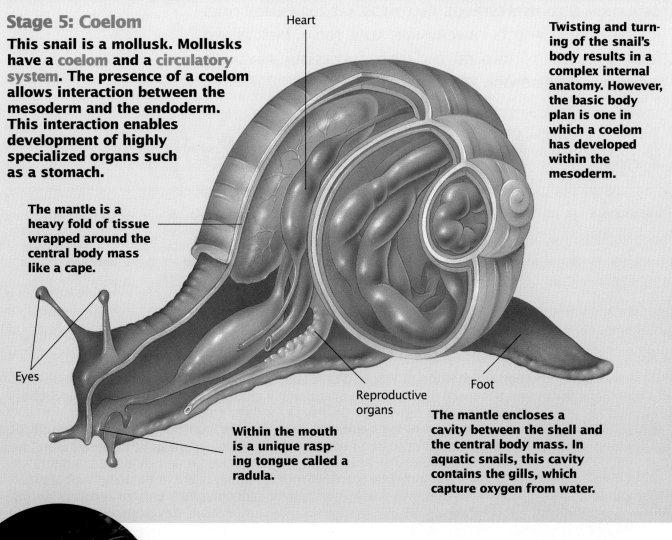

Heart

The mantle is a heavy fold of tissue wrapped around the central body mass like a cape.

Eyes

Twisting and turning of the snail's body results in a complex internal anatomy. However, the basic body plan is one in which a coelom has developed within the mesoderm.

Reproductive organs

Foot

Within the mouth is a unique rasping tongue called a radula.

The mantle encloses a cavity between the shell and the central body mass. In aquatic snails, this cavity contains the gills, which capture oxygen from water.

Figure 21.11
Mollusks, such as this snail, are more complex than nematodes.

Mollusks are coelomates with a circulatory system

A coelom and a circulatory system first evolved in the mollusks, phylum Mollusca. Snails (such as the one in **Figure 21.11**), clams, squids, and mussels belong to this phylum. The mollusk body plan is illustrated in the **Exploration** above.

Section Review

❶ Explain the difference between bilateral symmetry and radial symmetry.

❷ What structures in your body are derived from mesoderm?

❸ Summarize the advantage of a one-way gut.

❹ Diagram the two types of body cavities.

THERE ARE MORE THAN 5 BILLION PEOPLE ON EARTH. THIS FIGURE MAY SEEM LARGE, BUT IT IS TINY COMPARED WITH THE INSECT POPULATION. SCIENTISTS ESTIMATE THAT THERE ARE 200 MILLION TIMES MORE INDIVIDUAL INSECTS THAN HUMANS ALIVE TODAY. INSECTS AND HUMANS BELONG TO TWO OF THE MOST SUCCESSFUL PHYLA OF ORGANISMS—ARTHROPODA AND CHORDATA.

21.3 Four Innovations in Body Plan

Objectives

1 Contrast the body plans of segmented and nonsegmented animals.

2 Compare the exoskeleton of arthropods with the endoskeleton of vertebrates.

3 Summarize the differences between protostomes and deuterostomes.

4 List two traits that reveal the relationship between chordates and echinoderms.

Segmented Worms: Annelids

Look at the worm in **Figure 21.12**. This worm is made up of many similar units linked together, like beads in a necklace. Animals showing **segmentation** are composed of repeated body units. Three very successful animal phyla are segmented: annelids (earthworms and their relatives), arthropods (insects, crustaceans, and spiders), and chordates (mostly vertebrates).

Can you point out an example of segmentation in your body? Don't be surprised if you cannot. In vertebrates, segments are not usually visible externally in adults but are apparent during embryonic development. Vertebrate muscles develop from repeated blocks of tissue that occur in the embryo. Another example of segmentation is the vertebral column, which is a stack of very similar vertebrae.

What great advantage does segmentation provide? Its main advantage is the evolutionary flexibility it offers. A small change in an existing segment can produce a new kind of segment with a different function. As illustrated in the **Exploration** shown on the next page, some segments of the earthworm are modified for reproduction, some for feeding, and some for eliminating wastes.

Figure 21.12
The bristle worm is an example of an annelid. Like the earthworm on the next page, its body is segmented. Specialized segments perform specific tasks.

EXPLORATION OF AN ANNELID

Stage 6: Segmentation

Annelids, such as this earthworm, were the first organisms to evolve a body plan that consisted of **segments**. Most segments are separated by partitions that cross the coelom. In each segment, parts of the excretory, circulatory, and nervous systems are repeated.

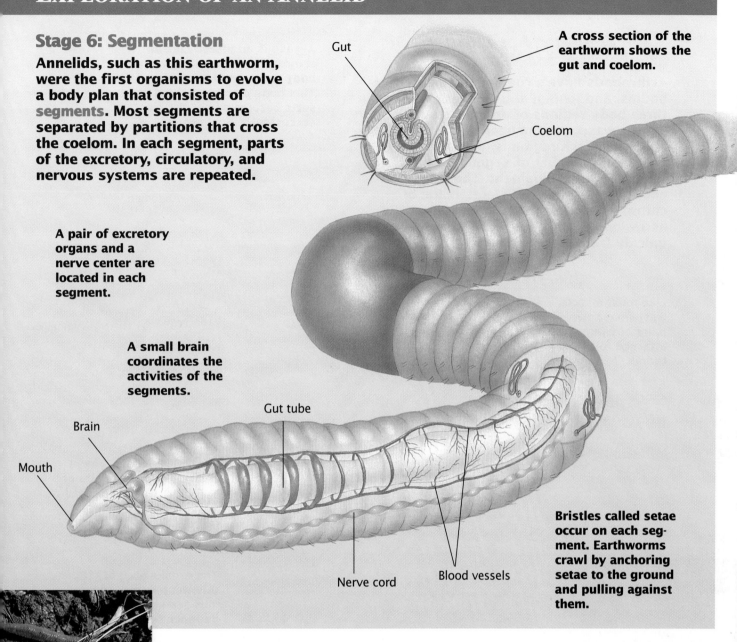

A cross section of the earthworm shows the gut and coelom.

Gut

Coelom

A pair of excretory organs and a nerve center are located in each segment.

A small brain coordinates the activities of the segments.

Gut tube

Brain

Mouth

Nerve cord

Blood vessels

Bristles called setae occur on each segment. Earthworms crawl by anchoring setae to the ground and pulling against them.

Figure 21.13
The earthworm is an annelid. Members of this phylum have segmented bodies.

Annelids have segments that are specialized

Annelids (phylum Annelida), such as the earthworm in **Figure 21.13**, were the first segmented animals to evolve. The earthworm body plan is shown in the **Exploration** above. The basic body plan of an annelid is a tube within a tube. The gut tube, which extends from mouth to anus, is suspended within the larger tube of the coelom. Note that the body is partitioned internally between segments. This partitioning limits diffusion of materials from segment to segment. A circulatory system overcomes this limitation by transporting materials between segments.

The front segments of annelids are modified to house a small brain and sense organs. Each segment along the body is controlled by an individual nerve center. A nerve cord running along the underside of the worm connects these nerve centers with the brain so that all of the body's activities can be coordinated.

EXPLORATION OF AN ARTHROPOD

Stage 7: Jointed Appendages, Exoskeleton, and Wings

Arthropods have a coelom, segmented bodies, and jointed appendages. The three body regions of an insect, such as this wasp, are the head, thorax, and abdomen. Each region is actually composed of a number of segments that fuse during development. The presence of a strong exoskeleton made of chitin, a complex muscular system, and wings permit this wasp to move quickly from place to place.

A wasp is an insect. Like all arthropods, it has a segmented body and jointed appendages. This body plan has helped insects become one of the most successful animal groups.

Like most insects, wasps have two pairs of wings attached to the thorax.

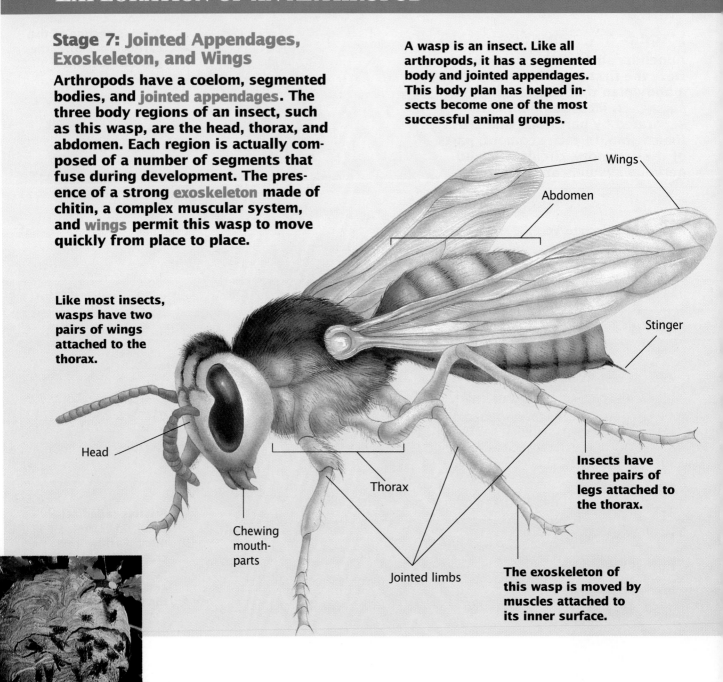

Wings

Abdomen

Stinger

Head

Chewing mouthparts

Thorax

Jointed limbs

Insects have three pairs of legs attached to the thorax.

The exoskeleton of this wasp is moved by muscles attached to its inner surface.

Figure 21.14
Wasps are highly social insects that build a nest where the colony lives and breeds. This nest is made of a paperlike material and may house hundreds of wasps.

Limbs and Skeletons: Arthropods

The name arthropod comes from the Greek words *arthros,* meaning jointed, and *podes,* meaning feet. The great success of arthropods, such as the wasps in **Figure 21.14**, is due largely to their jointed appendages. These appendages have evolved to perform a variety of tasks. Some appendages serve as limbs for walking or grasping. Others function as antennae for sensing the environment or as mouthparts for chewing, sucking, or poisoning prey. For example, a scorpion seizes and tears apart its prey with appendages that have been modified into large pincers capable of grasping and holding tightly.

Like annelids, the basic body plan of arthropods is segmented, as shown in the **Exploration** above. Individual segments of an arthropod often exist only during early development, however, and fuse into functional groups in adults.

Arthropods

and chordates

show segmen-

tation. Many

scientists think

segmentation

evolved indepen-

dently in each

group. What

evidence supports

this conclusion?

For example, caterpillars have many segments, while butterflies have only three main body units—head, thorax, and abdomen—each composed of several fused segments. Arthropods have an external skeleton or **exoskeleton**. The arthropod exoskeleton is made of chitin (*KYT uhn*), a tough polysaccharide. The muscles lie within the skeleton and attach to its inner surface. As it grows, an arthropod periodically sheds its rigid exoskeleton. Many arthropods change their body form as they develop. A butterfly begins life as a wormlike caterpillar, later transforming into the flying adult. A silverfish, on the other hand, does not change as it develops; it merely grows larger.

An Embryonic Revolution

Humans and the other chordates are more closely related to the echinoderms—the phylum that includes the sea star—than to the arthropods. There seems to be little resemblance between humans and sea stars (often called "starfish"). For one thing, sea stars are radially symmetrical, as you can see in the **Exploration** on the next page.

In animals with tissues, except for echinoderms and chordates, the first opening that forms in the embryo during gastrulation eventually becomes the mouth. Animals that develop in this fashion are known as **protostomes**. The word protostome means "first mouth," and refers to the fact that the initial depression that starts gastrulation becomes the mouth of the adult organism. In protostomes, the developmental fate of each cell of the early embryo is usually determined when that cell first appears.

Echinoderms and chordates are **deuterostomes**. The word deuterostome means "second mouth," and refers to the fact that the first opening in the embryo does not form the mouth—it becomes the anus of the adult animal. The mouth develops from an opening that appears later in development. In these animals, all the cells of the early embryo are identical. This means that any isolated cell can develop into a complete organism. **Figure 21.15** shows some protostomes and deuterostomes.

Figure 21.15
Below are examples of protostomes and deuterostomes. Recent studies using DNA nucleotide sequences support the division of the coelomates into these two groups.

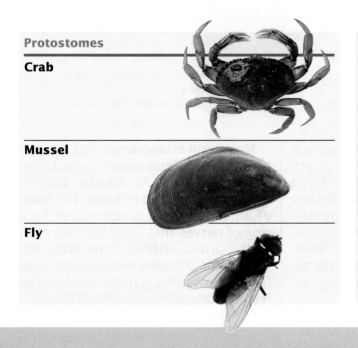

Protostomes

Crab

Mussel

Fly

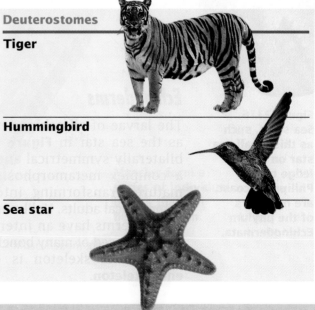

Deuterostomes

Tiger

Hummingbird

Sea star

Understanding Vocabulary

1. For each pair of terms, explain the difference in their meanings.
   a. endoderm, ectoderm
   b. tissue, organ
   c. radial symmetry, bilateral symmetry
   d. protostomes, deuterostomes

Relating Concepts

2. Copy the unfinished concept map below onto a sheet of paper. Then complete the concept map by writing the correct word or phrase in each oval containing a question mark.

Understanding Concepts

Multiple Choice

3. Which characteristic is shared by all animals?
   a. radial symmetry
   b. multicellularity
   c. segmentation
   d. pseudocoelom

4. Which body structures do *not* develop from mesoderm?
   a. brain and nerve
   b. muscles
   c. arteries and veins
   d. reproductive organs

5. Which animal is bilaterally symmetrical?
   a. sponge   c. snail
   b. sea star   d. hydra

6. Internal digestion, symmetry, and tissues are features found in
   a. annelids but not in cnidarians.
   b. both nematodes and sponges.
   c. cnidarians but not in sponges.
   d. both protists and sponges.

7. Which of these animals is cephalized?
   a. sponge   c. jellyfish
   b. liver fluke   d. sea star

8. Which of the following is *not* a characteristic of mollusks?
   a. pseudocoelom
   b. circulatory system
   c. radula
   d. mantle

9. Which is *not* an example of a segmented animal?
   a. wasp
   b. earthworm
   c. roundworm
   d. spider

10. What do chordates and echinoderms have in common?
    a. gene-controlled development
    b. a head and a brain
    c. radially symmetrical larvae
    d. hollow dorsal nerve cord

Completion

11. Members of the phylum _____ are multicellular, but their cells are not organized into tissues or organs.

12. The process that leads to the formation of tissue layers during development is called _____ .

13. The skeleton of arthropods is called a(n) _____ because the muscles lie inside the skeleton. However, the skeleton of vertebrates is covered by muscles and is called a(n) _____ .

14. The phylum most closely related to the chordates is the _____ .

15. In protostomes, the first opening in the embryo becomes the _____ and the second opening becomes the _____ .

Short Answer

16. What is meant by the phrase "division of labor"? Why is it advantageous for the cells of an animal to exhibit division of labor?

17. The brain and the sense organs in most bilaterally symmetrical animals are located in a head region. Why is this arrangement advantageous?

Kingdom Animalia — members are: heterotrophic, ? — allows — cell specialization — shown in — ? — can be part of — organs — such as — ? ; includes: sponges, ? — lack — jointed appendages — can be part of — ? — such as — digestive system ; have — ? — made of — chitin

18. How does the movement of food in a nematode compare with the movement of food in a flatworm?

19. What three features do all chordates have at some time during their lives?

Interpreting Graphics

20. Look at the diagrams below.

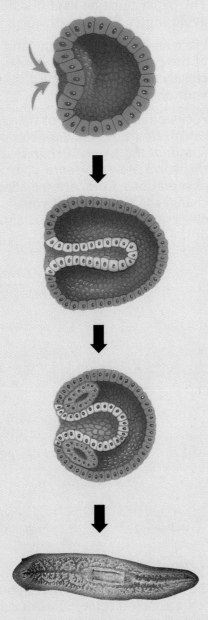

- What process is illustrated in these diagrams?
- Identify the mesoderm, endoderm, and ectoderm.

Reviewing Themes

21. *Patterns of Change*
    What evidence suggests that humans and sea stars are more closely related than humans and crabs?

22. *Evolution*
    How does segmentation provide for evolutionary flexibility?

23. *Structure and Function*
    List three functions performed by the jointed appendages of arthropods.

Thinking Critically

24. *Inferring Conclusions*
    Explain why animals that have a pseudocoelom are usually very small.

25. *Comparing and Contrasting*
    How is the gut of a jellyfish different from the gut of a roundworm?

26. *Building on What You Have Learned*
    In Chapter 15 you learned about how organisms are classified. Combine your knowledge of classification systems with your knowledge of body cavity types to explain why earthworms and sand worms are grouped in the phylum Annelida but flatworms are not.

Cross-Discipline Connection

27. *Biology and Health*
    Hookworms are parasitic roundworms that infect humans. Look for information in your library about the symptoms associated with hookworm infection. What can humans do to prevent becoming infected?

Discovering Through Reading

28. Read the article "Invasion of the Zebra Mussels," in *Discover*, January 1991, page 44. What problems are caused by zebra mussels in the Great Lakes? How are mussels being controlled?

Investigation

How Does Segmentation Help an Earthworm Move?

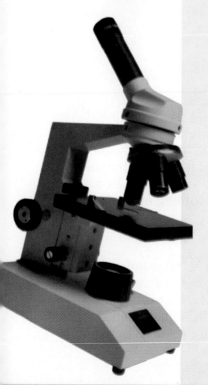

Objectives

In this investigation you will:
- *observe* a moving earthworm
- *relate* its pattern of locomotion to its segmented structure
- simulate an earthworm's movement

Materials

- pan
- paper towels
- water
- live earthworm
- long balloons
- graph paper

Prelab Preparation

- Review what you have learned about segmentation by answering the following questions:
- Define the word segmentation.
- Identify one example of segmentation in your body.
- What groups of animals are segmented?

Procedure: Observing Earthworm Movement

1. Form a cooperative team with another student to complete steps 2–9.

2. Line the bottom of a pan with six layers of paper towels. Thoroughly moisten the paper with tap water. The paper must be kept moist to avoid injury to the earthworm.

3. Place an earthworm at one end of the pan. Observe the worm closely as it moves. Record your observations.

4. Focus on one segment of the earthworm. *How does the diameter and length of the segment change as the worm moves?* Record your observations. Compare the shape changes in the segment you observe with those in the diagram of the earthworm below.

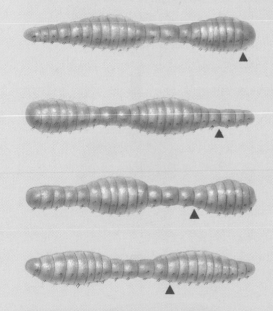

5. Carefully touch the underside of the earthworm. The bristles you feel are called setae. Setae occur on each segment of the earthworm. As the earthworm moves forward, setae anchor it to the soil so that it does not slip back and so that the segments have something to pull

against. Carefully return the earthworm to its container.

6. You will now simulate earthworm movement using two balloons containing water. From your teacher, get two balloons that have been tied together. Be careful with the balloons since rough treatment can easily break them. Each water-filled balloon represents one segment of an earthworm. Place the balloons in a straight line on a sheet of graph paper. Mark the position of each balloon on the graph paper.

7. Hold the rear balloon as shown in the figure below and gently squeeze it along its length. While squeezing the balloon, have your partner mark the position of the ends of each balloon on the graph paper. Your fingers simulate the action of muscles in the earthworm's body wall.

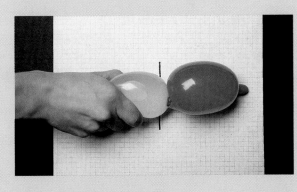

8. Now release the balloon. *What happens to the position of both balloons? How is this change different from what you saw in the earthworm's segments?*

9. Squeeze the rear balloon again. Hold the tip of the front balloon down on the graph paper. Now stop squeezing the balloon. *What happens to the position of both balloons?* Holding the tip of the balloon down simulates the effects of the earthworm's setae.

10. Clean up your materials and wash your hands before leaving the lab.

Analysis

1. *Evaluating Methods*
Like its exterior, the earthworm's coelom is segmented. Each segment of the coelom is separated from the others by membranes called septa. How is this pattern of internal division reflected in the balloon model of the earthworm?

2. *Inferring Relationships*
Relate the movement of fluid within a segment of the earthworm to the changes in shape you observed in that segment.

3. *Making Inferences*
Suppose the earthworm's coelom was not partitioned internally. Would the earthworm still be able to move in the same way? Explain.

4. *Making Inferences*
Explain why segments must be able to both elongate and shorten in order for the earthworm to move.

Thinking Critically

Explain why an earthworm's method of locomotion is not an effective way to move through water.

Adaptation to Land

The egg from which this young snapping turtle is emerging is one of the many adaptations that land animals evolved in their move from the sea onto land.

LIFE AROSE IN THE SEA AND STAYED THERE MORE THAN 3 BILLION YEARS UNTIL A PROTECTIVE SHIELD OF OZONE FORMED. WITHOUT THIS SHIELD, AQUATIC ANIMALS COULD NOT LIVE ON LAND. THE REQUIREMENTS FOR LIVING ON LAND ARE VERY DIFFERENT FROM THOSE FOR LIVING IN THE SEA. IN THIS CHAPTER, YOU WILL EXAMINE SOME KEY PROBLEMS ANIMALS FACED IN MOVING ONTO LAND.

22.1 Leaving the Sea

Objectives

1 List some animal groups that successfully made the transition to life on land.

2 Explain the importance of the skeleton in arthropods and vertebrates on land.

3 Describe the evolution of amphibian limbs from the fins of lobe-finned fishes.

4 Relate the structure of the vertebrate ear to the differences between air and water.

Which Animals Live on Land?

The animal body evolved over many millions of years in the sea. All of the major changes in body plan you read about in Chapter 21 took place in the sea. But the evolutionary journey of animals did not end there.

From the sea, animals invaded the land. Of the major animal phyla, only sponges, cnidarians, and echinoderms—animals that pump sea water through their bodies—were left behind. Members of every other major phylum can be found on land. Flatworms live in damp leaf litter. Nematodes and annelids burrow through the soil, and snails creep over damp ground at night. These animals, however, are found mainly in moist habitats. Only two groups of animals have fully adapted to life on dry land: arthropods and vertebrates such as the mudskipper shown in **Figure 22.1**. How these two groups evolved ways to survive the many challenges of living out of water is one of biology's most fascinating stories.

Figure 22.1
This mudskipper is an animal that has adapted to life in water and on land. Underwater, it breathes through gills. On land, it can absorb oxygen through the lining of its mouth. Mudskippers use their enlarged front fins to scurry across mud flats.

Figure 22.2
Several changes in limb structure took place as terrestrial vertebrates evolved from aquatic vertebrates.

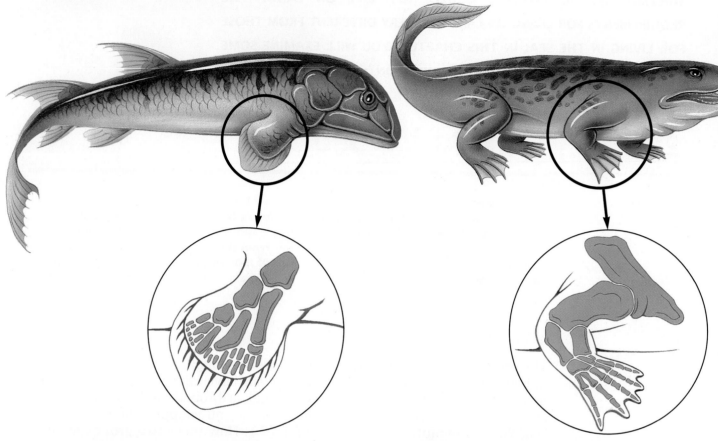

a Scientists think that a lobe-finned fish was the ancestor of the earliest land vertebrates. The drawing above shows the bones in the fin of an extinct lobe-finned fish.

b The earliest known land vertebrate is the amphibian *Ichthyostega*, which lived about 350 million years ago. It had four sturdy legs. Like its fish ancestors, it had a tail fin and some bony scales.

Supporting the Body

Water is about 1,000 times denser than air. Because of water's density, you can float on your back in a pool. For some aquatic animals, water provides much of the support necessary to keep their bodies from collapsing under the pull of gravity. That is why a jellyfish stranded on the beach cannot maintain its shape.

A variety of adaptations enabled animals that left the sea to overcome the loss of physical support. In nematodes and earthworms, the body cavity is filled with fluid under pressure that helps stiffen the body. In arthropods and vertebrates, support of the body is largely taken over by the skeleton. A land animal's skeleton holds up its body against gravity, much like beams and girders hold up a skyscraper.

Limbs play an important role in supporting vertebrates on land. When a terrestrial animal is standing, its legs bear the entire weight of its body. In this way, legs function like the pillars that hold up the roof of a building. Unlike pillars, however, animal legs have flexible joints where movement occurs.

In both arthropods and vertebrates, legs evolved from limbs adapted for movement in water. **Figure 22.2a–b** illustrates the evolution of amphibian limbs from the limbs of fishes. Recall

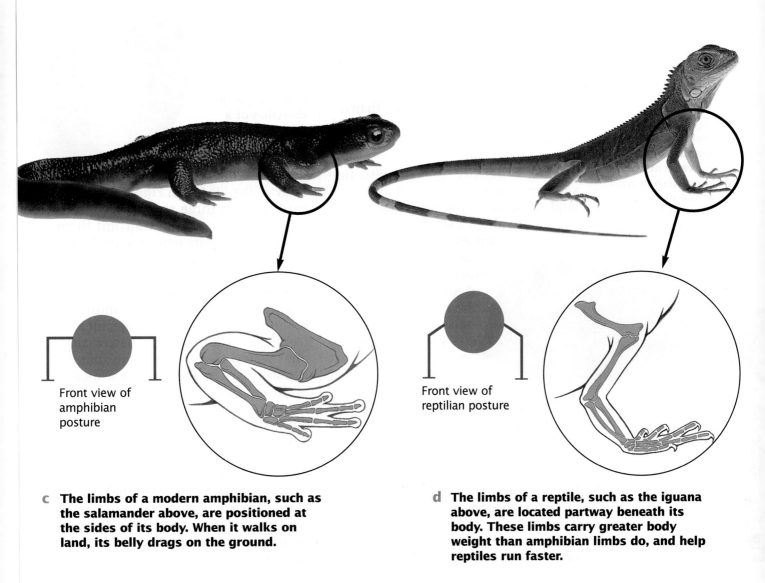

c The limbs of a modern amphibian, such as the salamander above, are positioned at the sides of its body. When it walks on land, its belly drags on the ground.

d The limbs of a reptile, such as the iguana above, are located partway beneath its body. These limbs carry greater body weight than amphibian limbs do, and help reptiles run faster.

Front view of amphibian posture

Front view of reptilian posture

from Chapter 10 that amphibians, the first vertebrates on land, evolved from fishes. The limbs of amphibians are homologous to fins of fishes. But most fishes have thin, paddlelike fins that would be useless for supporting and moving an animal on land. As you can see in **Figure 22.2a–b**, the arrangement and structure of bones in the limb of an early amphibian are similar to those in the fin of a lobe-finned fish. Because of these similarities, scientists think the first amphibians were descendants of the lobe-finned fishes, a group whose modern members include the coelacanth and the lungfishes.

Compare the posture of the amphibian in **Figure 22.2c** to that of the reptile in **Figure 22.2d**. Amphibian limbs are short and join the body horizontally; an amphibian's belly drags on the ground as it walks. Reptilian limbs are positioned slightly beneath the body. Limbs positioned in this way support more body weight than do the limbs of amphibians. Mammalian limbs are positioned directly beneath the body, raising the belly well off the ground. Compared with reptilian and amphibian limbs, limbs located directly beneath the body support the animal's weight more effectively. They also enable higher running speeds and give the animal a greater field of vision, helping it locate food and be alert to danger.

Vertebrate skin can be moist and thin, or dry and thick

Have you ever touched a frog? What did its skin feel like? Amphibians secrete a slippery mucus that is responsible for their slimy texture. Like the wax on an insect, this mucus coating helps limit evaporation. Despite this mucus coating, the skins of amphibians are not watertight. As you will see in the next section, amphibians absorb oxygen and release carbon dioxide directly through their skin. To do this, their skin must be moist and thin. Therefore, to keep from drying out, amphibians must remain in water or in moist environments.

A reptile is not slimy like an amphibian. Reptilian skin is dry and covered with tough scales, as shown in **Figure 22.7**. A watertight skin of scales was a significant evolutionary step, completely freeing reptiles from the necessity of living in a wet environment. Mammals and birds, which evolved from reptiles, also have skin that is dry and relatively watertight.

Figure 22.7
Reptiles have dry, scaly skin that is watertight. The skin of a banded rock rattlesnake is shown above.

Gas Exchange

You breathe about 12 times each minute. Have you ever considered why you breathe? From the perspective of water conservation, breathing is very costly. About 300 mL (9 oz.) of water evaporates from your lungs each day. You breathe, despite the water loss, because breathing is essential for gas exchange. Gas exchange is the process of absorbing oxygen from the environment and ridding the body of carbon dioxide. This "swap" of molecules is necessary for all animals, since they require oxygen to carry out cellular respiration, and they release carbon dioxide as a waste product.

Gas exchange occurs by diffusion, which you learned about in Chapter 4. Oxygen diffuses across the cell membrane into the cell, and carbon dioxide diffuses out. The cells that make carbon dioxide and use oxygen must be close to the environment, because diffusion works well only over short distances. Recall from Chapter 21 that flatworms and nematodes cannot be thick, since diffusion alone carries oxygen to their innermost tissues. The chordates, arthropods, mollusks, echinoderms, and some annelids can be large because they carry out gas exchange with specialized structures.

Gills are structures for aquatic gas exchange. Aquatic arthropods, fishes, and amphibian larvae have gills. Amphibian larvae have external gills. The gills of crabs and many other aquatic arthropods are located in chambers behind the head, as shown in **Figure 22.8**. The gills of fishes also lie directly behind the head, as you can see in **Figure 22.9**.

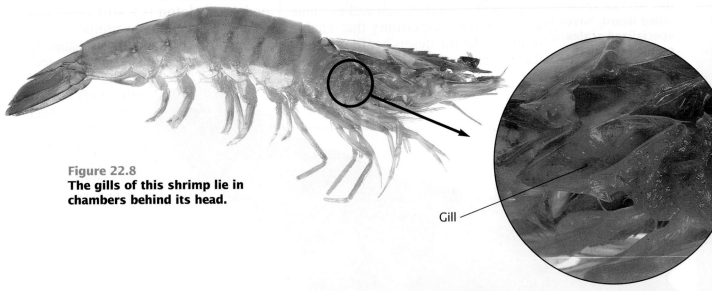

Figure 22.8
The gills of this shrimp lie in chambers behind its head.

Gill

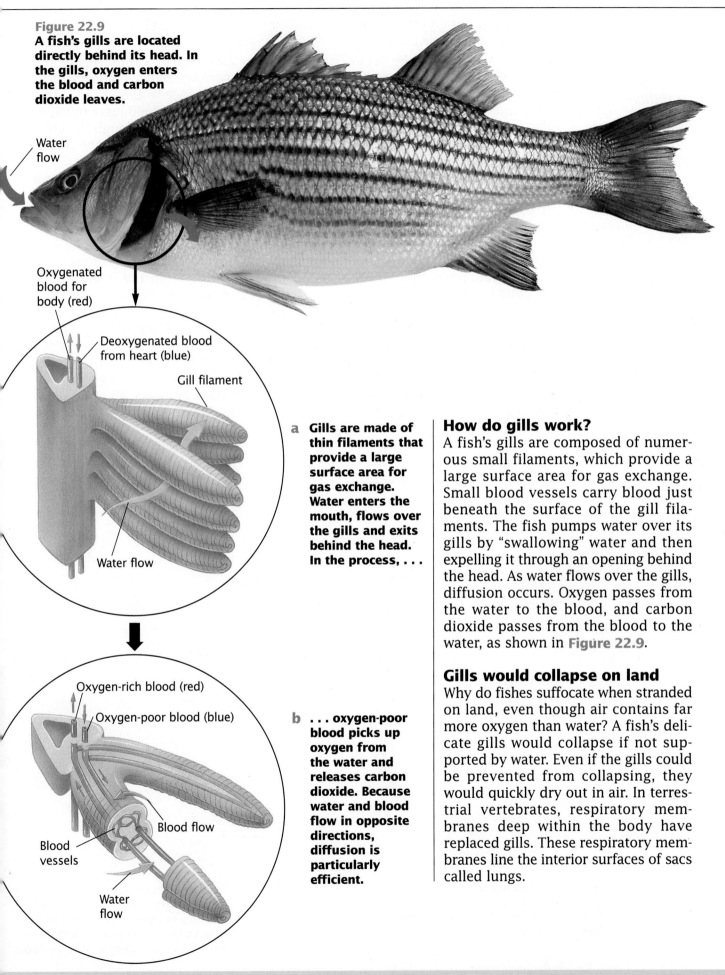

Figure 22.9
A fish's gills are located directly behind its head. In the gills, oxygen enters the blood and carbon dioxide leaves.

Water flow

Oxygenated blood for body (red)

Deoxygenated blood from heart (blue)

Gill filament

Water flow

a Gills are made of thin filaments that provide a large surface area for gas exchange. Water enters the mouth, flows over the gills and exits behind the head. In the process, . . .

Oxygen-rich blood (red)

Oxygen-poor blood (blue)

Blood flow

Blood vessels

Water flow

b . . . oxygen-poor blood picks up oxygen from the water and releases carbon dioxide. Because water and blood flow in opposite directions, diffusion is particularly efficient.

How do gills work?

A fish's gills are composed of numerous small filaments, which provide a large surface area for gas exchange. Small blood vessels carry blood just beneath the surface of the gill filaments. The fish pumps water over its gills by "swallowing" water and then expelling it through an opening behind the head. As water flows over the gills, diffusion occurs. Oxygen passes from the water to the blood, and carbon dioxide passes from the blood to the water, as shown in **Figure 22.9**.

Gills would collapse on land

Why do fishes suffocate when stranded on land, even though air contains far more oxygen than water? A fish's delicate gills would collapse if not supported by water. Even if the gills could be prevented from collapsing, they would quickly dry out in air. In terrestrial vertebrates, respiratory membranes deep within the body have replaced gills. These respiratory membranes line the interior surfaces of sacs called lungs.

Vertebrate Lungs

Stability

Vertebrates living

in the ocean also

lose water to their

environment.

Explain why this

occurs.

How does gas exchange occur in the lungs? Vertebrate lungs are lined with moist, thin tissues through which gas exchange occurs. A network of fine blood vessels called capillaries runs close to this lining. Oxygen-rich air is brought into the lungs by inhalation. Oxygen diffuses across the lining of the lungs into the capillaries. At the same time, carbon dioxide moves from the capillaries into the air in the lungs. Exhalation expels the "used" air from the lungs. You can follow the evolution of vertebrate lungs in the *Evolution of the Lung* feature on pages 490–491.

Terrestrial vertebrates have a double-loop circulatory system

In fishes, the heart pumps blood from the heart to the gills, where the blood picks up oxygen before flowing to the rest of the body. This creates the "single-loop" system, shown in **Figure 22.10a**. The capillaries of the gills are narrow, so they present a great deal of resistance to blood flow. As a result, the flow of blood loses much of its force as it passes through the gills. Circulation from the gills to the rest of the body is sluggish.

The fine capillaries of an amphibian's lung also slow down blood flow. In amphibians, however, the blood returns to the heart for repumping before circulating to the body's tissues. In effect, there are two circulatory systems here: blood flowing from the heart to the lungs and back, and blood flowing from the heart to the body and back. This type of dual circulatory system is found in all terrestrial vertebrates. It carries oxygenated blood much more rapidly through the body than does the circulatory system of fishes. **Figure 22.10b** illustrates the "double-loop" circulation of an amphibian.

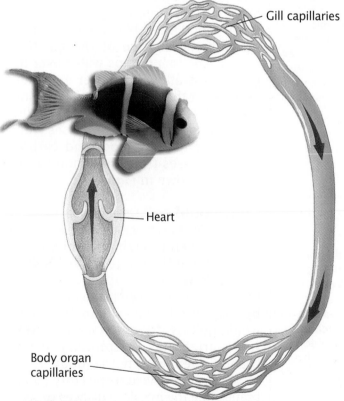

Gill capillaries

Heart

Body organ capillaries

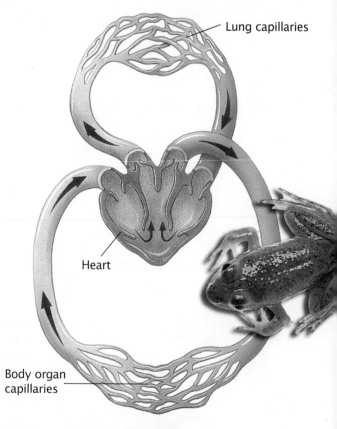

Lung capillaries

Heart

Body organ capillaries

Figure 22.10

a **In fishes, the heart pumps blood to the gills, where it picks up oxygen. The oxygenated blood then flows to the rest of the body. This creates a "single-loop" system.**

b **In amphibians and other land vertebrates, the heart first pumps blood to the lungs, where it picks up oxygen. The oxygenated blood returns to the heart to be pumped to the rest of the body, creating a "double loop."**

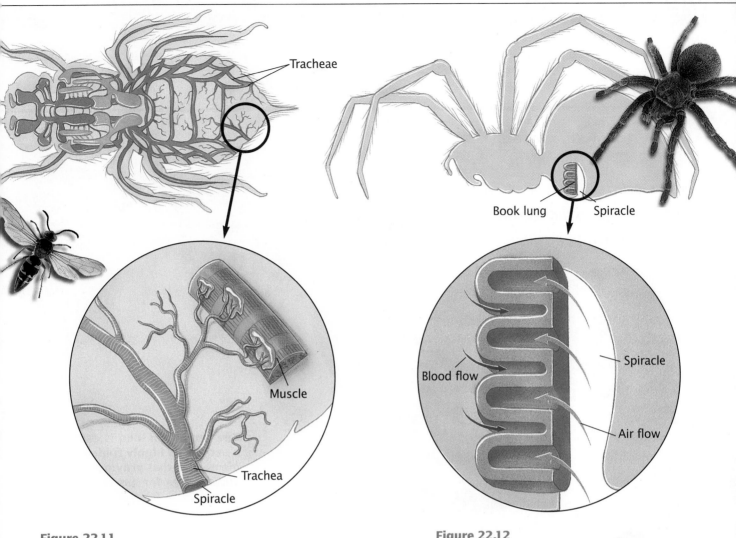

Figure 22.11
Tracheae leading into an insect's body from pores on the surface deliver oxygen directly to the cells and transport carbon dioxide outside the body.

Figure 22.12
In spiders, oxygen is absorbed through book lungs. The many folds of the book lungs provide a large surface area through which diffusion can occur.

How Terrestrial Arthropods Breathe

Land-dwelling arthropods, which are much smaller than most vertebrates, evolved different gas exchange systems. Insects breathe through **tracheae** (*TRAY kee ee*). Tracheae (singular, trachea) are tubes that lead into an insect's body cavity from pores called spiracles in the surface of the exoskeleton, as shown in **Figure 22.11**. Each trachea branches and rebranches, penetrating deep into the body. The tips of the tiny branches are open and carry oxygen directly to cells where it is needed. They also allow carbon dioxide to exit. This system is a very direct and efficient way of delivering oxygen and picking up carbon dioxide, and so is a good design to support the very active lifestyles of most insects. Notice that oxygen passes directly from the tracheae to the tissues. In your body, oxygen is first exchanged in the lungs, and then is carried to the tissues by blood. Insect blood does not transport oxygen.

Most spiders breathe through **book lungs**. As shown in **Figure 22.12**, book lungs are highly folded sacs, which are located inside the body, that open to the outside through a spiracle. Air moves into the sacs, where it is moistened so the oxygen in the air can dissolve and pass into the blood. Having respiratory membranes inside the body allows the surface where gas exchange occurs to remain moist without losing large amounts of water to evaporation.

Evolution of the Lung

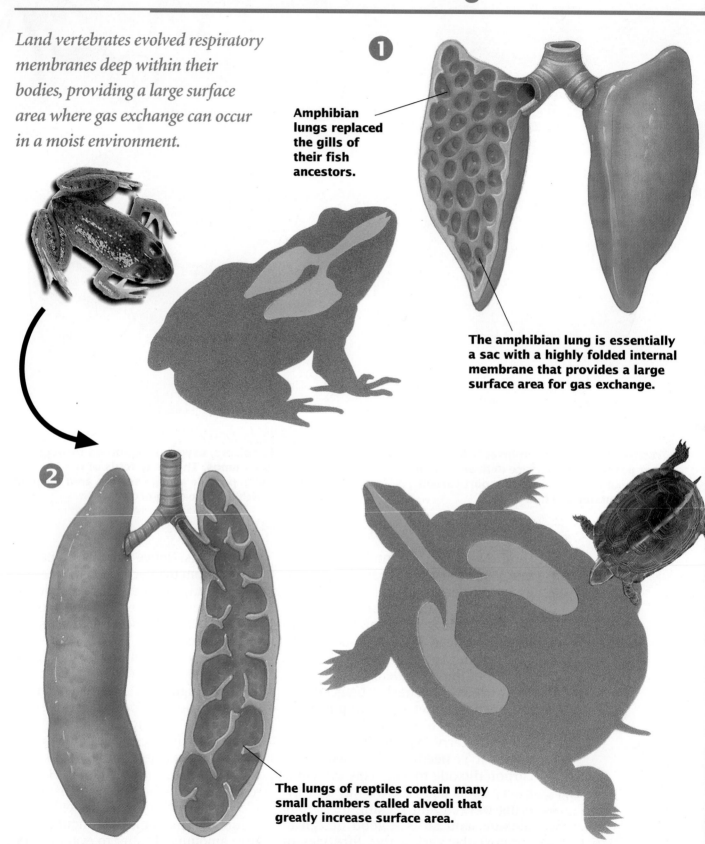

Land vertebrates evolved respiratory membranes deep within their bodies, providing a large surface area where gas exchange can occur in a moist environment.

1

Amphibian lungs replaced the gills of their fish ancestors.

The amphibian lung is essentially a sac with a highly folded internal membrane that provides a large surface area for gas exchange.

2

The lungs of reptiles contain many small chambers called alveoli that greatly increase surface area.

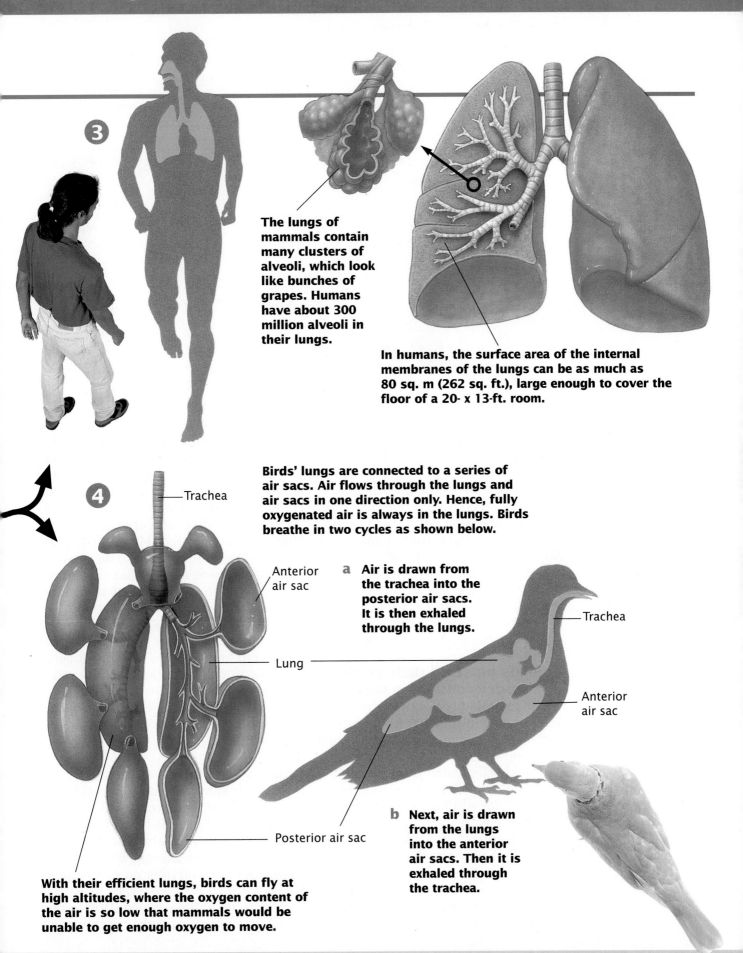

③

The lungs of mammals contain many clusters of alveoli, which look like bunches of grapes. Humans have about 300 million alveoli in their lungs.

In humans, the surface area of the internal membranes of the lungs can be as much as 80 sq. m (262 sq. ft.), large enough to cover the floor of a 20- x 13-ft. room.

④

Trachea

Birds' lungs are connected to a series of air sacs. Air flows through the lungs and air sacs in one direction only. Hence, fully oxygenated air is always in the lungs. Birds breathe in two cycles as shown below.

Anterior air sac

a Air is drawn from the trachea into the posterior air sacs. It is then exhaled through the lungs.

Trachea

Lung

Anterior air sac

Posterior air sac

b Next, air is drawn from the lungs into the anterior air sacs. Then it is exhaled through the trachea.

With their efficient lungs, birds can fly at high altitudes, where the oxygen content of the air is so low that mammals would be unable to get enough oxygen to move.

Adaptation to Land **491**

Table 22.1 Nitrogen-Containing Wastes

Type	Structure (nitrogen in yellow)	Solubility in water	Animal type
Ammonia		Soluble	Most aquatic animals
Urea		Soluble	Amphibians, mammals
Uric acid		Insoluble	Insects, reptiles, birds
Guanine		Insoluble	Spiders

Getting Rid of Wastes While Conserving Water

Stability

What role do the kidneys play in homeostasis?

Despite its covering of skin, your body is not completely watertight. Every day you lose about 2.5 L (2.4 qt.) of water—about 2 percent of the total amount of water in your body. How does that much water escape each day? About 300 mL (9 oz.) evaporates from your lungs. Another 500 mL (15 oz.) evaporates from your skin as sweat, cooling you in the process. The majority of the water you lose each day, about 1.5 L (45 oz.), is excreted as urine. Urine carries nitrogen-containing wastes from your body.

Animals must rid their bodies of nitrogen-containing wastes

When animals break down the amino acids found in foods containing protein, nitrogen is released as ammonia. Ammonia is toxic to animals and must be quickly eliminated from the body in very dilute form. Since the water required for dilution is plentiful for freshwater animals, they excrete their nitrogen directly as highly dilute ammonia. Land animals, however, cannot afford to lose so much body water. Instead, they spend some energy converting the ammonia to less toxic forms that need less water for dilution.

In vertebrates, ammonia is converted into one of two different compounds in the liver. Most mammals excrete nitrogen as **urea**, a less toxic compound of nitrogen. Elimination of urea requires some water, because urea is toxic if highly concentrated. Many birds and reptiles (and insects) excrete their waste nitrogen as **uric acid**. Unlike urea, uric acid is a nontoxic solid, and so can be eliminated with minimal water loss. (The white "deposits" birds leave on cars and statues is a combination of uric acid and feces.) Once produced, urea or uric acid are released by cells into the blood and are removed from the blood by the kidneys. The nitrogen-containing wastes are illustrated in **Table 22.1**.

Kidneys are blood filters

Vertebrate kidneys remove wastes from the blood and regulate the amounts of water and salts in the body. Blood circulates in the vertebrate body under pressure, and the kidneys act as filters, much as a car's oil filter does. As blood flows through the kidneys, its pressure forces water and dissolved substances such as sodium, potassium, chloride,

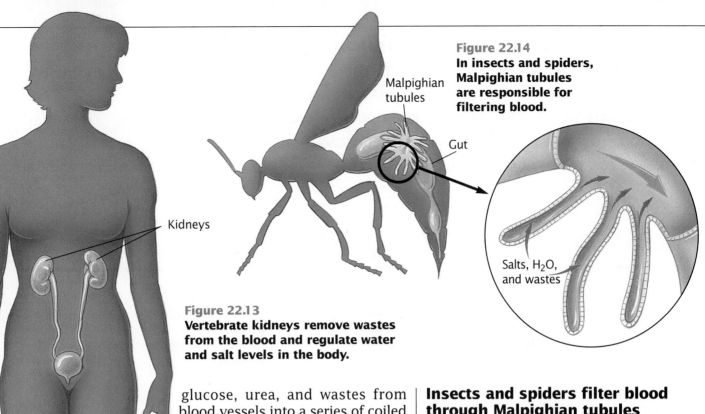

Malpighian tubules

Figure 22.14
In insects and spiders, Malpighian tubules are responsible for filtering blood.

Gut

Salts, H₂O, and wastes

Kidneys

Figure 22.13
Vertebrate kidneys remove wastes from the blood and regulate water and salt levels in the body.

glucose, urea, and wastes from blood vessels into a series of coiled tubes. Thus the kidneys are pressure filters, as is the oil filter in a car. Vertebrate kidneys are shown in **Figure 22.13**. Proteins and blood cells remain behind in the blood.

Most of the filtered substances are useful to the body, and it would be wasteful to discard them. Instead they are reabsorbed back into the bloodstream. Reabsorption is very selective. Urea, wastes, and some water are not reabsorbed, and pass out of the body as urine. The body regulates how much water is reabsorbed and therefore can control how much water leaves the body as urine. For instance, if you drink a large amount of water, your brain senses an increase in blood pressure and releases hormones that cause the kidneys to reabsorb less water. As a result, you produce more urine.

Insects and spiders filter blood through Malpighian tubules

Since blood does not circulate under pressure in insects and spiders, these terrestrial animals do not have pressure-filter kidneys. The organs responsible for filtering blood in insects and spiders are the **Malpighian** (*mal PIHG ee uhn*), **tubules** shown in **Figure 22.14**. Salts and wastes are actively transported from the blood into the Malpighian tubules. The high concentration of salts in the Malpighian tubules *draws* water from the blood. The contents of the Malpighian tubules are released into the gut, where useful salts such as potassium are reabsorbed. Water follows the salts by osmosis. Wastes such as uric acid and guanine are not reabsorbed and remain in the gut as a nearly dry paste that is excreted with the feces. Malpighian tubules are very efficient water conservation organs.

Section Review

❶ How would the exoskeleton of a land arthropod be different from that of an aquatic arthropod? Explain.

❷ Contrast the amount of evaporation from amphibian skins with that from reptilian skins.

❸ Describe the differences between amphibian lungs and mammalian lungs.

❹ List two adaptations that enable land vertebrates to get rid of wastes while still conserving water.

REPRODUCING ON LAND IS VERY DIFFERENT FROM REPRODUCING IN WATER. THE ADAPTATIONS TO LAND THAT YOU HAVE SEEN SO FAR WOULD NOT HAVE EVOLVED IF ANIMALS HAD NOT ALSO EVOLVED ADAPTATIONS THAT ENABLE THEM TO REPRODUCE ON LAND. IN THIS SECTION, YOU WILL SEE SOME OF THE ADAPTATIONS THAT MAKE IT POSSIBLE FOR TERRESTRIAL ANIMALS TO REPRODUCE FAR FROM WATER.

22.3 *Reproducing on Land*

Objectives

❶ **Recognize the advantages of internal fertilization for land animals.**

❷ **Compare a reptilian egg to an amphibian egg.**

❸ **Summarize two advantages of a placental mammal's development.**

Figure 22.15

Each animal shown below lays a different type of egg. The toads, for instance, lay strings of eggs in a gelatinous coating that is permeable to water.

Internal Fertilization

The toads you see in **Figure 22.15** are amphibians that have come from miles around to breed in the pond. The behavior of these toads reflects the strong ties between amphibians and water. Amphibians must reproduce in water or in damp environments. As you will see in this section, insects, mammals, birds, and reptiles have broken the reproductive tie to water.

When egg and sperm unite, they must do so in a moist environment. For many aquatic animals, including most fishes, finding a moist environment is easy: they release their gametes into the surrounding water. Fertilization takes place outside the body of either parent. This type of fertilization is called **external fertilization**. External fertilization is less common on land, because both sperm and egg run the risk of drying out and dying.

Most terrestrial animals reproduce by internal fertilization

How did terrestrial animals overcome the limitations of external fertilization? Recall that each animal carries its own supply of water around inside its body. Thus, if gametes can be transferred directly between individuals, the risk that gametes will dry out disappears. In almost all land animals, including flatworms, roundworms, annelids, mollusks, arthropods, mammals, reptiles, and birds, the male deposits his sperm directly inside the female. Fertilization occurs inside the female's body. This type of fertilization is called **internal fertilization**. Most amphibians lack internal fertilization. They reproduce by external fertilization, as did their fish ancestors. Now you can see one reason why amphibians such as toads must return to water to reproduce.

If you go to a pond a few days after toads breed, you will find that the water contains strings of small, round eggs, like those shown in **Figure 22.15** (far left). These eggs are the second reason that amphibians must reproduce in water or moist places. The embryonic toad develops within the egg's jellylike coating. This coating is freely permeable to carbon dioxide, oxygen, and water.

Amphibian eggs removed from moisture soon dry out and die.

The eggs of reptiles and birds are watertight

Unlike amphibian embryos, the embryos of reptiles are surrounded by a watertight protective membrane called the **chorion** (*KAWR ee ahn*). The chorion is impermeable to water, but it does permit oxygen to enter the egg and carbon dioxide to leave. Lying within the chorion is another membrane, the **amnion** (*AM nee uhn*). The amnion encloses the embryo within a watery environment, as shown in **Figure 22.16**. This kind of egg is called an **amniotic egg**. Amniotic eggs evolved first in reptiles and are also found in birds and mammals. In the eggs of reptiles and birds, a rich yolk supplies nutrition to the developing embryo. A tough shell surrounds and protects the eggs of birds, reptiles, and three species of egg-laying mammals.

Like the eggs of birds and reptiles, the eggs of most terrestrial arthropods are watertight and very yolky, providing nutrients for the young during development. Arthropod eggs are usually either thick-walled and resistant to drying or are encased in some sort of waxy protective covering.

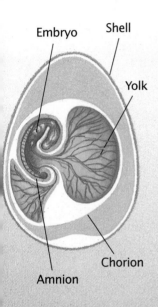

Figure 22.16
Bird and reptilian eggs, such as the turtle egg below, have a watertight protective coating called the chorion and an internal membrane called the amnion, which encloses the embryo in a watery environment.

Embryo

Shell

Yolk

Chorion

Amnion

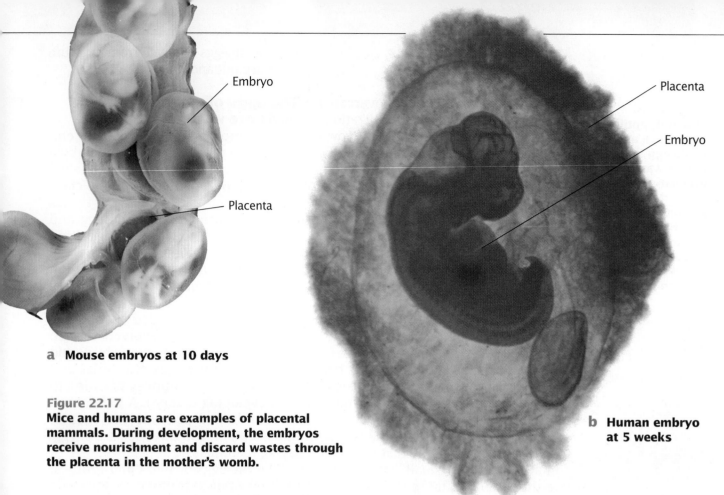

Embryo

Placenta

a **Mouse embryos at 10 days**

Placenta

Embryo

Figure 22.17
Mice and humans are examples of placental mammals. During development, the embryos receive nourishment and discard wastes through the placenta in the mother's womb.

b **Human embryo at 5 weeks**

Eggs Without Shells

Once the egg of a reptile or bird has been laid, the parent cannot provide further nourishment to the offspring until hatching. Furthermore, the egg is exposed to environmental hazards such as predators, overheating, or freezing.

Except for platypuses and echidnas (*ee KIHD nuhz*), mammals do not lay eggs. Instead, their eggs develop until birth inside the mother's body. No shell forms around the egg. Among marsupial mammals, such as kangaroos and opossums, birth occurs very early, and the offspring matures in the mother's pouch, nourished by suckling her milk.

Among placental mammals (mice and humans, for instance), birth occurs much later, when development is essentially complete. A human embryo develops for nine months inside its mother. During its development, the embryo of a placental mammal is nourished by the mother through a unique structure called the **placenta** (*pluh SEHN tuh*), as shown in **Figure 22.17**. The placenta is made of embryonic and maternal membranes. Nutrients and oxygen move from mother to embryo through the placenta, while carbon dioxide and other wastes leave the embryo.

Section Review

❶ **Why isn't external fertilization an effective way to reproduce on dry land?**

❷ **Describe the advantages of the shelled egg over the amphibian egg.**

❸ **Describe two hazards faced by reptilian eggs but not by eggs of placental mammals.**

"Gee, evolution is slow."

	Key Terms	Summary

22.1 Leaving the Sea

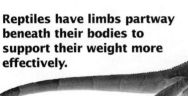

Reptiles have limbs partway beneath their bodies to support their weight more effectively.

Key Terms

lateral line (p. 484)

Summary

- Life on land places different demands on animals than life in water.
- The most successful land groups are the arthropods and the vertebrates. Both of these groups have sturdy skeletons that support their bodies out of water. They also have strong flexible limbs for moving on land.
- The high density of water makes sound easy to detect. Land animals hear well only with an amplifying system, such as the ear.

22.2 Staying Moist in a Dry World

Amphibian blood is pumped through a "double-loop" system.

Key Terms

gill (p. 486)

trachea (p. 489)

book lung (p. 489)

urea (p. 492)

uric acid (p. 492)

Malpighian tubule (p. 493)

Summary

- The exoskeleton of arthropods serves as a barrier to evaporation. Birds, reptiles, and mammals minimize water loss by means of their watertight skins.
- Animals obtain oxygen from the air and release carbon dioxide. In land vertebrates, this gas exchange occurs in the lungs. In insects, tracheae carry oxygen.
- Breakdown of amino acids produces the toxic byproduct ammonia. Aquatic animals excrete dilute ammonia. Terrestrial animals transform ammonia into urea or uric acid, which is eliminated with less water loss.

22.3 Reproducing on Land

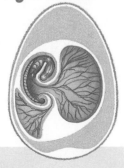

Reptilian eggs have a protective shell and a watery internal environment.

Key Terms

external fertilization (p. 494)

internal fertilization (p. 495)

chorion (p. 495)

amnion (p. 495)

amniotic egg (p. 495)

placenta (p. 496)

Summary

- Most amphibians reproduce in water by external fertilization. Most land animals reproduce by internal fertilization.
- The eggs of reptiles, birds, and mammals are surrounded by watertight membranes. In most mammals, development is completed within the mother's body.

Adaptation to Land **497**

Understanding Vocabulary

1. For each list of terms, identify the one that does not fit the pattern and describe why it does not fit.
   a. kidneys, lungs, tracheae, gills
   b. turtle, fish, toad, frog
   c. blood, ammonia, urea, uric acid
   d. chorion, amnion, shell, Malpighian tubules

Relating Concepts

2. Copy the unfinished concept map below onto a sheet of paper. Then complete the concept map by writing the correct word or phrase in each oval containing a question mark.

Understanding Concepts

Multiple Choice

3. Which animal group lives on land?
   a. sponges
   b. cnidarians
   c. roundworms
   d. echinoderms

4. Water serves to counteract the force of gravity for aquatic animals. What body structure takes the place of water for terrestrial vertebrates and arthropods?
   a. blood plasma
   b. skin
   c. skeleton
   d. lungs

5. Amphibian limbs are thought to be modified
   a. fish scales.
   b. lateral lines.
   c. reptile limbs.
   d. fish fins.

6. The waterproof coating that covers vertebrates and helps prevent water loss is called
   a. skin.
   b. endoskeleton.
   c. body armor.
   d. exoskeleton.

7. Which animal has the most efficient respiratory system?
   a. cow
   b. bird
   c. lizard
   d. crab

8. Compared with the single-loop circulation of a fish, the double-loop circulation of an amphibian has
   a. increased rate of blood flow.
   b. decreased ammonia removal.
   c. decreased rate of blood flow.
   d. increased oxygen removal from tissues.

9. In which group of land vertebrates does fertilization usually occur outside the female's body?
   a. mammals
   b. reptiles
   c. birds
   d. amphibians

10. Which of the following is a placental mammal?
    a. opossum
    b. mouse
    c. kangaroo
    d. toad

11. Gas exchange occurs in the gills of crabs and fishes by
    a. osmosis.
    b. diffusion.
    c. active transport.
    d. proton pumps.

12. The nitrogen-containing waste excreted by birds and reptiles is
    a. ammonia.
    b. urine.
    c. uric acid.
    d. urea.

Completion

13. Fishes detect disturbances in water with their _____ .

14. In mammals, gas exchange occurs in the _____ , while in spiders it occurs in the _____ .

15. Most of the water lost from the human body is eliminated as _____ .

16. In mammals, wastes are removed from the blood by the _____ . In insects, the same function is performed by the _____ .

Land animals include
do not include
arthropods
evolved from
?
?
lobe-finned fishes
lack
such as
amniotic egg
?
contain
have
?
produced by
tracheae — carry — ?
which nourishes
?
?
offspring
such as
lizards

Short Answer

17. Your friend caught an animal in her backyard. How could you tell if it is a reptile or an amphibian without harming the animal?

18. Explain the relationship between the density of a medium and the speed and distance that sound travels through it.

19. Fish eliminate nitrogen-containing wastes as ammonia. What complications would be associated with land animals also eliminating waste in this form?

20. Describe how kidneys enable a mammal to remove wastes and regulate how much water is retained by the body.

21. Explain why gills don't function out of water.

Interpreting Graphics

22. Look at the figure shown below.

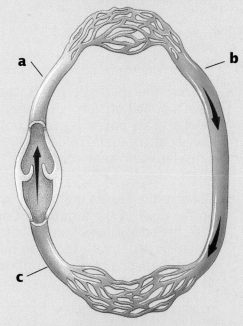

- In what kind of animal does this type of circulatory system occur?
- At which point would you expect to find oxygenated blood?
- At which point would you expect to find the highest blood pressure?

Reviewing Themes

23. *Evolution*
    What evidence has led scientists to conclude that the earliest land vertebrate evolved from lobe-finned fishes?

24. *Energy and Life*
    Land animals expend energy to convert ammonia into urea or uric acid. What do these animals gain from this conversion?

Thinking Critically

25. *Inferring Conclusions*
    The world-record time for running the mile is just under four minutes. How might this record be different if humans had air sacs like birds do?

26. *Comparing and Contrasting*
    What makes the lungs of mammals more efficient than the lungs of amphibians?

27. *Building on What You Have Learned*
    In Chapter 4 you learned about diffusion. What role does diffusion play in the exchange of gases in the gills of fish?

Cross-Discipline Connection

28. *Biology and Art*
    Draw a picture of a fictional animal that is adapted for survival in one of the seven terrestrial ecosystems studied in Chapter 12. Then write a description of how the animal reproduces, breathes, and eliminates wastes.

Discovering Through Reading

29. Read the article "Grasshoppers Change Coats to Beat the Heat," in *Science News*, August 24, 1991, page 119. What differences did scientists find between the lipid coats of southern and northern grasshoppers?

Investigation

How Are Anole Lizards Well Adapted to Their Environment?

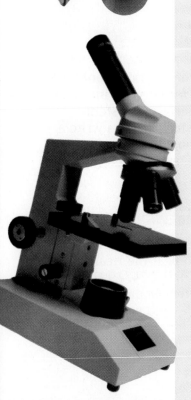

Objectives

In this investigation you will:
- *observe* structural and behavioral characteristics of an anole lizard
- *relate* these characteristics to the ability of the lizard to survive on land

Materials

- terrarium
- live anole lizards
- one piece of orange construction paper large enough to cover one-half of the terrarium floor
- one piece of green construction paper the same size as the orange paper
- live *Tenebrio* larvae or crickets
- small pieces of apple
- small dish
- desk lamp or heat lamp
- 1,000-mL beaker or large glass jar
- 600-mL beaker
- crushed ice

Prelab Preparation

Review what you have learned about reptiles by answering the following questions.
- What features of reptiles enable them to live on land?
- How are the limbs of a reptile different from the limbs of an amphibian and a mammal?

Procedure: Observing Anole Lizards

1. Form a cooperative team with another student to complete steps 2–9.

2. Observe one lizard in the terrarium. Record its coloration. Describe the texture of its skin. *Does its skin appear to be moist?*

3. Closely observe the lizard's limbs and posture. *How does the orientation of the lizard's hind limbs compare with the orientation of your legs?*

4. Beneath the jaw, male anole lizards have a pouch of skin called the dewlap. This pouch is pink. Look closely at the anole to determine if a dewlap is present and record your observations.

5. Observe the anoles that are sitting on the floor of the terrarium. Compare the coloration of anoles sitting on green paper with those on the orange paper. *Is there any difference in coloration?*

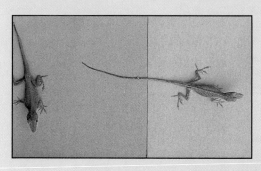

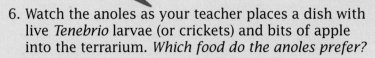

6. Watch the anoles as your teacher places a dish with live *Tenebrio* larvae (or crickets) and bits of apple into the terrarium. *Which food do the anoles prefer?*

7. Place a lamp near the glass at one corner of the terrarium. *What is the response of the anoles to the lamp?* Record your observations.

8. Observe as your teacher removes one anole and places it in a 600-mL beaker filled with crushed ice. *What is the anole's response?* Be sure to note any color change.

9. Your teacher will now introduce a new anole into the terrarium. Closely observe the behavior of the residents and the introduced anole. Record your observations.

Analysis

1. *Analyzing Observations*
   How do the limbs and skin of the lizard differ from those of an amphibian?

2. *Making Inferences*
   How are a lizard's limbs and skin better adapted for life on land than those of an amphibian?

3. *Analyzing Observations*
   What causes color change in anoles? Explain your answer.

Thinking Critically

1. Lizards are ectotherms that absorb heat from their surroundings. Yet lizards are able to maintain a fairly constant body temperature. Explain how lizards are able to control their body temperature.

2. What is the advantage of the anole's ability to change color, if its primary stimulus is not a change in background coloration? What evidence from your observations supports your answer?

3. What function does the male anole's dewlap serve? Why is it advantageous for the male to be able to retract the dewlap?

Chapter 23

Animal Diversity

Review

- animal tissues (Section 21.1)

- origin of body cavities
  (Section 21.2)

- innovations in body plan
  (Section 21.3)

This ocellated nudibranch is a marine mollusk. Unlike its relative the snail, it lacks a shell. This species of nudibranch can be found in both the Pacific and Indian Oceans.

CHAPTER 21 TOLD THE STORY OF THE EVOLUTION OF ANIMAL BODY
ARCHITECTURE. IN THIS CHAPTER, YOU WILL RETURN TO THE MAJOR
PHYLA OF ANIMALS FOR A LONGER VISIT. YOU WILL DISCOVER WHERE
MANY OF THESE ANIMALS LIVE, HOW THEY FEED, AND HOW THEY
REPRODUCE. YOU WILL ALSO LEARN HOW SOME OF THESE ANIMALS
AFFECT HUMANS, BOTH POSITIVELY AND NEGATIVELY.

23.1 Sponges, Cnidarians, and Simple Worms

Objectives

❶ Contrast the lifestyles of sponges and cnidarians.

❷ Compare the polyp and medusa stages of the cnidarian life cycle.

❸ Describe the life cycle of the beef tapeworm.

❹ List two parasitic nematodes that can live in humans.

Sponges

Sponges are probably most familiar to you as absorbent pads used for wiping up spills or cleaning dishes. The sponges you buy in the store are usually manufactured, not derived from the animals known as sponges. There are more than 9,000 species of sponges (phylum Porifera), most of which are marine. About 150 species live in fresh water. An adult sponge, like the one in **Figure 23.1**, spends its life attached to a hard surface. For this reason, it was not until 1765 that zoologists realized that sponges are animals, not plants. Most sponges are asymmetrical and grow to conform to the surface on which they live.

A sponge's body is perforated by holes (Porifera means "pore-bearer") that lead to an inner water chamber. Sponges pump water through these pores and expel it through a large opening at the top of the chamber. While water is passing through the body, nutrients are engulfed, oxygen is absorbed, and wastes are eliminated.

As you learned in Chapter 21, between the inner and outer layer of many sponges is a jellylike layer. Embedded within this layer is a network of needle-like **spicules**. Spicules form the skeleton of the sponge. The middle layer of some large sponges contains a mesh of tough protein called spongin. These kinds of sponges have been harvested for centuries because of the ability of their spongin skeletons to soak up water and release it when squeezed. Today, most of the sponges you can buy are copies of this spongin mesh manufactured from plastic or cellulose.

Figure 23.1
Sponges vary widely in size, shape, and color. This purple tube sponge lives in the ocean surrounding Bonaire, an island about 95 km (60 mi.) north of Venezuela.

Spiders

There are about 35,000 kinds of spiders. While a few live in fresh water, the majority are land dwellers. Indeed, spiders are thought to have been among the first arthropods to successfully invade the land, soon after their close relatives the scorpions. Spiders are predators, feeding on nearly any other animal of manageable size.

In addition to the characteristics that make arthropods so successful, spiders have several unique features that make them particularly well adapted to their lifestyles. First, they have efficient organs of excretion and water balance, so they can rid their bodies of wastes while still conserving precious water. Second, spiders are able to breathe air by using book lungs. Recall from Chapter 22 that book lungs allow air to come close to the spider's blood in a humid chamber where gas exchange occurs. Third, nearly all spiders are poisonous. The chelicerae of spiders are fangs equipped with poison glands. Fourth, and perhaps most important, spiders have the ability to produce silken threads, which they use in a great variety of ways.

Silk and fangs are a deadly combination

You may have watched a spider building its web. At the end of a spider's abdomen are small nozzle-like structures called **spinnerets**. Spinnerets direct the flow of silk from silk-producing glands in the abdomen. Spiders can produce different kinds of silk threads in various diameters.

Spider silk is composed of a complex structural protein that is produced in the glands as a liquid. As the liquid silk flows out through holes in the spinnerets, it solidifies into the threads that make up the web. As illustrated in **Figure 24.2**, there are as many different types of webs as there are types of spiders.

Figure 24.2

Each kind of spider has a specialized web that enables it to catch its prey. Some spiders build webs in high tree branches, while others build horizontal sheet webs that catch insects that drop from above or are carried by the wind. Some spiders build silk-lined burrows beneath the ground. These spiders extend threads that act as trip lines around the burrow entrance. Then they sit inside, waiting for an unsuspecting insect to become entangled.

Spider silk is elastic and extremely strong, about as strong as a nylon thread of the same diameter. Spiders use silk for many purposes. Threads are used as safety lines when spiders dangle from branches or drop to the ground. Many spiders also use silk to line their nests or burrows, to wrap prey, or to fashion cocoons for their young. Some newly hatched spiderlings spin long, thin threads on which they ride the winds over great distances.

A spider's web is a sticky trap to ensnare prey, such as the bee caught in the web shown in **Figure 24.3**. Constructing a web is a complex architectural feat that requires several steps. When an insect strikes the web and struggles, the spider is alerted by the threads' vibrations. The spider then rushes to the victim, bites it, and injects poison and digestive enzymes. The ability to stun or kill the prey quickly with poison from the chelicerae is a great advantage. It saves the spider from having to wrestle with large or potentially dangerous prey, such as bees and wasps.

Although nearly all spiders produce poison, only about 20 kinds are considered dangerous to people. Two such species are found in the United States: the American black widow spider and the brown recluse spider, both shown in **Figure 24.4**. Most spiders are more beneficial than harmful to humans. They are the world's champion pest controllers, consuming millions of insects in their daily activities. Many of these insects are agricultural pests.

Spiders have courtship rituals

Spiders are solitary animals. When they do seek a mate, it is important that males and females of the same species are able to recognize each other. Males are usually smaller than the females and could easily be mistaken for a meal if some signal did not identify the male to his prospective partner. Spiders have evolved complex behaviors that ensure

Figure 24.3
After wrapping a bee in a silk cocoon and injecting digestive enzymes, this *Argiope* spider leaves. Later the spider returns to its web and sucks out the now-liquid prey, and discards the empty carcass.

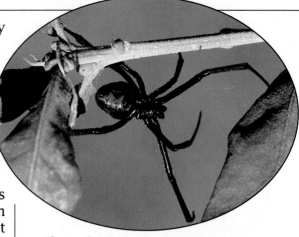

Figure 24.4
a **The American black widow spider is most common in the southern United States. The bite of the female injects a powerful venom that acts on the nervous system.**

b **The brown recluse spider can be recognized by the violin shape on its upper surface. They live in dark places such as basements and woodpiles. Their venom causes tissue damage in the area of the bite.**

successful mating and fertilization of the female's eggs.

Mating behavior in the black widow spider provides an example of the dangers a male spider faces. As the male approaches the web of a female, he drums a specific rhythm on the strands of the web to alert the female that he is a potential mate and not a meal. If he vibrates the female's web inappropriately, she will kill and eat him. After mating, the female often kills the male anyway, wrapping and saving him for her offspring's first dinner.

Pedipalps

Chelicerae

Cephalo-
thorax

Abdomen

Stinger

Figure 24.5
**This African
scorpion is about
18 cm (7 in.) long.
Scorpions found
in the United
States range
from 5–13 cm
(2–5 in.) in length.**

Other Arachnids

More than 400 million years ago, scorpions were the first arthropods to invade land. They share many of the adaptations that enable spiders to survive on land, including book lungs and efficient excretory and water conservation organs. Unlike the poison-injecting chelicerae of spiders, scorpions' chelicerae are ripping claws. Their pedipalps have evolved into enlarged pincers used to capture prey. Scorpions feed mostly on insects. As you can see in **Figure 24.5**, the scorpion's stinger is found at the tip of its tail. While the scorpion holds its victim in its pincers, it brings its tail forward to jab into the prey, and then tears the prey into pieces.

Scorpions live in the warmer regions of the world. About 20 scorpion species are found in the United States, mostly in the southwest. Scorpions hide during the day and hunt at night. Although it is painful, a scorpion's sting is rarely lethal to humans. Like spiders, scorpions perform elaborate mating rituals.

Most mites and ticks are parasites

If dust makes you sneeze, you are reacting not to the dust itself, but to the tiny mites found in the dust. Over 30,000 species of mites and ticks have been described. A number of these species are free-living on land or in water, but many are parasites on the bodies of other animals. The free-living forms are mostly scavengers or predators on other tiny creatures. If you have done much camping or hiking in brushy terrain, chances are you have spent some time removing ticks from your body. Most ticks are bloodsuckers that feed on vertebrates. Their sharp chelicerae are specialized for slicing skin. Although normally quite small, some ticks can swell

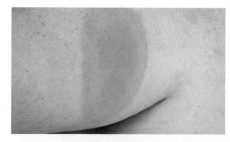

Figure 24.6
About one-half of all the people bitten by the tick that causes Lyme disease develop a "bulls-eye rash" within a few days.

to 3 cm (just over 1 in.) in length after a full meal of blood. Recall from Chapter 16 that ticks transmit Lyme disease and Rocky Mountain spotted fever to humans, and Texas cattle fever to livestock. **Figure 24.6** shows the rash that usually occurs a few hours after a person has been bitten by the tick that causes Lyme disease.

Mites, such as the red water mite in **Figure 24.7**, are much more diverse than ticks. They parasitize virtually all groups of animals and many plants. The incredible specialization of mites is shown by two species, *Demodex folliculorum* and *Demodex brevis*. One

Figure 24.7
The red water mite is an active swimmer. The larval stages of water mites may be parasitic on insects or fishes.

species lives only in hair follicles, the other in oil glands, respectively, of the human forehead. Mites have serious direct and indirect effects on humans. Many, such as chiggers, burrow just beneath the skin, where they cause severe itching. Mites also carry viruses that plague food crops, such as mosaic viruses of rye and wheat. The red-legged mite and the winter grain mite both feed on stored crops, destroying many tons of grain each year. Other mites cause feather loss in birds, decreased wool production in sheep, and mange in dogs. If you find yourself itching a bit after reading this section, don't be alarmed; we are all host to some of these parasites.

Horseshoe crabs are ancient arthropods

The most ancient arthropod with chelicerae is not an arachnid, but the horseshoe crab shown in **Figure 24.8**. There are only five species of horseshoe crabs alive today. They live in shallow oceans, where they plow through sandy bottoms and use their clawlike legs to grasp small animals on which they feed. They also scavenge on dead animals.

Horseshoe crabs are closely related to the extinct giant "water scorpions," or eurypterids, such as the one shown in **Figure 24.9**. The eurypterids flourished about 300 million years ago. Some eurypterids reached lengths of 3 m (10 ft.).

Figure 24.8
This horseshoe crab is not really a crab at all, but a close relative of the scorpion.

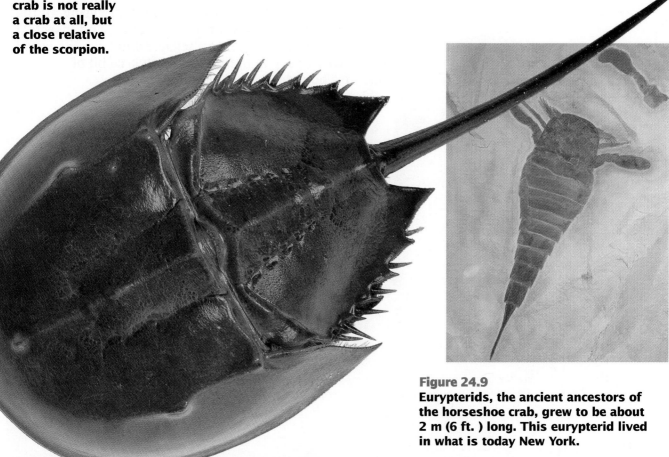

Figure 24.9
Eurypterids, the ancient ancestors of the horseshoe crab, grew to be about 2 m (6 ft.) long. This eurypterid lived in what is today New York.

Section Review

❶ **Name three arachnids that are not spiders.**

❷ **Contrast the functions of pedipalps and chelicerae in spiders.**

❸ **Describe two ways spiders use silk other than for building webs.**

❹ **How is your life affected by arachnids?**

Genaro Lopez: Entomologist

How I Became an Entomologist

"I grew up in Brownsville, Texas, where I spoke only Spanish until the first grade. I have always loved the outdoors. As a child, I went camping, hiking, and fishing whenever I could. In high school, my science classes were my favorite.

"Although neither of my parents finished high school, they encouraged me to go to college. A bunch of my friends talked about going, and I thought if they could do it, I certainly could. I started working for Robert Baker, a professor who was only five years older than I was. It's amazing how a good teacher can influence a person. Although I was only an undergraduate, he had me doing stuff that only graduate students did—like traveling to the Amazon River in Colombia, South America, to collect bats. On spring break, we'd go to Mexico to collect animals and plants. I acted as an interpreter and scientist. I'd never seen such peculiar life forms before.

"I loved working in the tropics—just being outside and discovering all the different trees, animals, and insects. I like to say science and I see the world in the same way—a little bit of fact and a whole lot of mystery."

Dr. Lopez is also a bicycle racer. "When you're going along at 25–28 mph and the guy in front of you is only three inches away, you know you're just a split second away from crashing. You have to make quick decisions. It's exciting. You have to be finely tuned."

Name:	**Genaro Lopez**
Home:	**Brownsville, Texas**
Employer:	**University of Texas at Brownsville, Brownsville, TX**
Personal Traits:	**• Adventurous**
	• Curious
	• Hard-working
	• Ambitious
	• Fun-loving

Applying Science to Life

Career Path

High School:
- Science
- Literature

College:
- Zoology
- Chemistry

Graduate School:
- Entomology
- Ecology

"During the academic year, I teach general biology, human anatomy, and physiology to college students. In the summer, I work as a consultant urban entomologist for a pest control company. I act as their Hispanic spokesperson. Every summer they fumigate entire apartment complexes in poor areas of New York City, Miami, and Los Angeles. Several hours after spraying, we go back into the buildings and look at all the dead roaches to see what species are more prevalent in different areas. I explain what's happening to the residents and media in Spanish.

"On one of my trips to Miami, my live roaches and I appeared on the highest rated Spanish variety television show, which is broadcast in 17 countries. When the host saw my Madagascar roaches, he turned white. Later he regained his composure and shook my hand."

Research Focus

Dr. Lopez has an ongoing project studying the tiger beetles along the edge of salt marshes in the lower Laguna Madre of Texas. He has found three species of beetles there. He samples their populations to learn how they coexist. It's an ecological question: How can three similar species share the same resources? He's trying to learn how cooperation, rather than competition, works in nature.

His hypothesis is that one species is most abundant in the spring, a second is most abundant in the summer, and a third is most abundant in the fall. That way they divide the available food equally.

Through his research, Dr. Lopez hopes to show that living organisms such as the beetle he is studying often separate their competition, so that they can successfully coexist.

AT THIS MOMENT, THERE ARE OVER 200 MILLION TIMES MORE INSECTS THAN HUMANS LIVING ON EARTH. THE NUMBER OF INSECT SPECIES IS FAR GREATER THAN THE NUMBER IN OTHER ANIMAL GROUPS. IN FACT, THERE ARE MORE SPECIES OF BEETLES THAN SPECIES OF ALL NON-INSECT ANIMALS COMBINED. INSECTS AND THEIR CLOSE RELATIVES, THE MILLIPEDES AND CENTIPEDES, HAVE MANDIBLES INSTEAD OF CHELICERAE.

24.2 Insects, Millipedes, and Centipedes

Objectives

❶ Contrast the anatomy of spiders with that of insects.

❷ Compare incomplete and complete metamorphosis.

❸ Describe five ways that insects affect your life.

❹ Identify two differences between millipedes and centipedes.

Insects

Figure 24.10
The bodies of all insects, including this grasshopper, consist of three distinct regions: head, thorax, and abdomen. Three pairs of legs are attached to the thorax. Most insects have wings, which are also attached to the thorax. The evolution of wings was a major reason for the diversification of insects.

Like other arthropods, insects are segmented. In insects, these segments have fused into three body regions: the head, the thorax, and the abdomen, as shown in **Figure 24.10**.

A hallmark of insects is the diversity of the specialized appendages attached to their heads. Unlike arachnids, insects have antennae, appendages specialized for sensing the environment. Insect mouthparts are also extremely varied, enabling insects to feed on a greater variety of foods than other arthropods. Among the insects are predators and parasites, bloodsuckers and plant-sapsuckers, scavengers and wood-eaters. The chelicerae and pedipalps of spiders are simple compared with the highly specialized mouthparts of insects. Instead of chelicerae,

insects have jaws, or **mandibles**, and other mouthparts that have evolved into different shapes for grinding, scraping, piercing, and sucking.

Insect legs have also evolved into a variety of shapes. The legs of beetles are modified for walking, while the legs of grasshoppers are adapted for jumping. Ant legs are suited for burrowing.

Like other terrestrial arthropods, insects have evolved mechanisms for gas exchange and water conservation. Insect tracheae are extremely efficient in delivering oxygen directly to the body's cells. Insects have a thick, waxy exoskeleton that prevents water loss. Their Malpighian tubules enable them to excrete nitrogenous wastes with little loss of water. **Table 24.1** shows some of the major orders of insects.

Forewing

Thorax

Antennae

Hind wing

Head

Jumping legs

Abdomen

Table 24.1 Major Orders of Insects

Order	Approximate Number of Species	Main Characteristics	Examples	
Coleoptera "Shield-winged"	350,000	Two pairs of wings (front pair covers transparent hind pair); heavy, armored exoskeleton; biting and chewing mouthparts; complete metamorphosis	Beetles, weevils	
Diptera "Two-winged"	120,000	Transparent front wings; hind wings reduced to knobby balancing organs; sucking, piercing, lapping mouthparts; complete metamorphosis	Flies, mosquitoes	
Lepidoptera "Scale-winged"	120,000	Two pairs of broad, scaly wings; hairy bodies; tubelike sucking mouthparts; complete metamorphosis	Butterflies, moths	
Hymenoptera "Membrane-winged"	100,000	Two pairs of transparent wings; mobile head; well-developed eyes; chewing and sucking mouthparts; stinging; many species social; complete metamorphosis	Ants, bees, wasps	
Hemiptera "Half-winged"	60,000	Two pairs of wings or wingless; piercing, sucking mouthparts; incomplete metamorphosis	Giant water bug, bedbug, chinch bug	
Orthoptera "Straight-winged"	20,000	Two pairs of wings or wingless; biting and chewing mouthparts in adults; incomplete metamorphosis	Grasshoppers, crickets, cockroaches, mantids	
Odonata "Toothed"	5,000	Two pairs of transparent wings; chewing mouthparts; incomplete metamorphosis	Dragonflies, damselflies	
Isoptera "Equal-winged"	2,000	Two pairs of wings, but some stages are wingless; chewing mouthparts; social insects with division of labor; incomplete metamorphosis	Termites	
Siphonaptera "Tube-wingless"	1,200	Small, wingless, flattened body; piercing and sucking mouthparts; jumping legs; complete metamorphosis	Fleas	

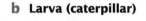

Most insects change their body form as they mature

Spiders, scorpions, mites, and ticks all undergo **direct development**. In this kind of development, a miniature copy of the adult form hatches from the egg. Some species of insects also undergo direct development.

Most insects, however, hatch as a form that is not identical to the adult. As it grows, the young insect changes to become more and more like its parents. Each of these changes is called a **metamorphosis** (*meht uh MAWR fuh sihs*). If the metamorphosis into the adult form involves a series of gradual changes, it is called **incomplete metamorphosis**. Insects that undergo incomplete metamorphosis include grasshoppers, dragonflies, mayflies, and cockroaches. In this type of development, the insect that emerges from the egg is called a **nymph**

(*NIHMF*). The nymph is a smaller version of the adult insect species, similar in structure but without wings or mature reproductive organs.

In **complete metamorphosis**, the juvenile form usually does not resemble the adult. When the egg of an insect hatches, an immature form called a larva emerges. The larva looks nothing like the adult. Larvae, like nymphs, cannot yet fly. When the larval stage is complete, the insect enters a stage called the pupa (*PYOO puh*) or chrysalis (*KRIHS uh lihs*), depending on the species. Many pupae form a covering around themselves. This covering is called a cocoon. During this stage, most insects remain immobile while larval tissues and organs are replaced with new tissues and organs. At the end of the pupal stage, a fully formed mature adult emerges, as shown in **Figure 24.11e**. About 90 percent of insect species undergo complete

a   Egg

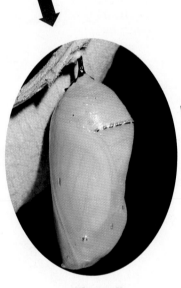

b   Larva (caterpillar)

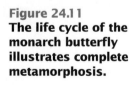

d   Adult butterfly about to emerge

c   Chrysalis

Figure 24.11
The life cycle of the monarch butterfly illustrates complete metamorphosis.

e   Adult butterfly

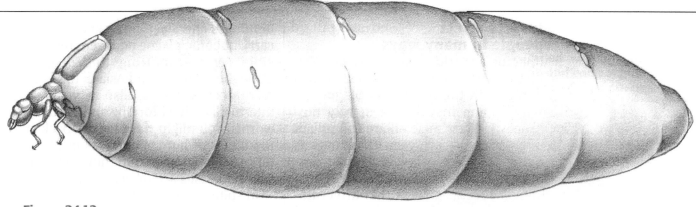

Figure 24.12

a A termite nest usually has a single queen who lays all the eggs. A queen is about 6 cm (2.5 in.) long. A nest of termites may contain as many as 5 million individuals.

metamorphosis, including flies, beetles, ants, bees, wasps, butterflies, and moths.

What is the evolutionary advantage of the complicated insect life cycle? During the life cycle, the various stages are not only physically different from each other, they are ecologically different as well. As a result, at different times in its life a single animal can exploit different habitats and different food resources. For example, the leaf-chewing caterpillar will eventually transform into a nectar-drinking butterfly. This ecological separation of young from adults eliminates competition between the two life-cycle phases.

Ants, bees, wasps, and termites live in complex societies

Ants, termites, and certain kinds of bees and wasps are highly social. These insects live in highly organized "societies" of related individuals, with division of labor among the society members. Different individuals serve the group by performing different

functions. In a termite colony, workers gather food, tend offspring, and excavate the tunnels and chambers of the colony. Soldiers, as the name implies, defend the colony against attack. Enlarged jaws and heads enable them to carry out this role. Both soldiers and workers are sterile. Only the king and queen of the colony reproduce. The role played by an individual in the colony is its caste. **Figure 24.12** shows representatives of the four castes of a termite colony. All four caste members are shown to scale and are pictured slightly more than three times their actual size. Insects have no choice about their caste in the society. Their genes determine the caste to which they will belong.

b Worker termites search for food, which they supply to the queen, king, and soldiers.

c Soldier termites guard the passages in the nest and the covered trails that radiate out from the nest.

d The king termite fertilizes all the eggs laid by the queen.

Tour of a Crustacean

Lobsters are the most advanced crustaceans. They have five fused head segments, eight thorax segments, and six abdomen segments.

Lobsters and other crustaceans are the only arthropods with two pairs of antennae.

Lobsters have five pairs of appendages for walking, and so are called "decapods," from the Latin word for ten.

The massive pincers of lobsters are used to crush shellfish and slice fish flesh.

The head and thorax are covered by a cuticle called the carapace, which is made of chitin.

Periodically, lobsters molt, shedding their hard shell so that their bodies can grow larger.

540 Chapter 24

Crustacean Diversity

Figure 24.17 **Crustaceans exhibit a wide variety of body forms. They have many different kinds of appendages.**

Lobsters, shrimp, crabs, and crayfish are all **decapods**. They have five pairs of thoracic legs (the name "decapod" means "ten feet"). The first pair of legs is often an enlarged set of claws used for food gathering and for defense, as you can see in **Figure 24.17**. Lobsters, shrimp, and crayfish all resemble one another in that the abdomen is large and muscular.

Barnacles don't look very much like other crustaceans, or even other arthropods, for that matter. However, as larvae they clearly resemble their crustacean relatives. As adults, barnacles live inside the walls of their calcium shells, which are attached to solid objects like rocks, pilings, ships' hulls, and even whales. If you look along almost any rocky seashore, you will probably find two different kinds of barnacles. Acorn, or volcano, barnacles have their shells cemented directly to the rock on which they live. The second kind, the goose barnacles, sit atop a fleshy stalk, which attaches to the rock.

Pill bugs and their relatives are called isopods, which means "same feet." They have seven pairs of similar walking legs on the thorax. They are closely related to another group of crustaceans called the amphipods. The name "amphipod" refers to having two different kinds of feet. If you turn over a stone along the beach or lake shore, you are sure to find some examples of amphipods. Beach hoppers are very common on sandy coastlines. Often, you can find these animals by looking for their burrows or by turning over piles of seaweed lying on the sand.

a Shrimp use their legs and long "feelers" to bury themselves in the sand when a predator approaches.

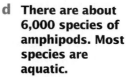

c Pill bugs are often found under rotting logs in forests throughout the continental United States.

b The abdomen of a crab is small and is folded beneath the thorax. Crabs can be found along most major coastlines in the world. Almost 100 different species live in the waters surrounding Australia.

d There are about 6,000 species of amphipods. Most species are aquatic.

e These common acorn barnacles feed by extending their feathery legs into the water to filter small food particles.

Figure 24.18
Large numbers of *Daphnia* can be found in almost any sample of water taken from a nutrient-rich lake or pond. There are nearly 400 species of *Daphnia* in freshwater environments. They are between 0.2 and 3 mm (.008 and 0.1 in.) in length.

the most important animal on Earth. It is a **copepod** (*KOH puh pahd*). While some kinds of copepods live on the sea bottom, and a few are parasitic, the vast majority are part of the zooplankton, the heterotrophic organisms that feed on phytoplankton. Copepods occur both in the sea and in fresh water, often in incredibly high concentrations. Their abundance follows the seasonal changes in concentrations of phytoplankton. Like the water fleas in lakes, the copepods link the ocean's photosynthetic life to the rest of the ocean's food web. Copepods are consumed by a variety of small predators, which are eaten by larger predators, and so on. Virtually all animal life in the open sea depends on the copepods, either directly or indirectly. Although humans do not eat copepods directly, our sources of food from the ocean would disappear without the copepods.

Figure 24.19
Although most copepods are pale and transparent, some species are brilliant red, orange, purple, blue, or black.

Among the most important crustaceans in freshwater environments are the so-called water fleas or cladocerans (*kluh DAHS uhr uhns*), like *Daphnia* shown in **Figure 24.18**. *Daphnia* feed on phytoplankton. In high concentrations, water fleas consume huge quantities of photosynthetic organisms and in turn are fed upon by small predators, such as young fishes. In many freshwater habitats, these little crustaceans are a vital link between the producers and the rest of the food web.

Look at the photograph in **Figure 24.19**. This tiny crustacean is perhaps

---

Section Review

❶ **Name three kinds of crustaceans.**

❷ **Describe three differences between insects and crustaceans.**

❸ **How would the ecology of the sea and the human food supply be affected if all crustaceans died?**

"Ernie! Look what you're doing —take those shoes off!"

Key Terms	Summary

24.1 Spiders and Their Relatives

chelicera (p. 525)

arachnid (p. 525)

pedipalp (p. 525)

spinneret (p. 526)

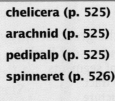

Spiders and their relatives are characterized by chelicerae.

- **Spiders, scorpions, mites, ticks, and horseshoe crabs have chelicerae, a specialized pair of appendages at the front of the body. They have two main body units: the cephalothorax and the abdomen. Spiders, scorpions, mites, and ticks are arachnids.**

- **The chelicerae of spiders are poison-delivering fangs. Spider silk is used as a safety line, to line burrows, to protect the young, and to trap food.**

- **Scorpions' chelicerae are adapted for tearing prey. The pedipalps of scorpions are pincers for grasping prey.**

- **Many ticks and mites are parasites. Some transmit diseases to humans and domestic animals.**

24.2 Insects, Millipedes, and Centipedes

mandible (p. 532)

direct development (p. 534)

metamorphosis (p. 534)

incomplete metamorphosis (p. 534)

nymph (p. 534)

complete metamorphosis (p. 534)

There are more species of insects on Earth than any other group of animals.

- **There are more species of insects than any other group of animals. Insects have three main body units: the head (with three sets of mouthparts), the thorax (with three sets of legs), and the abdomen.**

- **Insect development involves either complete or incomplete metamorphosis.**

- **Millipedes have two pairs of legs per segment and feed on decaying matter. Centipedes have only one pair of legs per segment and are predators.**

24.3 Crustaceans

decapod (p. 541)

copepod (p. 542)

Most crustaceans live in or close to the ocean.

- **Crabs, lobsters, shrimp, pill bugs, and barnacles are crustaceans, subphylum Crustacea. Most crustaceans are aquatic.**

- **The crustaceans known as copepods are extremely important links in marine food webs.**

Can Pill Bugs Detect Differences in Moisture and pH?

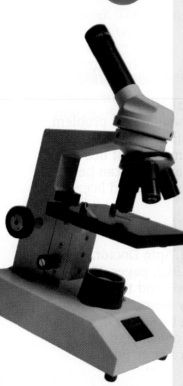

Objectives

In this investigation you will:
- *observe* a terrestrial crustacean
- *identify* a pill bug's responses to pH and moisture differences

Materials

- ruler
- paper towels
- pan
- 10 pill bugs
- 3 medicine droppers
- distilled water
- dilute sodium hydroxide solution
- vinegar
- 3 filter paper disks
- safety goggles
- lab apron
- pH paper

Prelab Preparation

1. Review what you have learned about crustaceans by answering the following questions.
   - To which class of arthropods do pill bugs belong?
   - In what habitats are pill bugs found?
2. Review what you have learned about acids in Chapter 14 by answering the following questions.
   - What does a solution's pH indicate?
   - What range of pH values indicates an acidic solution? A basic solution? A neutral solution?

Procedure: Observing Pill Bugs

1. Form a cooperative group of four students. Work with one member of your group to complete steps 2–10. Make three tables similar to the one shown below. Mark one table "Control," one "Water," and one "Acidic/Basic."

30-second intervals	Trial 1			Trial 2			Trial 3		
	Number of pill bugs in quadrant								
	1 2 3 4			1 2 3 4			1 2 3 4		
1									
2									
3									
4									
5									
6									
7									

2. Use a ruler and a pencil to divide a paper towel into four equal quadrants. Then, in the center of the paper towel, draw a circle about 8 cm (3 in.) in diameter. Place the marked paper towel on top of five unmarked sheets of paper towel and place the stack in a clean pan.

3. Place 10 pill bugs in the center of the circle. Observe their responses over a period of four minutes.

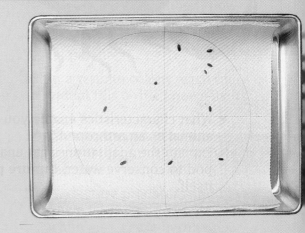

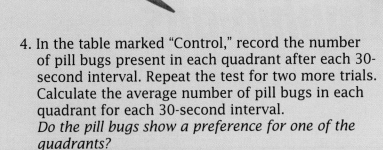

4. In the table marked "Control," record the number of pill bugs present in each quadrant after each 30-second interval. Repeat the test for two more trials. Calculate the average number of pill bugs in each quadrant for each 30-second interval.
Do the pill bugs show a preference for one of the quadrants?

5. Using one medicine dropper, moisten a disk of filter paper with distilled water. Place the disk in the middle of one quadrant. Repeat step 3, but record the results in the table marked "Water." *Do the pill bugs now show a preference for a particular quadrant? Why do they prefer this quadrant?*

6. **CAUTION: Wear safety goggles and a lab apron while working with the sodium hydroxide and vinegar solutions, as they can injure the skin and eyes. Thoroughly wash any area that has been contaminated by these solutions.**

7. Use small strips of pH paper to find the pH of the distilled water, vinegar, and sodium hydroxide solutions. Record your results.

8. Using a clean medicine dropper, soak a fresh disk of filter paper with vinegar. Using a different medicine dropper, soak another disk with dilute sodium hydroxide. Place each disk in a separate quadrant of the paper towel.

9. Repeat step 3, recording the results in the table marked "Acidic/Basic."

10. Return the pill bugs to their storage container. Clean up your materials and wash your hands before leaving the lab.

Analysis

1. *Summarizing Observations*
Describe the behavior of the pill bugs when placed on clean, dry paper.

2. *Summarizing Observations*
How did pill bug behavior change when the water-soaked disk was added to the paper towel?

3. *Analyzing Observations*
Are pill bugs attracted directly to moisture, like a moth to light, or do they move randomly until moisture is encountered? Explain how your observations support your statement.

4. *Evaluating Methods*
Why was it necessary to first observe the animals on a dry paper towel?

5. *Analyzing Observations*
How do pill bugs respond to differences in pH?

6. *Making Inferences*
How does the pill bug's response to moisture and pH increase its chance of survival?

7. *Making Predictions*
How do you think a pill bug would respond to light? Explain the reasons for your prediction.

Thinking Critically

How does the fact that pill bugs are crustaceans help explain why they are attracted to moist conditions?

Fishes and Amphibians

Review

- characteristics of chordates (Section 21.3)
- characteristics of vertebrates (Section 21.3)
- evolution of the amphibian heart (Section 22.2)

This frog in Botswana lays its eggs in water. Amphibians rely on the presence of water for fertilization and survival.

YOU AND MOST OF THE ANIMALS YOU COMMONLY SEE ARE VERTEBRATES. AS YOU LEARNED IN CHAPTER 10, VERTEBRATES ARE CHORDATES THAT HAVE A VERTEBRAL COLUMN, OR SPINE. THERE ARE APPROXIMATELY 40,000 SPECIES OF VERTEBRATES TODAY AND OVER HALF OF THEM ARE FISHES. THE FIRST VERTEBRATES EVOLVED MORE THAN 500 MILLION YEARS AGO.

25.1 Early Fishes

Objectives

❶ List three characteristics of agnathans.

❷ Describe how a lamprey feeds.

❸ Describe how jaws are thought to have evolved.

❹ Contrast sharks and rays with agnathans.

The First Fishes: Class Agnatha

The first vertebrates evolved about 510 million years ago. They were jawless fishes belonging to the class Agnatha (Agnatha means "without jaws"). Thick bony plates covered the bodies of these ancient agnathans (*AG na thuns*), as you can see in **Figure 25.1**. Bone and cartilage first evolved in this group of fishes. Agnathans were an abundant and diverse group for about 150 million years, until they were largely replaced by jawed fishes about 360 million years ago. Only 63 species of agnathans exist today.

The living agnathans are the lampreys and hagfishes, which are shown on the next page. These eel-like creatures have scaleless, slimy skin and lack paired fins. The skeleton is mostly composed of cartilage, and there is no well-developed vertebral column. The notochord, which is replaced by vertebrae in most vertebrates, functions as the major support structure in adult lampreys and hagfishes. The gills of agnathans lie within pouches that branch from the pharynx. Water exits these pouches through several openings behind the head.

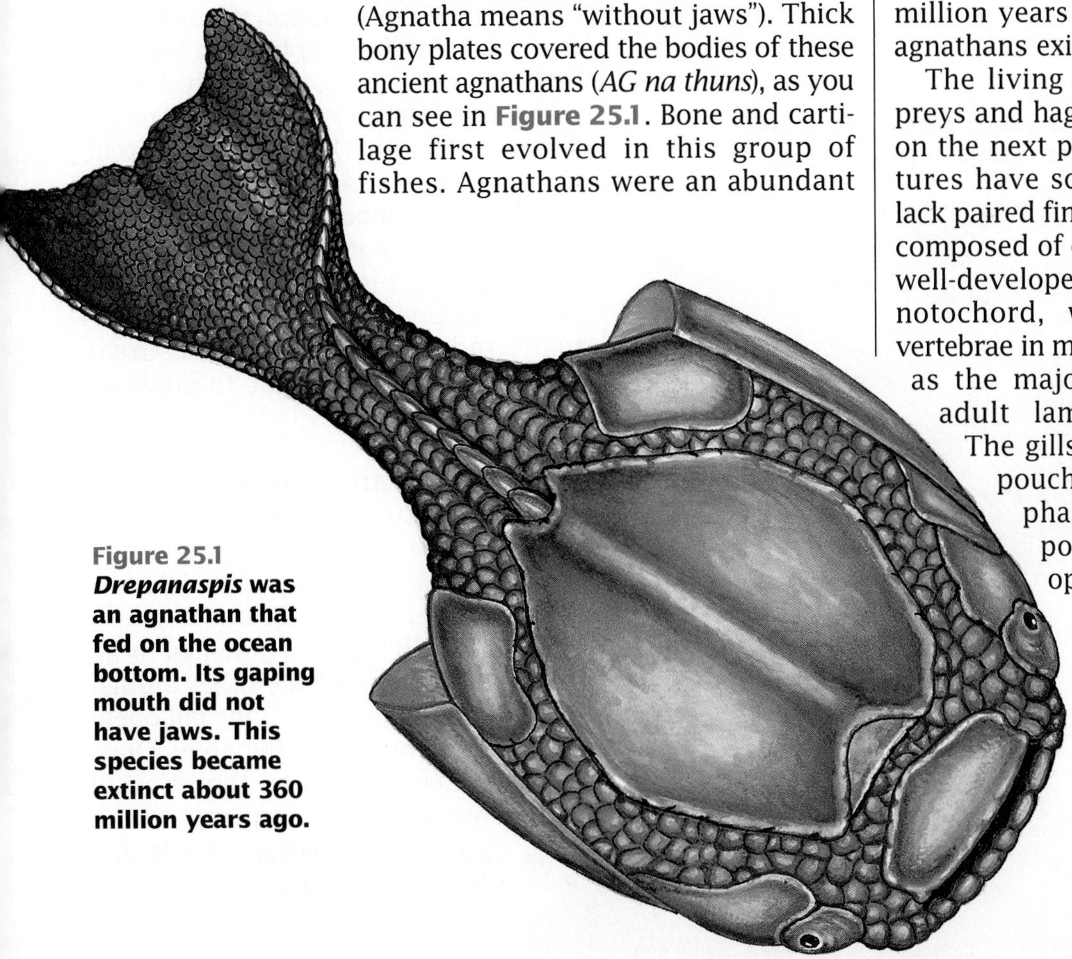

Figure 25.1
Drepanaspis was an agnathan that fed on the ocean bottom. Its gaping mouth did not have jaws. This species became extinct about 360 million years ago.

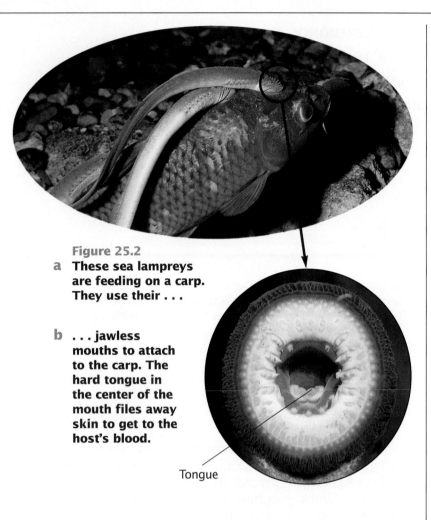

Figure 25.2

a **These sea lampreys are feeding on a carp. They use their . . .**

b **. . . jawless mouths to attach to the carp. The hard tongue in the center of the mouth files away skin to get to the host's blood.**

Tongue

Figure 25.3
Hagfishes are jawless scavengers. They invade the bodies of dying or dead animals and feed on the internal organs.

Lampreys are parasites

Most kinds of lampreys spend their entire lives in fresh water. A few species live in the sea as adults but migrate into fresh water to breed. In a way, an adult lamprey's lifestyle is similar to that of a leech. As shown in **Figure 25.2**, lampreys are external parasites that feed on other fishes. A lamprey's mouth is recessed within a funnel-like structure. Sharp teeth in the funnel help the lamprey hook onto its host. The rim of the funnel functions as a suction cup, which fastens the lamprey to its host. A rough tongue scrapes off small particles of the host's skin and flesh. The lamprey sucks in these particles along with the host's blood. Like some leeches, the lamprey secretes an anti-coagulant that prevents its host's blood from clotting.

After a lamprey has fed, it drops off of its host. Damage to the host can be severe since the wound left by the lamprey may become infected or cause the host to bleed to death. For this reason, large lamprey populations can cause great damage to populations of other fishes. For instance, when a canal that allowed ships to bypass Niagara Falls was deepened in the early 1900s, the ocean lamprey was able to move from the Atlantic Ocean into the Great Lakes. By the 1940s and 1950s, this lamprey was abundant enough to cause a serious decline in the commercial and sport fishing industries of the Great Lakes. Lamprey populations were eventually reduced by treating the Great Lakes with poisons toxic to lamprey larvae.

Hagfishes are scavengers

Hagfishes, such as the one in **Figure 25.3**, are scavengers that generally feed on dead or dying animals, such as large invertebrates or other fishes. When a hagfish locates a potential meal, it enters the body of the other animal by squirming in through the gill openings, the mouth, or the anus. Once inside, it feeds on the internal organs of the animal, biting with jawlike folds of muscle that close side to side.

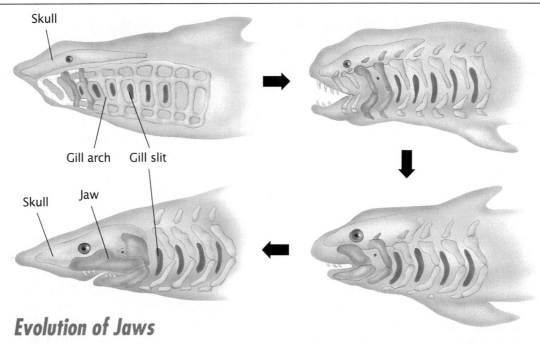

Figure 25.4
In the process of evolution, the gill arches present in jawless fishes probably moved forward to form a a jaw.

Skull

Gill arch   Gill slit

Skull   Jaw

Evolution of Jaws

Because they lack jaws, agnathans can only eat food that can be sucked into their mouths. Apart from the agnathans, all vertebrates have movable jaws. Animals with jaws can exploit a much wider range of foods than can jawless animals. Scientists think that jaws evolved from one or more of the gill arches that support the pharynx in agnathans, possibly in the way shown in **Figure 25.4**.

The earliest jawed fishes were acanthodians (*uh KAN thoh dee uhns*), members of the class Acanthodii. Although superficially resembling sharks, the acanthodians or "spiny fishes" had a bony internal skeleton and were definitely not sharks. Acanthodians evolved about 435 million years ago. Another group of early jawed fishes was the placoderms (*PLAK uh durms*), class Placodermi, which evolved about 400 million years ago. Placoderms (meaning "plate skin")

were armored with heavy, bony plates and had strong jaws, as in the predator *Dunkleosteus* illustrated in **Figure 25.5**. Both placoderms and acanthodians show a characteristic found in most other vertebrates: paired appendages, in this case paired fins. Placoderms and acanthodians diversified rapidly. Perhaps because of jaws and paired fins, these fishes largely replaced the agnathans. Lampreys and hagfishes may have survived because of their specialized feeding habits. Placoderms became extinct about 345 million years ago, and the acanthodians died out about 270 million years ago. Scientists do not know why these once-diverse groups became extinct.

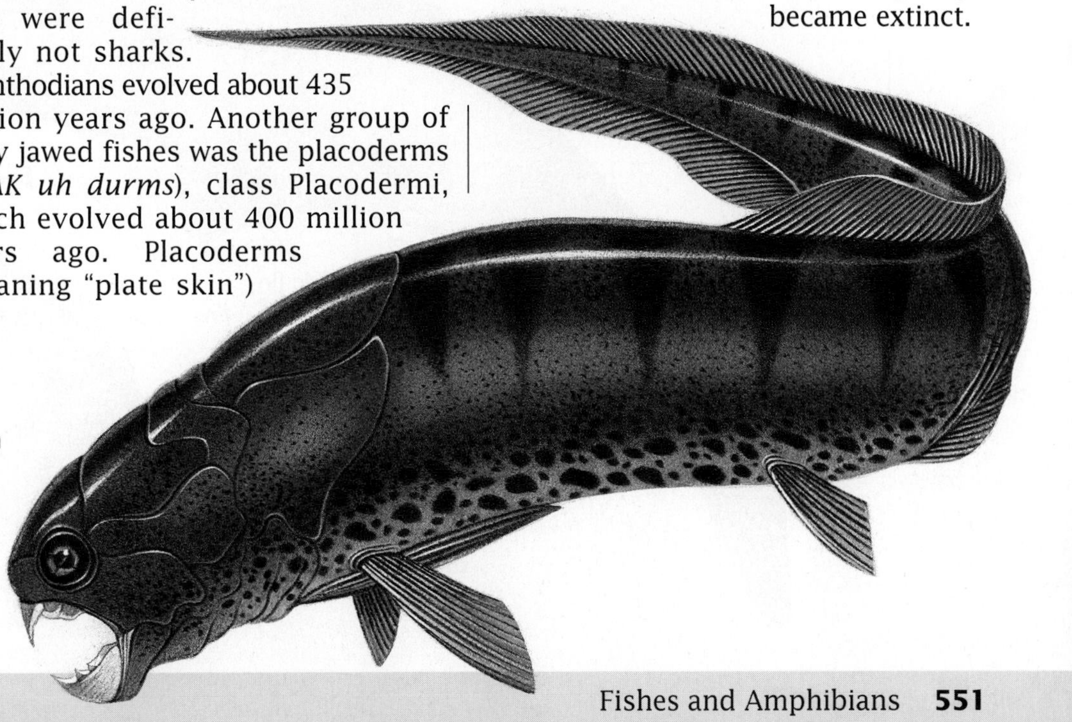

Figure 25.5
***Dunkleosteus,* a placoderm that was a predator in the ancient seas, might have looked like this drawing. Its skull and jaws were over 65 cm (2 ft.) long.**

Sharks and Rays: Class Chondrichthyes

Soon after placoderms evolved, another group of jawed fishes arose, the class Chondrichthyes (*kahn DRIHK thees*). The modern members of this class are the sharks, skates, and rays. Unlike the bony skeletons of acanthodians, the skeletons of sharks and rays are composed of a flexible substance called cartilage (*KAHRT'l ihj*). Chondrichthyes means "cartilage fishes." There are about 850 living species of cartilaginous fishes, the vast majority of which live in salt water. **Figure 25.6** shows a shark and a ray.

If you were to touch a shark or ray, you would notice rough, sandpaper-like skin. This texture results from the many small scales that are embedded in the skin. As shown in **Figure 25.7**, shark scales are very similar to shark teeth, which probably evolved from scales. The mouth of a cartilaginous fish includes upper and lower jaws, generally armed with rows of hard teeth. As the outer teeth are lost or broken, they are replaced by others moving up from behind.

The gills of sharks and rays open to the outside through a series of slits, as shown in **Figure 25.8**. One gill slit, the spiracle, opens directly to the outside on the side or top of the head. Water can be brought to the pharynx through the spiracle and then expelled through the gill slits. This arrangement not only frees the mouth for feeding but is a great advantage for sharks and rays that lie on the bottom of the ocean, where drawing water in through the mouth would also bring in sediment.

Figure 25.6
Sharks, skates, and rays have two sets of paired fins: pectoral fins and pelvic fins. The dorsal fin (on the back) and the caudal fin (the tail) are unpaired. In the skates and rays, the pectoral fins are greatly enlarged into a pair of large, wing-like fins.

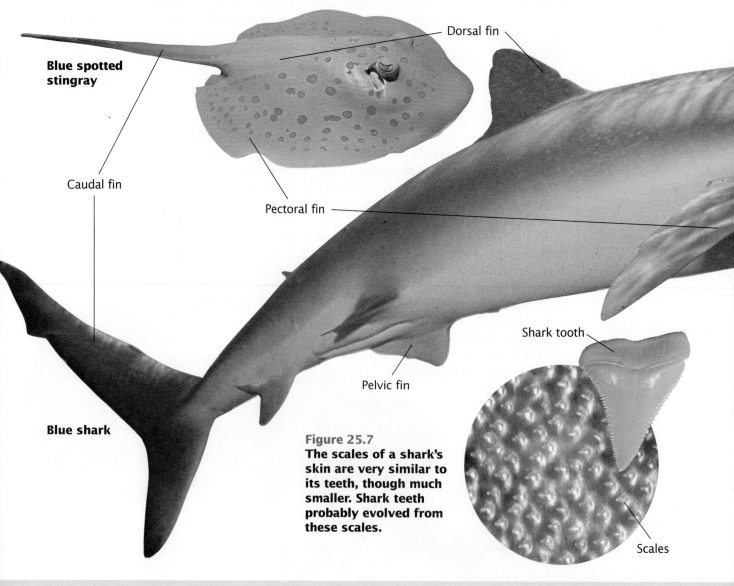

Blue spotted stingray

Dorsal fin

Caudal fin

Pectoral fin

Pelvic fin

Blue shark

Shark tooth

Scales

Figure 25.7
The scales of a shark's skin are very similar to its teeth, though much smaller. Shark teeth probably evolved from these scales.

How sharks detect and capture prey

Most sharks and rays are carnivores. Although sharks have the reputation of being man-eaters, only a few species of sharks are dangerous to humans. Three well-developed senses enable sharks and rays to detect their prey. First, they have an acute sense of smell. Second, they have the ability to sense electric currents in water. This sense is particularly useful in detecting the small electric currents generated by the muscle movements of animals. The electric ray *Torpedo* is also able to produce a powerful electric current, which it uses to deter predators and to stun prey. Third, sharks and rays have a lateral line system. As you learned in Chapter 22, the lateral line system is a series of pressure-sensitive cells that lie within canals along the sides of a fish. Changes in pressure caused by a fish or other animal swimming nearby can be detected by the cells in the lateral line.

Given the variety of shapes and sizes of cartilaginous fishes, it is not surprising that they exhibit a variety of feeding methods. Like the largest whales, the largest shark (the whale shark) and the largest ray (the manta ray) feed on plankton. The whale shark may exceed 13 m (45 ft.) in length, and the manta ray can have a "fin span" of nearly 6 m (20 ft.). Neither the whale shark nor the manta ray has teeth. Instead, their mouths are filled with a bony mesh that traps tiny organisms. Both of these large animals cruise the oceans and filter small crustaceans and protists from the water.

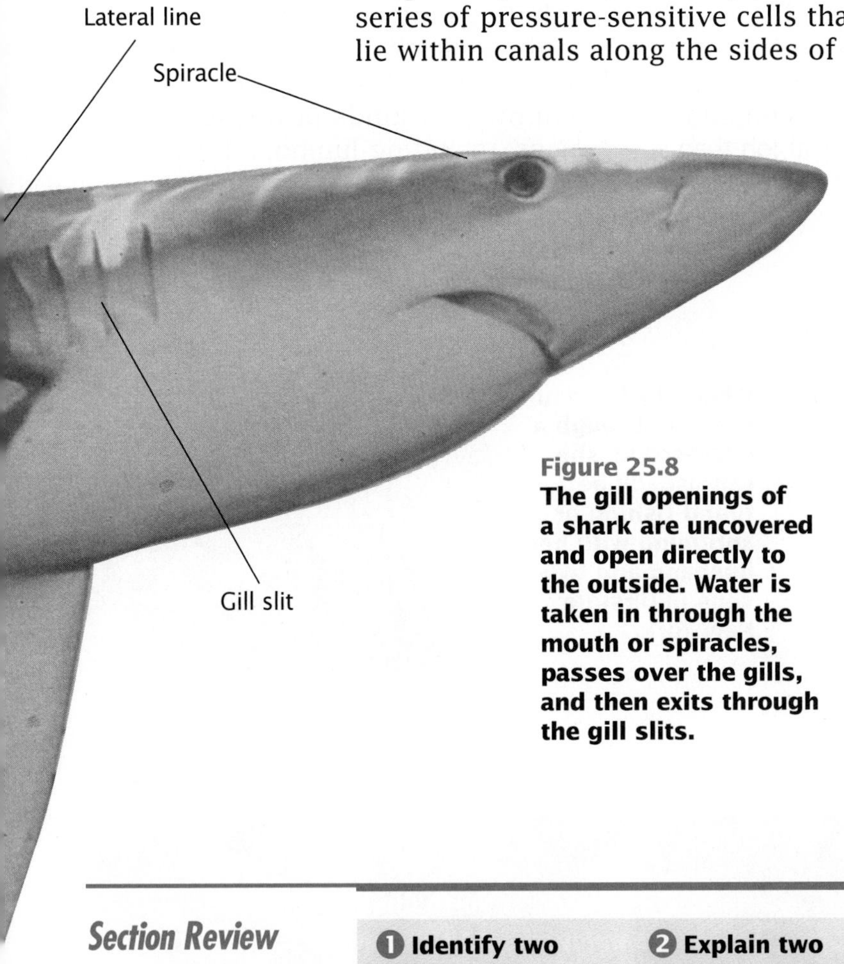

Sharks and rays have internal fertilization

The pelvic fins of male cartilaginous fishes are modified into a pair of claspers, which are used to transfer sperm to the female during mating. Thus, fertilization in these animals is internal. Skates, rays, and some sharks lay eggs. The extremely yolky fertilized eggs are usually housed in an elaborate leathery case. Many species of sharks do not lay eggs. Instead, the female keeps the yolky eggs inside her body until they hatch, and the young sharks are born alive. In some sharks, the female provides nutrients to her developing young through a membranous sac somewhat like the placenta.

Lateral line

Spiracle

Gill slit

Figure 25.8
The gill openings of a shark are uncovered and open directly to the outside. Water is taken in through the mouth or spiracles, passes over the gills, and then exits through the gill slits.

Section Review

❶ Identify two characteristics of agnathans.

❷ Explain two adaptations that enable a lamprey to be an external parasite of fishes.

❸ Diagram the stages in the evolution of jaws.

❹ List two differences between sharks and lampreys.

Phyllis Stout: Microscope Slide Technician

How I Became Interested In Science

"I can't remember a time in my life that I wasn't involved with science. I loved my high school biology classes, especially when we did lab work. My sister once needed a skull for a science project. My father knew about a biological supply company through his work and was able to find it for her there.

When I graduated from high school I wasn't sure if I could really pursue a career in science without a college degree, but I wanted to try.

"I started out as a Microscope Slide Order Filler at WARD'S, a biological supply company. It was more of a clerical job than science-related, but it was a great foot in the door. As soon as an opportunity for advancement came along, I took it— I wanted hands-on experience! For the next six months, I trained in the Osteology Department. I learned how to process, assemble and repair human and animal skeletons. My specialty became disarticulating and reassembling skulls using the Beauchene mounting technique, which lets students see and study all the individual parts of the skull. As far as I know, WARD'S has the only Osteology Lab of its kind in the country. Back around the turn of the century, WARD'S prepared and mounted all kinds of unusual specimens, including Jumbo, P.T. Barnum's circus elephant!"

When Phyllis is not peering through a microscope, she can usually be found fishing or antiquing with her husband and two sons in Upstate New York.

Name:	**Phyllis Stout**
Home:	**North Chili, New York**
Employer:	**WARD'S Natural Science, NY**
Personal Traits:	**• Sense of Humor**
	• Dedicated
	• Patient
	• Detail Oriented
	• Enthusiastic
	• Outgoing

"When my two children came along, I was able to work part-time. It would have been almost impossible to try and juggle family, job and college at that time. When an opportunity to work in the Microscope Slides Department came along, I decided to go back to work full-time. It was a complete switch from working with large skeletons to dissecting tiny specimens under a microscope! I've now been trained to prepare slides from the beginning to the final steps. I still often work with bones, only now I slice them paper thin for students to study individual cells. Last year, our Micro Slide team prepared over 200,000 slides that were distributed to students all over the world.

"I personally feel that if an individual has the motivation and desire to do something, no barriers can stop them. I am proud of my accomplishments even without a college degree. Now the results of my on-the-job education are in turn used to teach others."

The Craft of Preparing Micro Slides

Career Path

High School:
- Science
- Art

Work Experience:
- Trained in Osteology/ Human and Ligamentary
- Microscope Slide Technician

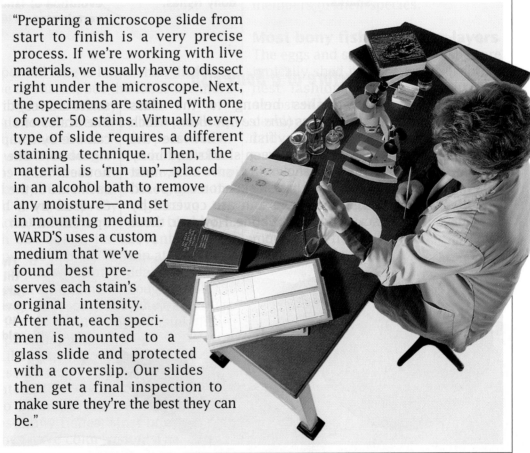

"Preparing a microscope slide from start to finish is a very precise process. If we're working with live materials, we usually have to dissect right under the microscope. Next, the specimens are stained with one of over 50 stains. Virtually every type of slide requires a different staining technique. Then, the material is 'run up'—placed in an alcohol bath to remove any moisture—and set in mounting medium. WARD'S uses a custom medium that we've found best preserves each stain's original intensity. After that, each specimen is mounted to a glass slide and protected with a coverslip. Our slides then get a final inspection to make sure they're the best they can be."

Research Focus

Phyllis Stout has participated in a number of interesting projects at WARD'S Natural Science in Rochester, New York. One of her most memorable was in the Osteology Lab, preparing a huge bison skeleton for a college in Ontario, Canada. After processing and assembling the skeleton in a six-week time frame, she hand-delivered the 6' X 10' bison to the college, where it then had to be transported to its home in a tiny "museum" room on an upper floor of the building! Most recently, Phyllis worked as part of an effort to re-outfit the university system of Kuwait after the Gulf War.

Tour of a Fish

There are more than 18,000 species of bony fishes. This fish is a salmon, one of the few fishes that live in both salt water and fresh water.

Salmon hatch in fresh water. The young migrate to the sea, where they mature. Adult salmon live in the ocean but spawn (reproduce) in the river in which they were born. Most scientists believe salmon locate their native river by using their sense of smell.

All fishes have gills that are composed of tiny filaments richly supplied with blood.

The "slimy" surface of a fish reduces water friction by more than 66 percent.

Most fishes continue to grow throughout their lives.

Bony fishes have skeletons and swim bladders to maintain buoyancy.

Bony fishes have paired pectoral and pelvic fins.

Fishes swim by beating their tails back and forth. Most fishes swim no faster than about 10 body lengths per second.

To reach their spawning grounds, salmon must swim upstream, often trying to leap over any waterfalls they encounter.

Major Groups of Bony Fishes

Look carefully at the fins of the salmon shown in the *Tour of a Fish* on page 558. Salmon belong to the group of bony fishes known as the ray-finned fishes. All but seven species of bony fishes are **ray-finned fishes**. In these fishes, the fins are fan-shaped and are supported by thin bony rays. Ray-finned fishes are the most successful and diverse group of vertebrates.

The other group of bony fishes is the **lobe-finned fishes**. In this group, the fins are fleshy and are supported by central bones. The existing lobe-finned fishes consist of six species of lungfishes and

one species of **coelacanth** (*SEE luh kanth*). **Table 25.1** lists the major groups of fishes.

As their name suggests, lungfishes have functional lungs. These fishes live in Africa, South America, and Australia. A lungfish from Africa is shown in **Figure 25.12**. Some lungfishes inhabit

Figure 25.12
Lungfishes, like this African lungfish, have fleshy, lobed fins. Lungs enable these fishes to live in water that is low in oxygen.

Table 25.1 Major Groups of Fishes

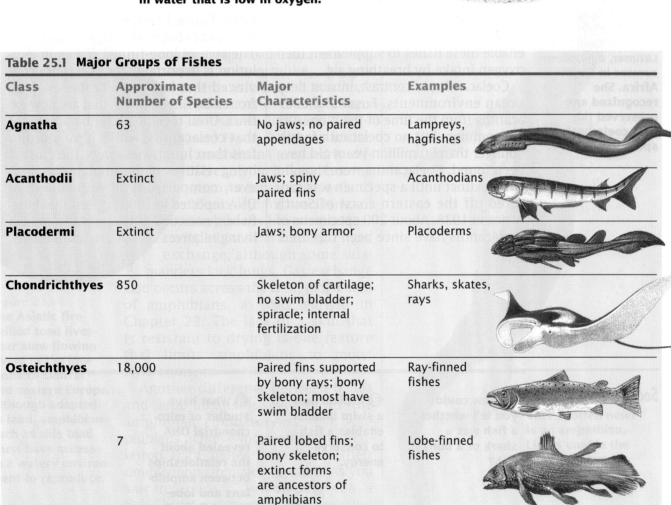

Class	Approximate Number of Species	Major Characteristics	Examples
Agnatha	63	No jaws; no paired appendages	Lampreys, hagfishes
Acanthodii	Extinct	Jaws; spiny paired fins	Acanthodians
Placodermi	Extinct	Jaws; bony armor	Placoderms
Chondrichthyes	850	Skeleton of cartilage; no swim bladder; spiracle; internal fertilization	Sharks, skates, rays
Osteichthyes	18,000	Paired fins supported by bony rays; bony skeleton; most have swim bladder	Ray-finned fishes
	7	Paired lobed fins; bony skeleton; extinct forms are ancestors of amphibians	Lobe-finned fishes

blood, sending it to the gills where it picks up oxygen. When the blood reaches the gills, it slows down while passing through narrow capillaries. The amphibian heart, in contrast, pumps both deoxygenated and oxygenated blood. Deoxygenated blood returning from the body is sent back out by the heart to the lungs to absorb oxygen. This oxygen-rich blood, which also slows as it passes through the narrow capillaries of the lungs, returns to the heart to be pumped to the rest of the body. Thus, amphibians are able to pump oxygenated blood at higher pressures and faster rates of flow than are fishes. The amphibian heart is partly divided, and some mixing of oxygenated and deoxygenated blood occurs as the blood flows through the heart.

The Tie to Water

As you learned in Chapter 22, amphibians do not lay watertight eggs and so must reproduce in water or in moist environments. Many species of frogs and toads, for example, congregate in ponds during the spring in preparation for mating. The males generally arrive first and begin noisy mating calls that attract nearby females. You have probably heard the raucous symphony of male frogs on warm spring evenings. Once mating begins, the male grasps a female and holds her. While she releases eggs, he simultaneously releases sperm. Fertilization takes place externally. Follow the development of a frog in **Figure 25.16**. You can read more about frogs in the *Tour of a Frog* on the next page.

Salamander reproduction follows a slightly different pattern than that of frogs and toads. Salamanders that live on land return to the water to breed. The males deposit packets of sperm in the water and then perform complex courtship behaviors to attract the females. The females draw the sperm packets into their reproductive openings, and fertilization occurs internally. Eggs are then deposited in the water where they continue developing. Salamander larvae are not as different from the adults as tadpoles are from adult frogs. For example, salamander larvae are carnivorous, like their parents, and they retain external gills until the time of metamorphosis.

Figure 25.16

a **The life cycle of a frog involves large-scale changes in body form. First, a mass of eggs is laid in a wet or moist environment.**

b **The young tadpole emerges from the egg with external gills, which are later replaced by internal gills. After feeding and growing, the tadpole begins to transform into an adult frog.**

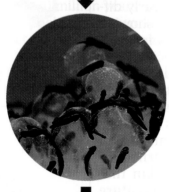

c **Dramatic changes occur in the tadpole. The tail and gills recede. Lungs and front and hind limbs grow. Feeding habits may also change. Herbivorous tadpoles change into carnivorous adults.**

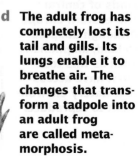

d **The adult frog has completely lost its tail and gills. Its lungs enable it to breathe air. The changes that transform a tadpole into an adult frog are called metamorphosis.**

Tour of a Frog

There are about 2,000 species of frogs and toads, all of them carnivores. All frogs, including this red-eyed tree frog, go through complete metamorphosis, changing from water-living tadpoles with gills to land-dwelling adults with lungs.

During winter, most frogs hibernate in the soft mud at the bottom of pools and streams. Tree frogs hibernate in the decaying material of the forest floor.

Toads are a kind of frog with short legs, stout bodies, and wart-covered skin.

Northern frogs can survive prolonged freezes; they prepare for winter by adding a chemical to their blood that serves as anti-freeze.

Most frogs are solitary except during the breeding season, when males call noisily to attract females.

Adult frogs have long, powerful hind legs and no tails; that is why they are called anurans, from the Greek for "without a tail."

Frog fossils have been found that are 150 million years old.

Frogs have moist, hairless, scaleless skin, through which they carry out much of their gas exchange. All frogs produce skin poison. The skin poison of South American *Dendrobates* frogs is more toxic than that of any spider or snake.

The world's largest frog is *Conraua goliath*, which is more than 30.5 cm (1 ft.) long from mouth to anus and weighs over 3.5 kg (7 lb.). This species lives in west Africa and eats animals as large as rats and ducks.

Kinds of Amphibians

Figure 25.17 shows a tropical burrowing amphibian called a caecilian. Caecilians are members of the order Gymnophiona (from the ancient Greek words for "naked" and "snakelike"). Caecilians burrow through the soil and feed on earthworms and other small animals.

Frogs and toads are **anurans**. They belong to the order Anura (meaning "without a tail" in Greek). This order is the largest amphibian order, containing over 3,600 species. As adults, frogs and toads are insect-eaters. They have large mouths, often with long tongues, and hind legs specialized for jumping. The body form of frogs and toads is distinct: the head and trunk are fused and there is no tail. Frogs and toads are found in a variety of habitats. Some species are completely aquatic, while others spend some time on land. Toads, which you might see in a garden or park, are mostly terrestrial, only returning to water to breed. The skin of a toad is dry and warty and is more resistant to evaporation than the skin of other amphibians.

Salamanders belong to the order Urodela ("visible tail"). The body shape of a salamander is more like that of reptiles. A salamander has a distinct head, trunk, and tail. The limbs are set at right angles to the body. Like frogs, toads, and caecilians, salamanders are carnivores. **Table 25.2** lists three orders of amphibians and their main characteristics.

Figure 25.17
Caecilians are limbless, one of their many adaptations for burrowing.

Table 25.2 Orders of Amphibians

Order	Approximate Number of Species	Major Characteristics	Examples
Gymnophiona	160	Wormlike body with no limbs; tail short or absent; restricted to tropics	Caecilians 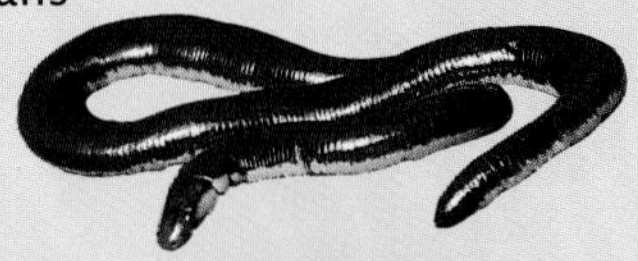
Anura	3,680	Head and trunk fused, no tail; lungs; limbs specialized for jumping	Frogs, toads 
Urodela	360	Body has distinct head, trunk, and tail; limbs set at right angles to body	Salamanders, newts 

Section Review

1 Compare an amphibian to a fish. In what important ways are they different?

2 Describe the stages in the life cycle of a frog.

3 Name one representative of each amphibian order.

A male Darwin's frog carries developing offspring in his vocal sacs. Here, a young frog has just been released.

Key Terms

Summary

25.1 Early Fishes

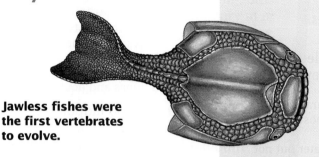

Jawless fishes were the first vertebrates to evolve.

- Jawless fishes were the first vertebrates to evolve.
- The jawless fishes include lampreys, which are external parasites on other fishes, and hagfishes, which are scavengers.
- Acanthodians were the first vertebrates with jaws.
- Sharks and rays have skeletons of cartilage.

25.2 Bony Fishes

swim bladder (p. 557)

ray-finned fishes (p. 559)

lobe-finned fishes (p. 559)

coelacanth (p. 559)

Most bony fishes are ray-finned fishes.

- Unlike sharks and rays, most bony fishes have skeletons of bone.
- Bony fishes have a swim bladder, a gas-filled sac that helps them to maintain position in the water.
- Most bony fishes are ray-finned fishes.
- A group of bony fishes known as the lobe-finned fishes are thought to be closely related to the amphibians. Studies of mitochondrial DNA suggest that lungfishes are the closest relatives of amphibians.

25.3 Amphibians

anuran (p. 564)

Most amphibians, including this frog, return to water to reproduce.

- Amphibians were the first terrestrial vertebrates.
- Frogs, toads, salamanders, and caecilians are amphibians.
- Because their skins are not watertight, most amphibians live in water or in damp environments.
- Amphibian eggs are not watertight and will dry out if not kept wet or moist.
- Caecilians are legless, tropical amphibians. They burrow in moist soils and eat worms and other small animals.
- Frogs and toads, the anurans, are adapted for jumping.
- Salamanders have four limbs and a tail. They are carnivorous and have internal fertilization.

Fishes and Amphibians **565**

Investigation

How Do Goldfish Respond To Light?

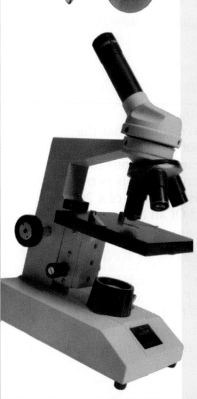

Objectives

In this investigation you will:
- *observe* a goldfish's behavior
- *test* the responses of goldfish to white light and red light

Materials

- cardboard box to fit over large glass jar or 1,000-mL beaker
- scissors
- cellophane tape
- black paper
- dechlorinated water
- large glass jar or 1,000-mL beaker
- aquarium fish net
- goldfish
- flashlight
- red cellophane
- wristwatch or clock with a second hand

Prelab Preparation

Review what you have learned about fishes by answering the following questions.
- To which class of fishes does a goldfish belong?
- Do fishes have color vision?

Procedure: Observing Goldfish Behavior

1. Form a cooperative team of four students. Work with your team to complete steps 2–10.

2. Cut a hole slightly smaller than the bulb end of the flashlight in the center of one side of the cardboard box. Make a similar hole in the bottom of the box. Tape one edge of a piece of black paper over each hole to form a flap.

3. **Caution: Use care when handling live animals.** Add 500 mL of dechlorinated water to a clean jar or 1,000-mL beaker. Use a net to transfer a goldfish to the jar. Allow the fish to become accustomed to its surroundings. Do not tap on the jar. Watch the fish for a few minutes. Record your observations.

4. Place the cardboard box over the jar so that the flap on the bottom of the box is directly above the jar. With the room lights down or off, shine the flashlight through the hole above the jar while watching the fish through the other hole. Record your observations. *Why is the box necessary?*

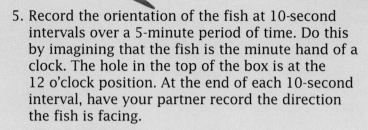

5. Record the orientation of the fish at 10-second intervals over a 5-minute period of time. Do this by imagining that the fish is the minute hand of a clock. The hole in the top of the box is at the 12 o'clock position. At the end of each 10-second interval, have your partner record the direction the fish is facing.

6. Repeat step 5 with the light shining through the hole in the side of the box while you watch through the hole in the top. Record your observations.

7. Tape a piece of red cellophane over the flashlight glass. Then, repeat steps 5 and 6.

8. Return the fish to the aquarium.

9. Combine your data with the other team in your group.

10. Make bar graphs of the combined data. Have each bar represent the number of times the fish was observed in each orientation around the imaginary face of the clock. Make a separate graph for each lighting situation.

11. Clean up your materials and wash your hands before leaving the lab.

Analysis

1. *Summarizing Data* Describe the response of the goldfish to white light.

2. *Summarizing Data* Describe the response of the goldfish to red light.

3. *Making Predictions* Based on your observations, how might the fish respond to white light coming from below the beaker?

Thinking Critically

What is the adaptive advantage of the goldfish's response to white light?

Endangered Species: How Far Do We Go to Save Them?

To what lengths should we go to save an endangered species? Should efforts to preserve the species take precedence over economic development? Or are the goals of species conservation and development compatible?

The Endangered Species Act, passed by Congress in 1973, is the cornerstone of efforts to preserve endangered species. This act requires the federal government to compile and maintain a list of endangered species (those in immediate danger of extinction) and threatened species (those declining in abundance and likely to be endangered soon). Once added to this list, a species is protected under federal law. It is illegal to kill, injure, capture, import, or export a listed species. Disrupting the habitat of a listed species is also prohibited. Furthermore, the government is forbidden from constructing, funding, or authorizing projects that would threaten listed species.

Under the protection of the Endangered Species Act, the American alligator, the bald eagle, the peregrine falcon, and the brown pelican recovered from near extinction. When efforts to save species have clashed directly with development, however, bitter controversies have resulted.

For instance, in 1973 a small (about 8 cm, or 3 in.), brownish fish was captured in the Little Tennessee River in eastern Tennessee. This fish belonged to an undescribed species; it was named the snail darter, after its primary food. In 1975 the snail darter was declared endangered. The snail darter's future looked bleak because its only known habitat was to be flooded after completion of the Tellico dam on the Little Tennessee River. A lawsuit was filed to stop construction of the dam. Tellico dam supporters argued that the economic gains from the dam, such as increased electric

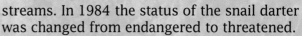

power generation and enhanced recreational opportunities, far exceeded the value of a small fish. Opponents countered that in addition to eliminating the snail darter, the dam would destroy thousands of acres of rich farmland and would flood important historical sites, such as an early Cherokee village. In 1978 the Supreme Court ruled against the dam, and construction was stopped.

Congress responded to this decision by amending the Endangered Species Act to create a committee that could grant exceptions to the act's provisions. This committee, nicknamed the "God Committee" because it held the power of life and death over species, also decided against the dam. After Congress voted to exempt the dam from all federal laws—including the Endangered Species Act— the project was completed. Economic considerations had won, and Congress had decided to sacrifice the snail darter.

Although the snail darter seemed to teeter on the verge of extinction, it survived all legal and bureaucratic battles over its preservation. Several hundred snail darters that had been transplanted to a nearby river thrived in their new habitat.

Additional natural populations of snail darters were discovered in several other rivers and streams. In 1984 the status of the snail darter was changed from endangered to threatened.

Is compromise possible when species preservation and development clash? In California's Coachella Valley, about 160 km (100 mi.) east of Los Angeles, such a compromise appears to have been reached. By the mid-1980s, much of the land in the valley had been converted from desert to golf courses, subdivisions, and hotels. Development was rapidly consuming the habitat of the Coachella Valley fringe-toed lizard, which lives only in this valley. This lizard, about 23 cm (9 in.) from nose to tip of tail, is adapted for life on fine, wind-blown sand. Instead of prohibiting development, bulldozing the last remnants of the lizard's habitat, or fighting a long and costly court battle, environmentalists and developers agreed to a compromise. Using $25 million in federal, state, and private funds, 5,300 hectares (13,000 acres) of land were purchased and set aside as a preserve. Development was permitted on the remaining land in the valley.

Thinking Critically

❶ If scientists find a plant with important medical uses, should they be allowed to remove the plant from its natural habitat? Give three reasons why or why not.

❷ If an underdeveloped country were to destroy a species in order to build up technologically, should the United States government interfere? Why or why not?

❸ During the Gulf War, Iraq intentionally spilled crude oil into the Persian Gulf. How should world powers deal with a country that maliciously damages the environment?

❹ You are the attorney for the environmentalists in the snail-darter case. Give three arguments in support of your case to stop construction of the dam.

❺ How do you think biology education has affected the public's reaction to endangered species?

Acting on the Issue

❶ Obtain a book from your local library on endangered species, and list those species that are endangered in your area.

❷ Find a copy of the Endangered Species Act and summarize it for your class.

❸ Write to your representatives in Congress to find out what legislation is pending that is related to the environment.

❹ Write to your representatives in Congress in support of legislation that you feel has a positive impact on the environment.

Reptiles, Birds, and Mammals

Review

- placenta
  (Section 22.3)

- amniotic egg
  (Section 22.3)

- the terms *ectotherm* and
  endotherm (Glossary)

It looks as if this hawk plans to make a meal of this tortoise, but he is really looking for a place to perch. Birds and mammals evolved from reptiles, which evolved from amphibians.

REPTILES (CLASS REPTILIA) WERE THE FIRST FULLY TERRESTRIAL VERTEBRATES. THEIR FREEDOM FROM AQUATIC ENVIRONMENTS WAS MADE POSSIBLE BY SEVERAL ADAPTATIONS THAT MADE THEM AND THEIR EGGS ESSENTIALLY WATERTIGHT. MODERN REPTILES INCLUDE SNAKES, LIZARDS, TURTLES, CROCODILES, AND ALLIGATORS. THEY REPRESENT THE LIVING MEMBERS OF A GROUP THAT ONCE DOMINATED THE LAND.

26.1 Reptiles

Objectives

1. **Identify three adaptations that make reptiles well suited to terrestrial life.**

2. **Contrast ectothermy and endothermy.**

3. **Describe one hypothesis that explains the disappearance of the dinosaurs.**

4. **List the four orders of living reptiles.**

Reptilian Adaptations to Terrestrial Life

Hold a snake or lizard and you will discover that the skin of a reptile, contrary to what many people think, is not wet and slimy. Instead, reptilian skin is dry and covered with tough, hard, platelike scales, as shown in **Figure 26.1**. This dry skin forms a barrier to water loss in land environments. Reptilian skin is resistant to water loss because it contains large amounts of lipids and the protein keratin. Keratin is the tough, wear-resistant material that composes your hair and fingernails. In some reptiles, thick bony plates develop beneath the scales, such as those that form the shells of turtles.

Figure 26.1
This Philippine sail-finned lizard has dry, scaly skin that protects its body from drying out and from being cut or scratched while moving over the ground.

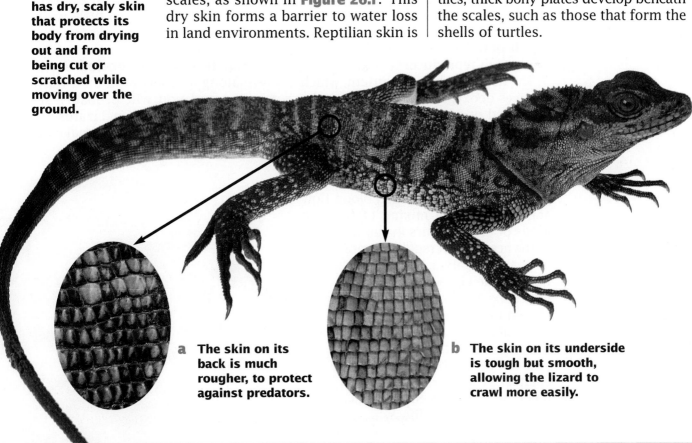

a **The skin on its back is much rougher, to protect against predators.**

b **The skin on its underside is tough but smooth, allowing the lizard to crawl more easily.**

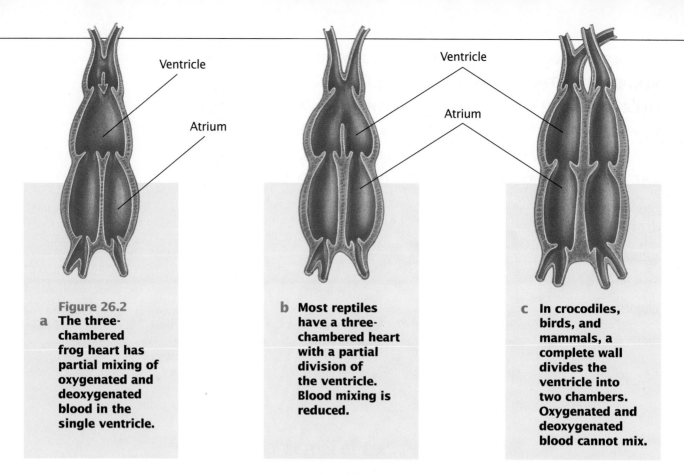

Figure 26.2

a The three-chambered frog heart has partial mixing of oxygenated and deoxygenated blood in the single ventricle.

b Most reptiles have a three-chambered heart with a partial division of the ventricle. Blood mixing is reduced.

c In crocodiles, birds, and mammals, a complete wall divides the ventricle into two chambers. Oxygenated and deoxygenated blood cannot mix.

An important adaptation of reptiles to life on dry land is the amniotic egg, which is resistant to water loss. As you learned in Chapter 22, the reptilian egg protects the vulnerable embryo and provides for its needs as it develops. Since the amniotic egg contains its own supply of water, reptiles need not travel to water to reproduce.

The fish heart has one atrium, which receives deoxygenated blood from the body, and one ventricle, which pumps blood to the gills. In the amphibian heart, such as the frog heart shown in **Figure 26.2a**, a wall separates the atrium into two chambers, so that oxygenated and deoxygenated blood flow separately. Some mixing of the two kinds of blood occurs in the ventricle. In the hearts of all reptiles except crocodiles and alligators, the ventricle is partially divided by a wall of tissue that reduces the amount of blood mixing, as shown in **Figure 26.2b**. In crocodiles, alligators, birds, and mammals, this partition is complete, as you can see in **Figure 26.2c**. The circulatory system of reptiles helps them meet the increased energy demands of an active terrestrial lifestyle.

You read about another reptilian adaptation for water conservation in Chapter 22. Reptiles excrete nitrogenous waste as uric acid, a form that requires very little water for dilution. Reptile urine contains so little water that it is paste rather than liquid.

Are reptiles "cold-blooded"?

Drive along a desert road on a warm morning and you will probably see lizards basking in the sun. Lizards and other reptiles, such as the snakes in

Figure 26.3, raise their body temperature by absorbing heat from their surroundings. Reptiles, like fishes and amphibians, are ectotherms. Ectotherms cannot regulate their body temperature through metabolism. Consequently, their body temperature changes with the temperature of their surroundings. Birds and mammals, in contrast, produce their body heat internally through metabolism. They are endotherms. Endotherms maintain their body temperature within narrow limits.

Throughout the day, the body temperature of an ectotherm often follows the temperature of its surroundings. Body temperature falls at night, when the air is cool. In the morning, many reptiles seek sunny places, letting the sun's rays warm their bodies. Once warmed, many reptiles maintain a relatively constant body temperature by moving into and out of the sunshine, their bodies warming and cooling, as described in **Figure 26.4**. You might hear ectotherms called "cold-blooded," but this description is inaccurate. On warm days, some desert lizards have body temperatures higher than yours.

Endotherms such as birds and mammals can sustain activity for longer periods of time than can reptiles and amphibians. Birds and mammals can also be active on colder days and live in colder climates than can reptiles and amphibians. There is a high cost to endothermy, however. It requires large amounts of food. A mouse must eat about 10 times as much food as a similarly sized lizard.

Figure 26.3
These red-sided garter snakes are emerging from their den, where they have spent the winter. By clustering together, they are able to conserve heat.

Figure 26.4

a Heat exchange is very important for ectotherms. This swift is warmed by heat from the rock on which it sits and by sunlight.

b During the hottest part of the day, the swift may retreat into the shade to lower its body temperature.

Tour of a Lizard

The largest group of reptiles today is the lizards. This gecko is a very fast runner. Most living lizards are small; few are bigger than a squirrel.

All reptiles have a tough, dry, scaly skin. The outer layer of scales is shed periodically.

Most lizards have external ears; snakes do not.

Most lizards have four legs, but some are legless. These lizards are not snakes because they lack the "floating jaws" that allow snakes to swallow large prey.

A gecko has adhesive toe pads that enable it to walk upside down and on vertical surfaces.

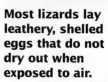

Most lizards lay leathery, shelled eggs that do not dry out when exposed to air.

Many lizards drop their tails when seized by a predator; the tail regrows in about seven weeks.

Modern lizards evolved late in the age of dinosaurs.

Figure 26.5

Figure 26.5
This phylogenetic tree shows the relationships of some of the major vertebrate groups that evolved from the early reptiles.

Birds

Lizards

Snakes

Turtles

Dinosaurs

Crocodiles

Mammals

Early reptiles

The Age of Reptiles

The first reptiles evolved about 320 million years ago, when the world was entering a dry period. Well suited to dry conditions, reptiles diversified rapidly after their initial appearance, giving rise not only to the ancestors of modern reptiles, but also to the wide variety of dinosaurs and other reptiles that were the dominant animals on Earth for over 170 million years. The period of reptile dominance, which lasted from 250 million to 65 million years ago, is called the Age of Reptiles. The evolutionary relationships of the reptiles and their descendants are shown in **Figure 26.5**.

Two groups of reptiles inhabited the oceans while the dinosaurs lived on land. Ichthyosaurs (*IHK thee oh sawrs*) were fully adapted for an aquatic existence. Ichthyosaurs ("fish lizards" in Greek) resembled dolphins, having a pointed snout, streamlined body, fins, and a flattened tail. The long-necked plesiosaurs (*PLEE see oh sawrs*) had barrel-shaped bodies with paddlelike fins. Plesiosaurs and ichthyosaurs probably fed on fish. These creatures are shown in **Figure 26.6**.

Dinosaurs, meaning "terrible lizards" in Greek, evolved from reptiles that were only 0.5 m to 1 m (2 ft. to 3 ft.) in

Figure 26.6
The 2-m (6-ft. 6-in.) *Ichthyosaurus* gave birth to live young, like modern day whales. The 2.3-m (7-ft. 6-in.) body of the *Plesiosaurus* was adapted for maneuverability to catch fish. Both existed during the Jurassic period.

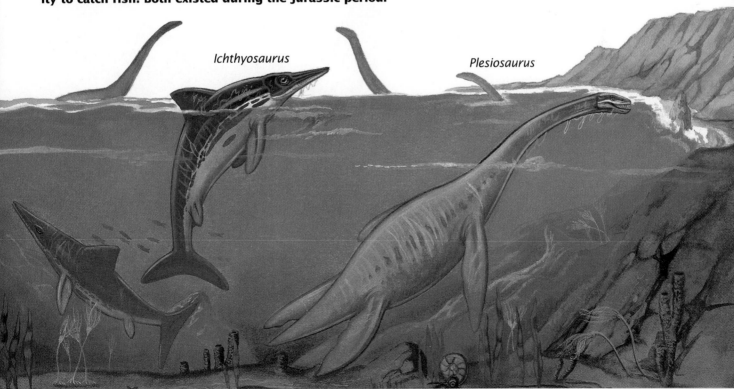

Ichthyosaurus

Plesiosaurus

length. Among the dinosaurs were the largest land animals, such as the herbivores *Apatosaurus*, *Diplodocus*, and *Brachiosaurus*. *Brachiosaurus* was about 23 m (75 ft.) long and about 17 m (56 ft.) tall. The dinosaurs also included the largest land carnivore, *Tyrannosaurus rex*. This great predator was over 5 m (16 ft.) tall when standing on its hind legs. Dinosaurs, known for their large size, were varied in size and habits, as you can see in **Figure 26.7**.

Pterosaurs were the first vertebrates to fly

No reptiles can fly today, but one group of flying reptiles existed alongside the dinosaurs: the pterosaurs (*TEHR oh sawrs*), such as *Rhamphorhynchus* shown in **Figure 26.8**. These reptiles were flying 75 million years before the first birds. During the 160 million years they existed, pterosaurs were a diverse

Figure 26.8
Pterosaurs such as *Rhamphorhynchus* were the first group of vertebrates to evolve the ability to fly. They evolved 75 million years before the first bird.

group. Some were as small as a sparrow. Others had wingspans of 11 m (35 ft.), greater than the wingspan of many small airplanes.

How did pterosaurs fly? Did they only glide? Or were they capable of flapping their wings? By comparing pterosaur fossils to birds and bats, scientists have been able to draw some

Figure 26.7
The popular conception of dinosaurs is of huge lumbering animals like *Stegosaurus*, which was 6 m (20 ft.) long and weighed up to 1,500 kg (2 tons). But many dinosaurs, such as *Struthiomimus*, were small and fast. Although the species shown here did not all exist at the same time, this illustration shows some of the great diversity among dinosaurs.

Tyrannosaurus rex

Diplodocus

Deinonychus

Protoceratops

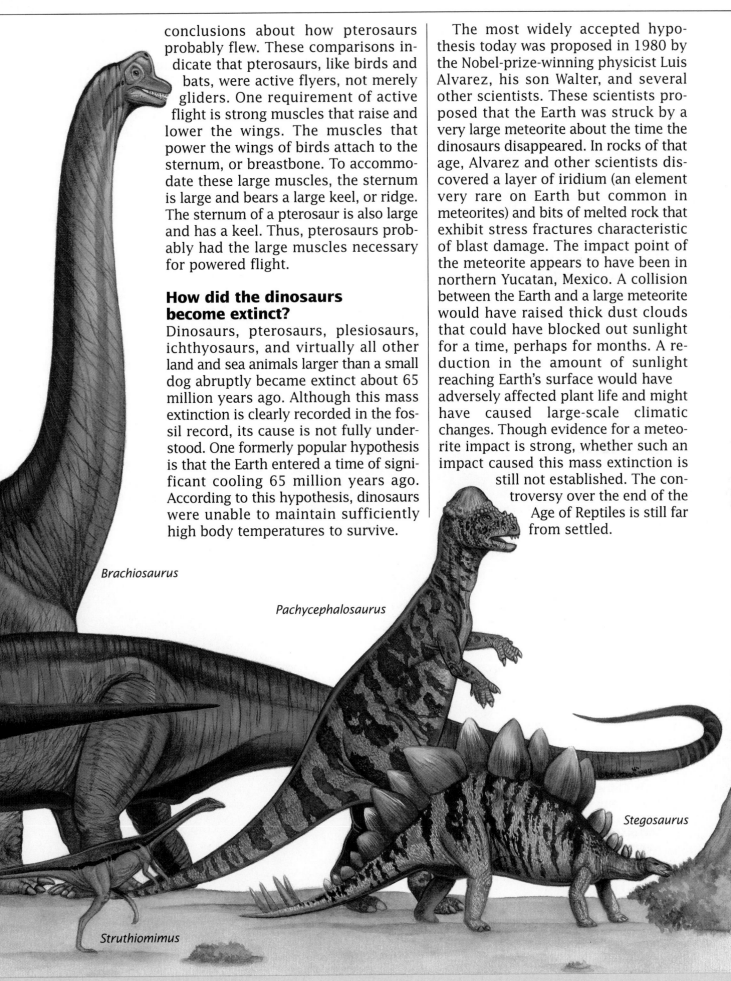

conclusions about how pterosaurs probably flew. These comparisons indicate that pterosaurs, like birds and bats, were active flyers, not merely gliders. One requirement of active flight is strong muscles that raise and lower the wings. The muscles that power the wings of birds attach to the sternum, or breastbone. To accommodate these large muscles, the sternum is large and bears a large keel, or ridge. The sternum of a pterosaur is also large and has a keel. Thus, pterosaurs probably had the large muscles necessary for powered flight.

How did the dinosaurs become extinct?

Dinosaurs, pterosaurs, plesiosaurs, ichthyosaurs, and virtually all other land and sea animals larger than a small dog abruptly became extinct about 65 million years ago. Although this mass extinction is clearly recorded in the fossil record, its cause is not fully understood. One formerly popular hypothesis is that the Earth entered a time of significant cooling 65 million years ago. According to this hypothesis, dinosaurs were unable to maintain sufficiently high body temperatures to survive.

The most widely accepted hypothesis today was proposed in 1980 by the Nobel-prize-winning physicist Luis Alvarez, his son Walter, and several other scientists. These scientists proposed that the Earth was struck by a very large meteorite about the time the dinosaurs disappeared. In rocks of that age, Alvarez and other scientists discovered a layer of iridium (an element very rare on Earth but common in meteorites) and bits of melted rock that exhibit stress fractures characteristic of blast damage. The impact point of the meteorite appears to have been in northern Yucatan, Mexico. A collision between the Earth and a large meteorite would have raised thick dust clouds that could have blocked out sunlight for a time, perhaps for months. A reduction in the amount of sunlight reaching Earth's surface would have adversely affected plant life and might have caused large-scale climatic changes. Though evidence for a meteorite impact is strong, whether such an impact caused this mass extinction is still not established. The controversy over the end of the Age of Reptiles is still far from settled.

Brachiosaurus

Pachycephalosaurus

Stegosaurus

Struthiomimus

The Survivors

The mass extinction 65 million years ago spared four groups of reptiles: lizards and snakes (order Squamata), turtles and tortoises (order Testudines), crocodiles and alligators (order Crocodylia), and the tuatara (order Rhynchocephalia).

Of these four groups, alligators, such as the one in **Figure 26.9**, and crocodiles are the most closely related to dinosaurs. Twenty-two species of crocodiles and alligators live in tropical and sub-tropical regions of the world. Two species, the American alligator and the American crocodile, occur in the United States. Crocodiles and alligators lead a largely aquatic life, feeding on aquatic animals such as fishes and turtles and on terrestrial animals that come to drink or feed in the water, occasionally including humans. Crocodiles and alligators have been hunted intensely for their hides, which are used to make handbags, shoes, and other leather products. Because of overhunting, three species of crocodiles are now endangered.

All turtles, such as the box turtle shown in **Figure 26.10a**, have a shell. A turtle's shell is composed of bony plates that are fused together. The vertebrae and ribs are fused to the interior of the shell, as shown in **Figure 26.10b**. Turtles lack teeth but have a sharp beak. Most turtles spend some of their

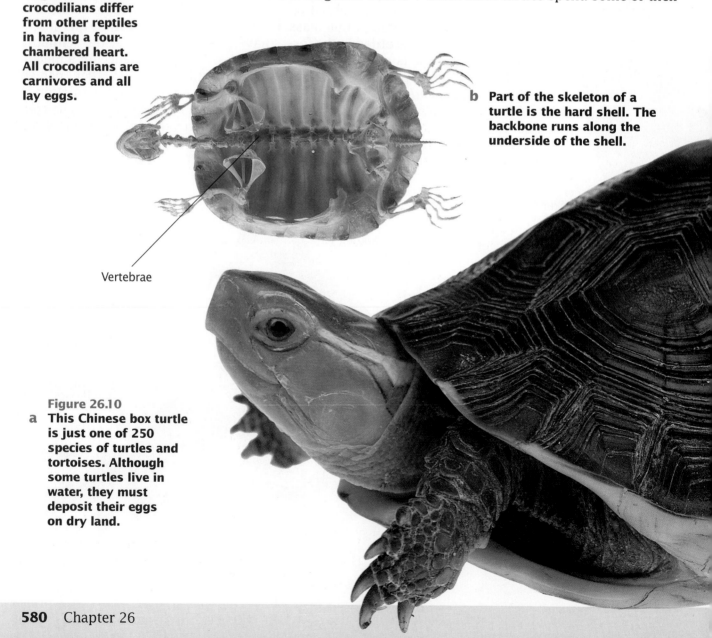

Figure 26.9
The American alligator and other crocodilians differ from other reptiles in having a four-chambered heart. All crocodilians are carnivores and all lay eggs.

b Part of the skeleton of a turtle is the hard shell. The backbone runs along the underside of the shell.

Vertebrae

Figure 26.10

a **This Chinese box turtle is just one of 250 species of turtles and tortoises. Although some turtles live in water, they must deposit their eggs on dry land.**

Figure 26.11
The tuatara is the sole survivor of the order Rhynchocephalia. The name tuatara means "spiny crest."

feeding on animals much larger than themselves. A snake's jaw has five joints (your jaw has only one), which give the jaw great flexibility. This flexibility enables the snake to engulf prey several times its own diameter, as shown in **Figure 26.12**. Most species of snakes are nonpoisonous, but a few species can deliver a poisonous bite. Rattlesnakes, coral snakes, copperheads, and cobras are examples of poisonous snakes.

Figure 26.12
The structure of the snake's jaw and the flexible nature of its skin and skeleton allow the snake to eat animals larger in diameter than its own body. It may take this snake an hour or more to swallow its prey.

time in water. An extreme case is the sea turtles, which only leave the ocean to lay their eggs. The limbs of sea turtles are flattened and paddlelike for steering and propulsion in the ocean. Tortoises, on the other hand, are almost completely terrestrial and have elephant-like limbs for walking.

The tuatara, like the one shown in **Figure 26.11**, is the only surviving member of the order Rhynchocephalia. This lizardlike reptile closely resembles its relatives that were common 150 million years ago. Tuataras live only on a few islands near New Zealand.

Snakes and lizards belong to the largest order of reptiles, order Squamata. There are about 6,800 species of snakes and lizards, and they live on every continent except Antarctica. All snakes lack limbs, movable eyelids, and external ears. Snakes are carnivores,

a **After suffocating its prey by constriction, . . .**

b **. . . the snake begins the long process of swallowing.**

Unlike snakes, most lizards have four limbs and external ears. Lizards are most abundant in the tropics and in deserts. Most species are carnivorous, feeding on insects, small mammals, or other lizards. A few species are herbivorous. Only two species of lizards are poisonous: the Gila monster, shown in **Figure 26.13**, of the southwestern United States and northern Mexico, and the Mexican beaded lizard of western Mexico. **Table 26.1** summarizes the major reptile orders.

Figure 26.13
The glands that produce the Gila monster's poison are located in the lower jaw. The lizards chew it into their victims.

Table 26.1 Orders of Reptiles

Order	Approximate Number of Living Species	Main Characteristics	Examples
Ornithischia	Extinct	Mostly plant-eating dinosaurs with two pelvic bones facing backwards, as in a bird's pelvis; hole in the skull in front of eye socket; legs positioned beneath the body; Over 150 genera	*Triceratops, Stegosaurus, Iguanodon* *Stegosaurus*
Saurischia	Extinct	Flesh-eating and plant-eating dinosaurs with one pelvic bone facing forward, the other backward, as in a lizard's pelvis; terrestrial with three or sometimes five toes; hole in skull in front of eye socket; legs positioned beneath body; over 200 genera	*Tyrannosaurus, Brontosaurus, Brachiosaurus* *Tyrannosaurus rex*
Pterosauria	Extinct	Flying reptiles with wings of skin between fourth finger and body; wing span of early (Jurassic) *Rhamphorhynchus* was typically 60 cm (2 ft.), of later (Cretaceous) *Pteranodon* over 7.5 m (25 ft.)	*Pteranodon, Pterodactylus, Rhamphorynchus* *Rhamphorynchus*
Plesiosauria	Extinct	Marine reptiles with very large paddle-shaped fins, a barrel shaped body, and long jaws with sharp teeth; some had a snakelike neck twice as long as the body; others had a short neck and elongated skull about 3.7 m (12 ft.) in length	*Plesiosaurus, Elasmosaurus, Kronosaurus* *Plesiosaurus*

Ichthyosauria	Extinct	Marine reptiles with streamlined bodies up to 3 m (10 ft.) in length; the four legs modified into balancing fins; apparently fast swimmers, with many body similarities to modern fishes such as tuna or mackerel	*Ichthyosaurus*
Squamata suborder Sauria	3,800	Lizards; largely terrestrial with limbs set at right angles to body; dry skin of scales; teeth; anus is in transverse (sideways) slit	Anoles, geckos, horned lizards
Squamata suborder Serpentes	3,000	Snakes; largely terrestrial; no legs; scaly skin shed periodically; teeth	Rattlesnakes, garter snakes
Testudines	250	Body encased in shell of bony plates; sharp, horny jaw edges without teeth; vertebrae and ribs fused to shell	Turtles, tortoises, terrapins
Crocodylia	25	Four-chambered heart; extended jaw with socketed teeth; five digits on forelimbs, four digits on hind limbs; anus is a longitudinal (lengthwise) slit	Crocodiles, alligators
Rhynchocephalia	1	Sole survivor of a group that largely disappeared about 100 million years ago. Skull like those of early Permian reptiles; fused, wedgelike, socketless teeth; primitive eye under skin of forehead	Tuatara

Section Review

❶ Describe two adaptations to land shown by reptiles.

❷ What is one disadvantage of endothermy?

❸ What evidence indicates a meteorite collided with Earth about the time the dinosaurs became extinct?

❹ List a representative of each of the four living orders of reptiles.

Reptiles, Birds, and Mammals **583**

ALTHOUGH THE DINOSAURS DIED OUT 65 MILLION YEARS AGO AT THE END OF THE CRETACEOUS PERIOD, THEIR DESCENDANTS, THE BIRDS, SURVIVED. TODAY BIRDS ARE THE MOST DIVERSE GROUP OF LAND VERTEBRATES, WITH ABOUT 8,800 SPECIES. ALL BIRDS HAVE FEATHERS, AND ALMOST ALL BIRDS ARE CAPABLE OF POWERED, SUSTAINED FLIGHT.

26.2 Birds

Objectives

❶ List two similarities between birds and reptiles.

❷ Identify two differences between birds and reptiles.

❸ Identify two functions of feathers.

❹ Describe two bird adaptations, other than feathers, for flight.

Birds Evolved From Reptiles

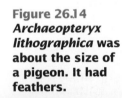

Figure 26.14
Archaeopteryx lithographica **was about the size of a pigeon. It had feathers.**

Birds (class Aves) evolved about 150 million years ago. The oldest known bird fossils were found in fine limestone in Bavaria. These specimens were named *Archaeopteryx*, which means "ancient wing." *Archaeopteryx* shared many characteristics with the small dinosaurs from which it evolved, including teeth in sockets and a long, bony tail. But as you can easily see in **Figure 26.14**, *Archaeopteryx* had feathers, a characteristic unique to birds.

Modern birds lack teeth and have only a vestigial tail, but they still retain many reptilian characteristics. For instance, birds lay amniotic eggs, although the shell of bird eggs is hard rather than leathery. Scales are also present on the feet and lower legs of birds.

Feathers are unique to birds

Feathers have replaced scales as the body covering of birds. Feathers are flexible, strong structures that can be regrown and that make an excellent wing for flying. Feathers provide most of the surface area of a bird's wing. A bird can change the surface area of its wing and alter its flight patterns by spreading or collapsing its wing feathers and the wing itself.

Like wing feathers, a bird's tail feathers can be spread or collapsed, changing their effective surface area. Tail feathers can also be used for braking and steering during flight. Watch the action of the tail of a bird in flight and during landing to see some of the actions of these feathers.

As shown in **Figure 26.15**, birds have two major types of feathers. Most of a bird's feathers are **contour feathers**. These feathers cover the body of the bird and give the wings and tail their shape. Contour feathers also insulate against heat loss. Fine **down feathers** growing underneath or among the contour feathers are specialized for insulation. The down feathers of eider ducks are used in sleeping bags because they are a lightweight, effective insulation.

Bird skeletons are lightweight

The skeleton of a bird is adapted for flight. The bones are thin and hollow. Many are reinforced by internal struts, like the wings of an airplane. The sternum is large and has a keel, providing solid anchorage for some of the large flight muscles.

Bird wings are modified forelimbs. The bones of the forelimbs fully support and move the wings. The finger bones are very tiny, but the arm and hand bones are long, providing strength and enabling complex movements of the wings. As you learned in Chapter 9, the bones of a bird's wing are homologous to the bones of your arm and hand.

Birds are endothermic and active

Flight is an energy-demanding activity. Birds, like mammals, are endothermic. Endothermy enables birds to meet the energetic demands of flight. In addition, birds have a four-chambered heart with separate circulatory loops to the lungs and to the body. Therefore, oxygen-rich blood is rapidly delivered to tissues where it is needed, without mixing with deoxygenated blood. Bird respiration is very efficient because their system of air sacs permits air to flow in only one direction through the lungs, as explained in Chapter 22. You can read more about birds in the *Tour of a Bird* on page 587.

Figure 26.15

a **Contour feathers and down feathers have different functions. Down feathers serve primarily as insulation. Contour feathers provide insulation, steering, balance, and coloration.**

Vane

Shaft

Quill

b **The individual filaments, or barbs, of a contour feather are linked together by hooked barbules to form a continuous surface, thereby decreasing wind resistance.**

Barbules

Barbs

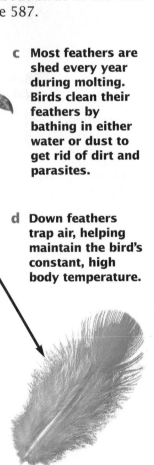

c **Most feathers are shed every year during molting. Birds clean their feathers by bathing in either water or dust to get rid of dirt and parasites.**

d **Down feathers trap air, helping maintain the bird's constant, high body temperature.**

How Birds Fly

Birds must overcome gravity in order to achieve and maintain flight. The wing is the feature that enables birds (and airplanes) to fly. An upward force known as lift is generated when air passes across the surface of a wing. Lift is produced when the pressure of the air passing over the top of the wing is less than that of the air passing under the wing. This pressure difference can be created in two ways. First, the wing can be arched, creating a pocket on the underside of the wing. Second, the front edge of the wing can be held higher than the back edge, increasing what is called the angle of attack.

Lift is produced only when air is moving across the wing, either when the wind blows toward the bird or when the bird travels forward. Since wind direction and speed are unpredictable, efficient flight demands that birds generate force to propel themselves forward. This forward force is known as thrust. Birds typically move forward by beating their wings. Thrust is produced largely by the downstroke during flapping of the wings.

Active flight requires strong muscles to raise and lower the wings. The muscles that power the wings of birds during flight attach to the sternum, or breastbone. The sternum is large and keeled, like the hull of a boat, to accommodate these large muscles.

There are different kinds of bird flight

Birds fly in a variety of ways. The most common method is called flapping flight. It is also the most energy-demanding form of flight. The bird must actively lower and raise the wings in a complex twisting pattern that produces both lift and forward thrust.

If you watch birds in the air, you will notice that not all of them flap their wings all of the time. The simplest form of nonflapping flight is gliding. A bird always loses altitude as it glides because gravity overcomes the lift provided by its forward descent. Birds dropping from treetops to the ground often do so by simple gliding; it requires very little energy.

Some birds exhibit a different kind of nonflapping flight called soaring, which allows them to overcome gravity and stay aloft for long periods while spending only minimal energy. Soaring birds, such as the tern in **Figure 26.16**, are generally large-bodied and have relatively small flight muscles. The large body results in forward momentum once the bird is flying, thus maintaining lift. It is this ability to maintain lift that makes soaring different from simple gliding.

Figure 26.16
This soaring Caspian tern is a member of the gull family and is the largest North American tern.

Tour of a Bird

Birds , such as this cedar waxwing, evolved from dinosaurs. The similarities between dinosaurs and birds prompted Thomas Huxley, a colleague of Darwin, to call birds "glorified reptiles."

Ranging from 40°C–42°C (104°F–108°F), bird body temperatures are higher than the lethal limit for mammals.

All special adaptations found in flying birds contribute to two ends: more power and less weight.

Birds are like airplanes in many ways. They have wings for lift, a tail for steering, and wing slots to avoid stalling at slow speeds.

A bird's neck is highly flexible, with elaborately interwoven and subdivided muscles.

All flying birds have a keeled sternum to which powerful flight muscles are attached.

Most birds molt at least once a year, a few feathers at a time. They survive winter cold by ruffling their feathers to provide insulation.

Feathers evolved from reptilian scales. A feather consists of a shaft and several hundred parallel barbs.

There are 8,800 species of living birds. All birds have forelimbs modified into wings, are covered with feathers, have horny beaks, and lay hard-shelled eggs.

The bones of birds are laced with air cavities, making them light and strong at the same time. The skeleton weighs less than the feathers.

Many birds migrate seasonally. The bobolink commutes 6,400 km (4,000 mi.) each year between North America's Great Lakes region and central South America.

Major Orders of Birds

You can often tell a great deal about the habits and food of a bird by examining its beak and feet. For instance, carnivorous birds such as eagles have curved talons for seizing prey and a sharp beak for tearing apart their meal. The beaks of seed-eating birds such as finches are short, thick, seed-crushers.

What birds are common in your neighborhood? What sorts of adaptations do they show? Twenty-eight orders of living birds have been described. **Table 26.2** lists 16 of the most common orders of birds, showing examples. Pay particular attention to the feet and beaks of these birds.

Table 26.2 Major Orders of Birds

Order	Approximate Number of Living Species	Main Characteristics	Examples	
Passeriformes	5,276	Songbirds; perching feet; well-developed vocal organs; dependent young; largest bird order, containing 60 percent of all bird species	Sparrows, robins, warblers, crows, starlings, mockingbirds	
Apodiformes	428	Small bodies, short legs, rapid wing beat; hummingbirds are the smallest birds	Hummingbirds, swifts	
Piciformes	383	Sharp, chisel-like bills for pounding through wood in search of insects; grasping feet	Woodpeckers, toucans, honeyguides	
Psittaciformes	340	Well-developed vocal organs; large powerful bills for crushing seeds	Parrots, cockatoos	
Charadriiformes	331	Shorebirds; typically with long, slender, probing bills and long, stiltlike legs	Gulls, terns, plovers, auks, sandpipers,	
Columbiformes	303	Stout bodies; perching feet	Pigeons, doves	
Falconiformes	288	Birds of prey; day-active carnivores; sharp pointed beaks for tearing flesh; keen vision; strong fliers	Eagles, hawks, falcons, vultures	
Galliformes	268	Rounded bodies; often limited flying ability	Chickens, quail, grouse, pheasants	

Gruiformes	209	Marsh dwellers; diverse body shapes; long, stiltlike legs	Rails, coots, bitterns, cranes
Anseriformes	150	Waterfowl; webbed toes; broad bill with filtering ridges at margins	Swans, geese, ducks
Ciconiiformes	114	Long-legged waders; often with large bodies	Storks, herons, ibises
Strigiformes	146	Nocturnal birds of prey; large eyes; powerful beaks and feet	Owls
Procellariiformes	104	Sea birds; tube-shaped bills; many can fly for long periods of time	Albatrosses, petrels
Sphenisciformes	18	Marine; flightless; confined to Southern Hemisphere; thick coat of insulating feathers; wings modified as paddles for swift swimming	Penguins
Dinornithiformes	2	Primitive; small and flightless; found only in New Zealand	Kiwis
Struthioniformes	1	Large; flightless; only two toes; long, strong running legs	Ostrich

Section Review

❶ **What evidence indicates that birds evolved from reptiles?**

❷ **Name two differences between a lizard and a bird.**

❸ **Ostriches have feathers but cannot fly. What functions do feathers perform for an ostrich?**

❹ **List two features of the bird's skeleton that suit it for flight.**

YOU ARE A MAMMAL. SO ARE MOST OF THE ANIMALS THAT HUMANS EAT, USE AS WORK ANIMALS, AND KEEP AS PETS. ALL MAMMALS HAVE HAIR OR FUR ON THEIR BODIES, A CHARACTERISTIC NOT FOUND IN ANY OTHER KIND OF ANIMAL. MAMMALS ARE ALSO ENDOTHERMIC, LIKE BIRDS. FEMALE MAMMALS PRODUCE MILK WITH WHICH THEY NURSE THEIR OFFSPRING.

26.3 Introduction to Mammals

Objectives

❶ List the unique characteristics of mammals.

❷ Identify two features of monotremes.

❸ Contrast the manner of development of marsupials and placentals.

Evolution of Mammals

Figure 26.17
The animals illustrated below are members of a group of reptiles called therapsids. At first glance, these therapsids seem to be mammals. The similarity is not surprising: mammals evolved from therapsids.

Mammals (class Mammalia) arose from early reptiles called therapsids (*thur AP sihdz*), some of which are illustrated in **Figure 26.17**. The fossil record provides a well-documented transition between these reptiles and mammals, with fossil forms ranging from reptiles with a few mammalian characteristics to true mammals. True mammals appeared about 230 million years ago, about the same time as the first dinosaurs. These early mammals were small, about the size of mice. For 165 million years—all of the time the dinosaurs flourished— mammals were a minor group that changed little. In the 65 million years since the dinosaurs became extinct, mammals have rapidly diversified to fill the ecological opportunities made available by the disappearance of the dinosaurs. There are now about 4,500 species of mammals.

a *Massetognathus* (below) was a small, early Triassic herbivore, able to pack food into its cheeks like a rodent.

b *Cynognathus* (left) was about 1 m (3 ft. 3 in.) long and lived during the early Triassic. It was one of the largest therapsids.

c *Oligokyphus* (right) was small, about 50 cm (20 in.) long. These Jurassic animals could have been mistaken for mammals were it not for their reptilian jaws.

Figure 26.18
The grizzly bear, *Ursus horribilis*, is a large mammal, standing nearly 2.8 m (9 ft.) tall and weighing close to 800 kg (1,760 lbs.). Like all mammals, it has fur and nurses its young on milk.

a Grizzly bears are omnivores, eating mostly plants and fruits. But they do have sharp teeth for pulling apart flesh.

b The grizzly bear's name comes from the fact that its fur is silver tipped, giving it a grizzled appearance.

c One or two tiny and nearly naked cubs are born in midwinter, during hibernation. The mother is able to produce milk even though she does not eat or drink until spring.

d The four-chambered heart of the bear is very efficient, rapidly delivering oxygen to the body's muscles and organs.

Ventricles

Atria

Mammalian Characteristics

Mammals are distinguished by two characteristics: the presence of hair or fur on the body, and the ability to produce milk. Even the apparently naked whales and dolphins grow sensitive bristles on their snouts. Evolution of fur and the ability to regulate body temperature through metabolism enabled mammals to inhabit colder environments than could be tolerated by ectothermic reptiles and amphibians.

Like birds, mammals have a four-chambered heart. There is no mixing of deoxygenated and oxygenated blood in the heart.

Female mammals have **mammary glands** that secrete milk. Newborn mammals, which are born without teeth, suckle this rich milk until able to feed on their own. The grizzly bear shown in **Figure 26.18** shows all of these key mammalian characteristics.

Table 26.3 Major Orders of Mammals

Order	Approximate Number of Living Species	Main Characteristics	Examples
Monotremata	3	The only egg-laying mammals; once widespread, now found only in Australia and New Guinea	Platypus, echidnas
Marsupialia	280	Primitive mammals; have an abdominal pouch in which young are reared	Kangaroos, koalas, opossums
Rodentia	1,814	Small herbivores with chisel-like, incisor teeth that grow continuously	Squirrels, rats, mice, beavers, porcupines
Chiroptera	986	The only flying mammals; elongated fingers that support a thin wing membrane; mainly fruit or insect eaters; many fly at night, navigating by sonar	Bats
Insectivora	390	Small, chiefly night-active mammals; feed on insects; sharp-snouted; spend most of their time underground; the most primitive placental mammals	Moles, shrews, hedgehogs
Carnivora	240	Land-living predators; teeth adapted for seizing prey and shearing flesh; there are no native families in Australia	Dogs, bears, cats, wolves, otters, weasels
Primates	233	Largely tree dwellers; binocular vision and an opposable thumb; large brains; the end product of a line that branched off early from other mammals; retains many primitive characteristics	Prosimians, apes, monkeys, humans

Artiodactyla	211	Hoofed mammals with two or four toes; large herbivores; most are grass eaters	Sheep, pigs, cattle, deer, giraffes
Cetacea	79	Aquatic, streamlined bodies; front limbs modified into broad flippers; no hind limbs; nostrils are blowholes on top of head; hairless except on muzzle	Whales, dolphins, porpoises
Lagomorpha	69	Rodentlike mammals with four upper incisors, rather than the two seen in rodents; hind legs often longer than forelegs, an adaptation for jumping	Rabbits, hares, pikas
Pinnipedia	34	Marine carnivores with limbs modified for swimming; feed mainly on fish	Seals, sea lions, walruses
Edentata	30	Mostly insect eaters; many are toothless, but some have degenerate, peglike teeth	Sloths, anteaters, armadillos
Perissodactyla	17	Hoofed mammals with one or three toes; herbivores with teeth adapted for chewing	Horses, zebras, rhinoceroses, tapirs
Proboscidea	2	Enormous herbivores with long trunks; two upper incisors elongated as tusks; the largest living land animals	Elephants

Section Review

1 Why are monotremes classified as mammals even though they lay eggs?

2 List two differences between monotremes and other mammals.

3 Contrast the pattern of development of a marsupial with that of a placental mammal.

Food and Feeding

Mammals feed on a variety of foods. Horses, giraffes, antelopes, and elephants are herbivores. Lions, wolves, seals, and sperm whales are carnivores. The blue whale, the largest animal, filters small crustaceans from the ocean. It is usually possible to determine a mammal's diet by examining its teeth. For example, look at the skull of the coyote (a carnivore) and the deer (a herbivore) shown in **Figure 26.24**. The coyote's long canine teeth are suited for biting and holding prey. Its premolar and molar teeth are triangular and sharp for shearing off chunks of flesh. The deer's canines, in contrast, are small. It clips off mouthfuls of plants with its flat incisors. The deer's molars are large, and their surfaces are covered with ridges to form an effective grinding surface that can break up tough plant tissues.

Cellulose is the major component of plant cell walls, and thus is a major constituent of the plant body. Mammals do not have enzymes that can digest cellulose. Herbivorous mammals rely on a mutualistic partnership with bacteria that can produce cellulose-splitting enzymes. Mammals such as cows, buffaloes, antelopes, goats, deer, and giraffes have huge four-chambered stomachs that function as storage and fermentation vats. The first chamber is the largest and holds a large population of bacteria. When the animal swallows, chewed plant material passes into this chamber. Bacteria partly digest the material, which is then regurgitated and chewed again. A cow chewing its cud is rechewing this partly digested food. After another thorough grinding, the cud is swallowed and further digested in the stomach. It then passes from the stomach into the intestines.

Rodents, horses, elephants, and rabbits are herbivores but have relatively small stomachs lacking mutualistic bacteria. These animals do not chew a cud. Bacteria that aid in digestion live in a pouch that branches from the large intestine.

Even with these complex adaptations for breaking down cellulose, a mouthful of plants is less nutritious than a mouthful of flesh. Herbivores must consume large amounts of plant material to gain sufficient nutrition. An elephant eats 135–150 kg (300–400 lbs.) of food per day. You can read more about mammalian adaptations in the *Tour of a Mammal* on page 599.

Figure 26.24
The structure of a mammal's jaw and teeth usually reveals its diet.

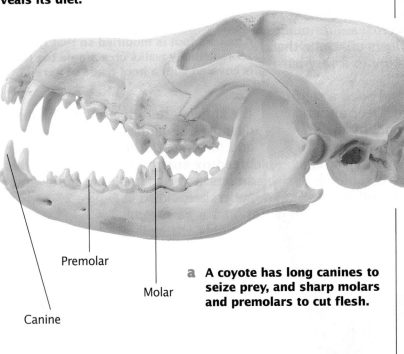

Premolar

Molar

Canine

a **A coyote has long canines to seize prey, and sharp molars and premolars to cut flesh.**

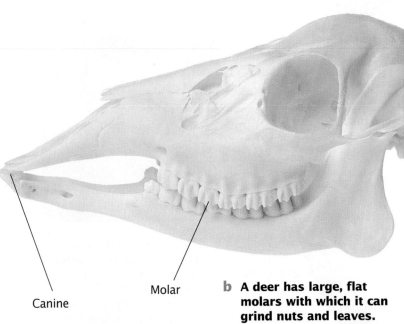

Molar

Canine

b **A deer has large, flat molars with which it can grind nuts and leaves.**

Tour of a Mammal

Mammals, such as this field mouse, are endotherms and maintain a constant body temperature. Mammals have highly developed brains.

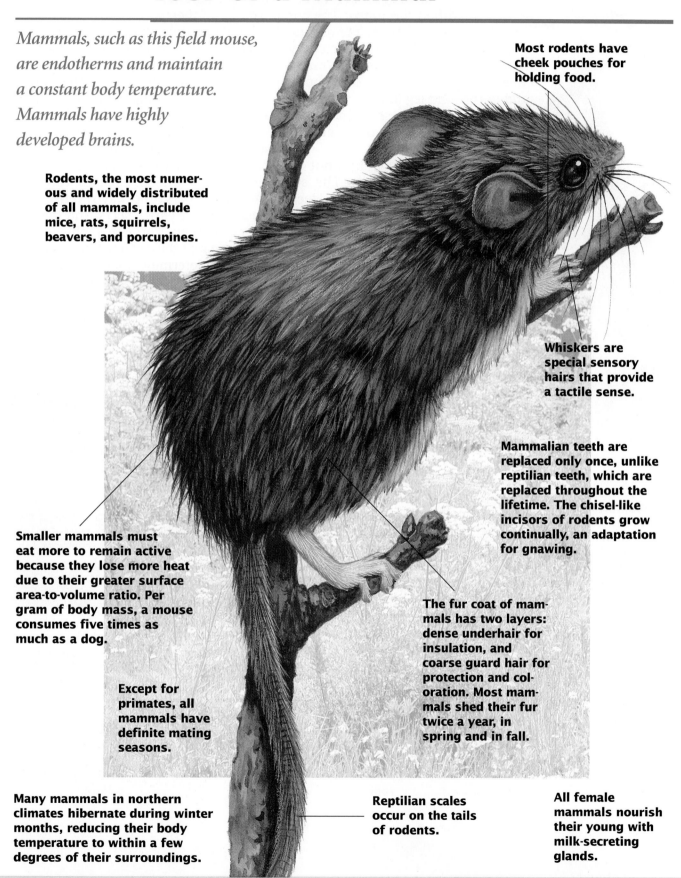

Rodents, the most numerous and widely distributed of all mammals, include mice, rats, squirrels, beavers, and porcupines.

Most rodents have cheek pouches for holding food.

Whiskers are special sensory hairs that provide a tactile sense.

Mammalian teeth are replaced only once, unlike reptilian teeth, which are replaced throughout the lifetime. The chisel-like incisors of rodents grow continually, an adaptation for gnawing.

Smaller mammals must eat more to remain active because they lose more heat due to their greater surface area-to-volume ratio. Per gram of body mass, a mouse consumes five times as much as a dog.

The fur coat of mammals has two layers: dense underhair for insulation, and coarse guard hair for protection and coloration. Most mammals shed their fur twice a year, in spring and in fall.

Except for primates, all mammals have definite mating seasons.

Many mammals in northern climates hibernate during winter months, reducing their body temperature to within a few degrees of their surroundings.

Reptilian scales occur on the tails of rodents.

All female mammals nourish their young with milk-secreting glands.

Flying Mammals

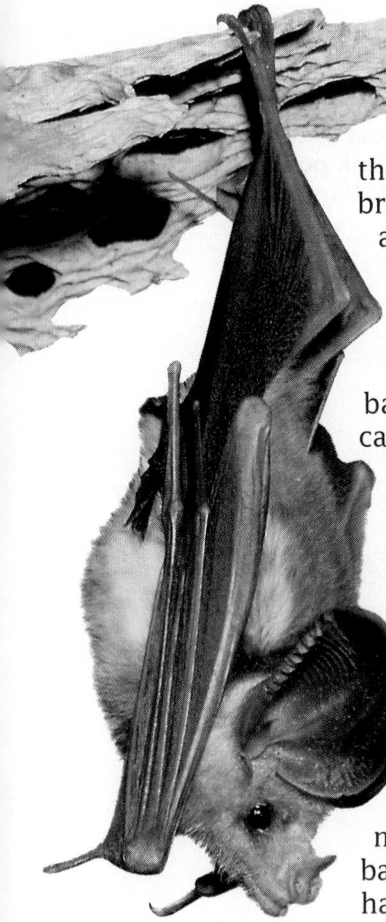

Bats are the only mammals capable of powered flight. Like the wings of birds, the wings of bats are modified forelimbs. The bat wing is a leathery membrane of skin and muscle supported by the bones of four fingers. The membrane attaches to the side of the body and to the hind leg (and to the tail in some bats). When resting, most bats prefer to hang from their legs, as shown in **Figure 26.25**.

Contrary to popular belief, not all bats emerge only at night. The so-called flying foxes, for instance, feed on fruit during the day. Most bats, however, are active at dusk or at night. Some eat fish and frogs, which they pluck from the water's surface. Others feed on nectar from flowers or prey on small mammals, including other bats. Vampire bats of Central America and South America drink blood from large mammals. Flying insects are the main food source for most nocturnal bats. Since few birds fly at night, bats have almost exclusive access to this rich food supply.

Figure 26.25
Bats, such as this California leaf-nosed bat, are social animals. They roost together in groups of thousands, and some species hunt cooperatively.

How do bats navigate in the dark?

How can bats fly in dark caves and hunt their food at dusk or even at night? Do bats have particularly sensitive vision? Or do they use another sense to find their way around? Late in the eighteenth century, the Italian scientist Lazzaro Spallanzani showed that a blinded bat could fly without crashing into things and could still capture flying insects. However, when Spallanzani plugged the ears of the bat, it was unable to navigate and collided with objects. Spallanzani concluded that bats "hear" their way through the world. It was not until the late 1930s that bats' ability to "fly blind" was explained.

Bats have evolved a sonar system that functions much like the sonar devices used by ships to locate underwater objects. As a bat flies, it emits extremely high-pitched sounds well above our range of hearing. These high-frequency pulses are emitted through the mouth or, in some cases, through the nose. The sound waves reflect off obstacles or flying insects, and the bat hears the echo. Through sophisticated processing of this echo within its brain, a bat can determine not only the direction of an object but also the distance to the object. This sonar system enables bats to navigate and to capture their prey at night, as shown in **Figure 26.26**.

Figure 26.26
"Blind as a bat" is an inaccurate phrase. This bat was able to use its sonar to catch this moth in midair in the dark.

Section Review

❶ List two functions of hair.

❷ Name two keratin-containing structures on your body.

❸ Contrast the teeth of a coyote with those of a deer.

❹ Explain why a blinded bat can catch prey but a deafened one cannot.

Chapter 26 Highlights

"Well, of course I did it in cold blood, you idiot! . . . I'm a reptile!"

	Key Terms	**Summary**

26.1 Reptiles

Reptiles control their body temperatures through their behavior.

		• Reptiles have dry, largely watertight skin, lay watertight eggs, and are ectotherms.
		• Reptiles evolved about 320 million years ago.
		• Pterosaurs were the only reptiles that evolved the ability to fly.
		• Dinosaurs became extinct about 65 million years ago.
		• The surviving reptiles include crocodiles and alligators, turtles, the tuatara, lizards, and snakes.

26.2 Birds

Some birds may have more than 25,000 feathers on their body.

	contour feather (p. 585)	• The first birds evolved about 150 million years ago.
	down feather (p. 585)	• All birds have feathers, are endothermic, and lay eggs.
		• Birds have down feathers and contour feathers.
		• The skeletons of birds are lightweight. The bones are hollow and thin.

26.3 Introduction to Mammals

Mammals, such as these bears, receive nourishment from their mother's mammary glands while young.

	mammary gland (p. 591)	• Mammals evolved from reptiles about 230 million years ago.
	monotreme (p. 592)	• Mammals have hair and are endothermic. Female mammals have mammary glands.
	marsupial (p. 592)	
	placental mammal (p. 592)	• Monotremes are mammals that lay eggs.
		• Marsupial mammals are born early and complete their development in the mother's pouch.
		• Placental mammals nourish their young via the placenta throughout development.

26.4 Mammalian Adaptations

The skull of a coyote has long canines and strong jaws.

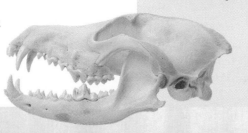

	guard hair (p. 596)	• Hair serves as insulation, as camouflage, as a sensory structure, and as a defense.
	underhair (p. 596)	• Bats are the only flying mammals. Bats are able to navigate and capture prey in the dark by means of a sonar system.

Understanding Vocabulary

1. For each pair of terms, explain the differences in their meanings.
   a. endotherm, ectotherm
   b. contour feathers, down feathers
   c. monotremes, marsupials
   d. guard hair, underhair

Relating Concepts

2. Copy the unfinished concept map below onto a sheet of paper. Then complete the concept map by writing the correct word or phrase in each oval containing a question mark.

Understanding Concepts

Multiple Choice

3. The skin of reptiles forms a watertight barrier because it
   a. is sticky and covered with slime.
   b. is covered with scales and hair.
   c. contains fats and the protein keratin.
   d. is covered with feathers.

4. Which animal is not an ectotherm?
   a. lizard   c. toad
   b. alligator   d. chicken

5. What evidence suggests that reptiles and birds are closely related?
   a. both lay amniotic eggs
   b. scales and feathers are found on birds and reptiles
   c. both are endothermic
   d. birds and reptiles live where mammals cannot

6. Which of the following are not bird adaptations for flight?
   a. a cartilaginous skeleton
   b. an efficient respiratory system
   c. contour feathers
   d. the ability to regrow lost feathers

7. Mammals are different from any other vertebrates in that they
   a. lay eggs.
   b. produce milk.
   c. are endotherms.
   d. have structures that contain keratin.

8. In placental mammals,
   a. embryos are nourished through a placenta throughout development.
   b. embryos are nourished by yolk before attaching to the placenta.
   c. the young finish developing in a pouch.
   d. young develop in hard-shelled eggs.

9. Which is not a function of hair?
   a. insulation   c. camouflage
   b. navigation   d. sensory structures

10. Which structures do not contain keratin?
    a. fingernails   c. teeth
    b. hooves   d. hair

11. In Spallanzani's experiment to determine how bats navigate in the dark, what variable did he manipulate?
    a. the type of insect used as food for the bat
    b. sense organs: he blocked them one at a time
    c. the time of day the experiment was carried out
    d. bat feeding preferences

Completion

12. The first flying vertebrates were the _____ . They were not birds but _____ .

13. Snakes are members of the order _____ , while turtles belong to the order _____ .

14. Birds are most closely related to _____ , which became extinct _____ years ago.

15. Kangaroos are _____ . Newborn kangaroos crawl from the mother's birth canal to her _____ , where they finish developing.

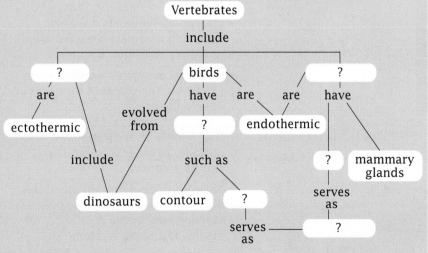

Short Answer

16. How could use of the term "cold-blooded" lead to a misunderstanding about how reptiles regulate their body temperature? Why is it more appropriate to call them ectothermic?

17. Describe the hypothesis proposed by Luis Alvarez and his colleagues about the end of the Age of Reptiles.

18. Explain how a bat is able to locate a flying insect on a dark night.

19. Why do scientists think that monotremes are more closely related to reptiles than are either marsupials or placental mammals?

20. Explain why birds can live in colder climates than reptiles.

Interpreting Graphics

21. Examine the animals pictured below.

a

b

c

- Identify the class to which each animal belongs.
- Which animal is endothermic?
- In which animal is the ventricle of the heart incompletely divided?

Reviewing Themes

22. *Patterns of Change*
    If you were a marsupial, how would you have spent the early months of your life?

23. *Evolution*
    What evidence suggests that reptiles and birds have a common ancestor?

24. *Scale and Structure*
    How could you tell if a mammalian skull you found was that of a herbivore or a carnivore?

Thinking Critically

25. *Inferring Conclusions*
    How would losing 50 percent of its down feathers affect a bird?

26. *Compare and Contrast*
    How are underhair and down feathers alike? How are they different?

27. *Comparing and Contrasting*
    How are an echidna and a crocodile alike? How are they different?

28. *Building on What You Have Learned*
    In Chapter 16 you learned about bacteria. How are bacteria helpful to herbivorous mammals?

Cross-Discipline Connection

29. *Biology and English*
    Identify fears that people have about bats by surveying students in your school. Then learn if their fears are justified by writing a letter to Bat Conservation International, P.O. Box 162603, Austin, Texas 78716.

Discovering Through Reading

30. Read the article "Listening to the Mockingbird," in *National Wildlife*, June/July 1992, pages 12–16. How did the mockingbird get its name? According to scientists who have studied mockingbirds, why do male mockingbirds sing? What effect does the male's singing have on females?

Investigation

How Do Down and Contour Feathers Differ?

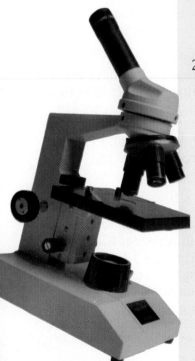

Objectives

In this investigation you will:
- *observe* down and contour feathers
- *contrast* the structure of down and contour feathers
- *relate* the structure of each kind of feather to its function

Materials

- contour feather
- down feather
- compound light microscope
- prepared slide of contour feather
- prepared slide of down feather

Prelab Preparation

1. Review what you have learned about feathers by answering the following questions:
   - What is the function of down feathers?
   - Where are down feathers found on a bird?
   - What functions do contour feathers perform?
   - Where would you find contour feathers on a bird?

2. Review the procedures in the Appendix for using the compound microscope.

Procedure: Comparing Down and Contour Feathers

1. Form a cooperative team with another student to complete steps 2–7.

2. Examine a down feather. The stiff base of the feather is known as the quill. The quill extends into the flesh to anchor the feather. Hold the feather by the quill and wave the feather in the air. Describe the texture of the feather. Record your observations.

3. Examine a contour feather. Using the photograph below as a guide, identify the quill, shaft, vane, and barbs.

4. Hold the feather by the quill and gently bend the tip with your other hand. Be careful not to break the feather. Next, hold the feather by the quill and wave the feather through the air. Record your observations. *What makes the contour feather stiffer than the down feather?*

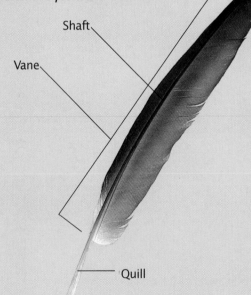

Shaft

Vane

Quill

5. Examine the vane of the contour feather. *How does the texture of the vane compare with the texture of the down feather?*

6. Observe the prepared slide of the down feather under low power. Now observe it under high power. Make a drawing of the feather as it appears under under high power.

7. Repeat step 6 for the contour feather. Using the photograph above as a guide, you should be able to identify the barbules that branch from each barb and the hooklets that interlock the barbules. *Were these structures visible on the down feather?*

8. Clean up your materials and wash your hands before leaving the lab.

Analysis

1. *Identifying Relationships*
Which kind of feather has a longer quill? How does the difference in quill length relate to the function of each kind of feather?

2. *Identifying Relationships*
The contour feathers of the tail and back edge of the wings are often called flight feathers. What features make contour feathers better than down feathers for flight?

3. *Inferring Conclusions*
Birds spend much time preening their feathers. They rearrange and straighten their feathers and align the barbs on individual feathers. Explain why preening is crucial for the feathers to function effectively.

Thinking Critically

Unlike a bird's wing, a bat's wing is a thin membrane of skin stretched across elongated fingers. By contrasting the structure of feathers and hair, explain why a wing could not be made of hair.

DECIPHERING THE FOSSIL RECORD

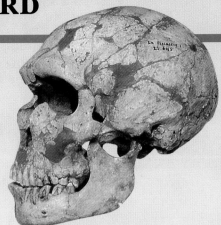

Neanderthal skull

1800

Baron Georges Cuvier

1800s French naturalist **Baron Georges Cuvier** excels in comparative anatomy (comparing animal structures) and pioneers the founding of paleontology (the study of fossils). Cuvier reconstructed extinct animals by comparing their fossils to the skeletons of modern animals.

1842 Fossil finds of large, extinct reptiles lead **Sir Richard Owen**, an English scientist, to suggest that these reptiles belong to a group of reptiles that were unlike any living animals, the dinosaurs.

Sir Richard Owen

1855 Geologists **Sir John Dawson** (Canadian) and **Sir Charles Lyell** (British) discover bones of amphibians in layers of the earth associated with tree ferns and early gymnosperms.

1856 **Neanderthal skull** is found in Feldhofer Cave near Dusseldorf, Germany.

1861 *Archaeopteryx*, the earliest known bird, is discovered in a limestone quarry in Germany.

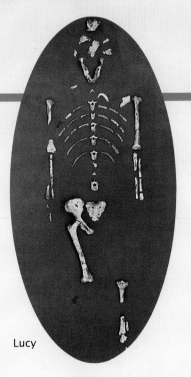

Lucy

1933

Mary Douglas Leakey

1959 Anthropologist **Louis S. B. Leakey** and his wife, **Mary Leakey**, find the skull of *Australopithecus boisei* in Tanzania. It dates back to more than one million years.

1960 British anthropologist **Mary Leakey** and son **Jonathan** discover the first *Homo habilis* fossils at Olduvai Gorge in Tanzania. *Homo habilis* is the oldest member of our genus, and lived in Africa 2 million years ago.

1972 **Richard Leakey**, son of Louis and Mary Leakey, discovers a 1.75-million-year-old skull of *Homo habilis* in western Kenya.

1974 In Ethiopia, American anthropologist **Donald Johanson** discovers "Lucy," the 3 million-year-old skeleton of a bipedal female. Lucy is classified as *A. afarensis* and is the oldest known hominid.

Louis S. B. Leakey

Dinosaur eggs

1932

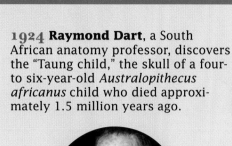

1924 **Raymond Dart**, a South African anatomy professor, discovers the "Taung child," the skull of a four- to six-year-old *Australopithecus africanus* child who died approximately 1.5 million years ago.

Raymond Dart

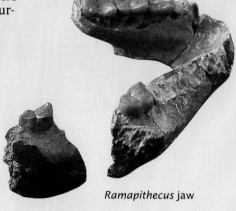

Ramapithecus jaw

1903 American explorer **Roy Chapman Andrews** leads expeditions into the Gobi Desert. His team finds dinosaur eggs and the remains of Earth's largest land animals, dinosaurs thought to be 95 million years old.

1927 Canadian anthropologist **Davidson Black** and French philosopher and paleontologist **Pierre Teilhard de Chardin** discover fossilized bones of *Homo erectus* near Beijing, China.

1932 American anthropologist **George E. Lewis** discovers jaw fragments and teeth of the ape *Ramapithecus,* which lived 8 million to 14 million years ago in northern India.

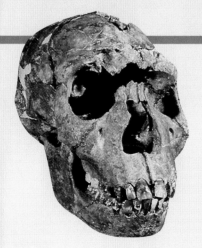

1990

1987 **Wendy Slobada**, a recent Canadian high school graduate, discovers a new dinosaur eggshell site near Milk River, Alberta, Canada; this site yields a fossil of an unhatched dinosaur.

1988 In Argentina, University of Chicago paleontologist **Paul Sereno** (American) discovers what could be the oldest known dinosaur fossil, 230 million-year-old *Herrarasaurus.*

Homo erectus skull

1985 **Kamoya Kimeu**, Richard Leakey's Kenyan assistant, finds a nearly complete *Homo erectus* skeleton in western Kenya. The skeleton is that of a 12-year-old boy who died 1.6 million years ago.

Wendy Slobada

1992 **Brian Anderson**, an American graduate student at the University of California, Riverside, discovers rare fossilized impressions of dinosaur skin in the Books Cliffs of Utah.

Human Life

Every human life begins when a single sperm cell penetrates an egg cell and fertilization occurs. As the fetus grows, the organ systems needed to maintain life develop. Those systems are put to work daily in everything we do. As we age, organ systems slow down and cease to function properly. Although every person, like every living thing, must die, the cycle of life continues. In this unit, you will learn how life begins and how the body's organ systems work together.

The Human Body

The swift strokes of a crew team require the coordination of the body's skin, bones, muscles, and other tissues.

YOUR BODY CONSISTS OF MANY SPECIALIZED TISSUES SUCH AS THOSE FOUND IN SKIN, BONES, MUSCLES, NERVES, AND BLOOD VESSELS. THESE TISSUES, EACH MADE OF SPECIALIZED CELLS, ARE ARRANGED DIFFERENTLY IN EACH PART OF YOUR BODY. THIS PROCESS OF SPECIALIZATION AND ARRANGEMENT BEGAN BEFORE YOU WERE BORN WHEN A SINGLE CELL BEGAN TO DIVIDE.

27.1 Tour of the Human Body

Objectives

❶ **Explain the roles of the four kinds of tissues in the human body.**

❷ **Describe four types of connective tissues.**

❸ **List four organ systems of the human body.**

Similar Cells Form Tissues

Figure 27.1
The human body is made of four types of tissue, as shown below. All work together to make the body a functional unit capable of all life activities.

All of the trillions of cells in the human body arise from a single cell—the fertilized egg. The first few cells that form by cell division after fertilization look alike. Soon, however, the new cells become specialized to carry out particular tasks.

As cells specialize, they begin to group together to form tissues. A **tissue** is a group of similar cells that work together to perform a specific function, such as movement or protection. Your body contains four types of tissues—epithelial tissue, muscle tissue, nerve tissue, and connective tissue, as shown in **Figure 27.1**.

Epithelial tissue covers and protects
Epithelial (*ehp uh THEE lee uhl*) **tissue** is made of tightly connected cells that are arranged in flat sheets, often only a few cells thick. Just as canvas protects machinery that must be stored outdoors, epithelial tissue prevents damage to the cells that lie beneath it. Skin is an example of epithelial tissue. Epithelial tissue also lines spaces within the body and covers the inner and outer surfaces of your internal organs.

Some epithelial tissue has an entirely different function. This epithelial tissue contains exocrine glands. An **exocrine gland** is a cell or group of cells that produces and releases secretions onto a body surface. One type of exocrine gland in your skin secretes sweat, which helps cool your body. Exocrine glands in the epithelial tissue lining your digestive system produce enzymes that break down food.

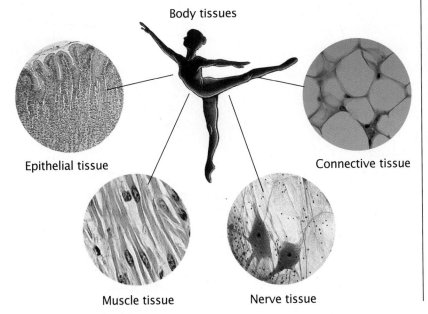

Body tissues

Epithelial tissue

Connective tissue

Muscle tissue

Nerve tissue

Muscle tissue moves the body

Have you ever awakened with sore muscles? If so, you probably became aware of how often you use those particular muscles. **Muscle tissue** moves the parts of your body. Even when you are sitting still, muscles are at work moving food through your digestive system, maintaining your posture, and moving blood through your heart.

Muscle tissue is made of cells that contract, or shorten, and then return to their normal length. Another characteristic of muscle tissue is that it responds to electrical stimulation. Electrical signals control when a muscle will contract.

Nerve tissue sends electrical signals through the body

As you walk up a flight of stairs, the electrical signals that cause the muscles in your legs to contract are produced by nerve tissue. **Nerve tissue** is found in your brain, nerves, and sense organs. It contains cells called neurons that are able to generate electrical impulses and transfer the impulses to other cells. Neurons have long extensions that can carry electrical impulses long distances. Nerve tissue also contains other cells that nourish and protect the neurons that carry the electrical signals.

Connective tissue joins, supports, and transports

The bones of your skeleton and the ligaments that join them are examples of **connective tissue**. Connective tissue actually contains few cells. Most of the tissue consists of fluid and fibers secreted by the cells in connective tissue. **Figure 27.2** shows the four basic types of connective tissue, all of which can be found in your knee. They are cartilage, bone, connective tissue proper, and blood.

Figure 27.2
The graceful leap of a dancer depends on many types of tissues including muscle, nerve, and epithelial tissues. These tissues and all four types of connective tissue can be seen in this longitudinal section of a human knee.

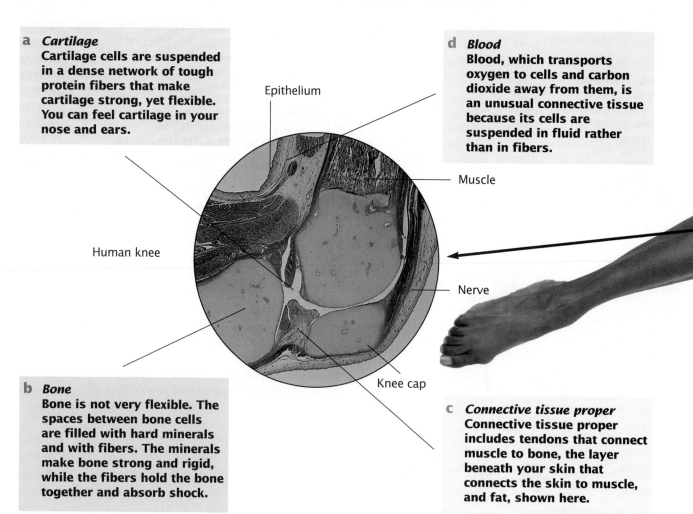

a Cartilage
Cartilage cells are suspended in a dense network of tough protein fibers that make cartilage strong, yet flexible. You can feel cartilage in your nose and ears.

d Blood
Blood, which transports oxygen to cells and carbon dioxide away from them, is an unusual connective tissue because its cells are suspended in fluid rather than in fibers.

Epithelium

Muscle

Human knee

Nerve

Knee cap

b Bone
Bone is not very flexible. The spaces between bone cells are filled with hard minerals and with fibers. The minerals make bone strong and rigid, while the fibers hold the bone together and absorb shock.

c Connective tissue proper
Connective tissue proper includes tendons that connect muscle to bone, the layer beneath your skin that connects the skin to muscle, and fat, shown here.

Tissues Form Organs

How can only four kinds of tissues make up all the different organs in your body? The answer is simple: different tissues are combined in different ways to form each organ. An **organ** is a structure composed of a number of tissues that work together to perform a specific job in the body.

For example, the largest organ of your body—your skin—contains all four types of tissues. Your skin is made of many watertight sheets of epithelial tissue covering a cushioning layer of connective tissue. Nerve tissue in your skin makes it sensitive to touch. Tiny muscles can make the fine hairs that cover most of your body stand on end when you are cold. All of these tissues cooperate to protect your body.

Organs that work together form an **organ system**. An example of an organ system is the circulatory system. The heart and the blood vessels are separate organs that work together to carry substances through the body. **Figure 27.3** shows four organ systems you will learn more about in this chapter and in chapters to come.

Figure 27.3
Each organ in this dancer's body performs a specific task related to other organs in its system. For instance, in the circulatory system, the heart pumps blood through vessels that deliver oxygen to the lungs and to other organs.

a **Circulatory system**

b **Skeletal system**

c **Reproductive system**

d **Digestive system**

Section Review

❶ How does each of the four tissue types function in the human body?

❷ How might a disease that slowly destroys nerve cells affect muscle tissue in the body?

❸ Name four connective tissues found in the knee.

❹ List four organ systems of the human body.

Videoscope Surgery

Futuristic Surgery? The Time Is Now

In a popular futuristic science-fiction series, doctors perform surgery without spilling a drop of blood. Knives and scalpels are obsolete, and patients are up and around in no time. Now, thanks to fiber optics and video technology, bloodless surgery may not be too far in the future.

Increasingly, physicians are using a technique called videoscope surgery to diagnose and treat injured and diseased body organs. In videoscope surgery, a surgeon makes a small incision and inserts a long slender fiber-optic tube into the body. The tube has a tiny camera lens at the end that transmits images to a nearby television screen in the operating room. Surgical instruments for grasping, snipping, and stapling can be inserted through other small incisions. The surgeon controls the instruments from outside the body while watching the screen.

In 1987, videoscope surgery was used for the first time to remove a diseased gallbladder. In traditional gallbladder surgery, a 7-cm to 13-cm (3-in. to 6-in.) cut is made along the patient's abdomen. With the new surgery, four small incisions are made that heal faster and hurt less. Videoscope surgery is now used on about three-quarters of the American patients having their gallbladders removed.

Widespread Applications

In addition to gallbladder surgery, the new technique is rapidly being adapted to other kinds of surgery. For instance, one surgeon has removed a diseased kidney this way, after using instruments to maneuver the organ into a small sac and chop it into smaller pieces. Another surgeon recently began using videoscope surgery to perform lung biopsies to identify cancerous tissue. In the past, biopsy surgery involved slicing through chest and rib muscles. Afterward, a patient would spend several days in intensive care and then have weeks of painful recovery. With the new technique, blunt-tipped instruments are pushed gently through the muscles, which stretch temporarily. When finished, the surgeon closes up the small incision with a few stitches and a bandage.

Although videoscope surgery is likely to have a great impact in surgery of the chest and abdomen, other operations have already benefited from fiber-optic instruments. For example, the torn cartilage associated with many joint injuries, especially in the knee, can be repaired using an instrument called an arthroscope. Arthroscopic surgery has become widely used on athletes who used to have to undergo extensive surgery to treat their knee injuries.

Will these surgical instruments soon be outdated? One study predicts that 40 percent of all surgery will be performed with videoscopes by the next century.

Gallbladder removal

Surgery:	Traditional	Videoscope
Incision(s):	one 7-cm to 13-cm (3-in. to 6-in.) cut	four 0.5-cm (0.25-in.) punctures
Average cost:	$4,250	$3,500
Hospital stay:	5 to 8 days	overnight
Recovery time:	4 to 6 weeks	5 to 7 days
Complications:	1% to 9.4%	5.1%
Deaths:	0.2% to 0.6%	0.1%

Taking Precautions

As videoscope surgery becomes more common, however, some experts have become concerned. Operating by video camera requires a lot of practice. Surgeons have to acquire a new set of skills. For example, when looking at a television screen, a surgeon sees only a two-dimensional image. So it is easier for the surgeon to misjudge the distance to important organs or major blood vessels. A tiny cut in an artery, for example, could cause serious bleeding that would be hard to stop.

In videoscopic gallbladder surgery, if the surgeon is relatively new at the technique, patients stand a greater risk of injury to the bile duct, the tube connecting the gallbladder to the small intestine. Consequently, some medical societies are drafting guidelines to ensure that surgeons are well trained in this new technique. One recommendation is to have doctors practice on animals first. Then, surgeons who are already experts in the technique will supervise the newly trained surgeons' first operations on humans.

In addition to better training and more practice, video surgeons also need improvements in their equipment. Stiff necks from straining to watch a television screen and stiff elbows from using awkwardly designed instruments are common problems. Nevertheless, the field is progressing quickly. Eventually, say some doctors, cutting a patient open for certain kinds of surgery might go the way of the dinosaurs.

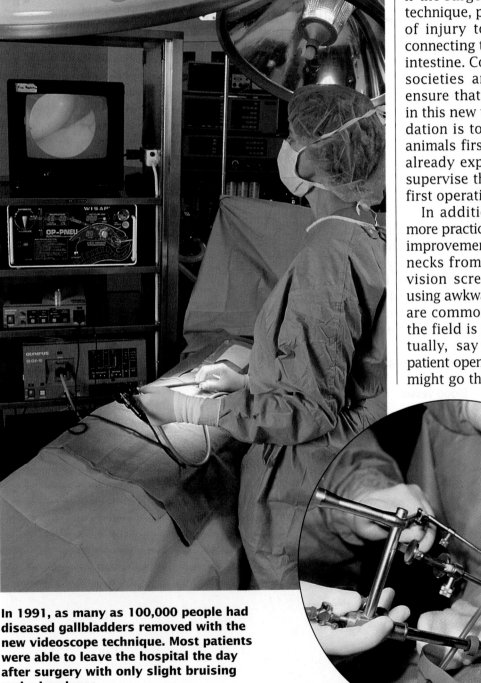

In 1991, as many as 100,000 people had diseased gallbladders removed with the new videoscope technique. Most patients were able to leave the hospital the day after surgery with only slight bruising and a bandage.

The Epidermis

Not all of your skin is alive. Your **epidermis** is made of inner layers of living epithelial cells and outer layers of dead cells. These dead cells, shown in **Figure 27.6a**, are packed with a waterproof protein called keratin.

Each epidermal cell arises from the division of a cell in the innermost layer of the epidermis. This layer is one of the most active regions of cell division in your body. As the new cell begins to make keratin, it is pushed outward by even newer cells forming underneath. The epidermal cell moves farther from the nutrients that diffuse from the blood flowing through the dermis. When the cell is pushed too far from its nutrient supply and has filled with keratin, it dies. In effect, it becomes part of the skin's "surface armor," absorbing the wear and tear of life until it is eventually shed to make way for its replacement.

Pigments give skin color

Human skin color ranges from pale pink to almost blue-black. Skin is mainly colored by cells that produce a brown pigment called **melanin** (*MEHL uh nihn*). These cells are found in the lowest layer of the epidermis. The amount of melanin these cells produce varies from person to person. Melanin provides some protection from the sun's ultraviolet rays. The more sun you are exposed to, the more melanin your skin produces. That is why pale skin "tans." Two other pigments contribute to skin color: hemoglobin, which gives pale skin a pinkish cast, and carotene, a yellow-orange pigment.

Hair and nail are dead cells

A strand of hair is made of dead, keratin-filled cells that overlap like roof shingles, as shown in **Figure 27.6b**. Hairs grow from specialized epidermal structures called hair follicles. Each hair on your head grows for several years before its follicle enters a resting phase for a few months. While the follicle is dormant, the hair falls out. The length of time that a hair follicle is active before the hair falls out is a genetic trait. If you have inherited a short cycle of hair growth, you will never be able to grow very long hair, no matter how carefully you care for it.

Like hair, nails are also produced by specialized epidermal cells, as described in **Figure 27.6c**.

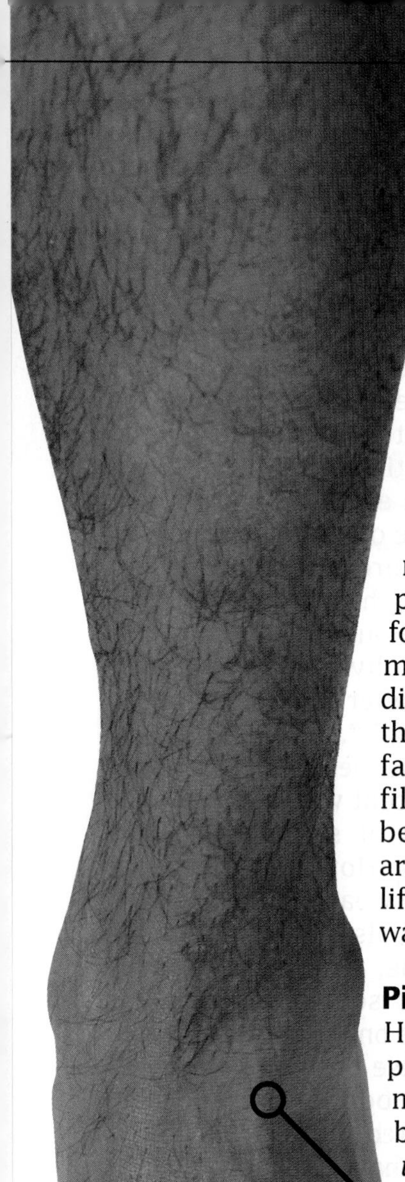

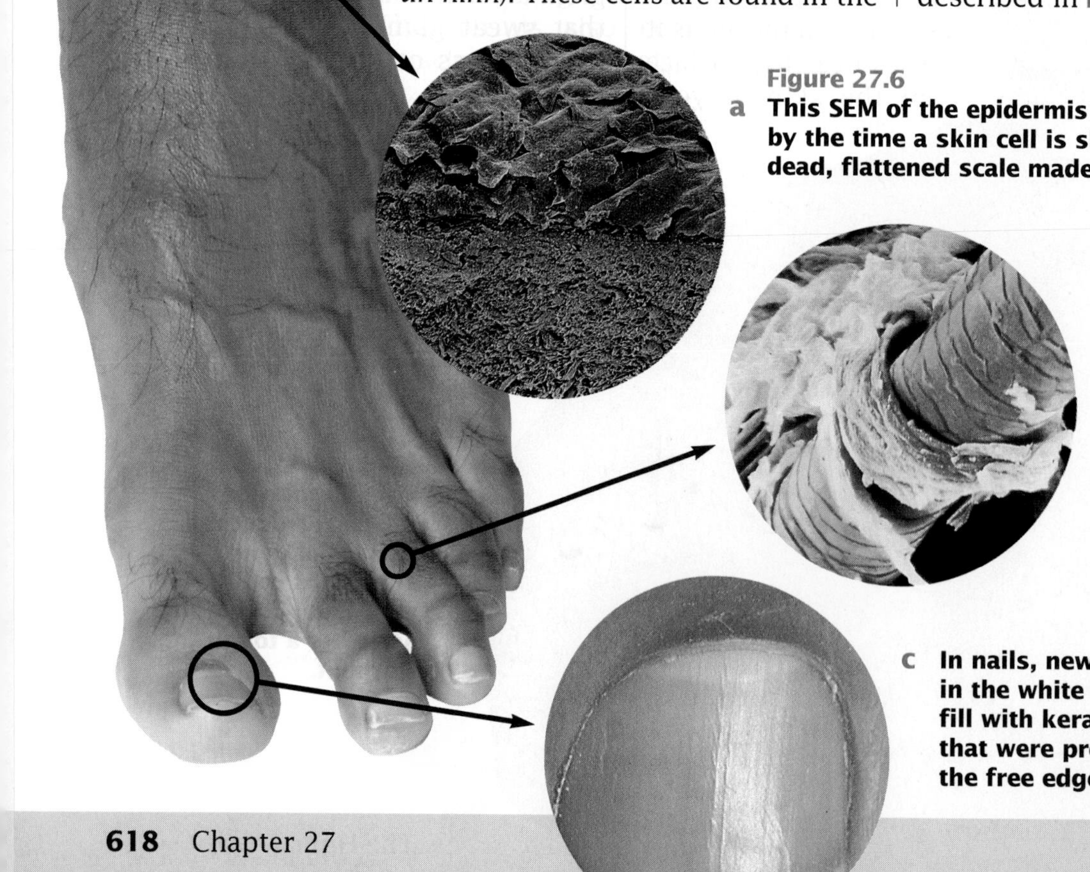

Figure 27.6

a **This SEM of the epidermis reveals that by the time a skin cell is shed, it is a dead, flattened scale made of keratin.**

b **The center of each hair is filled with melanin and air bubbles. Dark hair contains more melanin than blond hair.**

c **In nails, new cells are produced in the white half-moons. These cells fill with keratin while they push cells that were produced earlier toward the free edge of the nail.**

Skin Disorders

Since your skin is the most exposed part of your body, it is often damaged. The damage may be minor—a blister, an insect bite, or a small cut. Other damage is more serious, as in the case of skin cancer. Your skin may also be affected by changes within your body.

Acne is caused by overactive oil glands

Acne is a common, undesirable side effect of adolescence. Normally, oil glands in your skin produce just enough oil to waterproof your hair and to seal moisture into the skin. However, high levels of sex hormones produced in the body during adolescence increase the oil production of the oil glands.

Excessive oil production can clog the oil glands, causing a buildup of oil. This accumulation of oil can appear as a whitehead. If this material is exposed to air, it darkens, forming a blackhead. If the oil builds up to the point that the oil gland actually bursts, the area around the gland becomes red and inflamed. Bacteria from the skin may also infect the damaged area, making the inflammation worse. The pimple that forms is evidence that the body is fighting the tissue damage and the invading bacteria. Although acne cannot be prevented, it can usually be controlled with proper skin care.

Skin cancer is caused by DNA mutations in skin cells

Cancers occur when mutations in DNA cause cells to lose the ability to stop dividing when they are supposed to. The sun's ultraviolet light is known to cause mutations. Years of exposure

Figure 27.7
Sunburn can be painful and can lead to skin cancer and premature aging of the skin. To prevent sunburn, use a sunscreen and avoid direct sunlight.

to sunlight may result in the mutations that can cause cancer of the skin.

The danger of a particular skin cancer depends on the type of skin cell that becomes cancerous. The most common skin cancers have a very high cure rate when detected early. These include cancers that arise from the rapidly dividing cells of the lower layer of the epidermis, or the cells in the first stages of keratin production. About 1 percent of all skin cancers, however, result from mutations in the melanin-producing cells. These cancers, called malignant melanomas, are very dangerous and have a very low cure rate.

The most effective ways to reduce your risk of skin cancer are to minimize your exposure to sunlight and to use a sunscreen. Sunbathers, like the person in **Figure 27.7**, increase their chances of skin cancer. They also age their skin prematurely.

Section Review

❶ **Which part of the skin has the simpler structure, the dermis or the epidermis? Explain your answer.**

❷ **What do nerves, blood vessels, glands, and connective tissue contribute to the skin?**

❸ **How could keeping your skin very clean help minimize the effects of acne?**

❹ **What are two effective ways to reduce your risk of skin cancer?**

YOU MAY THINK OF YOUR SKELETON AS JUST A RIGID FRAMEWORK OF
BONE THAT ENABLES YOU TO SIT, STAND, OR RUN. BONE ITSELF MAY
APPEAR AS LIFELESS AS ROCK. HOWEVER, BONE IS IN FACT A DYNAMIC
CONNECTIVE TISSUE MADE OF LIVING CELLS. THROUGHOUT YOUR LIFE,
THESE CELLS CONTINUE TO PRODUCE THE MANY FIBERS AND MINERALS
THAT FILL THE SPACES BETWEEN THE CELLS.

27.3   Bones

Objectives

| ❶ Draw a diagram of a typical long bone and label its parts. | ❷ Differentiate a fracture from a sprain. | ❸ Discuss the causes and effects of osteoporosis. | ❹ Identify five types of joints in your body and explain the movement each permits. |

Figure 27.8
The skull of an infant has more separate bones than that of an adult. As the child grows, many of these bones will fuse together. The "channels" you see in the infant skull will disappear by the age of four.

Bone Structure and Growth

Throughout your childhood and adolescence, bone cells build more and more bone as your body grows. Compare the size and shape of the infant's skull with the adult's skull in **Figure 27.8**. Even in adults, specialized bone cells continue to break down and rebuild bone tissue. For example, the bone tissue at the end of the thighbone where it joins the knee is completely replaced every six months.

The minerals produced by the bone cells—mostly calcium and phosphorus—make bones strong. These minerals also regulate an amazing variety of activities in your body. For example, nerves and muscles cannot work without the proper level of calcium. Bones act as a warehouse for storing calcium. If the level of calcium falls, bone cells release more calcium into the blood. The storage and release of calcium in your bones help maintain a precise level of calcium inside your body.

The human body contains bones of all shapes and sizes. The long bones of your arms and legs are shaped like cylinders. Curved, flat plates of bone form the part of the skull that protects your brain. Wrists and ankles contain many small bones that look like pebbles, and bones of the face and spine have unusual, irregular shapes.

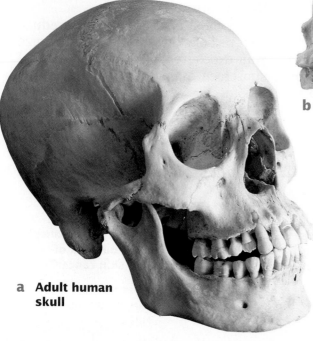

b Infant skull

a Adult human skull

Bone growth begins with cartilage

Bone growth begins long before birth. The basic shape of a long bone, such as an arm bone, is first formed in cartilage. Later, the cartilage cells begin to be replaced by cells that form bone. While you are growing, long bones still have a region of cartilage near each end that allows bones to grow longer. When you reach your adult size, these regions, too, are converted to bone.

Figure 27.9 shows a cross section of the femur in the thigh, a typical long bone. The outer layer of a long bone consists mainly of minerals and mature bone cells "trapped" by the minerals that they have deposited. To form this dense outer layer, bone cells deposit minerals in concentric rings, leaving a canal in the center of each group of rings. Each canal contains a small blood vessel that carries nutrients to the bone cells.

Inside the ends of a long bone are bone cells and minerals with large spaces in between, like a sponge. These spaces, and the entire center of the long middle part of the bone, are filled with **marrow**. The marrow inside long bones produces blood cells in newborns. As you get older, this marrow gradually changes its job to storing fat. In adults, most blood cells are produced in the marrow of flat bones like the sternum, or breastbone.

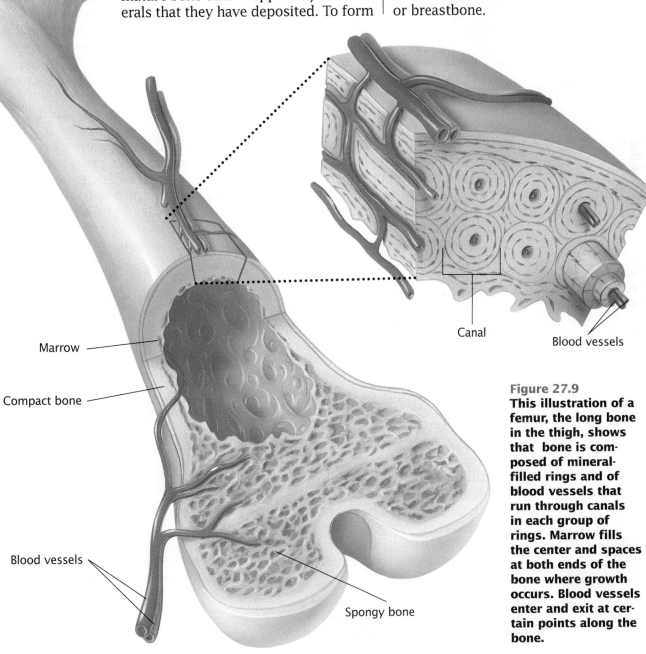

Marrow

Compact bone

Blood vessels

Spongy bone

Canal

Blood vessels

Figure 27.9
This illustration of a femur, the long bone in the thigh, shows that bone is composed of mineral-filled rings and of blood vessels that run through canals in each group of rings. Marrow fills the center and spaces at both ends of the bone where growth occurs. Blood vessels enter and exit at certain points along the bone.

Bones break under stress

The minerals deposited in bone tissue make bones hard and rigid. These characteristics enable bone to protect and support the body. Although the fibers in bones make them much less brittle than a piece of chalk, severe stress placed on a bone may cause it to break. For example, certain types of falls while snow skiing can cause the bones in the lower leg to break at the top of the ski boot. In contact sports like football, a bone may break because a limb has been twisted.

A broken bone is called a **fracture**. A bone fracture may be a simple crack, or the bone may actually break into two or more pieces as shown in **Figure 27.10**. The most serious type of fracture is a compound fracture, in which pieces of broken bone often protrude through the skin. One reason compound fractures are so serious is that they may result in infection of the bone. Because of this, treatment of compound fractures usually includes large doses of antibiotics.

When a bone breaks, there is considerable bleeding caused by damage to the blood vessels in the bone itself and in the surrounding tissues. Healing of a fracture begins as the blood in this swollen region around the broken bone begins to clot. The bone tissue is then rebuilt between the two broken ends of bone in much the same way as bone tissue forms before birth. The rebuilding is not always perfect, however. Many people who have broken a bone can still feel a thicker region of bone where the fracture occurred.

Bone fractures heal at different rates. Large bones heal more slowly than small bones, and the bones of young people heal more quickly than those of older people. Holding the broken ends close to each other and keeping them completely still speeds healing of bones. This is why bone fractures are often treated by encasing the fractured limb in a cast.

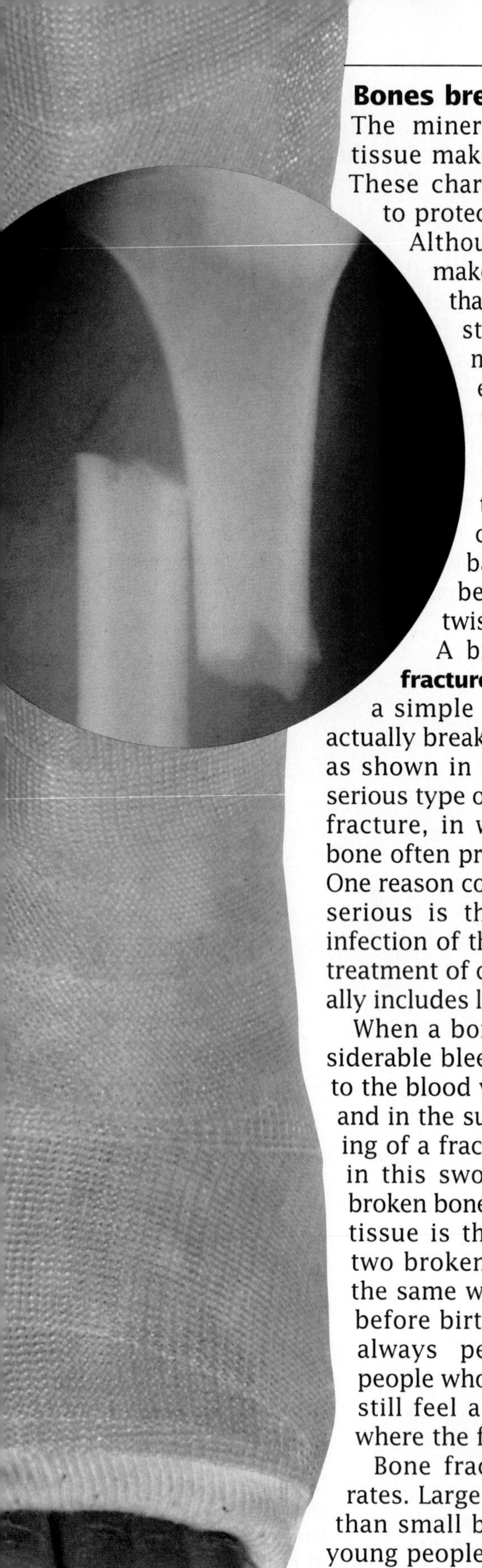

Osteoporosis causes bones to become brittle

As bones grow longer, they also grow thicker and denser. In young adults, the density of bone usually remains relatively constant as bone tissue is broken down and replaced at a steady rate. During middle age, bone replacement gradually becomes less efficient, and bones become less dense. The loss of bone density that may eventually result is called **osteoporosis** (*ahs tee oh puh ROH sihs*). Compare the healthy bone with the bone that has undergone severe mineral loss after the onset of osteoporosis, shown in **Figure 27.11**.

Osteoporosis can cause bones to become light, brittle, and easily broken. In the United States, more than 600,000 bone fractures a year result from osteoporosis. Severe osteoporosis in the bones of the spine often changes the posture of very old people. Although both men and women lose bone as they

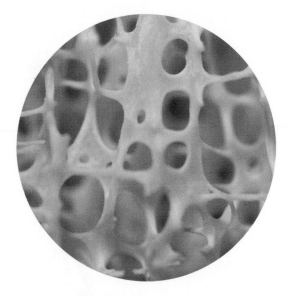

Figure 27.11

a **In healthy bone tissue (magnified 10X in this photograph) minerals are continuously replaced, so the bone remains strong.**

Figure 27.10

The X ray above shows a fracture of the tibia, a long bone in the lower leg. This fracture can be repaired without surgery by setting the fracture with a cast.

age, women are at a greater risk for osteoporosis for two reasons. First, women's bones are usually smaller and lighter than men's bones. Therefore, the loss of the same amount of bone in a man and a woman could result in the woman having thin, fragile bones, while the man's bones might still be quite strong. Second, the production of female sex hormones declines rapidly during menopause. Because sex hormones help to maintain bone density, this decline in hormone production increases the rate of bone loss.

Researchers have discovered that exercise can increase the amount of minerals deposited in bone. For people in their teens and twenties, regular exercise and a balanced diet that includes plenty of calcium can actually increase bone density. And for older people, regular exercise can slow the bone loss that can lead to osteoporosis. That's good news for the tennis player in **Figure 27.12**.

You may think that osteoporosis is something that you won't have to think about for a long time. However, bone density can be increased only during your teens and twenties. Regular exercise and a healthy diet will make you healthier now and will also pay off later. The stronger your bones are now, the less likely you are to be affected by osteoporosis later.

Figure 27.12
This woman is over 60 years old and is still very active. Regular exercise at any age can help slow the bone loss that can lead to osteoporosis.

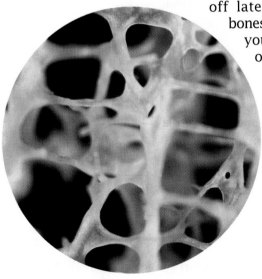

b After the onset of osteoporosis, minerals are not rapidly replaced. As a result, bones become brittle and can break easily.

What Exercise Does for Muscles

Most people can increase the flexibility of their joints, build endurance, and increase their strength and agility through exercise and conditioning. Different kinds of exercise affect your muscles in different ways, as shown in **Figure 27.17**.

Skeletal muscle cells can be fast-twitch or slow-twitch

Every skeletal muscle in your body contains two types of muscle cells. Fast-twitch muscle cells are called into action when a person, such as a pole-vaulter, requires speed and quick movements over a short period. Slow-twitch muscle cells respond more slowly but do not tire as quickly as fast-twitch muscle cells. When an activity such as cycling requires endurance, slow-twitch

muscle cells are working to provide it. You can read more about fast- and slow-twitch muscles in **Figure 27.17a**.

Anaerobic exercises use fast-twitch muscle cells

Some activities demand that your muscles work at high intensity for a very brief period. Such activities are called **anaerobic** (*an uh ROH bihk*) **exercise**. These short bursts of activity use fast-twitch muscle cells. Running up stairs, sprinting, and making a dash for home plate in a baseball game are examples of anaerobic exercise. **Figure 27.17b** explains how anaerobic exercise works.

Muscle cells rapidly use large amounts of energy during anaerobic exercise. As you learned in Chapter 5, cells must have oxygen to produce

Figure 27.17
Each exercise, from cycling to sprinting to weight training, affects muscles in different ways.

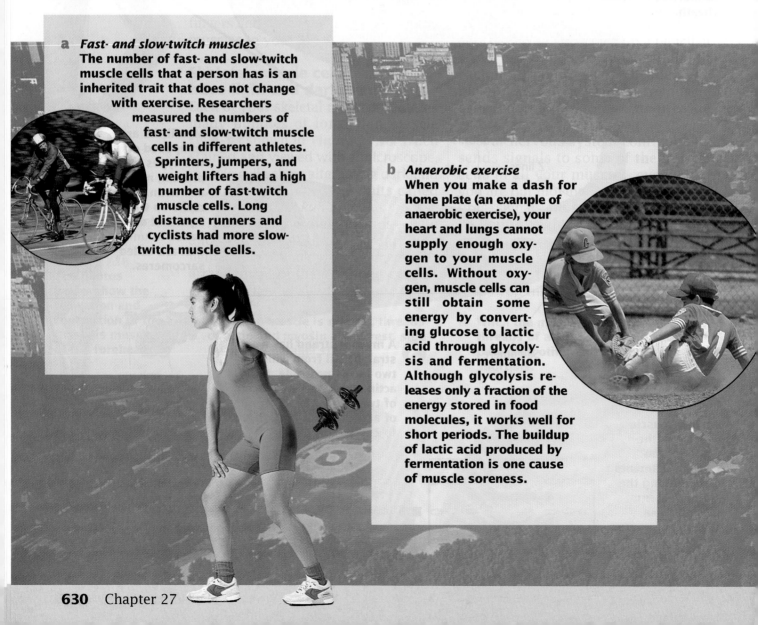

a *Fast- and slow-twitch muscles*
The number of fast- and slow-twitch muscle cells that a person has is an inherited trait that does not change with exercise. Researchers measured the numbers of fast- and slow-twitch muscle cells in different athletes. Sprinters, jumpers, and weight lifters had a high number of fast-twitch muscle cells. Long distance runners and cyclists had more slow-twitch muscle cells.

b *Anaerobic exercise*
When you make a dash for home plate (an example of anaerobic exercise), your heart and lungs cannot supply enough oxygen to your muscle cells. Without oxygen, muscle cells can still obtain some energy by converting glucose to lactic acid through glycolysis and fermentation. Although glycolysis releases only a fraction of the energy stored in food molecules, it works well for short periods. The buildup of lactic acid produced by fermentation is one cause of muscle soreness.

the maximum amount of energy from food molecules.

After you stop anaerobic exercise, your body needs extra oxygen to burn up the excess lactic acid and return your energy reserves to normal. This need for extra oxygen is called oxygen debt. Oxygen debt is the reason you must breathe rapidly and deeply for a few minutes after a hard run.

Aerobic exercises demand a continuous supply of oxygen

Steady, low-intensity exercise like jogging, in-line skating, or swimming laps is called **aerobic** (*ehr OH bihk*) **exercise**. The slow, steady pace of aerobic exercise ensures that your lungs and heart can deliver oxygen to your muscles at the same rate at which the muscle cells are using it. Your muscles can use this continuous supply of oxygen to extract the maximum amount of energy from food molecules and continue the exercise for as long as 20 minutes or more. **Figure 27.17c** describes some of the benefits of aerobic exercise.

Resistance exercises increase muscle size and strength

Aerobic exercise will increase the size and strength of your muscles somewhat, but it won't make you look like a bodybuilder. To significantly increase the size of your muscles, resistance exercises are the most effective form of exercise. **Resistance exercises** are exercises that require exerting a great deal of force against a heavy or immovable object. Weight lifting and using exercise machines that provide resistance are forms of resistance exercises. You can read more about resistance exercise in **Figure 27.17d**.

c *Aerobic exercise*
The amount of oxygen that can be taken into the body and delivered to the muscles can be increased, in most people, through aerobic training. Training strengthens the chest muscles so that more air—and oxygen—enters the body with each breath. Aerobic training helps the heart pump blood more efficiently and increases the number of blood vessels in the muscles. This increased ability to supply oxygen to the muscles results in an increase in endurance.

d *Resistance exercise*
Exercising will not increase the number of your muscle cells, but resistance exercises will increase muscle size because each muscle cell gets bigger. However, resistance exercises are much less effective than aerobic exercises in strengthening the heart and lungs, and in increasing the number of blood vessels in the muscle. Because resistance exercises do not significantly improve the ability of your body to deliver oxygen to your muscles, they do not increase your endurance.

Anabolic steroids are dangerous

Some people are tempted to experiment with anabolic steroids to increase the size of their muscles. Anabolic steroids are powerful synthetic compounds that chemically resemble the male sex hormone, testosterone. The use of steroids may produce serious side effects—including severe acne, cancer, failure to reach full height, heart disease, and psychological disorders. Some males who use anabolic steroids develop female-like breasts and shriveled testes. Females who use these chemicals may develop facial hair, deepening of the voice, and male-pattern baldness. Many of these symptoms of steroid abuse are irreversible. The use of anabolic steroids may have caused death in some cases, as you can see in **Figure 27.18**.

Figure 27.18
Lyle Alzado, former defensive lineman for the Denver Broncos, Los Angeles Raiders, and Cleveland Browns died of a rare type of brain cancer. Alzado attributed the cause of his cancer to steroid use.

"If I win this battle, it will be the best one. . . . If I lose this battle, whatever I've done has been real wrong."
Lyle Alzado
1949–1992

Overuse causes muscle injuries

Overusing muscles by exercising too much or without proper conditioning can lead to muscle injury. A muscle strain, commonly called a "pulled muscle," is the overstretching or even tearing of a muscle. Muscle strain may occur when a muscle is overused or when strenuous exercise is done before doing warm-up exercises. **Tendinitis** is a painful inflammation of a tendon caused by too much friction or stress on the tendon.

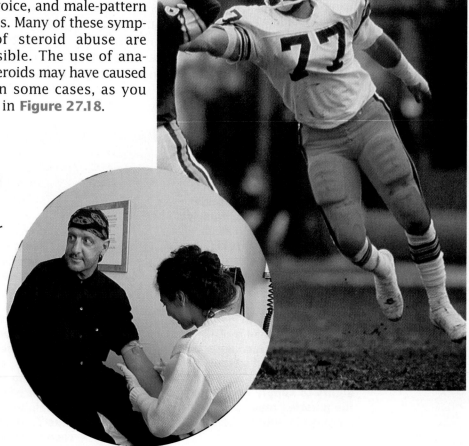

Section Review

❶ **Describe the three main types of muscle.**

❷ **How might your ability to move your arm be affected by an injury to the biceps muscle? To the triceps muscle?**

❸ **If you wanted to increase your endurance, which kind of exercise would be most effective?**

❹ **List three of the dangers associated with anabolic steroids.**

8. Using scissors, remove any tissues covering the muscle. Use a blunt probe to separate the individual muscles from each other without tearing them. On a sheet of unlined paper, draw your chicken wing, showing all of its separate muscles.

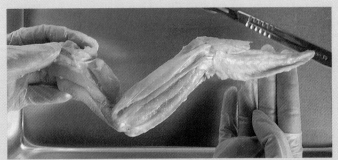

9. Straighten the chicken wing and hold it horizontally above the tray. Pull on each of the muscles and note the movement that each muscle causes. Turn the wing upside down and bend the joints. Again pull on each muscle and note how the bones move. On your drawing, color red each muscle that bends a joint. Color yellow each muscle that straightens a joint.

10. Find as many tendons as possible in your chicken wing. Determine where each tendon connects to a bone. *How will you decide if the structure that you have found is a tendon or a ligament?* With the blue pencil, add tendons to your drawing.

11. Cut crosswise through each muscle. Then follow each half of the muscle to the point it attaches to the bone and cut only the tendon. *How did removing the muscles and tendons from the wing affect the stability of this joint?*

12. Closely examine the joint between the upper wing and the lower wing and identify the ligaments. Using a purple pencil, add ligaments to your drawing.

13. Bend and straighten the joint and observe how the bones fit together. The shiny, white covering of the joint surfaces is made of cartilage. *What characteristics of the joint enable it to move smoothly?*

14. Clean up your materials and wash your hands before leaving the lab. **Follow your teacher's directions for proper disposal of the chicken wing.**

Analysis

1. *Analyzing Structure*
   How does tendon tissue differ from muscle tissue?

2. *Recognizing Relationships*
   Describe how bones and cartilage form joints.

3. *Relating Ideas*
   Explain how muscles, tendons, bones, ligaments, and joints combine to allow the back-and-forth movement of the lower chicken wing.

Thinking Critically

1. Compare the arrangement of bones in a chicken wing with the bones found in the human arm shown in **Figure 27.13.** Describe the similarities and differences between the limbs of these two species.

2. What can you infer about the evolution of limb structure in vertebrates?

3. The chicken is not capable of sustained flight. How do you think the structure of the chicken wing differs from the wing structure of a bird that flies?

The Nervous System

Review

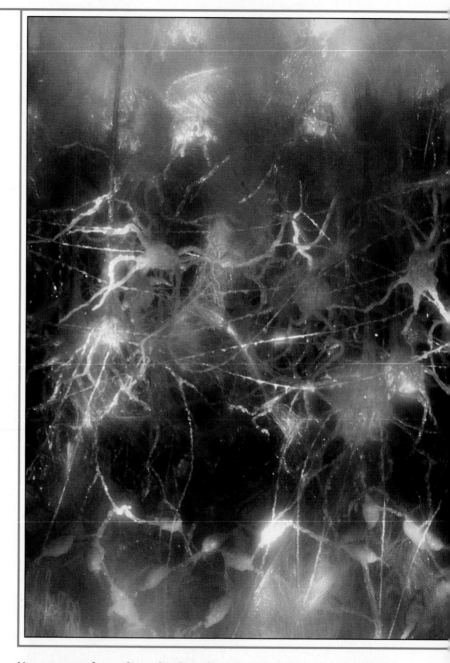

Neurons such as these brain cells are specialized to carry electrical messages. Their membranes are rich in gated ion channels that can open and close, routing messages along nerves in the body.

WHEN YOU TOUCH A HOT STOVE, YOUR HAND INSTANTLY JERKS BACK BECAUSE OF A MESSAGE FROM YOUR SPINAL CORD. WHEN YOU SWAT A FLY, THE MUSCLES OF YOUR ARM CONTRACT QUICKLY BECAUSE OF A MESSAGE FROM YOUR BRAIN. THESE MESSAGES TRAVEL ALONG NERVES, WHICH ARE BUNDLES OF CELLS THAT CARRY ELECTRICAL SIGNALS THROUGHOUT YOUR BODY.

28.1 How a Nerve Carries a Message

Objectives

❶ Describe the structure of a neuron.

❷ Explain how a nerve impulse travels along a myelinated neuron.

❸ Explain how a nerve impulse is carried across a synapse.

❹ Explain how nerves and muscles interact.

The Neuron

Figure 28.1
Neurons such as these brain cells receive information and send it to other parts of the body.

The basic units of communication in the nervous system are cells called **neurons** (*NOO rahnz*). Neurons carry electric signals like the electric wires in a house. When you flip a light switch, an electric current travels through a wire and a light bulb comes on. When your finger touches an uncomfortably hot surface, a message travels along neurons to your spine. Your spine sends out orders along other neurons that cause your muscles to withdraw your hand.

Neurons are specialized to carry signals throughout the body

Although they differ greatly in the details of their structure, neurons with the same basic architecture occur in the brain, the spinal cord, and the nerves that travel throughout the body. As you can see in **Figure 28.1**, short, slender branches called **dendrites** (*DEHN dryts*) extend from one end of the neuron's body. Dendrites are input channels. They receive information from other neurons or from sensory cells. At the other end of the cell body there is a single, long, tubelike extension called an **axon**. Axons are output channels. It is along the axon that the neuron sends out messages to other neurons or directly to the muscles.

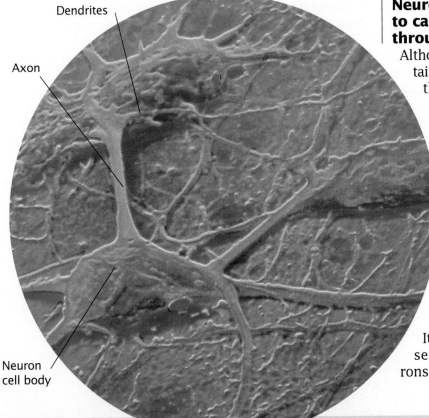

Dendrites

Axon

Neuron cell body

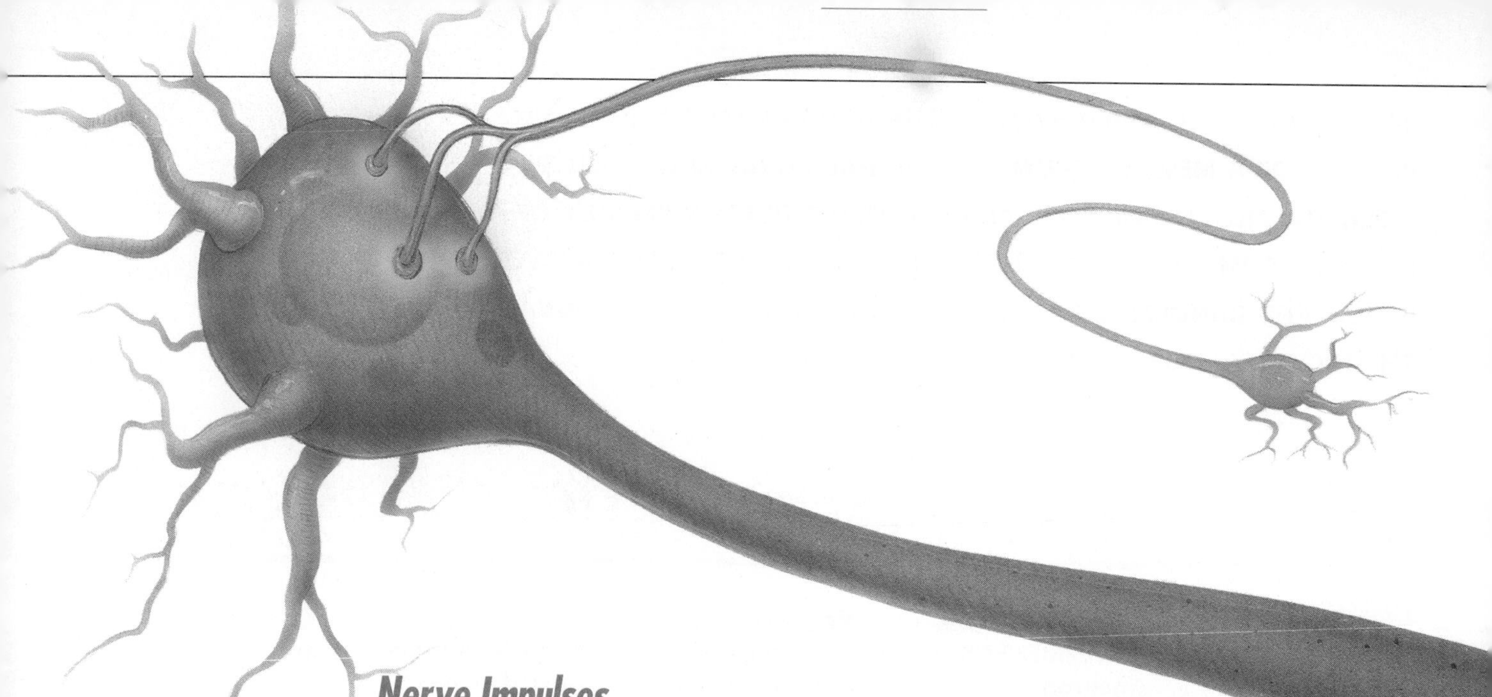

Nerve Impulses

How does a nerve work? A neuron is able to transmit a signal because its plasma membrane contains **gated ion channels** that can open or close. Some of these channels allow only sodium ions to pass through them, others only potassium ions. Information passes along a neuron as an electrical current generated by ions moving in and out across the neuron's plasma membrane through gated channels. Each channel is shaped like a tube of protein sticking through the membrane's lipid bilayer. Some of these channels open and close in response to an electrical voltage; these "voltage-gated" channels are responsible for transmitting the nerve impulse along the neuron. Other channels are opened by chemicals; these "chemical-gated" channels are responsible for passing a nerve impulse from one cell to another.

The inside of a neuron at rest is negatively charged

When a neuron is not carrying an impulse and is at "rest," active transport channels in the neuron's plasma membrane, called sodium-potassium pumps (you met them in Chapter 4), transport sodium ions out of the cell and potassium ions in, as shown in **Figure 28.2a**. Sodium ions cannot easily move back into the cell once they are pumped out, so the concentration of sodium ions builds up outside the cell. Similarly, potassium ions build up inside the cell, although not as many since many potassium ions are able to diffuse out through open channels. You can see that the effect of this is to make the outside of the neuron more positive than the inside. This difference in voltage is called the **resting potential**. The resting potential is the starting point for the transmission of a nerve impulse.

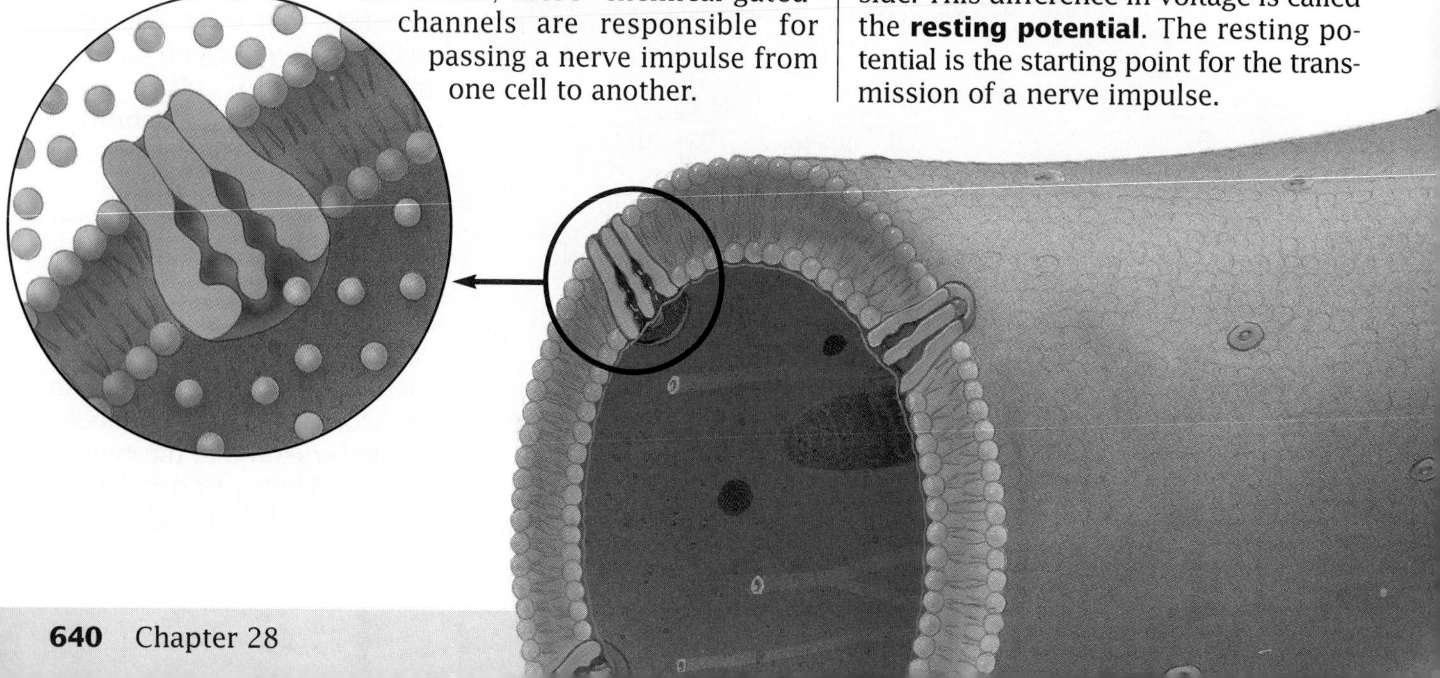

Figure 28.2

a At rest, sodium-potassium pumps keep sodium ions outside the membrane and potassium ions inside. This results in a difference in voltage across the membrane. This difference is called the resting potential.

A nerve impulse disrupts the resting potential

A nerve impulse starts when pressure, or other sensory inputs, disturbs a neuron's plasma membrane, causing sodium channels on a dendrite to open. As a result, sodium ions flood into the neuron from outside, and for a brief moment the inside of the membrane becomes more positive in that immediate area of the cell. This sudden local reversal of electric voltage across the neuron membrane is called an **action potential**.

The sodium channels in the small patch of membrane with the action potential remain open for only about half a millisecond. However, the change in voltage causes nearby voltage-gated sodium ion channels to open, which starts the action potential moving down the neuron, as the opening of the gated channels causes nearby voltage-gated channels to open, like a chain of falling dominoes.

When the action potential has passed, the voltage-gated sodium channels snap closed again and the resting potential is restored. So a nerve impulse is actually the movement of an action potential along a neuron as a series of voltage-gated ion channels open and close. Examine the details of how nerve impulses move along an axon in **Figure 28.2b**.

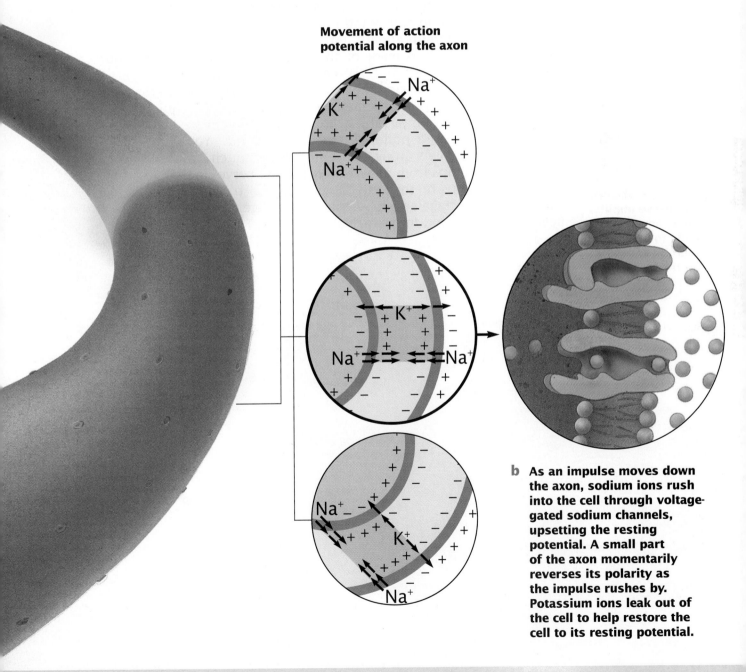

Movement of action potential along the axon

b As an impulse moves down the axon, sodium ions rush into the cell through voltage-gated sodium channels, upsetting the resting potential. A small part of the axon momentarily reverses its polarity as the impulse rushes by. Potassium ions leak out of the cell to help restore the cell to its resting potential.

Myelin sheaths speed signals through the nervous system

A neuron that reaches from the tip of your index finger to your spinal cord is very long. Some vertebrates have neurons that are even longer. Think of one stretching from a giraffe's foot to its spine—a distance of 5 m (16 ft.)! It takes time for a nerve impulse to travel down a long neuron. If the nerve impulse is too slow, you risk burning your finger on the hot stove. Similarly, a giraffe would be in serious trouble if the message from its brain to its legs were slower than a charging lion.

Humans, giraffes, and other vertebrates have evolved ways of speeding up nerve signals. Many of their neurons have axons encased in myelin (*MEY uh lihn*) sheaths. A **myelin sheath**, illustrated in **Figure 28.3**, is made of special cells that wrap their fatty cell membranes in layers around the axon. Between zones of myelin, the neuron is exposed for a short interval. The exposed gap is called a node. When a nerve impulse travels along a myelinated axon, it jumps from one node to the next. Nodes are the only sites with exposed voltage-sensitive channels that can respond to the arrival of

an electrical charge. Jumping from node to node like this is much faster than traveling along the full length of the bare axon.

Destruction of large patches of myelin characterize a disease called multiple sclerosis. In multiple sclerosis, small, hard plaques appear throughout the myelin. Normal nerve function is impaired, causing symptoms such as double vision, muscular weakness, loss of memory, and paralysis.

Some animals, like cockroaches, have neurons with giant axons that are extra thick and carry impulses faster than normal axons. These giant axons run from receptors in the roach's abdomen to its brain. If you swat at a cockroach, the giant axons carry the message that there is a current of air caused by your hand. When the brain gets the message, the cockroach scurries away, quickly evading your descending hand. You can read much more about nerves in the *Tour of a Neuron* on page 643.

Evolution

Why are myelin sheaths considered an evolutionary advancement?

Figure 28.3
Myelin sheaths wrap the axon membrane many times, forming an insulating sheath. The nerve impulse then "hops" from node to node, thereby speeding up the transmission of electrical signals.

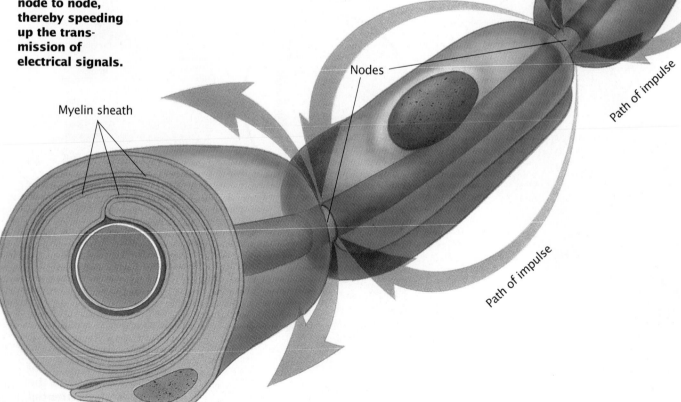

Myelin sheath

Nodes

Path of impulse

Path of impulse

Tour of a Neuron

Nerves are made of cells called neurons. A motor neuron like this one is specialized to transmit messages rapidly to muscle cells.

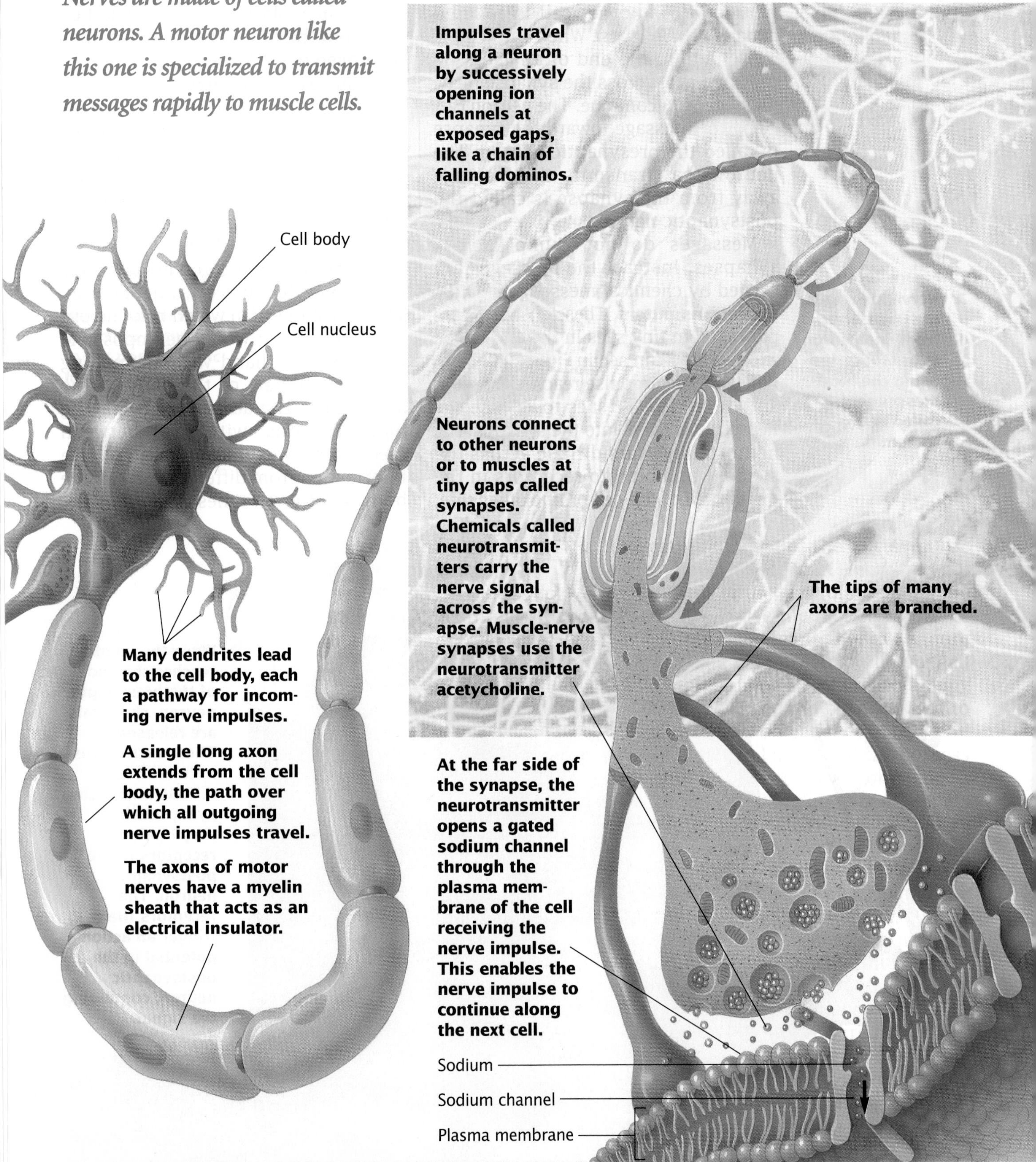

Impulses travel along a neuron by successively opening ion channels at exposed gaps, like a chain of falling dominos.

Cell body

Cell nucleus

Neurons connect to other neurons or to muscles at tiny gaps called synapses. Chemicals called neurotransmitters carry the nerve signal across the synapse. Muscle-nerve synapses use the neurotransmitter acetycholine.

The tips of many axons are branched.

Many dendrites lead to the cell body, each a pathway for incoming nerve impulses.

A single long axon extends from the cell body, the path over which all outgoing nerve impulses travel.

The axons of motor nerves have a myelin sheath that acts as an electrical insulator.

At the far side of the synapse, the neurotransmitter opens a gated sodium channel through the plasma membrane of the cell receiving the nerve impulse. This enables the nerve impulse to continue along the next cell.

Sodium

Sodium channel

Plasma membrane

Evolution of the Brain

Every vertebrate brain has a forebrain, a midbrain, and a hindbrain. As land vertebrates have evolved, the forebrain has become increasingly dominant; in humans it envelops the rest of the brain.

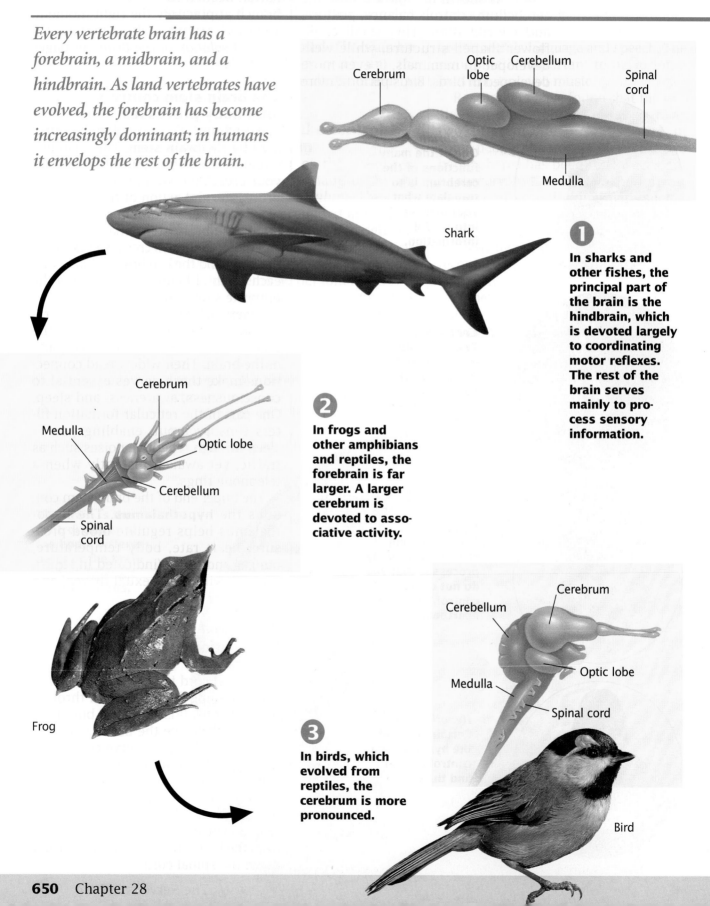

Cerebrum

Optic lobe

Cerebellum

Spinal cord

Medulla

Shark

Cerebrum

Medulla

Optic lobe

Cerebellum

Spinal cord

Frog

Cerebellum

Cerebrum

Optic lobe

Medulla

Spinal cord

Bird

1 In sharks and other fishes, the principal part of the brain is the hindbrain, which is devoted largely to coordinating motor reflexes. The rest of the brain serves mainly to process sensory information.

2 In frogs and other amphibians and reptiles, the forebrain is far larger. A larger cerebrum is devoted to associative activity.

3 In birds, which evolved from reptiles, the cerebrum is more pronounced.

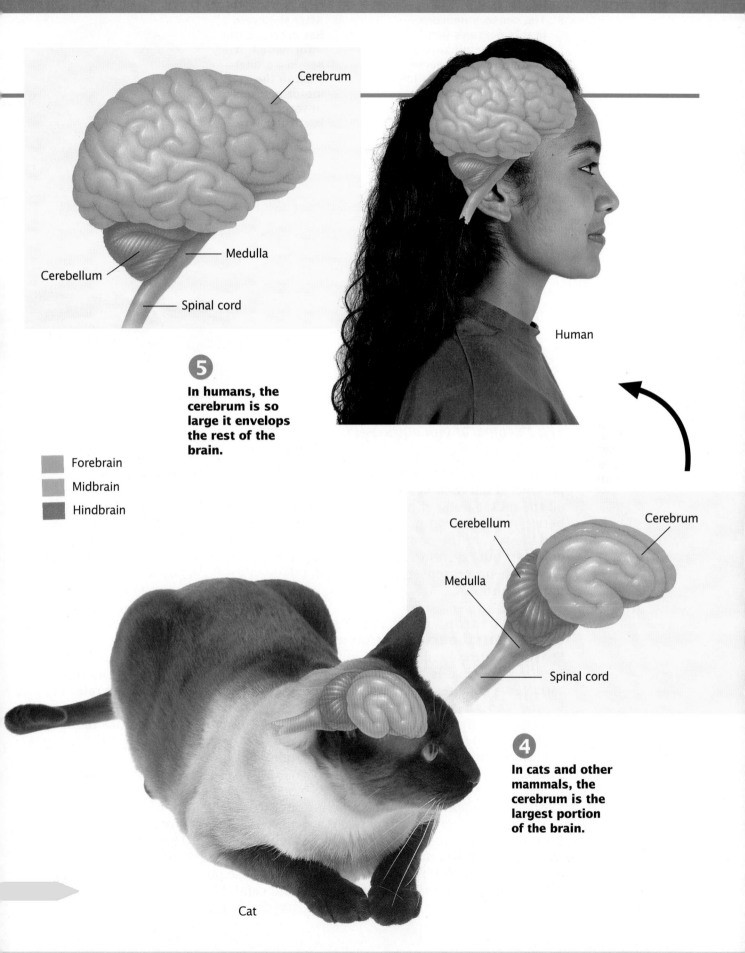

Cerebrum

Medulla

Cerebellum

Spinal cord

Human

5

In humans, the cerebrum is so large it envelops the rest of the brain.

Forebrain

Midbrain

Hindbrain

Cerebellum

Cerebrum

Medulla

Spinal cord

4

In cats and other mammals, the cerebrum is the largest portion of the brain.

Cat

The Nervous System   **651**

Figure 28.9

a The sensory neurons in this person's hand react to stimuli such as heat and initiate nerve signals that are sent to the spine and brain.

b After the brain has received the information, it sends a signal back to the motor neurons to react in a particular way.

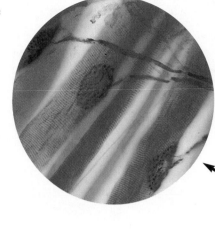

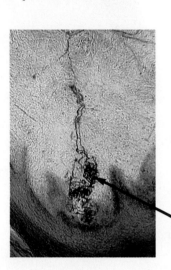

The Peripheral Nervous System

All of the nervous system outside the spinal cord and brain is known as the **peripheral nervous system**. It carries all the messages sent back and forth between the central nervous system and the rest of the body. The peripheral nervous system has two main types of neurons: sensory neurons and motor neurons.

Sensory neurons relay signals to the central nervous system

Sensory neurons tell the central nervous system what is happening. They carry nerve impulses from sense organs to the central nervous system, as shown in **Figure 28.9a**.

Sense organs are organs that react to changes inside and outside the body. They detect many different things, including changes in blood pressure, strain on ligaments, and smells in the air. Sense organs include complex organs, such as eyes and ears. Your skin has many small structures called sensory receptors, which enable you to sense pressure and temperature.

Motor neurons deliver information to muscles and glands

Motor neurons are partners of sensory neurons. Motor neurons carry information from the central nervous system to a muscle or gland, as shown in **Figure 28.9b**. They act on the information delivered by the sensory neurons. If your eyes see a runaway truck speeding toward you, the central nervous system sends messages through motor neurons to glands that secrete adrenaline. The adrenaline increases your heartbeat and breathing rate. The central nervous system also sends messages through motor neurons to many muscles, which contract and get the body out of there—fast!

In each segment of the spine, sensory nerves go into the cord and motor nerves come out of it. The motor nerves control most of the muscles below the head. This is why injuries to the spinal cord often paralyze the lower part of the body. A muscle is paralyzed and cannot move if its motor neurons are damaged.

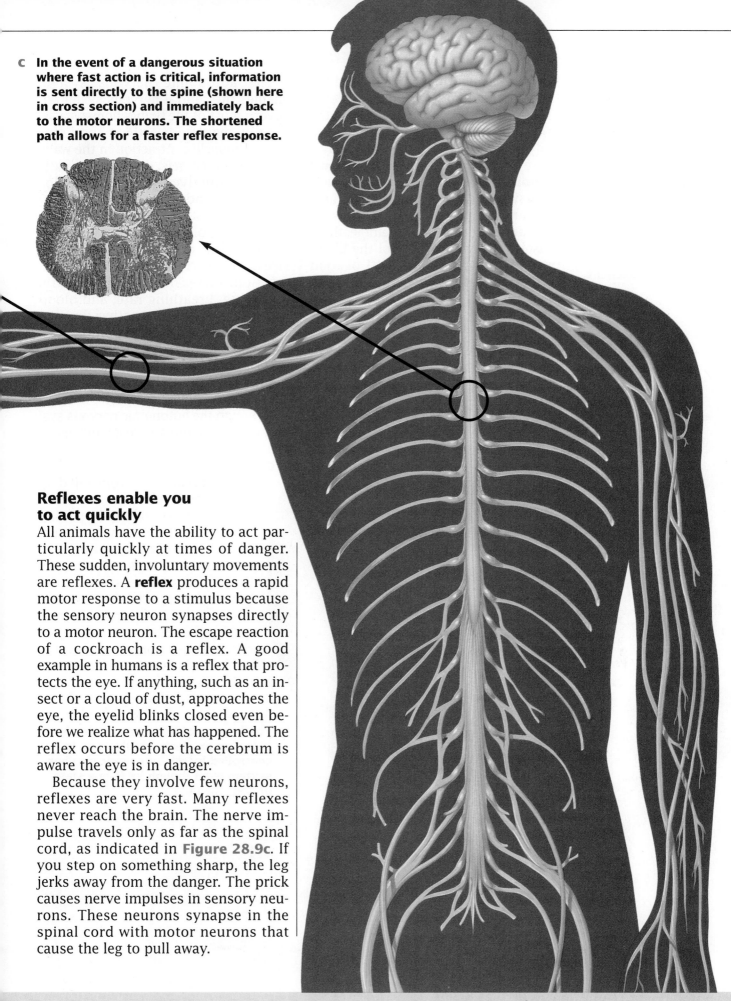

c In the event of a dangerous situation where fast action is critical, information is sent directly to the spine (shown here in cross section) and immediately back to the motor neurons. The shortened path allows for a faster reflex response.

Reflexes enable you to act quickly

All animals have the ability to act particularly quickly at times of danger. These sudden, involuntary movements are reflexes. A **reflex** produces a rapid motor response to a stimulus because the sensory neuron synapses directly to a motor neuron. The escape reaction of a cockroach is a reflex. A good example in humans is a reflex that protects the eye. If anything, such as an insect or a cloud of dust, approaches the eye, the eyelid blinks closed even before we realize what has happened. The reflex occurs before the cerebrum is aware the eye is in danger.

Because they involve few neurons, reflexes are very fast. Many reflexes never reach the brain. The nerve impulse travels only as far as the spinal cord, as indicated in **Figure 28.9c**. If you step on something sharp, the leg jerks away from the danger. The prick causes nerve impulses in sensory neurons. These neurons synapse in the spinal cord with motor neurons that cause the leg to pull away.

The Autonomic Nervous System

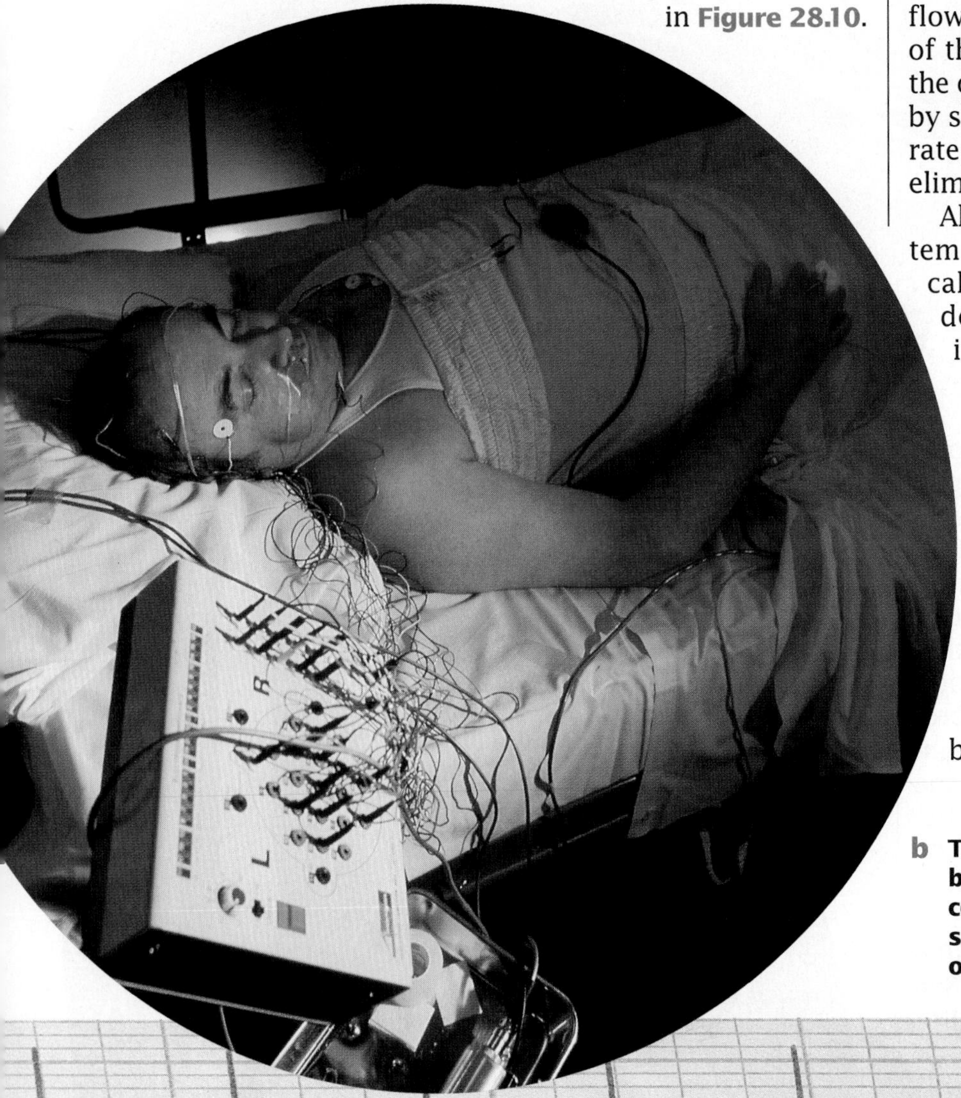

Figure 28.10

a **This sleeping man may appear to be inactive, but his brain is really quite active. Not only is he dreaming, but his brain is also maintaining his vital body functions.**

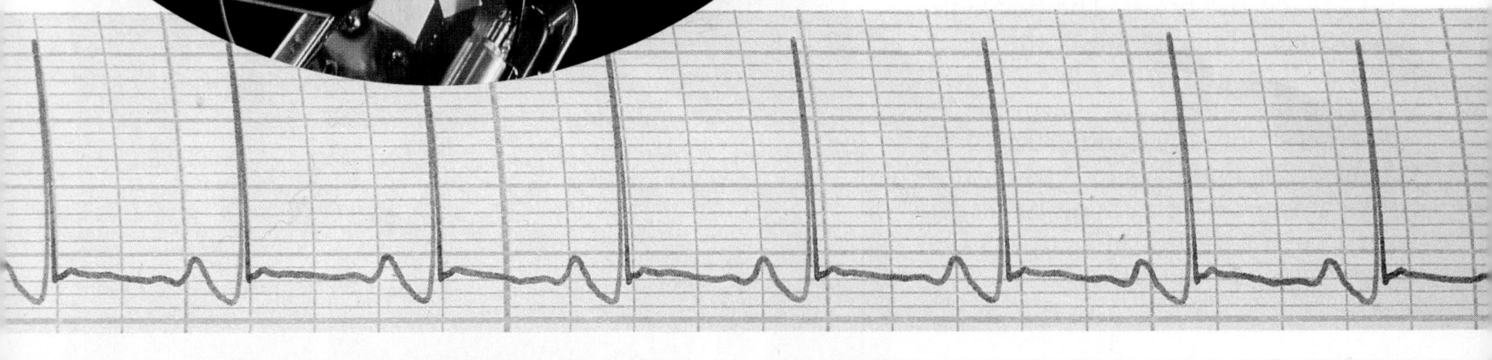

Some motor neurons are active all the time, even when the body is asleep. These neurons carry messages from the central nervous system that keep the body going even when it is not active. These are the neurons of the **autonomic nervous system**. The autonomic nervous system carries messages to muscles and glands that usually work without our noticing. For example, these muscles and glands control the blood pressure and the movement of food through the digestive system even during sleep, as shown in **Figure 28.10**.

The autonomic nervous system enables the central nervous system to govern most of the body's homeostasis. It helps regulate heartbeat and helps control muscle contraction in the walls of the blood vessels, digestive, urinary, and reproductive tracts. It also helps stimulate glands to secrete tears, mucus, and digestive enzymes.

One division of the autonomic nervous system dominates in times of stress. It controls the "fight-or-flight" reaction, increasing blood pressure, heart rate, breathing rate, and blood flow to the muscles. Another division of the autonomic nervous system has the opposite effect. It conserves energy by slowing the heartbeat and breathing rate, and by promoting digestion and elimination.

Although the autonomic nervous system can carry out its tasks automatically, it is not completely independent of voluntary control. For instance, breathing is controlled by the autonomic nervous system, but one can decide to stop breathing for a short time. However, any voluntary control of the autonomic nervous system that endangers life disturbs homeostasis of the brain tissue, causing unconsciousness. The autonomic nervous system then takes over again and restores normal functions. This is why you cannot hold your breath indefinitely.

b **The beating of his heart, as shown below on an electrocardiogram, is controlled by the autonomic nervous system even though he is in a state of deep sleep.**

How Scientists Study the Brain

Figure 28.11
Using new imaging techniques, scientists have been able to study the brain while it does certain jobs, like recalling information. In the images below, the colored areas are the most active.

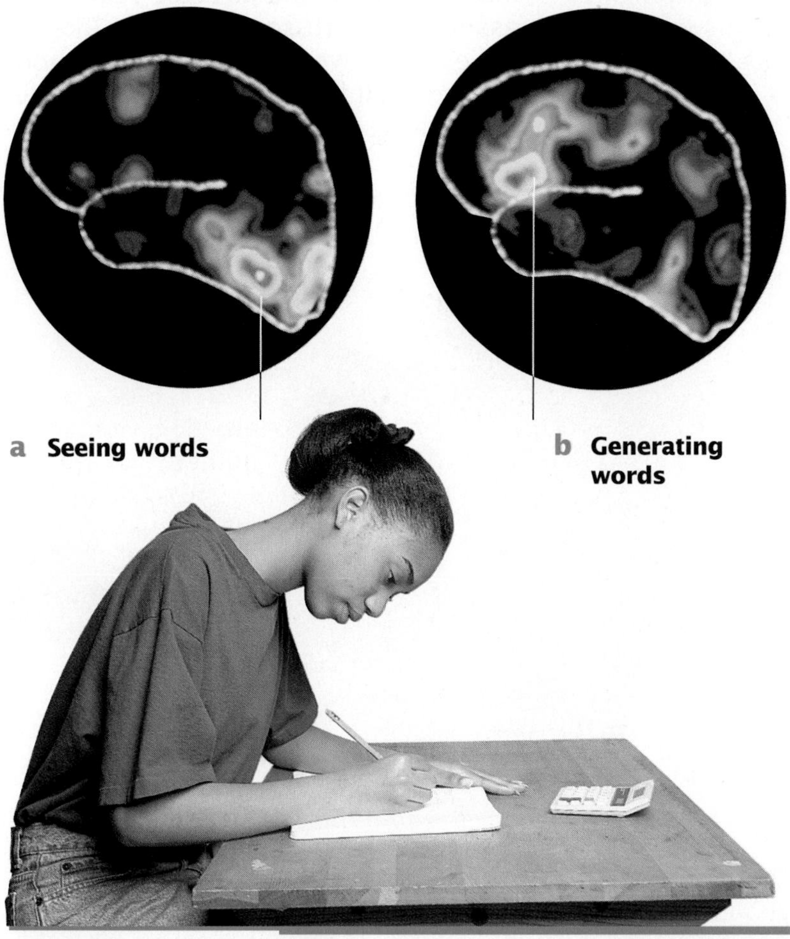

How do we know the functions of the different parts of the brain? The different regions of the brain have been investigated in several ways. Researchers can map the activities of each area by stimulating different parts of the brain with electricity. Or they may observe which functions of the body or thought processes are lost when parts of the brain are deliberately or accidentally destroyed. A newer method is to find out which parts of the brain use the most ATP while a person performs some act, such as recalling information, as shown in Figure 28.11. The neurons that use the most ATP are the most active.

a Seeing words

b Generating words

Learning differs from memory

Learning and memory are brain functions that are especially well developed in humans but remain poorly understood by scientists. Learning and memory are not the same thing. As young children we learn many skills, such as talking, but we do not remember learning them.

Memory lasts for different lengths of time. When you take notes in class, you often remember what the teacher says only long enough to write it down. This is short-term memory, which can last for a few seconds or a few hours. We use it for things like remembering to do an errand or cramming for a test. If information is stored for any length of time, it is transferred to long-term memory. Here it may remain for life. Much long-term memory is stored subconsciously, meaning that we do not know we know it.

Scientists still do not fully understand sleep

Sleep is another mysterious function of the nervous system. Even after 50 years of research, scientists still have no solid ideas about why you must sleep each night, nor about why sleeping affects your temper, alertness, and emotions. And scientists do not understand why some people need more sleep than others. Although you may think of sleep as giving the brain a rest, the brain is active during sleep periods. It must maintain breathing and the rate of heartbeat. Scientists hypothesize that sleep permits the brain to restore biochemical functions depleted by the day's activities and to process and reorganize information taken in during the day.

Section Review

❶ Summarize the functions of each division of the nervous system.

❷ Discuss the different functions of the cerebrum.

❸ Compare the functions of the central nervous system, sensory neurons, and motor neurons.

❹ What are two ways scientists can study the brain?

James Moreland:
National Pharmaceutical Sales Manager

How I Became Interested in Pharmaceuticals

"Though I majored in political science and history both in high school and in college, science has always intrigued me. I took the required courses—biology, chemistry, and physics. When I got out of college, I started reading about life science. That's how I got into the industry. I liked the idea of helping hospitals, doctors, and patients with medications to relieve illness.

"My pharmaceutical company put me through a very extensive training program. I studied anatomy, pharmacology, chemistry, and biology. I learned how drugs are put together, how they interact in the body, what they're good for, and what they're not good for. We also looked at the adverse reactions associated with some drugs. I like to believe that we provide products and services that improve the quality of life for patients."

James is married and has two children. His son, Kareem, is 20 and his daughter, Jessica, is 10.

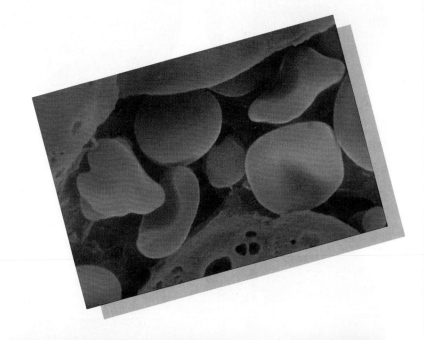

Name:	**James Moreland**
Home:	**Buckingham, Pennsylvania**
Employer:	**Ortho Biotech, division of Johnson & Johnson, Raritan, NJ**
Personal Traits:	**• People-oriented**
	• Likes to travel
	• Articulate
	• Curious

The Excitement of the Pharmaceutical Industry

"The company I work for manufactures genetically-engineered drugs. For example, we make Orthoclone OKT-3. This drug is especially helpful to patients about to receive a kidney transplant. When the body receives a new organ, the immune system initially rejects the organ. The body knows the organ is foreign and builds antibodies for that organ. Orthoclone OKT-3 shuts down the immune system for two weeks, giving the new organ an opportunity to get adjusted to the new body. Thus when the immune system comes back on, it doesn't recognize the new organ as foreign. There is a much better chance of accepting that organ and not experiencing rejection, which could be complicated and even result in death. This same drug is also useful with liver and heart transplants.

"Another major product I am responsible for is a synthetically-produced version of erythyropoietin, a hormone produced by the kidneys that stimulates the production of red blood cells. Red blood cells carry oxygen to all parts of the body. All people need red blood cells in order for their heart, brain, and lungs to function. If your body does not produce enough red blood cells, you will become anemic. You begin to feel tired and cannot function normally.

"Our synthetically-produced version of erythyropoietin is helpful for people on dialysis, a process that purifies the blood during kidney failure. Dialysis patients can become very anemic. They can take synthetically-produced erythyropoietin to compensate for their inability to naturally produce erythyropoietin. As a result, the amount of blood and iron in the body is increased, thereby providing the patient with energy and enabling him or her to function at a near normal level.

"Of the approximately 120,000 patients on dialysis therapy, 90,000 are on this drug. It makes a tremendous difference in their treatment.

"Anemia also occurs in patients who are being treated for AIDS or HIV infection. As part of their therapy, they take an antiviral drug called AZT in order to arrest the virus. As a result of taking AZT, their bodies become highly immune-suppressed. AZT can affect the bone marrow, which is responsible for generating red blood cells. The patient will probably become anemic because the body is not producing enough red blood cells.

"Erythyropoietin therapy can increase the production of red blood cells, let the patient continue his or her AZT drugs, and even allow the doctor to increase the AZT dosage."

Career Path

High School:
- Language Arts
- Math
- Biology

College:
- Political Science
- History
- Biology

VIVID REPORTS ABOUT YOUR ENVIRONMENT STREAM INTO YOUR BRAIN THROUGH YOUR EYES, EARS, SKIN, NOSE, AND MOUTH. THE FACT IS, THERE ARE MORE THAN A DOZEN DIFFERENT TYPES OF SENSORY CELLS THAT DETECT CHANGES OUTSIDE AND INSIDE YOUR BODY. TOGETHER, THEY HELP YOU INTERACT WITH THE WORLD THAT SURROUNDS YOU.

28.3 *The Sense Organs*

Objectives

❶ List three stimuli to which sense organs react.

❷ Explain how the inner ear helps you maintain balance.

❸ Explain how your ear detects sound.

❹ Compare and contrast the functions of rods and cones.

Sensing Internal Information

Many receptors that enable the body to receive information from the environment are located in highly specialized organs called **sense organs**. The most familiar sense organs are the eyes, ears, nose, and tongue. These organs have receptors that can respond to stimuli by producing nerve impulses in a sensory neuron. Your body's receptors detect stimuli such as light, heat, or pressure. The receptor converts the energy of a stimulus into electrical energy that can travel in the nervous system.

Receptors inside the body inform the central nervous system about the condition of the body. For instance, temperature receptors throughout the body detect changes in temperature. This information travels to the hypothalamus, which helps control body temperature.

Receptors in the joints, tendons, and muscles detect changes in the position of parts of the body. Receptors in the ear tell us the position of the head. Together, all these receptors inform the brain where the body is in three dimensions. This knowledge is essential to move freely and maintain your balance, like the man in **Figure 28.12.**

Receptors in the inner ear sense position in space

The ear is really two sense organs in one. It not only detects sound waves,

Figure 28.12

a **In-line skating requires balance and coordination. Without sense organs in this man's inner ear telling him how to balance, where his legs are, and where he is in space, he would be in big trouble.**

b The semicircular canals are three looped rings in the inner ear.

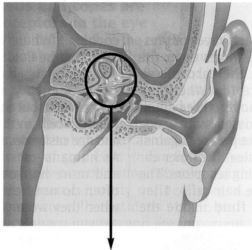

c Each canal has a swelling at one end, which is lined with receptor cells inside. Tiny hairs protrude from the receptor cells into a jellylike fluid.

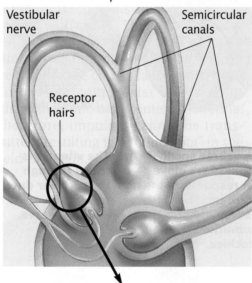

Vestibular nerve

Semicircular canals

Receptor hairs

d When your head is upright, the fluid is still and the hairs are upright.

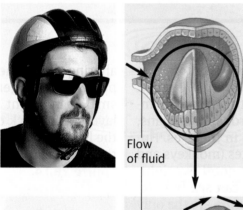

Flow of fluid

e Any movement causes the fluid to slide over the hairs and bend them in the opposite direction. The hair cells send messages to the brain about your position in space.

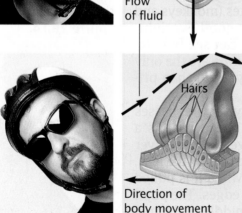

Hairs

Direction of body movement

it also senses the position of the head, whether it is still, moving in a straight line, or rotating.

The receptor cells for your sense of position are located in canals and chambers of the inner ear, shown in **Figure 28.12c**. Each receptor cell has a bundle of tiny hairs protruding from it. In the semicircular canals, the receptor cells are covered with a jellylike cap. Each canal is full of fluid and contains hair cells. When the fluid moves, it stimulates the hair cells, as shown in **Figure 28.12d-e**. The three canals lie in three different planes, so they can detect head movement in any direction.

The semicircular canals detect when the head changes speed or direction. They do not react if the head rotates at a constant speed or moves in a straight line. Traveling in a car or airplane at constant speed gives no sense of movement. The semicircular canals detect movement only if the car or airplane turns. This is because when the head is moving in a straight line, the fluid in the canals does not move. When the head changes direction, however, the fluid tends to keep going in its original direction, and so it bends the hair cells.

The brain interprets stimuli from sense organs

Messages from sense organs to the central nervous system are all in the form of nerve impulses. How does the brain know whether an incoming nerve impulse indicates a shining light, a melodious sound, or a familiar odor? This information is built into the "wiring" in the pathways of neurons that synapse with each other, and into the location in the brain where the information arrives. The brain "knows" it is responding to light when it gets a message from a sensory neuron that comes from light receptor cells. Neurons from the eye may send impulses for other reasons. For instance, when you press your fingertips gently against the corners of your eyes, you "see stars." This is because the brain treats any impulse from the eye as light, even though the eye received no light.

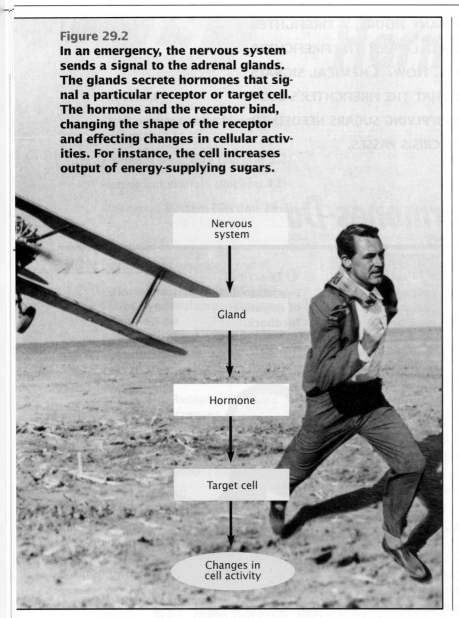

Figure 29.2
In an emergency, the nervous system sends a signal to the adrenal glands. The glands secrete hormones that signal a particular receptor or target cell. The hormone and the receptor bind, changing the shape of the receptor and effecting changes in cellular activities. For instance, the cell increases output of energy-supplying sugars.

Nervous system

↓

Gland

↓

Hormone

↓

Target cell

↓

Changes in cell activity

The effects of hormones last longer than nerve impulses

Imagine that you start to cross a street and suddenly a car speeds around the corner. Instantly, nerve impulses deliver the message—"Danger!"—from your eyes to your brain. In a matter of seconds, you sprint back to the safety of the curb. Nerve impulses prompt a nearly instantaneous response to a change in the environment.

Hormones, on the other hand, are released more slowly than nerve impulses, but their effects usually last longer. Imagine that the reckless driver crashes into another car and then speeds away. Within seconds, hormones from your adrenal glands cause your heart to beat faster, pumping extra oxygen through your bloodstream. Your brain becomes more alert; you might think quickly enough to look at and remember the license plate number of the hit-and-run driver. You race down the street to see if anyone is injured. After this is over, you may find yourself shaking a little. **Figure 29.2** summarizes the events involved in an emergency.

The effects of some hormones will probably last 10 or 20 minutes. But when the body has to deal with an emergency that lasts longer, other adrenal hormones are released. These hormones can maintain extra energy levels for several hours after an emergency occurs.

The Hypothalamus-Pituitary Connection

The endocrine system and the nervous system are the two main control systems of the body. They are so closely linked that they often are considered a single system—the **neuroendocrine system**. In Chapter 28 you learned that the hypothalamus is the part of your brain that regulates body temperature, breathing, hunger, and thirst. The hypothalamus can also be considered the master switchboard of the endocrine system. Your hypothalamus is continuously checking conditions inside your body. Are you too hot or too cold? Are you running out of fuel? How about your blood pressure? Is it too high or too low? If your internal environment starts to get out of balance, your hypothalamus has several ways to set things right again. For example, the hypothalamus can send a nerve signal to another part of the brain—the medulla—to speed up or slow down your heart rate. The hypothalamus also sends out commands in the form of hormones, thus acting like an endocrine gland.

The pituitary gland secretes and stores hormones

All of the hormones produced by the hypothalamus move through a slender thread of tissue to the **pituitary** *(puh TOO uh tehr ee)* **gland**. The pituitary gland produces at least six different hormones in response to the hormones released from the hypothalamus. As described in **Figure 29.3 a–b**, two of these hormones, growth hormone and prolactin, have direct effects on tissues in the body. Growth hormone, for instance, stimulates protein synthesis and cell division in target cells. It also profoundly influences the growth of cartilage and bone. Because these pituitary hormones are controlled by "releasing" hormones from the hypothalamus, the brain exercises direct control over the endocrine system.

The pituitary gland secretes four other hormones that control the activities of a number of other endocrine glands such as the thyroid gland and the adrenal glands. Some of these glands are discussed in more detail later in this chapter.

In addition to producing hormones, the pituitary gland stores hormones made in the hypothalamus. Two of these hormones, oxytocin *(ahk see TOHS ihn)* and vasopressin *(vay so PREHS ihn)*, discussed in **Figure 29.3c–d**, are released from the pituitary gland when needed by the body.

Figure 29.3
The pituitary gland, located at the base of the hypothalamus, was formerly called the "master gland" because so many of its hormones regulate other endocrine functions.

a *Growth hormone* stimulates general growth of the body, particularly of the skeleton. When too little growth hormone is produced, a condition called dwarfism can result. With too much growth hormone, a condition called gigantism can occur.

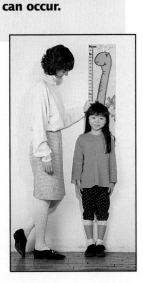

b *Prolactin* affects breast tissue in nursing mothers, causing glands in the breasts to produce milk.

c *Oxytocin* is the hormone that causes contractions of the uterus during labor.

d *Vasopressin* causes the kidneys to form more concentrated urine, thereby conserving water in the body.

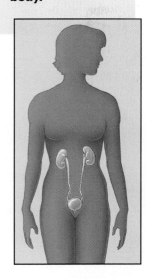

Regulating Hormone Release

Because the body produces more than 30 hormones, it must be able to regulate the release of these hormones. For example, endocrine glands usually do not secrete their hormones at a constant rate. A nerve impulse may cause a gland to increase or decrease its rate of hormone production. In most cases, however, chemical signals, including other hormones, regulate the function of the endocrine system through **negative feedback**. Negative feedback is a process by which a change in an environment causes a response that returns conditions to their original state. For example, negative feedback enables a thermostat to keep your classroom at a constant temperature. Suppose the thermostat is set for 22°C (72°F). The temperature in the classroom is not always exactly that temperature. If it is a chilly winter day and the room temperature falls below 22°C, the furnace will come on and blow warm air into the room until the temperature rises. Then the furnace shuts off, and the temperature slowly begins to fall. Just about the time you start feeling a little chilly—at about 20°C or 21°C (68°F or 69°F)—the furnace switches on again.

Parathyroid glands regulate levels of calcium in the blood

The way a thermostat maintains the temperature in a room is similar to the way most endocrine glands help a body maintain homeostasis. For example, the parathyroid glands, four tiny oval glands embedded in the back of the thyroid, secrete a hormone that regulates the level of calcium in the bloodstream. Calcium is a mineral necessary for proper growth, healthy teeth and bones, nerve function, and muscle contraction. While some calcium is stored inside cells, most of it is stored in bones. When the level of calcium in the blood drops even slightly, the parathyroid glands are stimulated to secrete parathyroid hormone. This hormone causes the release of calcium from bone into the blood. When the level of calcium in the blood rises to a certain level, the parathyroid glands halt production of the hormone. The diagram in **Figure 29.4** summarizes these events.

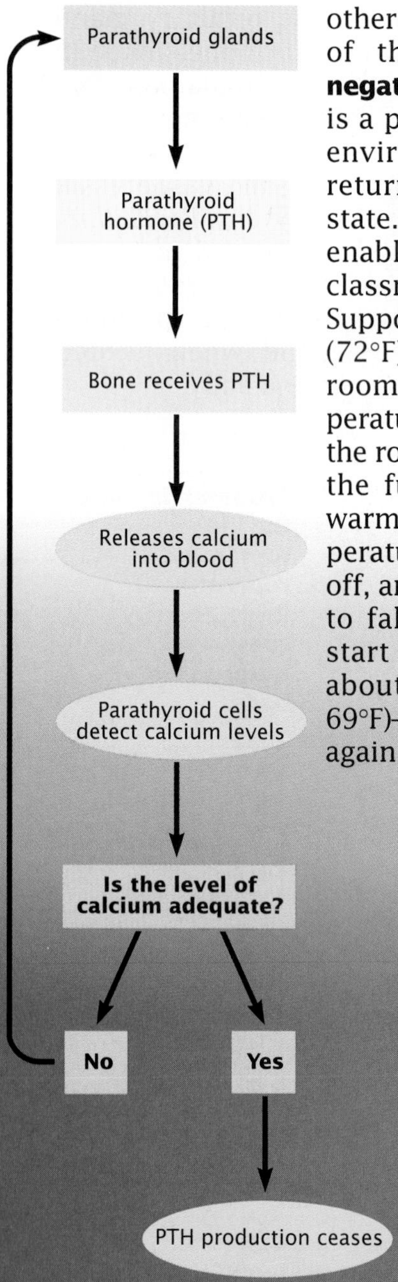

Figure 29.4
Calcium is necessary for the growth and maintenance of the strong, healthy bones of this skier. When calcium levels in the blood drop, a negative feedback process, summarized in the diagram at left, causes calcium levels in the blood to rise.

Parathyroid glands

↓

Parathyroid hormone (PTH)

↓

Bone receives PTH

↓

Releases calcium into blood

↓

Parathyroid cells detect calcium levels

↓

Is the level of calcium adequate?

No   Yes

↓

PTH production ceases

Negative feedback also maintains body temperature

Imagine that you are waiting for the bus on a windy winter day, shivering in the cold. Even though you are not aware of it, your hypothalamus is monitoring your body temperature. As you get colder, your hypothalamus produces a hormone that stimulates your pituitary gland. Your pituitary gland then secretes a hormone that stimulates the thyroid gland. The thyroid gland, in turn, secretes thyroxine, a hormone that speeds up the rate of metabolism in your body, keeping your body temperature at a normal level. When the thyroxine level reaches a certain point, the secretion of the thyroid-stimulating hormone by the pituitary is reduced, and the thyroid stops secreting thyroxine. This negative feedback process is summarized in **Figure 29.5**.

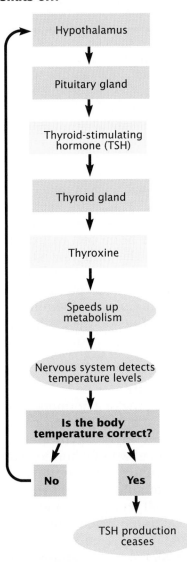

Hypothalamus

↓

Pituitary gland

↓

Thyroid-stimulating hormone (TSH)

↓

Thyroid gland

↓

Thyroxine

↓

Speeds up metabolism

↓

Nervous system detects temperature levels

↓

Is the body temperature correct?

No → Yes

↓

TSH production ceases

Section Review

1. **Define the term "hormone" and name four endocrine glands in the body.**

2. **Why are the nervous system and endocrine system often called the neuro-endocrine system?**

3. **Explain how the parathyroid glands regulate levels of calcium in the blood.**

4. **Explain how the hypothalamus helps maintain a constant body temperature.**

HORMONES ARE CARRIED THROUGHOUT THE HUMAN BODY IN THE BLOODSTREAM. ONLY TARGET CELLS ACTUALLY RESPOND TO THE MESSAGE THAT HAS BEEN SENT. AFTER A HORMONE ARRIVES AT ITS TARGET, THE HORMONE CAUSES CERTAIN ACTIONS TO OCCUR. TWO TYPES OF HORMONES CAUSE CHANGES IN CELLS IN DIFFERENT WAYS, AS YOU WILL LEARN IN THIS SECTION.

29.2   *How Hormones Work*

Objectives

❶ Describe how a target cell responds to a hormone.

❷ Explain how a steroid hormone affects a target cell.

❸ Explain how a peptide hormone affects a target cell.

Hormone Receptor Proteins

As you read in Chapter 4, the plasma membrane of the cell contains receptor proteins. A receptor protein has a unique shape that will only hold a particular type of molecule. If a cell has a receptor protein that will hold a particular hormone molecule, then the cell will respond to that hormone. For example, bone cells have receptor proteins that will hold parathyroid hormone molecules. A parathyroid hormone molecule will fit into one of these receptor proteins like a key fitting into a lock, as shown in **Figure 29.6**. Fitting the hormone molecule into the receptor changes the receptor's shape, which causes the cell's activities to change. In this case, the bone cell will begin to break down the minerals stored in the bone so that calcium can be released. While many receptor proteins are located in the plasma membrane, others are found inside the cytoplasm of the cell.

Hormones affect enzymes in cells

How can the binding of a hormone molecule with a receptor protein cause drastic changes in the activities of a cell? The main effect of a hormone on a cell is to change the activity or amounts of enzymes present in that cell. Recall that enzymes speed up chemical reactions. When a hormone causes changes in the enzymes in a cell, it causes changes in the chemical reactions that are happening inside the cell.

Figure 29.6

a When a hormone comes into contact with a plasma-membrane receptor protein that will hold it, the hormone binds to the receptor protein like a key fitting into a lock.

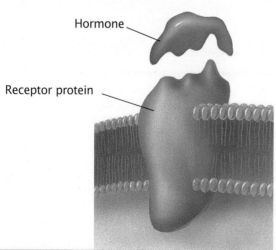

Hormone

Receptor protein

b The receptor protein changes shape, causing changes within the cell.

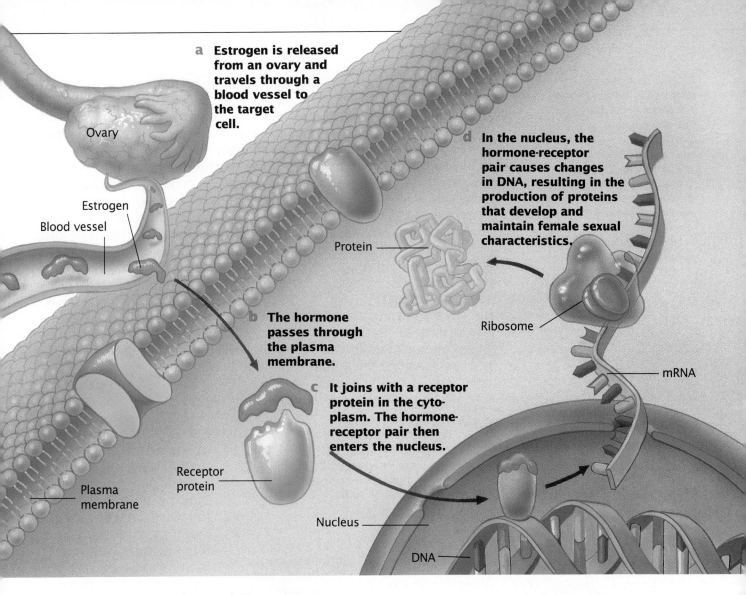

a Estrogen is released from an ovary and travels through a blood vessel to the target cell.

Ovary

Estrogen

Blood vessel

d In the nucleus, the hormone-receptor pair causes changes in DNA, resulting in the production of proteins that develop and maintain female sexual characteristics.

Protein

Ribosome

mRNA

b The hormone passes through the plasma membrane.

c It joins with a receptor protein in the cytoplasm. The hormone-receptor pair then enters the nucleus.

Receptor protein

Plasma membrane

Nucleus

DNA

Steroid Hormones

Figure 29.7
The steroid hormone estrogen influences development of female sexual traits. The way steroid hormones, such as estrogen, cause changes in cellular activity is described above.

Hormones assembled from cholesterol are called **steroid hormones**. Since cholesterol is a lipid, steroid hormones can pass through the lipid bilayer of the plasma membrane. Within the cytoplasm, a steroid hormone can bind to a receptor protein. The hormone-receptor complex can then pass into the nucleus, where it will trigger changes in the chromosomes. The male sex hormone testosterone and the female sex hormones estrogen and progesterone are examples of steroid hormones.

Steroid hormones activate genes

Recall that every cell in your body (except sperm or egg cells) contains all of the genes that you inherited from your parents. However, not all genes are actively being used to make proteins. Different sets of genes are switched on in different types of cells. For example, the genes that are active in liver cells are different from the genes that are active in skin cells.

One way to alter a cell's enzyme activity is to change which genes are switched on. Steroid hormones work by activating specific genes in the target cell, as shown in **Figure 29.7**. First, a steroid hormone passes through the plasma membrane and joins a receptor protein in the cytoplasm. Then the hormone-receptor unit passes into the nucleus and attaches itself to the DNA, as shown in **Figure 29.7d**. This attachment activates certain genes in that cell, causing particular proteins, including new protein enzymes, to be produced.

Peptide Hormones

Figure 29.8
Liver cells have receptor proteins for the peptide hormone glucagon, which is made by the pancreas. Glucagon uses the cyclic AMP as a second messenger to release glucose into the blood.

Hormones made of amino acids joined to form small peptides or larger proteins are called **peptide hormones**. Recall that amino acids dissolve easily in water. This is because amino acids and water molecules are both polar molecules. However, the positive and negative charges on peptide hormones prevent them from passing through the lipid bilayer of a plasma membrane. Thus, peptide hormones must send messages from outside the cell.

Second messengers carry information into the cytoplasm

Since peptide hormones cannot enter the cell, some other molecule must carry the message from the cell membrane into the cytoplasm. The peptide hormone acts as the first messenger, carrying the message from the endocrine gland to the cell surface. The molecule that carries the information into the cytoplasm is called a **second messenger**.

One of the most common second messengers is cyclic AMP. Cyclic AMP is made from ATP by an enzyme that removes two phosphate groups, forming AMP. The ends of the AMP join, forming a circle. Even though many hormones use cyclic AMP as a second messenger, these hormones cause many different effects in target cells. How? Each target cell has different enzymes in its cytoplasm that are activated by cyclic AMP. **Figure 29.8** shows how the peptide hormone glucagon (*GLOO kuh gahn*) signals liver cells to release glucose into the blood.

Second messengers are hormone amplifiers

A single hormone molecule binding to a receptor in the plasma membrane can result in the formation of many second messengers in the cytoplasm. Each second messenger, in turn, can activate many molecules of a certain enzyme. Sometimes this enzyme activates another enzyme, enabling each hormone molecule to have a tremendous effect inside a cell, even though it never actually enters the cell.

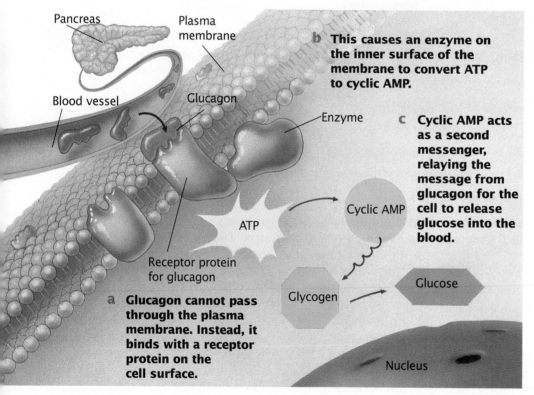

Pancreas

Plasma membrane

Blood vessel

Glucagon

Enzyme

b **This causes an enzyme on the inner surface of the membrane to convert ATP to cyclic AMP.**

c **Cyclic AMP acts as a second messenger, relaying the message from glucagon for the cell to release glucose into the blood.**

Cyclic AMP

ATP

Receptor protein for glucagon

Glycogen

Glucose

a **Glucagon cannot pass through the plasma membrane. Instead, it binds with a receptor protein on the cell surface.**

Nucleus

Section Review

❶ How does a receptor protein respond to a hormone molecule?

❷ Name the two ways hormones affect enzymes.

❸ How does the hormone estrogen deliver a message to a target cell?

❹ How does the hormone glucagon deliver a message to a target cell?

YOUR BODY FUNCTIONS DEPEND ON THE DELICATE INTERACTION BETWEEN YOUR ENDOCRINE SYSTEM AND YOUR BODY TISSUES. THE ENDOCRINE SYSTEM ORCHESTRATES SEVERAL PROCESSES NEEDED BY THE BODY TO COPE WITH STRESS, TO REGULATE METABOLISM, TO CONTROL BLOOD SUGAR LEVELS, TO CARRY ON RESPIRATION, AND EVEN TO REPRODUCE.

29.3 Glands and Their Functions

Objectives

❶ **Name** three endocrine glands and list the hormones each produces.

❷ **Explain** the effects of producing too much or too little thyroxine.

❸ **Describe** diabetes and explain how the condition can be treated.

❹ **List** two effects prostaglandins have on cells.

The Adrenal Glands

Eight different endocrine glands in the body produce dozens of different hormones. More than two dozen hormones are produced by the **adrenal glands**. The adrenal glands are two almond-sized glands located on top of the kidneys. Each adrenal gland is two endocrine glands in one. The inner part of each adrenal gland is called the adrenal medulla, and the outer part is called the adrenal cortex.

Figure 29.9
This is a famous movie scene. But if it were a real, life-threatening emergency, the adrenal medulla would secrete the hormone epinephrine. This hormone speeds up the heart rate and increases blood flow to the muscles.

The adrenal medulla helps the body react to a sudden crisis

The adrenal medulla is different from other glands because of the signal that activates it. Rather than being activated by hormones, the adrenal medulla is stimulated to release its hormones by nerves that run from the hypothalamus directly to the adrenal glands.

The adrenal medulla produces two hormones—epinephrine (*ep uh NEF rihn*), or adrenaline, and norepinephrine. These hormones produce what is called the "fight-or-flight" reaction. They are secreted in response to sudden stresses such as fear, anger, pain, or physical exertion. In a fraction of a second, the adrenal medulla can respond to any emergency that may arise, such as the scenario in **Figure 29.9**. The heart beats faster and blood flow increases to the heart and muscle cells. At the same time, air passages in the lungs relax and more oxygen is delivered throughout the body. For this reason, epinephrine is sometimes administered to people with asthma, a condition in which air passages in the lungs swell, making breathing difficult. During an acute asthma attack, epinephrine can help relax the swollen air passages.

Figure 29.10
In 1989, hurricane Hugo ripped through the island of St. Croix, leaving many people without water. As these hurricane victims can tell you, standing in a long line for many hours can be stressful. The diagram to the right illustrates how the endocrine system releases hormones to help the body deal with such situations.

Hypothalamus → Pituitary gland → Hormones → Adrenal cortex → Cortisol → Increases level of glucose for energy

Scale and Structure

Why are the

adrenal glands

considered two

separate glands?

The adrenal cortex helps the body deal with long-term stress

While the effects of epinephrine and norepinephrine wear off within a few minutes, hormones produced by the adrenal cortex enable the body to handle stress for hours or even days. Cortisol *(KAWRT uh sawl)* is one of the main hormones produced by the adrenal cortex. Cortisol increases the amount of energy available to the body by forming glucose from fats and proteins. Hormones released by the hypothalamus and the pituitary gland control the release of cortisol. For instance, standing in line for many hours can be stressful. The endocrine pathway in **Figure 29.10** shows how your body is supplied with energy during such circumstances. Even when you are not under stress, the level of cortisol fluctuates during the day. This is because you need different amounts of energy for different activities.

Cortisone is a compound that is similar to cortisol. Cortisone is produced synthetically and is used as a drug for the treatment of arthritis and a variety of other diseases. The prolonged use of cortisone, however, can be hazardous to your health. Possible side effects include the loss of resistance to infections, loss of bone and muscle protein, poor wound healing, and excess fat deposits, especially in the face. Because of these side effects, cortisol and other drugs chemically similar to cortisone are usually prescribed in the lowest effective dosages and are used only for short periods.

The Thyroid Gland

The thyroid gland is located in the neck, just below the Adam's apple. As you have read, the thyroid gland releases thyroxine, which regulates the body's metabolic rate. Thyroxine is also necessary for normal growth and development of the brain, bones, and muscles during childhood. In addition, thyroxine maintains a normal heart rate and affects reproductive functions.

Too much thyroxine causes Graves' disease

Some common disorders of the endocrine system are caused by problems with the thyroid. If too much thyroxine is produced, the condition that results is called **hyperthyroidism**, or Graves' disease. People with Graves' disease have a rapid, irregular heart rate, feel very nervous, and lose weight. Olympic gold medalist Gail Devers, pictured in **Figure 29.11**, began to experience the symptoms of Graves' disease just prior to the 1988 Summer Olympics. Concern about an irregular heartbeat also led to the diagnosis of Graves' disease in former President George Bush in 1991. Most people with Graves' disease can lead normal, productive lives once they are treated.

Graves' disease seems to be caused by a malfunction of the immune system. Recall that hormones work by binding to receptor proteins, which causes changes in the activity of the cell. The binding of a hormone to receptor proteins in thyroid cells causes the thyroid gland to release thyroxine. In a person with Graves' disease, however, antibodies made by the immune system fit into these receptor proteins instead. As a result, even though too much thyroxine already is present in the body, the thyroid gland still releases more.

Graves' disease is usually treated with a dose of radioactive iodine. The iodine collects in the thyroid gland and, over the course of several months, slowly destroys thyroid tissue. It is hard to control just how much of the gland is destroyed, however, so the patient often makes too little thyroxine after the treatment. Thyroxine levels are then brought up to normal by taking pills containing the hormone.

Figure 29.11
In 1988, American sprinter Gail Devers experienced severe weight loss, uncontrollable shaking fits, and loss of vision in her left eye. Nearly two years later she was diagnosed with Graves' disease, a condition in which too much thyroxine is produced by the thyroid gland. With proper treatment, however, Devers went on to win the gold medal in the women's 100-meter dash in the 1992 Summer Olympics.

Chapter 30

Drugs and the Nervous System

Review

- **protein receptors (Section 4.1)**
- **action potential (Section 28.1)**
- **neurotransmitters (Section 28.1)**

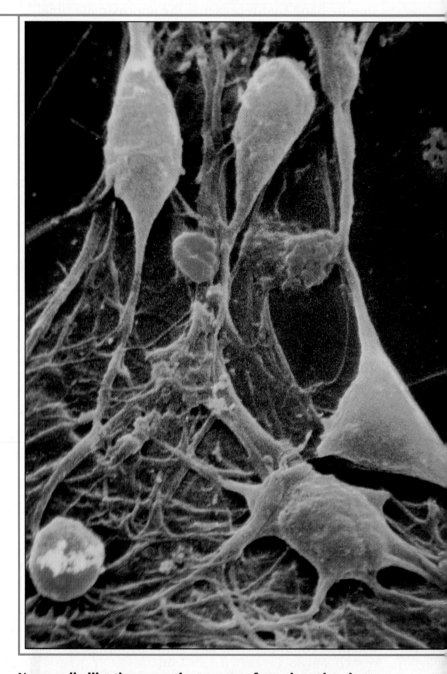

Nerve cells like these are the targets of psychoactive drugs. Drugs cause changes in the synapses between nerve cells, which can lead to addiction.

FEW SOCIAL PROBLEMS IN THIS COUNTRY HAVE A GREATER IMPACT ON PEOPLE'S LIVES THAN THE SPREADING ABUSE OF ADDICTIVE DRUGS. ADDICTION TO HEROIN, COCAINE, AND CRACK UTTERLY DESTROYS PEOPLE'S LIVES, WHILE ADDICTION TO DRUGS LIKE ALCOHOL AND NICOTINE CONDEMNS HUNDREDS OF THOUSANDS OF OTHER PEOPLE TO EARLY DEATHS.

30.1 Drugs and Addiction

Objectives

❶ Discuss information that research has yielded about psychoactive drugs.

❷ Summarize how drugs affect the nervous system.

❸ Describe addiction in terms of physiology.

❹ Explain the role of neuro-modulators in nerve impulse transmission.

What Are Psychoactive Drugs?

Figure 30.1
This woman is addicted to crack cocaine. Drug addiction is a serious personal and social problem that can destroy lives and communities. Understanding the biology of drug addiction can be a first step in tackling drug problems.

Drugs are chemicals put into the body that are not normally found there. Heroin, cocaine, and alcohol are drugs, as are aspirin, penicillin, and caffeine. **Psychoactive drugs** such as alcohol, cocaine, heroin, and nicotine affect the nervous system. Psycho-active drugs are often addictive. The woman in **Figure 30.1** is addicted to crack, a form of cocaine. Widespread use of addictive psychoactive drugs has created major social problems, particularly in our nation's inner cities.

Research has shown how addictive drugs work

Scientists have recognized an important fact: addiction to psychoactive drugs is a physiological response, one that involves drug molecules and nerve cell membranes. Addiction is the human body's attempt to cope with chemical disruption of its signaling systems. Psychoactive drugs, taken to fool the nervous system into providing pleasure or taken to control pain, cause the nervous system to adapt physiologically to their presence, leading to addiction. Scientists are beginning to understand that the body's response is a straightforward chemical response. There is no way to prevent addiction with willpower, any more than willpower can stop a bullet when playing Russian roulette with a loaded gun. To understand the nature of addiction, you must focus your attention on how nerves communicate with one another, for it is in attempting to disrupt this process that addictive drugs cause their damage.

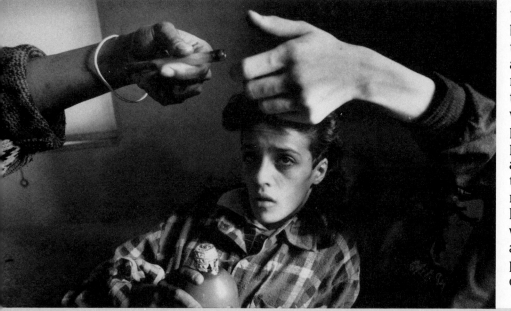

How Psychoactive Drugs Affect Nerves

Most addictive drugs act at the ends of nerve cells. In a house, all the wires are connected to each other by switches. A current passes from one wire to the next when a switch makes a physical connection between two wires. In your body, nerve cells are not connected at all. Instead they are separated from one another by tiny gaps called synapses that act as chemical switches. A current passes from one nerve cell to another when a synapse makes a chemical connection between the two nerve cells. Addictive drugs work by short-circuiting these chemical connections. In effect, they "hot-wire" certain nerve pathways in the brain.

A nerve impulse passes across a synapse carried by a chemical called a neurotransmitter, as shown in **Figure 30.2a**. Your body has dozens of different kinds of neurotransmitters. Each is present at the ends of particular kinds of nerves. The neurotransmitter does not actually carry an electric charge. Instead, it opens ion channels on the far side of the synapse, an action which has an electrical effect. Some neurotransmitters open sodium ion (Na^+) channels, letting sodium ions flood the receiving nerve cell. This action starts a nerve impulse that travels down that neuron. Other kinds of neurotransmitters open potassium ion (K^+) channels, letting potassium ions rush out of the receiving nerve cell and making it harder for the nerve to fire. This action is called inhibition. A nerve cell body can have many axons in close contact with it, some stimulating and others inhibiting. The summed effect is how your brain integrates signals.

Neurotransmitters work by binding to specific receptor proteins in the membrane on the far side of the synapse. Each kind of neurotransmitter fits into its own type of receptor, as shown in **Figure 30.2b**. The reason all the nerves in your brain don't fire all at once when a signal arrives is that different nerves have different combinations of neurotransmitters and receptors. A signal can only pass between two nerves if the sending nerve releases a neurotransmitter for which the receiving nerve has a receptor.

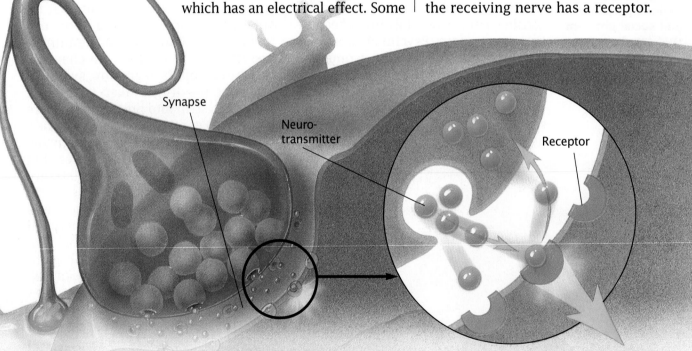

Synapse

Neuro-transmitter

Receptor

Figure 30.2

a *Normal transmission of an impulse across a synapse.*
When a nerve impulse reaches a synapse, the electrical signal is transferred across the gap to the next neuron by a chemical called a neurotransmitter.

b **Neurotransmitter molecules bind to receptor proteins in the receiving neuron's membrane, initiating a new nerve impulse. Then the neurotransmitter molecules are reabsorbed or broken down.**

When a neurotransmitter has passed a signal to the far side of a synapse, it is either destroyed or reabsorbed by the nerve cell that released it. If it were not destroyed or reabsorbed, it would just keep firing the nerve cell again and again. Nerve gas works by inhibiting an enzyme that normally destroys the neurotransmitter acetylcholine (*uh seet uhl KOH leen*) within synapses between nerve cells and muscle cells. When the acetylcholine cannot be broken down, muscle cells cannot cease contracting and death results.

Neuromodulators prolong the action of neurotransmitters

Your body often prolongs the transmission of a signal across a synapse by slowing the destruction of neurotransmitters. When this happens, the receptor proteins on the far side of the synapse continue to encounter the neurotransmitter for a longer period of time. Your body is able to do this naturally with certain chemicals called **neuromodulators**. Neuromodulators last longer, compared with neurotransmitters. Some of them aid in releasing a neurotransmitter into the synapse; others inhibit the reabsorption of a neurotransmitter. And still other neuromodulators delay the breakdown of neurotransmitters after their reabsorption. This leaves more of a neurotransmitter in the neuron ending to be released when the next signal arrives.

How drugs alter mood

Mood, pleasure, pain, and other mental states are determined by particular groups of nerves in the brain that use special sets of neurotransmitters and neuromodulators. Mood, for example, is strongly influenced by the neurotransmitter serotonin (*sihr uh TOH nuhn*). Many researchers think that depression results from a shortage of serotonin. It is difficult to treat depression directly with serotonin (it has too many other effects). However, depression can be treated successfully with drugs that act as serotonin neuromodulators, as shown in **Figure 30.3**. One class of serotonin modulators blocks a specific enzyme, which then slows the breakdown of serotonin after it is reabsorbed from synapses. Prozac®, the world's top-selling antidepressant, inhibits the reabsorption of serotonin. Antidepressants thus act to increase the amount of serotonin in the synapse by slowing its removal and destruction. This allows the nerve impulses necessary for transmission of proper mood.

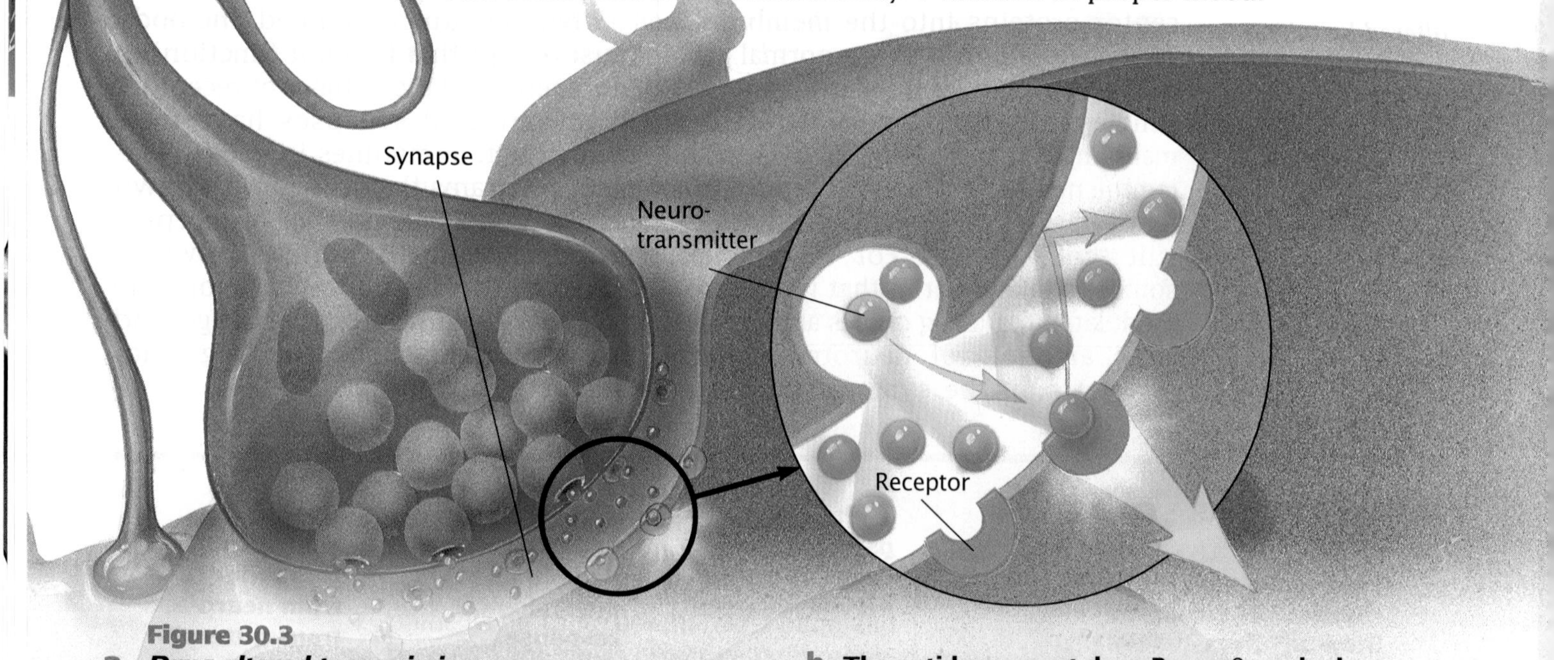

Synapse

Neuro-transmitter

Receptor

Figure 30.3

 a *Drug-altered transmission of impulse across the synapse* Depression can result from a shortage of the neurotransmitter serotonin (shown as green spheres).

b The antidepressant drug Prozac® works by blocking reabsorption of serotonin from the synapse. This action leaves more serotonin in the synapse, making up for the shortage.

Drugs and the Nervous System **693**

Circulation and Respiration

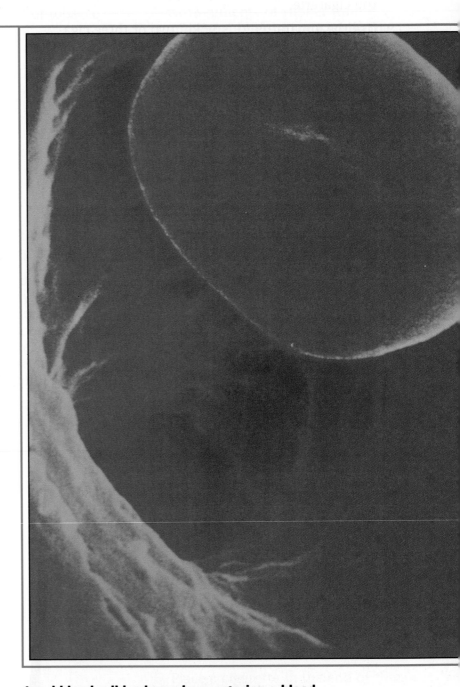

A red blood cell is shown here entering a blood vessel. The cell will travel on a 60,000-mile journey throughout the human body.

INSIDE YOUR BODY, BLOOD FLOWS ALONG A VAST INNER HIGHWAY ON A 60,000-MILE JOURNEY, FAR ENOUGH TO ENCIRCLE THE PLANET TWO AND ONE-HALF TIMES. BLOOD NOURISHES AND CLEANSES THE ENTIRE BODY, DELIVERING FOOD AND OXYGEN TO EVERY CELL AND PICKING UP WASTES. BLOOD TRAVELS THROUGH EVERY LIVING TISSUE, PROPELLED BY THE BEATING OF THE HEART.

31.1 Circulation

Objectives

❶ Explain the three functions of circulation.

❷ Describe the components of blood.

❸ Name the three major types of blood vessels.

❹ Describe the role of the lymphatic system.

Transporting Materials Through the Body

Figure 31.1
In 1628, William Harvey demonstrated that blood travels from the heart to the body's limbs and then back again. If an arm is bound above the elbow, swellings appear in the lower arm. Harvey used this as evidence that blood flow back to the heart was slowed.

All living things must capture materials from their environment to use in carrying on their life processes. Many bacteria, single-celled protists, and simpler multicellular animals live within a liquid environment, enabling materials to diffuse directly into and out of the organism through plasma membranes. However, larger animals with many cells stacked in layers, such as earthworms, armadillos, and humans, cannot rely solely on diffusion to supply needed materials and carry away wastes. In these organisms, a circulatory system transports oxygen, carbon dioxide, food molecules, hormones, and other materials to and from the cells of the body. In addition, the circulatory systems of mammals and birds help maintain their constant body temperatures. The circulatory system also carries cells that help protect the body from disease. All of these functions help the body maintain homeostasis and are essential to survival.

Our modern understanding of the human circulatory system began with the work of the seventeenth-century English physician William Harvey. Harvey demonstrated that blood circulates in one direction in a closed circuit throughout the body. An illustration of his experiment is shown in **Figure 31.1**.

Blood: A Liquid Tissue

Blood has been called the river of life. It is the tissue responsible for transporting nearly everything within the body. Among all the body's tissues, blood is the only liquid tissue. The material that makes blood a liquid is a protein-rich substance called **plasma** (*PLAZ muh*). Plasma makes up approximately 55 percent of blood. The other 45 percent is made mostly of cells—red blood cells, white blood cells, and platelets.

Red blood cells carry oxygen to other body tissues

Most of the cells that make up blood are red blood cells. There are approximately 5 million red blood cells per cubic millimeter of blood! The main function of a **red blood cell** is to carry oxygen from the lungs to all cells of the body. Red blood cells also carry carbon dioxide from cells back to the lungs to be exhaled. The structure of a red blood cell is well suited to this function.

As shown in **Figure 31.2a**, red blood cells are shaped like round cushions, squashed in the center. Red blood cells are filled with a protein called hemoglobin, an iron-containing molecule that gives blood its red color. Oxygen binds easily to the iron in hemoglobin, making red blood cells efficient oxygen carriers. A single red blood cell contains about 250 million hemoglobin molecules. Each red blood cell can carry about 1 billion molecules of oxygen.

A condition in which a person has too few red blood cells or not enough hemoglobin is called **anemia** (*uh NEE mee uh*). Anemia results in a shortage of oxygen in the cells of the body. Recall from Chapter 6 that people who have sickle cell anemia have an abnormal form of hemoglobin that causes cells to become sickle-shaped. The sickled red blood cells clog small blood vessels and interfere with oxygen delivery. Sickle cell anemia is a serious disease that cannot be cured and can be fatal.

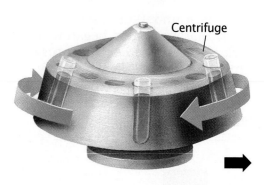

Vacuum syringe

Centrifuge

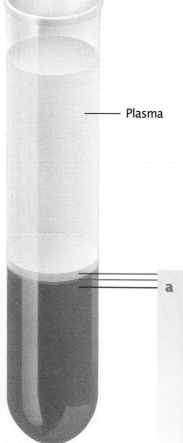

Plasma

Figure 31.2
After a blood sample is collected in a vacuum syringe, it is spun rapidly in a device called a centrifuge (above). This rotation forces the blood to settle in layers, with the heaviest cells (red blood cells) at the bottom of the tube, white blood cells and platelets in the middle, and plasma at the top.

a *Red blood cells*
Red blood cells carry oxygen from the lungs to all cells of the body, and carry carbon dioxide back to the lungs to be exhaled. Each red blood cell has an average life span of only 120 days because they mature without a nucleus. About 2 million new red blood cells are produced every second in the bone marrow to replace those that are worn out.

White blood cells defend the body

White blood cells form a mobile army that protects the body against invading bacteria, viruses, or other foreign cells. White blood cells called macrophages (*MAK roh fayj iz*) act as a clean-up crew, scavenging worn-out or dead cells. While red blood cells are confined to the bloodstream, white blood cells are able to squeeze in and out of blood vessels. They move toward areas of tissue damage and infection by responding to chemicals released by damaged cells. By following the chemical trail, macrophages are able to migrate to infected areas in the body. There, they gather in large numbers to engulf and digest foreign cells.

As shown in **Figure 31.2b**, white blood cells are colorless and irregularly shaped. Most of them are manufactured and stored in bone marrow until they are needed by the body. Although they are larger than red blood cells, they are considerably less numerous; normally there are 4,000 to 11,000 white blood cells per cubic millimeter of blood. Whenever white blood cells are mobilized for action, the body speeds up their production so that twice the normal number may appear within a few hours. A white blood cell count over 11,000 generally indicates the presence of a bacterial or viral infection in the body.

Platelets help repair damaged blood vessels

Platelets are not really cells; they are fragments of cells formed from large white blood cells within bone marrow. Each cubic millimeter of blood contains between 250,000 and 500,000 tiny platelets.

Platelets are essential for the clotting process that occurs when blood vessels are injured, as shown in **Figure 31.2c**. When you accidentally cut your finger, platelets come into contact with the ends of the broken blood vessels. There, the platelets swell and stick to the rough surfaces created by the injury as well as to each other. Certain of these platelets rupture and release a protein, which triggers a series of reactions that forms a clot. The clot stops the bleeding and hardens into a patch over the injured area. In time, the injury is repaired by the growth of cells that replace those that were damaged.

Patterns of Change

What can cause the number of white blood cells in the body to increase?

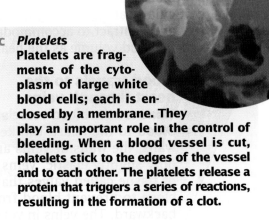

b *White blood cells*
White blood cells help defend the body against invading bacteria and other intruders. Some white blood cells move against the flow of the bloodstream. Others squeeze through the walls of blood vessels. They arrive at the damaged area by following chemical signals given off by infected cells. The white blood cells then gather and attack foreign intruders.

c *Platelets*
Platelets are fragments of the cytoplasm of large white blood cells; each is enclosed by a membrane. They play an important role in the control of bleeding. When a blood vessel is cut, platelets stick to the edges of the vessel and to each other. The platelets release a protein that triggers a series of reactions, resulting in the formation of a clot.

BLOOD COULD NOT MEET THE BODY'S NEEDS IF IT DID NOT FLOW.
BLOOD FLOWS BECAUSE THE CIRCULATORY SYSTEM INCLUDES A PUMP—
THE HEART—THAT FORCES BLOOD TO MOVE THROUGHOUT THE BODY.
KEEPING FIT, NOT SMOKING, AND EATING FOODS LOW IN ANIMAL FATS
WILL REDUCE YOUR RISK OF DISEASES AFFECTING THE HEART AND BLOOD
VESSELS.

31.2   How Blood Flows

Objectives

❶ **Describe the structure of the heart.**

❷ **Trace the two routes blood can take through the body.**

❸ **Define blood pressure and explain how it is measured.**

❹ **Describe the causes and symptoms of two cardiovascular diseases.**

The Heart

The heart is a muscular organ that pumps blood throughout the body. When you are sitting still, your heart pumps about 5 L (5.3 qt.) of blood each minute. If you are riding a bike, like the person in **Figure 31.6**, your heart may have to pump up to seven times that amount per minute.

Your heart is divided into left and right halves. The right half pumps blood to the lungs to pick up oxygen and release carbon dioxide. The left half of the heart pumps the oxygen-rich blood to the rest of the body—your head, arms, legs, and all the tissues and organs in between.

As shown in **Figure 31.7**, each half of the heart has an upper and a lower chamber. The upper chamber, called the **atrium**, receives blood coming into the heart. The lower chamber, called the **ventricle**, pumps blood out of the heart. The upper and lower chambers are separated from each other by flaplike valves that control the direction of the blood flow inside the heart. Blood flows into both atria at the same time, and the atria contract together. Similarly, the ventricles contract together. All this activity causes the heartbeat, a sound that is usually described as "lubb dup."

Figure 31.6
When you are active, your heart pumps up to 35 L (37 qt.) of blood each minute. That's seven times the amount your heart pumps when you are at rest.

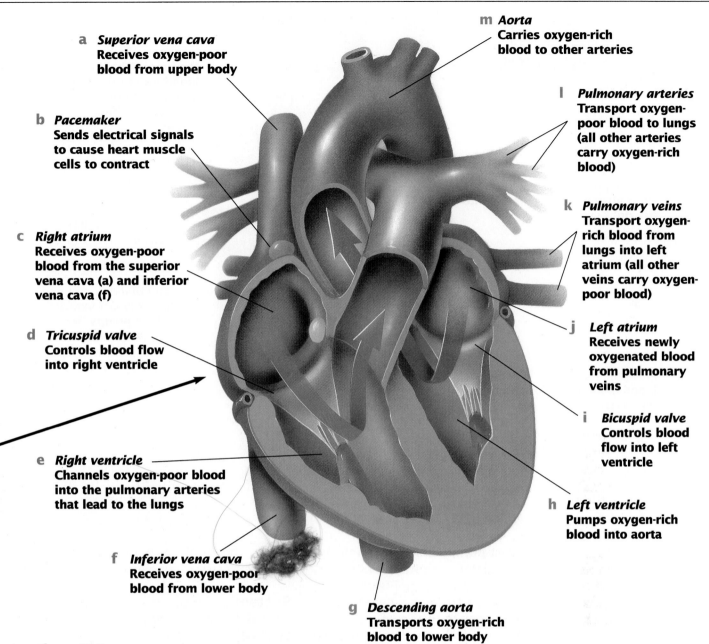

a *Superior vena cava*
Receives oxygen-poor blood from upper body

b *Pacemaker*
Sends electrical signals to cause heart muscle cells to contract

c *Right atrium*
Receives oxygen-poor blood from the superior vena cava (a) and inferior vena cava (f)

d *Tricuspid valve*
Controls blood flow into right ventricle

e *Right ventricle*
Channels oxygen-poor blood into the pulmonary arteries that lead to the lungs

f *Inferior vena cava*
Receives oxygen-poor blood from lower body

g *Descending aorta*
Transports oxygen-rich blood to lower body

m *Aorta*
Carries oxygen-rich blood to other arteries

l *Pulmonary arteries*
Transport oxygen-poor blood to lungs (all other arteries carry oxygen-rich blood)

k *Pulmonary veins*
Transport oxygen-rich blood from lungs into left atrium (all other veins carry oxygen-poor blood)

j *Left atrium*
Receives newly oxygenated blood from pulmonary veins

i *Bicuspid valve*
Controls blood flow into left ventricle

h *Left ventricle*
Pumps oxygen-rich blood into aorta

Figure 31.7
The human heart pumps oxygen-poor blood to the lungs and oxygen-rich blood to the rest of the body.

The "lubb" sound relates to the closing of the valves that lead from the atrium to the ventricle. The "dup" comes very shortly afterward and is related to the closing of the valves between the ventricles and the arteries that lead to the lungs and the rest of the body.

The four-chambered human heart, shown in **Figure 31.7**, is similar in design to the heart of birds. The *Evolution of the Heart* on pages 722–723 summarizes the evolution of the vertebrate heart.

What makes the heart beat?

The heartbeat originates in heart tissue. It begins to beat in the embryo, before any nerves connect it to the brain. It can continue to beat during transplant surgery, after all nerves have been cut. How is this possible?

Each heartbeat is started by the **pacemaker**, a small bundle of cells at the entrance to the right atrium. An electrical signal from the pacemaker, shown in **Figure 31.7b**, travels through the heart muscle cells in the right and left atria, causing them to tighten, or contract. When the signal reaches the right and left ventricles, they also contract. These contractions cause the chambers to squeeze the blood through the heart and push it to other parts of the body.

Figure 31.8
Pulmonary circulation carries blood from the heart to the lungs and back to the heart. Follow this pathway of blood flow beginning with oxygen-poor blood entering the right atrium.

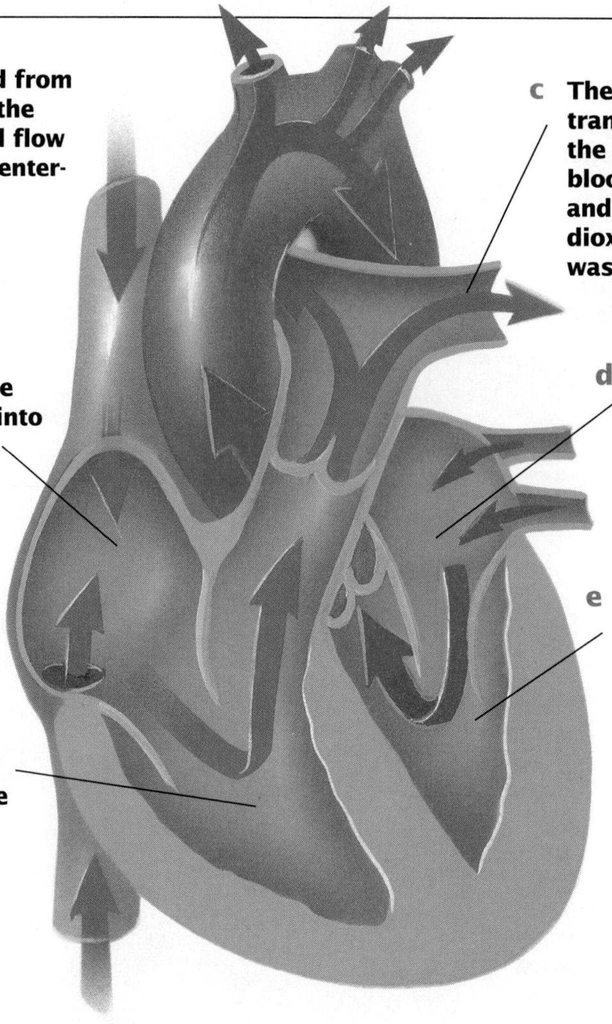

c The pulmonary arteries transport the blood to the lungs. There the blood picks up oxygen and gets rid of carbon dioxide and other wastes.

a Oxygen-poor blood enters the right atrium and is pumped into the right ventricle.

d The pulmonary veins transport the newly oxygenated blood to the left atrium.

e The blood is pumped into the left ventricle, which pumps the blood through the aorta (top) to the rest of the body.

b From the right ventricle, the blood is pumped through the pulmonary arteries.

Circulatory Pathways

Your heart pumps blood through two major circulatory pathways. Pulmonary circulation, shown in **Figure 31.8**, carries blood from the heart to the lungs and back to the heart. The pathway of blood from the heart to other parts of the body is called systemic circulation.

Blood picks up oxygen and releases carbon dioxide in the lungs

Blood returning to the heart from the body is low in oxygen and high in carbon dioxide. This blood is pumped through the right side of the heart into arteries that lead to the lungs. These arteries, the pulmonary arteries, are the only arteries that carry oxygen-poor blood. As blood flows through the capillaries in the lungs, it picks up oxygen and gets rid of carbon dioxide. Leaving the lungs, the blood travels through veins, back to the heart. These veins, the pulmonary veins, are the only veins that carry oxygen-rich blood.

Systemic circulation carries blood from the heart to the rest of the body

Blood returning to the heart from the lungs is ready to deliver oxygen to the rest of the body. Notice in **Figure 31.8e** that oxygen-rich blood is pumped through the left side of the heart and out the aorta, the largest artery in the body. From the aorta, blood flows to all parts of the body through a system of increasingly smaller arteries. Systemic circulation has three branches of special importance: the branch that carries blood to the heart, the branch that carries blood to the digestive tract and liver, and the branch that carries blood to and from the kidneys. Several of the major veins and arteries of systemic circulation are shown in **Figure 31.9**.

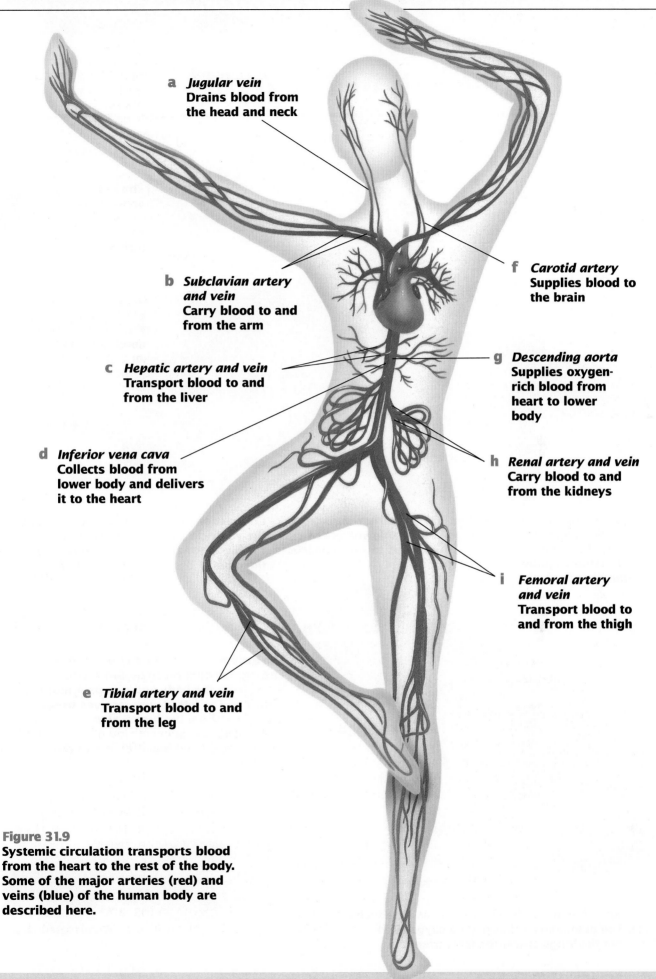

a *Jugular vein*
Drains blood from the head and neck

b *Subclavian artery and vein*
Carry blood to and from the arm

c *Hepatic artery and vein*
Transport blood to and from the liver

d *Inferior vena cava*
Collects blood from lower body and delivers it to the heart

e *Tibial artery and vein*
Transport blood to and from the leg

f *Carotid artery*
Supplies blood to the brain

g *Descending aorta*
Supplies oxygen-rich blood from heart to lower body

h *Renal artery and vein*
Carry blood to and from the kidneys

i *Femoral artery and vein*
Transport blood to and from the thigh

Figure 31.9
Systemic circulation transports blood from the heart to the rest of the body. Some of the major arteries (red) and veins (blue) of the human body are described here.

Evolution of the Heart

Contractions of the vertebrate heart propel blood through the body. Follow the evolution of heart structure below.

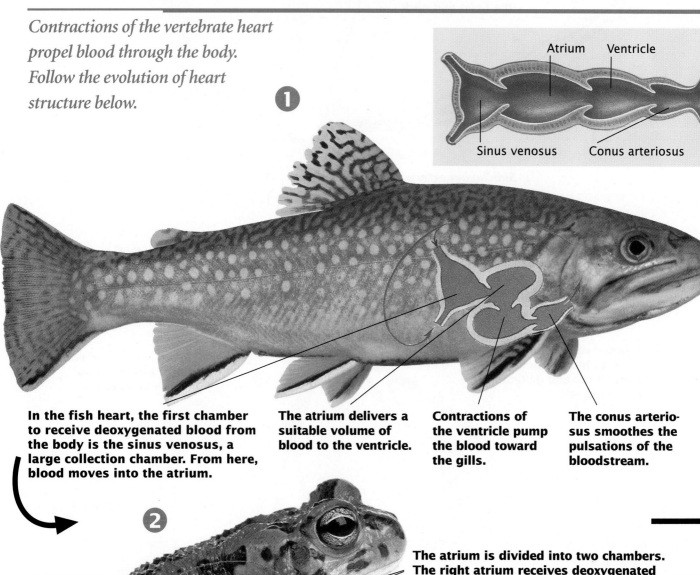

1

Atrium   Ventricle

Sinus venosus   Conus arteriosus

In the fish heart, the first chamber to receive deoxygenated blood from the body is the sinus venosus, a large collection chamber. From here, blood moves into the atrium.

The atrium delivers a suitable volume of blood to the ventricle.

Contractions of the ventricle pump the blood toward the gills.

The conus arteriosus smoothes the pulsations of the bloodstream.

2

The atrium is divided into two chambers. The right atrium receives deoxygenated blood from the body and pumps it into the ventricle. The left atrium receives oxygenated blood from the lungs, also sending it to the ventricle. Some mixing of oxygenated and deoxygenated blood occurs in the ventricle.

In the amphibian heart, the sinus venosus is reduced in size. The pulmonary vein carrying oxygenated blood from the lungs enters the left atrium .

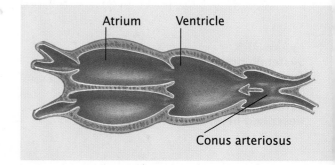

Atrium   Ventricle

Conus arteriosus

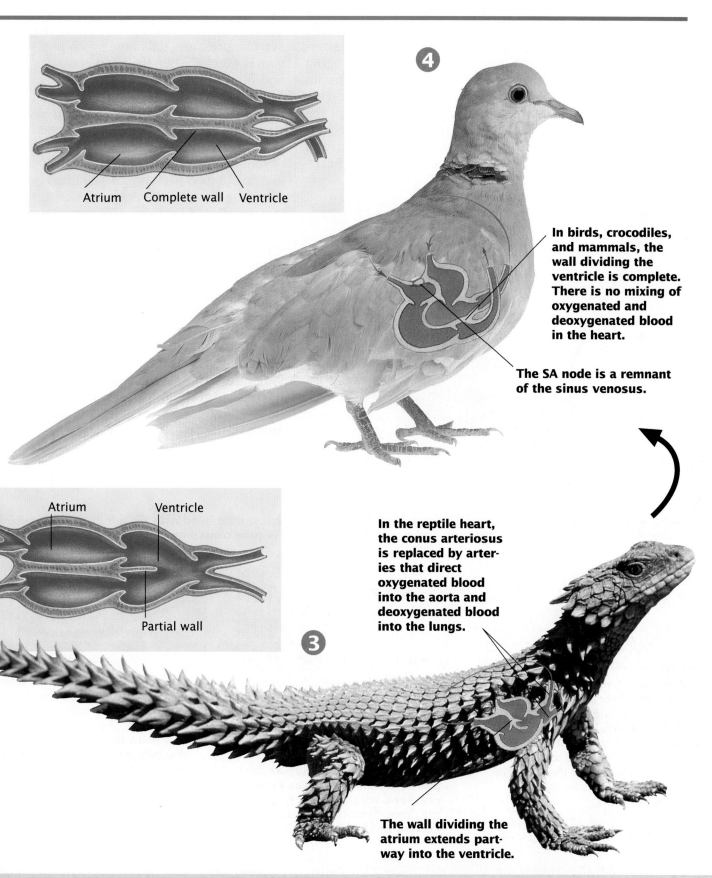

4

Atrium   Complete wall   Ventricle

In birds, crocodiles, and mammals, the wall dividing the ventricle is complete. There is no mixing of oxygenated and deoxygenated blood in the heart.

The SA node is a remnant of the sinus venosus.

Atrium   Ventricle

Partial wall

3

In the reptile heart, the conus arteriosus is replaced by arteries that direct oxygenated blood into the aorta and deoxygenated blood into the lungs.

The wall dividing the atrium extends partway into the ventricle.

Circulation and Respiration   **723**

Blood Pressure

When ventricles contract, blood is forced into the arteries, which exerts pressure on the walls of the blood vessel. This force is called **blood pressure**. When ventricles relax, the pressure decreases. The muscular, elastic walls of the arteries are able to adjust to the changing pressure. The elasticity helps maintain the pressure between heartbeats. This way blood is kept flowing through the body continuously. When you take your pulse, you are feeling the expansion and relaxation of an artery with each heartbeat.

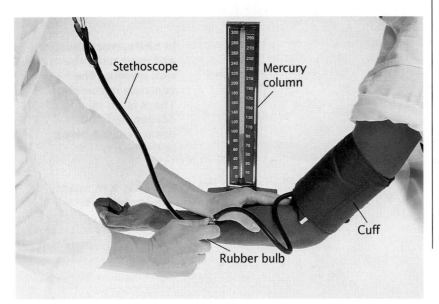

How is blood pressure measured?

As described in **Figure 31.10**, two numbers are used to register blood pressure. The first number, called the systolic pressure, tells how much pressure is exerted when the heart contracts and blood flows through the arteries. The second number is the diastolic pressure, which tells how much pressure is exerted when the heart relaxes. Blood pressure is expressed in terms of millimeters of mercury (mm Hg). For example, blood pressure of 120 mm Hg would be equal to the pressure exerted in a column of mercury 120 mm high. Blood pressure in a healthy adult is typically about 120/80 mm Hg. These figures indicate that the blood is pushing against the artery walls with a pressure of 120 mm Hg as the heart contracts and 80 mm Hg as the heart rests. Blood pressure steadily rises with increasing age as the arteries become less elastic. Blood pressure figures provide information about the conditions of the arteries and are useful in diagnosing high blood pressure and hardening of the arteries.

Figure 31.10
A kit for measuring blood pressure includes a stethoscope, a cuff that can be filled with air, a hollow rubber bulb that pumps air into the cuff, and a gauge or hollow glass tube containing a column of mercury.

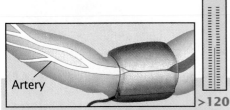

a The cuff is wrapped around the patient's upper arm. When the rubber bulb is pumped, air inflates the cuff. This squeezes an artery in the arm until no blood passes through. The stethoscope is used to listen for the sound of blood flow to ensure that the artery is closed.

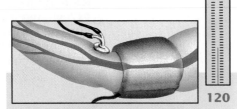

b The cuff is gradually deflated until blood begins to flow into the arm. A sound of blood pulsing can be heard, indicating that the blood pressure is greater than pressure exerted by the cuff. The pressure at this point (120 mm Hg) is the systolic pressure, which is read from the scale. Systolic pressure is exerted by ventricles contracting.

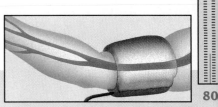

c The cuff is loosened until blood flows freely through the artery and the sounds below the cuff disappear. The pressure at this point (80 mm Hg) is the diastolic pressure, the pressure between heart contractions.

Diseases of the Heart and Blood Vessels

Diseases of the heart and blood vessels are referred to as **cardiovascular diseases**. Cardiovascular diseases are the leading cause of death in the United States, claiming about 1 million lives every year. An estimated 63 million Americans have some form of cardiovascular disease.

Heart attacks can kill

Heart attacks are the most common cause of death from cardiovascular disease. A heart attack results from an insufficient supply of blood to an area of heart muscle. Without the blood, heart muscle cells are starved for oxygen and die. A heart attack can result from blockage of a blood vessel due to atherosclerosis (*ath uhr oh skluh ROH sihs*), a condition in which fatty deposits form on the insides of arteries, as shown in **Figure 31.11b**. Recovery from a heart attack depends on how much heart tissue has been damaged, where the damage occurred, and whether other blood vessels can do the work of the damaged blood vessels.

Hypertension is high blood pressure

When the pressure of the blood against the artery wall is continually higher than normal, a condition called **hypertension**, or high blood pressure, results. Hypertension is dangerous because of the damage it can do to the heart, brain, and kidneys if not controlled. Because it has no warning symptoms, hypertension is often called the silent killer.

A healthy lifestyle can reduce your risk of cardiovascular disease

Certain factors can influence an individual's risk of cardiovascular disease. For example, a tendency toward some cardiovascular diseases appears to be hereditary. Nicotine, the drug in cigarette smoke, has been strongly linked to cardiovascular disease in many clinical studies. Scientists have also discovered that obesity, stress, and a lack of regular exercise can greatly increase the risk of high blood pressure, a heart attack, or a stroke. Reducing the amount of animal fat, cholesterol, and salt in your diet and exercising moderately at least three times a week will help maintain resistance to these and many other diseases. Not surprisingly, avoiding smoking greatly lowers your chances of cardiovascular disease.

Scale and Structure

How can

atherosclerosis

increase the risk of

heart attack?

Figure 31.11

a **In a normal artery, the passageway is clear for blood to pass through. However, . . .**

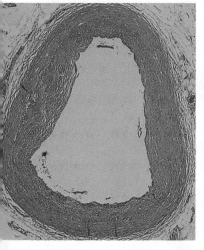
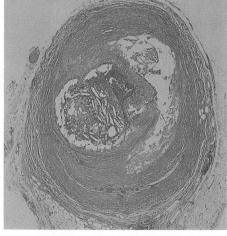

b **. . . in an artery that is partially blocked because of atherosclerosis, fat deposits decrease the amount of blood that can flow through the artery.**

Section Review

❶ **Describe the roles played by the left and right halves of the heart.**

❷ **Explain the two routes blood can take through the body.**

❸ **What does a blood pressure of 120/80 mean?**

❹ **What steps could you take to reduce your chances of developing cardiovascular disease?**

Cholesterol and Your Health

Cholesterol and Teens

You've read the claims "Cholesterol Free" and "Low Cholesterol" splashed across food packages. Clearly, food makers are hoping to tempt customers concerned about their health. But did you know that these messages are important not only for adults, but also for you as a teenager?

Studies show that a condition called atherosclerosis (a buildup of fat deposits in the arteries) can begin in childhood. The process progresses into adulthood and can lead to coronary heart disease, a major cause of death in the United States. According to many studies, preventing atherosclerosis in childhood or in your teenage years could extend your lifetime. By learning the facts about cholesterol and how it affects the human body, you can take the first step toward living a healthier lifestyle.

What Is Cholesterol?

Cholesterol is a fatty, waxlike substance that is an essential part of the membranes in cells of the human body. Humans need cholesterol to produce certain hormones, vitamin D, and the protective sheath around nerve fibers. Our bodies produce enough cholesterol for these requirements. We also take in cholesterol from some of the foods we eat.

Cholesterol is made in the liver and transported to all of the body's cells through the bloodstream. It is carried by lipoproteins, molecules containing fats and proteins. One type of lipoprotein is known as low-density lipoprotein (LDL). A second type is called high-density lipoprotein (HDL). LDL is often called "bad cholesterol" because it deposits cholesterol in body tissues, especially in the walls of arteries. As more and more cholesterol is deposited, the arteries become narrower, making it difficult for blood to flow through. The reduced blood flow may cause heart disease, heart attacks, or strokes. Often referred to as "good cholesterol," HDL removes cholesterol from the LDL and body tissues and transports it back to the liver for removal from the body. In general, a high LDL cholesterol level, or a low HDL cholesterol level, increases the risk for developing coronary heart disease.

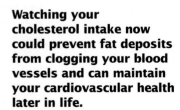

Watching your cholesterol intake now could prevent fat deposits from clogging your blood vessels and can maintain your cardiovascular health later in life.

Guidelines for Lowering Blood Cholesterol Levels

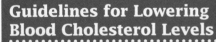

- **Eat fewer foods high in saturated fat (butter, cheese, whole milk, ice cream, meat, coconut and palm oil)**
- **Replace part of your saturated fat intake with unsaturated fat**
- **Eat fewer high-cholesterol foods (eggs, dairy products, liver)**
- **Choose foods high in complex carbohydrates (starch and fiber)**
- **Maintain a healthful diet and exercise regularly**

How to Modify Cholesterol Levels

Researchers believe that genetic factors may influence cholesterol levels in the blood. Although these factors cannot be controlled, a person can reduce blood cholesterol levels. One way is to eat foods that are low in cholesterol. Foods that come from animals, such as eggs, cream, and meat, are high in cholesterol. Fruits, vegetables, and grains contain no cholesterol. That's because animals produce cholesterol, whereas plants do not.

The kinds of fats you eat are important too. Saturated fats, such as butter and animal fat, tend to raise blood cholesterol levels. But unsaturated fats may actually help lower the amount of cholesterol carried in the blood. They do this by increasing the levels of HDL, or "good" cholesterol. Unsaturated fats include most of the liquid fats such as vegetable, olive, and fish oils, as well as margarine. Many of the saturated fats in your diet can be substituted with unsaturated fats. People who need to increase their HDL levels are encouraged to eat polyunsaturated fats. These fats are found mainly in safflower, corn, soybean, sesame, and sunflower oils.

Another way to reduce your cholesterol level is by exercising. Some research suggests that HDL levels are higher in people who exercise regularly. And being overweight seems to deplete the amount of HDL in the blood, as does smoking. So, by exercising regularly, not smoking, and eating a diet that is low in saturated fats and low in cholesterol, you can decrease your risk of developing heart disease.

Corn oil, a polyunsaturated fat, and olive oil, an unsaturated fat, should replace saturated fats, such as butter, in your diet.

Food labels should be read carefully. They provide a good source of information about the fat and cholesterol contents of food products.

How to Be a Smart Consumer

Eating to lower your blood cholesterol level means learning to select foods that are low in saturated fats and cholesterol. The packages in the supermarket with "cholesterol free" claims do contain food without cholesterol, but sometimes these statements can be misleading. A food made with highly saturated oil as the only fat source may be cholesterol-free but still be a poor choice if you're trying to reduce your blood cholesterol level.

Food labels can provide important information about the type and amount of fats and cholesterol. Ingredients are presented on the label according to their weight in the product. The ingredient found in the greatest amount is listed first, while the ingredient found in the least amount is listed last. One way to be sure that a certain food fits a low-fat, low-cholesterol diet is to limit your selection of foods in which ingredients high in saturated fats or cholesterol are among the first five on the list. Also, choose sparingly from foods that list many fats or oil.

The listing of fats and cholesterol is optional on food labels. However, under FDA regulations, any food product with a nutritional claim must have the nutritional content listed on the label.

HOW GASES ARE EXCHANGED IN ANIMALS DEPENDS ON THE SIZE OF THE ANIMAL. VERY SMALL ANIMALS CAN OBTAIN OXYGEN AND GIVE OFF CARBON DIOXIDE DIRECTLY THROUGH THE PLASMA MEMBRANES OF THEIR OUTERMOST LAYER OF CELLS. IN CONTRAST, LARGER ANIMALS HAVE EVOLVED SPECIAL RESPIRATORY SYSTEMS THAT PROVIDE A LARGE SURFACE AREA TO EXCHANGE GASES.

31.3 The Respiratory System

Objectives

❶ Explain how the diaphragm and rib muscles work to move air into and out of the lungs.

❷ Describe the pathway by which oxygen from the air travels to a cell in the body.

❸ Explain how your body regulates your breathing rate.

❹ Describe three diseases of the respiratory system.

Lungs and Breathing

Figure 31.12
On or off the soccer field, humans must obtain oxygen from the atmosphere and release excess carbon dioxide in the process of gas exchange.

Throughout this book, you have learned that most living organisms, like the soccer players in **Figure 31.12**, need to obtain oxygen from their environment and need to remove carbon dioxide from their bodies. As you read in Chapter 5, the oxygen is needed for cellular respiration, the chemical process that allows organisms to obtain energy from substances such as glucose. During aerobic respiration, carbon dioxide is produced as a waste product and must be removed from the cell. As you learned in Chapter 22, the evolution of lungs enables terrestrial animals to obtain oxygen from the atmosphere and release excess carbon dioxide.

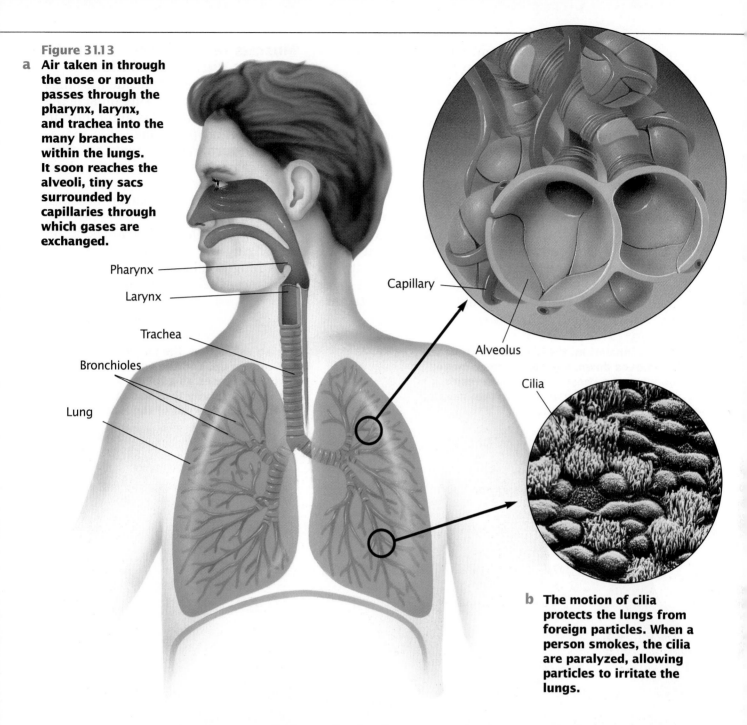

Figure 31.13

a Air taken in through the nose or mouth passes through the pharynx, larynx, and trachea into the many branches within the lungs. It soon reaches the alveoli, tiny sacs surrounded by capillaries through which gases are exchanged.

Pharynx

Larynx

Trachea

Bronchioles

Lung

Capillary

Alveolus

Cilia

b The motion of cilia protects the lungs from foreign particles. When a person smokes, the cilia are paralyzed, allowing particles to irritate the lungs.

Lungs contain branched tubes

Follow the passage of air through the respiratory system in **Figure 31.13**. Air enters the body through the nose or mouth and travels to the pharynx (*FAIR ihnks*), a tube at the back of the nose and mouth. From the pharynx, air enters the voice box, or larynx (*LAR ihnks*), and the windpipe, or trachea (*TRAY kee uh*). These passageways are lined with tissues that warm and moisten incoming air. The lower end of the trachea divides into two branches, which divide many times into smaller branches of tubes called bronchioles (*BRAHNG kee ohlz*). This branching network of tubes is lined with cilia, shown in **Figure 31.13b**, and a layer of protective mucus. The smallest of these tubes lead into **alveoli** (*al VEE uh leye*), clusters of tiny air sacs. Gases enter and leave the circulatory system through the alveoli. Each alveolus is surrounded by a network of capillaries. Blood in these capillaries picks up oxygen from the alveoli and releases carbon dioxide to be exhaled. Each lung contains about 150 million alveoli, providing a surface area larger than a small house for gas exchange.

Inhalation involves muscle contraction in the chest

Breathing begins when the diaphragm, the dome-shaped muscle below the chest cavity, contracts and moves downward. The muscles between the ribs also contract, causing the rib cage to move up and out. Together, these muscle contractions cause the chest cavity to enlarge. When the chest expands, the air pressure in the chest cavity drops. Air pressure outside the body is then greater than that inside the chest, causing air to rush into the lungs to equalize the pressure. This part of breathing is called **inhalation**.

Muscles return to their relaxed position during exhalation

When the air pressure inside the lungs is equal to the air pressure outside the lungs, the muscles relax and return to their original positions. This movement, in turn, reduces the size of the chest cavity. As the size of the chest cavity decreases, the air pressure inside the chest cavity gradually becomes greater than air pressure outside the body. Air then leaves the lungs, again equalizing the pressure. This part of breathing is called **exhalation**. Inhalation and exhalation are illustrated in **Figure 31.14**.

Figure 31.14

a During inhalation, the diaphragm contracts and moves down, and the rib cage moves up. When the chest expands as a result, air pressure in the chest cavity drops, causing air to rush into the lungs.

b During exhalation, the diaphragm relaxes and moves up, and the size of the chest cavity decreases as a result. Air pressure in the chest cavity thus increases, forcing air to be exhaled out of the lungs .

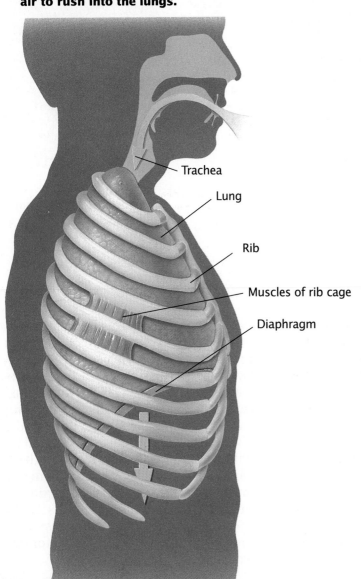

Trachea

Lung

Rib

Muscles of rib cage

Diaphragm

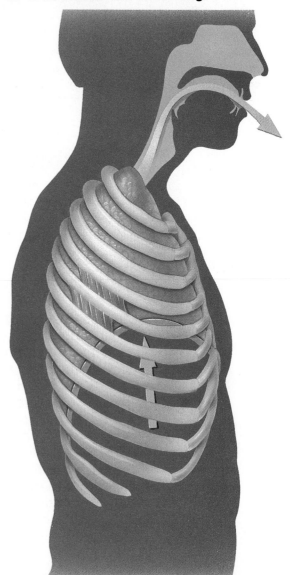

Figure 31.15
Follow the passage of carbon dioxide (CO_2) leaving body tissues (a) and being exhaled through the respiratory system (b) in the illustration below.

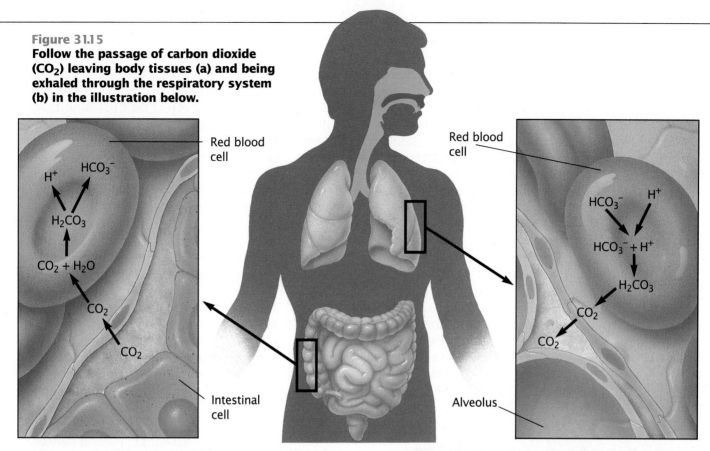

a When CO_2 leaves cells of the intestines, it can enter red blood cells and combine with hemoglobin. Or, it may combine with water to form carbonic acid, H_2CO_3, which breaks down to form hydrogen ions and bicarbonate ions.

b When blood reaches the lungs, the process is reversed and CO_2 is released. The CO_2 diffuses from the blood into the alveoli in the lungs. From there, it is exhaled along with water vapor.

Gas Exchange

The destination of the inhaled air is the alveoli, which are surrounded by capillaries. The tissue forming the walls of the alveoli and capillaries is only one cell thick. Gas exchange occurs when oxygen in the alveoli diffuses into the blood in the capillaries. In turn, the carbon dioxide in the blood diffuses into the air of the alveoli.

Oxygen binds to hemoglobin
In the blood, oxygen quickly binds with hemoglobin, the protein in red blood cells. Hemoglobin soaks up oxygen extremely effectively, which causes still more oxygen to enter red blood cells. The red blood cells then give up their oxygen to the cells of body tissues, where it is used in metabolism, the chemical activities of cells. As a result of metabolism, oxygen concentration in the body's cells is low, but carbon dioxide concentration is high.

Carbon dioxide is transported to the lungs to be exhaled
Carbon dioxide is a waste product that must be eliminated from cells. It is transported in the blood in three ways. About 5 percent of the carbon dioxide in the body dissolves in the plasma in blood. Another 25 percent enters the red blood cells and combines with hemoglobin. With help of an enzyme, the remaining 70 percent combines with water in the red blood cells to form carbonic acid, H_2CO_3. Because carbonic acid is unstable, hydrogen ions and bicarbonate ions quickly form, as shown in **Figure 31.15a**.

When blood reaches the lungs, chemical reactions occur that reverse the process, releasing carbon dioxide. As shown in **Figure 31.15b**, the carbon dioxide diffuses from the blood into the alveoli in the lungs. The carbon dioxide is exhaled with water vapor.

Regulation of Breathing

Receptors in the brain and circulatory system continuously monitor the levels of oxygen and carbon dioxide in the blood. These receptors enable the body to automatically regulate oxygen and carbon dioxide concentrations by sending signals to the brain. The brain responds by sending nerve signals to the diaphragm and rib muscles, speeding or slowing the rate of breathing. Perhaps surprisingly, carbon dioxide has more effect on breathing than does oxygen. For example, if the concentration of carbon dioxide in your blood increases, you breathe more deeply, ridding your body of excess carbon dioxide. When the carbon dioxide level drops, your breathing slows.

Diseases of the Respiratory System

Respiratory diseases affect millions of Americans. **Asthma** (*AZ muh*) is a respiratory disease in which certain airways in the lungs become constricted because of sensitivity to certain stimuli. The narrowing of the airways reduces the efficiency of respiration, which decreases the amount of oxygen reaching body cells.

Cigarette smoking is linked to emphysema and lung cancer, two respiratory diseases that claim millions of lives annually. In people who have **emphysema** (*em fuh SEE muh*), the lung tissue loses its elasticity, greatly reducing the efficiency of gas exchange. In **lung cancer**, carcinogens present in tobacco smoke trigger the growth of cancerous cells in lung tissue. More than 90 percent of lung cancer patients are smokers, as suggested by the antismoking posters in **Figure 31.16**. Lung cancer has an extremely low rate of cure. Fewer than 10 percent of its victims live more than five years after diagnosis. In Chapter 30 you learned about the effects of cigarette smoking on the body.

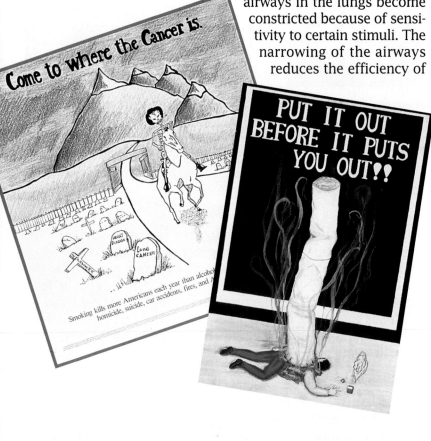

Figure 31.16
These posters, warning about some of the life-threatening illnesses associated with cigarette smoking, were designed by high school students.

Section Review

❶ What role do the diaphragm and rib muscles play in the processes of inhalation and exhalation?

❷ How does the exchange of gases occur in the lungs?

❸ When you exercise, you automatically begin to breathe faster. Explain how and why this occurs.

❹ How do asthma and emphysema affect respiration?

Strands of a protein called fibrin trap red and white blood cells. Soon a blood clot will form.

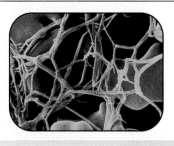

Key Terms

Summary

31.1 Circulation

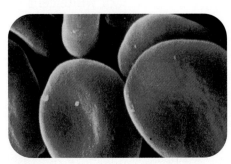

Red blood cells carry oxygen from the lungs to all cells of the body and carry carbon dioxide back to the lungs to be exhaled.

plasma (p. 712)

red blood cell (p. 712)

anemia (p. 712)

white blood cell (p. 713)

platelet (p. 713)

blood vessel (p. 714)

artery (p. 714)

vein (p. 714)

capillary (p. 714)

blood type (p. 716)

lymphatic system (p. 717)

- **The circulatory system transports materials to and from cells.**

- **Blood consists of plasma, red blood cells that transport oxygen, white blood cells that protect against infection, and platelets that help clotting.**

- **Blood types are defined by proteins on the surface of red blood cells.**

- **Blood vessels include arteries that carry blood away from the heart, veins that carry blood back to the heart, and capillaries that connect the arteries to the veins.**

- **The lymphatic system returns fluids back to the blood vessels.**

31.2 How Blood Flows

Blood pressure in a healthy adult is typically about 120/80 mm Hg.

atrium (p. 718)

ventricle (p. 718)

pacemaker (p. 719)

blood pressure (p. 724)

cardiovascular disease (p. 725)

hypertension (p. 725)

- **The heart has two sides: the right side moves blood to the lungs, and the left side moves blood to the rest of the body.**

- **Blood pressure consists of a high value (systolic) when the heart contracts and a lower value (diastolic) when the heart rests.**

- **Hypertension, or high blood pressure, is an example of a cardiovascular disease.**

31.3 The Respiratory System

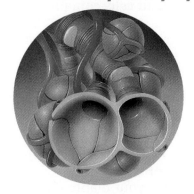

The walls of alveoli and the capillaries around them provide a large surface area for gas exchange.

alveolus (p. 729)

inhalation (p. 730)

exhalation (p. 730)

asthma (p. 732)

emphysema (p. 732)

lung cancer (p. 732)

- **When the diaphragm and the rib muscles contract, enlarging the chest cavity, inhalation occurs. When the muscles relax, air is forced out and exhalation occurs.**

- **Gas exchange occurs when oxygen in the alveoli diffuses into the blood in the capillaries. Carbon dioxide in the blood diffuses into the air of the alveoli.**

- **Breathing is regulated mainly by response to the level of carbon dioxide detected in the blood.**

- **Cigarette smoking is linked to emphysema and lung cancer.**

Understanding Vocabulary

1. For each pair of terms, explain the differences in their meanings.
   a. inhalation, exhalation
   b. red blood cell, white blood cell
   c. arteries, veins
   d. systolic pressure, diastolic pressure

Relating Concepts

2. Copy the unfinished concept map below onto a sheet of paper. Then complete the concept map by writing the correct word or phrase in each oval containing a question mark.

Understanding Concepts

Multiple Choice

3. When blood exits the right ventricle, it
   a. is not under pressure.
   b. is oxygenated.
   c. has more white cells than red cells.
   d. enters the pulmonary artery.

4. The maximum force exerted against the arterial walls occurs during
   a. ventricular contraction.
   b. systolic pressure.
   c. diastolic pressure.
   d. arterial relaxation.

5. By volume, blood is mostly
   a. white cells.   c. plasma.
   b. red cells.   d. platelets.

6. Most of the oxygen transported in the blood is
   a. dissolved in the plasma.
   b. bound to hemoglobin.
   c. in the form of carbonic acid.
   d. in the form of water.

7. Platelets act in
   a. blood clotting.
   b. red cell development.
   c. fighting infection.
   d. transporting oxygen.

8. Fluid balance in the blood is maintained by the
   a. lymphatic system.   c. arteries and veins.
   b. plasma.   d. enzymes.

9. Which chambers of the heart contract simultaneously?
   a. all four chambers
   b. right atrium and left atrium
   c. right atrium and right ventricle
   d. left atrium and left ventricle

10. A high count of white cells in the blood is a sign of
    a. bone marrow damage.
    b. infection.
    c. a bleeding wound.
    d. not enough hemoglobin.

11. If the amount of carbon dioxide in the blood increases, breathing
    a. stops.
    b. speeds up.
    c. becomes more infrequent.
    d. is controlled by hormones.

12. The site of gas exchange in the respiratory system is the
    a. lungs.   c. capillaries.
    b. bronchioles.   d. alveoli.

Completion

13. Bacteria and cancer cells that enter the body are destroyed in the _____ .

14. The pacemaker starts each _____ . It is _____ located at the entrance of the right _____ .

15. Blood is carried to the heart by the _____ and away from the heart by the _____ .

16. Red blood cells contain _____ that gives the cells their red color. A person who has too few red blood cells suffers from _____ .

17. A person with a blood pressure of 190/120 suffers from _____ , or high blood pressure.

Short Answer

18. Where is your diaphragm located? How does it function in breathing?

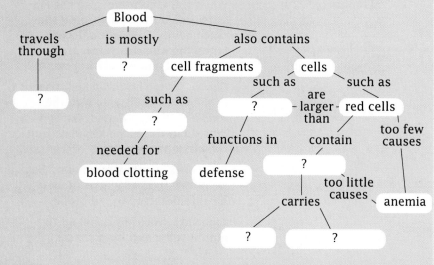

19. How are the "lubb" and "dup" sounds of the heart and the movement of the heart valves related?

20. Describe the three ways that carbon dioxide is transported in the blood.

21. Where do new red blood cells come from? What happens to old ones?

22. The circulatory system transports oxygen, carbon dioxide, and hormones. What else does it do?

Interpreting Graphics

23. Look at the cross sections of the two coronary arteries below and answer the questions that follow.

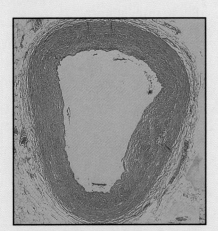

a

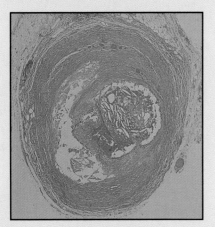

b

- How are the two arteries different?
- How can you explain the differences observed?

Reviewing Themes

24. *Stability*
   How do white blood cells help the body maintain homeostasis?

25. *Scale and Structure*
   Why can the four-chambered heart of mammals be thought of as two hearts in one?

26. *Evolution*
   Explain why the vertebrate lung is considered an adaptation to terrestrial life.

Thinking Critically

27. *Comparing and Contrasting*
   How are the pulmonary circulation and systemic circulation different?

28. *Comparing and Contrasting*
   How do red blood cells and white blood cells differ in terms of function, size, and number?

29. *Building on What You Have Learned*
   In Chapter 25 you learned about the open circulatory system of arthropods. How does the human circulatory system differ from that found in arthropods?

Cross-Discipline Connection

30. *Biology and Health*
   Asthma attacks are brought on by something ingested or inhaled. Do library research to discover what substances trigger asthma attacks.

Discovering Through Reading

31. Read the article "Working Out Under Pressure," in *Health*, July/August 1992, pages 98–99. What was wrong with Marc Cohen? What did Dr. Siegel prescribe for Marc? What information serves to justify Dr. Siegel's prescription?

Investigation

How Does Exercise Affect Pulse Rate?

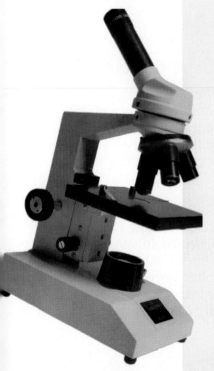

Objectives

In this investigation you will:
- *formulate* a hypothesis stating the effect that exercise has on pulse rate
- *design* an experiment that will compare pulse rate while at rest and immediately after exercise
- *collect* and *record* data

Materials

- unlined paper
- pencil
- watch with a second hand
- stable stool or chair about 30 cm (12 in.) high
- calculator

Prelab Preparation

Review what you have learned about circulation and respiration by answering the following questions:
- What is it that you are feeling when you take your pulse?
- What is the connection between the respiratory system and the circulatory system?

Procedure: Determining the Effect of Exercise on Pulse Rate

1. Form a cooperative team of two students. Work with your partner to complete steps 2–6.

2. Practice finding your partner's pulse by placing your index and middle fingers on the inner part of his or her wrist just below the base of the thumb. Use a watch with a second hand to determine the pulse rate for a 60-second interval.

3. Prepare a data table like the one shown below.

Pulse rate/minute

	Resting rate		After exercise	
Subject	1	2	1	2
Trial 1				
Trial 2				
Trial 3				
Individual average				
Team average				
Class average				

4. Discuss the question "How does exercise affect pulse rate?" with your partner. Formulate a hypothesis that answers the question.

5. Use the following guidelines to design an experiment to test your hypothesis.
   a. One member of the team will be the subject while the other member observes and records data. Team members will then switch roles and repeat the experiment.
   b. Resting pulse rate will be taken while the subject is sitting quietly.

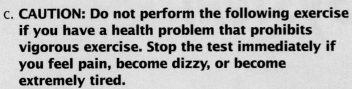

c. **CAUTION: Do not perform the following exercise if you have a health problem that prohibits vigorous exercise. Stop the test immediately if you feel pain, become dizzy, or become extremely tired.**

d. The subject will exercise by stepping onto and off of a stool that is about 30 cm (12 in.) high. The subject should step at a rate of about 30 times a minute for three minutes. Pulse rate is taken immediately afterward.

e. Data should be collected for more than one trial for each subject. Calculate the subject's average pulse rate.

f. Collect data from other teams. Calculate the average pulse rate for the class when appropriate.

g. All data should be organized in a table.

6. After your design is approved by your teacher, conduct your experiment with your partner.

Analysis

1. *Summarizing Data*
   Summarize your data. Explain whether the data support your hypothesis.

2. *Analyzing Data*
   State your conclusion about how pulse rate changes after exercise.

3. *Evaluating Methods*
   Why is it best to average data from many trials and from a number of subjects?

4. *Predicting Outcomes*
   What changes might occur in pulse rate after a person completes an eight-week physical fitness course? Explain the reasons for your answer.

Thinking Critically

Why does your breathing and pulse rate increase when you exercise?

Chapter 32

The Immune System

Review

- **membrane marker protein (Section 3.2)**
- **viral reproduction (Section 16.3)**
- **the term *pathogen* (Glossary)**

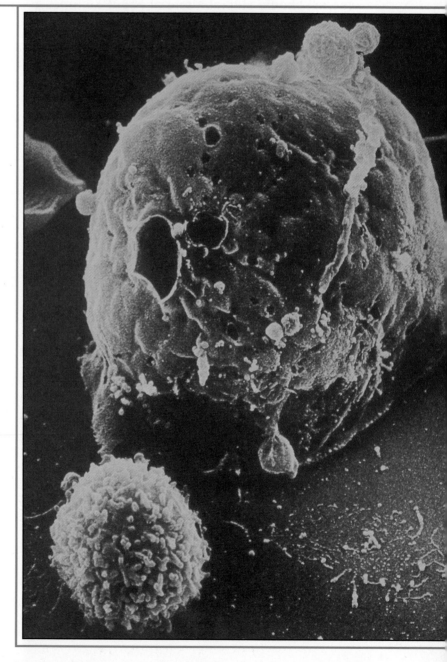

Immune system cells called killer T cells protect the body by destroying tumors. The large tumor cell above is being attacked by the smaller killer T cell.

LIKE A CITY UNDER SIEGE, YOUR BODY IS SURROUNDED BY LEGIONS OF WOULD-BE INVADERS: BACTERIA, VIRUSES, PROTISTS, PARASITIC ANIMALS, AND PARASITIC FUNGI. YET YOU ARE USUALLY WELL, BECAUSE YOUR BODY REPELS THESE PATHOGENS. IT ALSO DEFEATS THE FEW PATHOGENS THAT DO MANAGE TO ENTER YOUR BODY, USING A POWERFUL DEFENSE CALLED THE IMMUNE SYSTEM.

32.1   *First Line of Defense*

Objectives

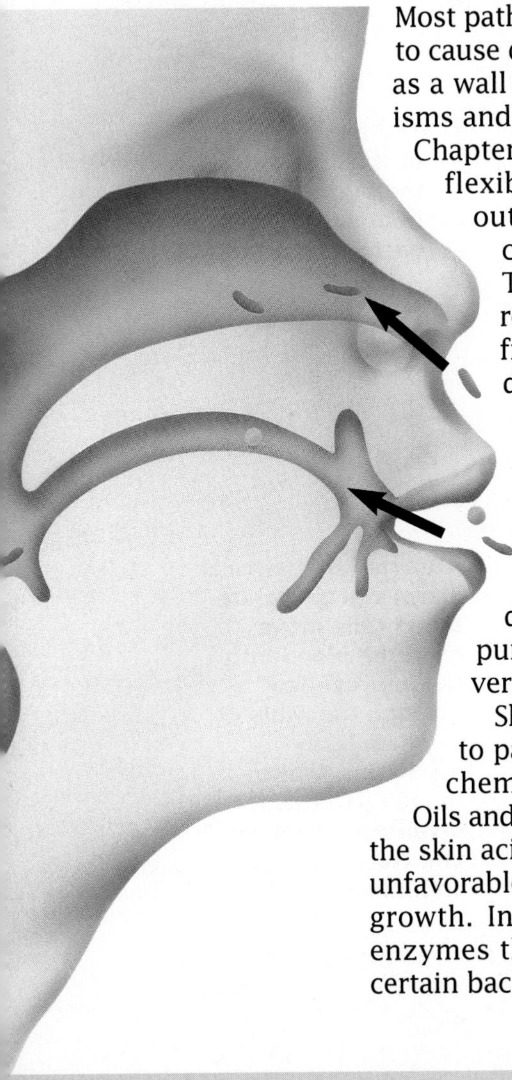

Figure 32.1
Every time you breathe, you inhale a variety of bacteria and viruses that could make you sick.

❶ **Identify two ways in which skin repels pathogens.**

❷ **Recognize the role of mucous membranes in defending the body.**

❸ **Summarize the reactions that make up the inflammatory response.**

❹ **Describe how your body distinguishes its own cells from invading pathogens.**

Keeping Pathogens Out

Most pathogens must enter the body to cause disease. Your skin functions as a wall to keep out foreign organisms and viruses. As you learned in Chapter 27, skin is the body's dry, flexible covering. Cells of the outer layers of epidermis are continually being worn away. These lost cells are quickly replaced by cells moving up from the lower layers of epidermis, where cell division is occurring. In only 40 minutes, your body loses and replaces approximately 1 million skin cells. Such rapid replacement of body cells keeps the skin from disintegrating and enables punctures or cuts to be sealed very quickly.

Skin not only acts as a barrier to pathogens, it also engages in chemical warfare against them. Oils and sweat secreted by glands in the skin acidify its surface, creating an unfavorable environment for bacterial growth. In addition, sweat contains enzymes that digest the cell walls of certain bacteria.

As shown in **Figure 32.1**, each time you inhale or eat, bacteria and viruses can bypass the protection provided by your skin. Openings in your skin—such as the mouth, nostrils, eyes, and anus—are essential to allow food, oxygen, and sensory input to enter the body, and to allow wastes such as carbon dioxide, urine, and feces to be eliminated. Your body guards these openings with another series of defenses. For instance, tears and saliva contain the same antibacterial enzyme found in sweat. And internal body surfaces that come into contact with the environment are covered with **mucous membranes**.

A mucous membrane is a moist epithelial layer that is impermeable to most pathogens. Mucous membranes line the nasal passages, mouth, lungs, digestive tract, urethra, and vagina. Mucous membranes contain glands that secrete mucus, a sticky fluid that traps pathogens. Pathogens inhaled into the respiratory tract become lodged in mucus and are swept into the mouth by cilia. They are then swallowed and pass into the stomach. Digestive enzymes and strong acids in the stomach destroy most pathogens that are swallowed.

Fighting Off a Local Infection

A splinter in your fingertip is only a minor injury, even though it punches a hole in your skin through which pathogens can enter. Your body quickly closes the puncture and activates a set of chemical and cellular defenses to destroy any invaders. Your finger swells, becomes red and painful, and feels hot. These are signs that your body is attacking pathogens. The redness and swelling are part of the **inflammatory response** to infection or injury. You can follow the events of an inflammatory response in **Figure 32.2**.

White blood cells destroy pathogens

White blood cells are the "soldiers" of the immune system and are crucial for the inflammatory response. **Phagocytes** (*FAG oh seyets*) are white blood cells that launch direct attacks on pathogens, ingesting them by phagocytosis.

Phagocytes consume bacteria, cells infected by viruses, dead cells, and foreign particles. **Macrophages** (*MAK roh fayjz*) are the most common phagocytes. Another type of phagocyte, called a neutrophil (*NOO troh fihl*), releases the same chemical found in household bleach, killing itself and any nearby bacteria. Phagocytes, dead cells, and pathogens are the components of pus, a yellowish or whitish fluid that often accumulates in a wound. The presence of pus signals that phagocytes are combating pathogens. The inflammatory response is sufficiently powerful to repel most small-scale infections.

Cells infected by viruses release a chemical messenger, a protein called interferon (*in tuhr FIHR ahn*). Interferon activates white blood cells that kill pathogens, and so increases the ability of noninfected cells to resist infection. Interferon may have medical uses.

Figure 32.2
When a splinter enters your finger, bacteria and viruses can gain access to your body. Your body initiates the inflammatory response to combat these pathogens.

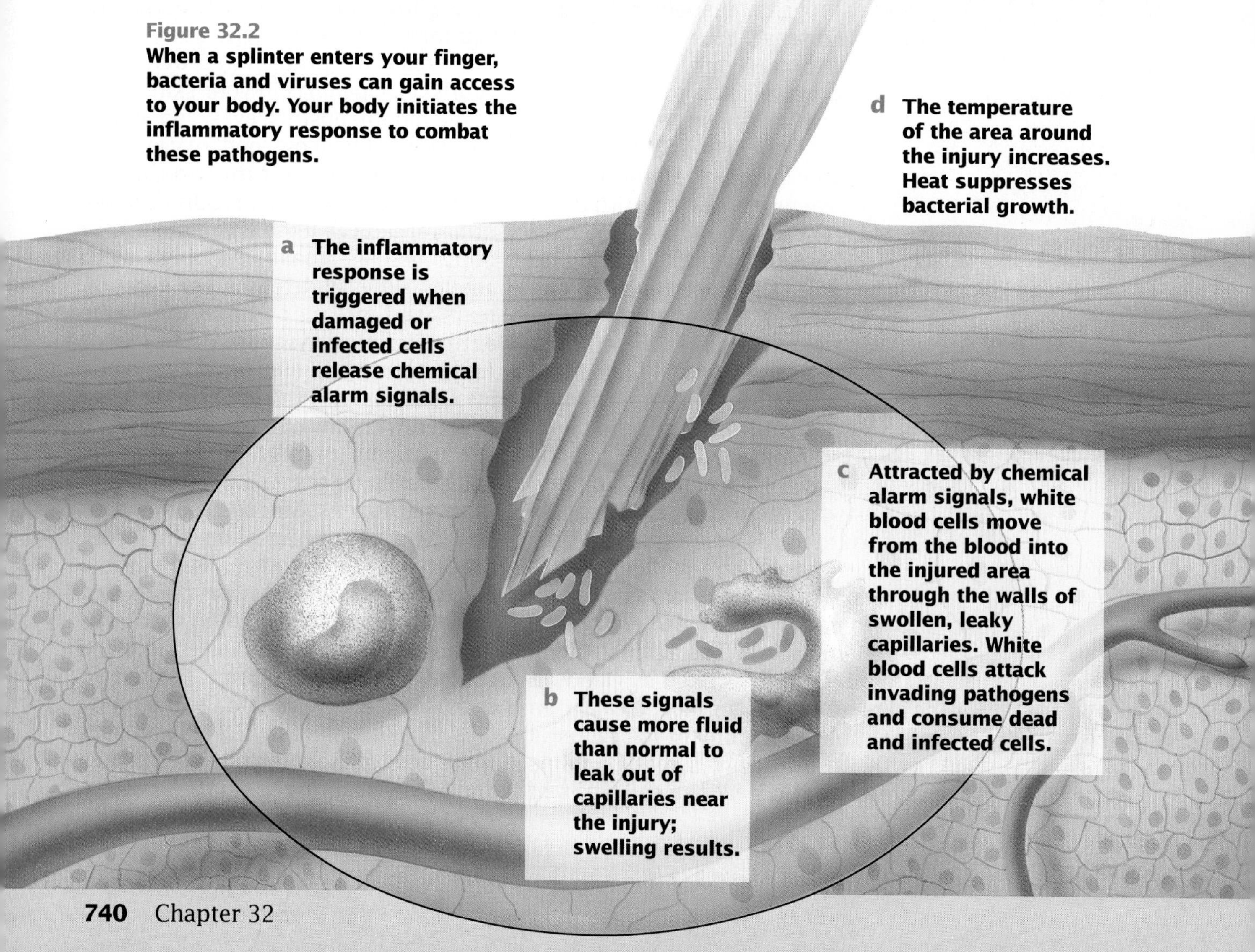

a The inflammatory response is triggered when damaged or infected cells release chemical alarm signals.

b These signals cause more fluid than normal to leak out of capillaries near the injury; swelling results.

c Attracted by chemical alarm signals, white blood cells move from the blood into the injured area through the walls of swollen, leaky capillaries. White blood cells attack invading pathogens and consume dead and infected cells.

d The temperature of the area around the injury increases. Heat suppresses bacterial growth.

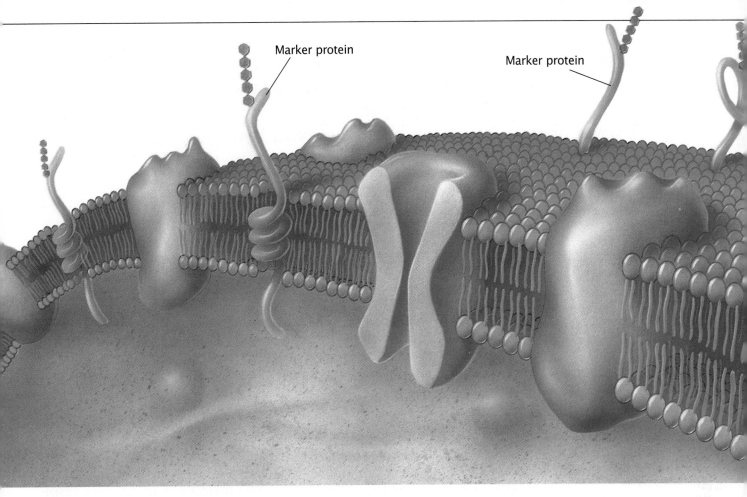

Marker protein

Marker protein

Figure 32.3

Because no two organisms have the same marker proteins, cells of the immune system are able to identify and destroy invading pathogens.

Recognizing Pathogens

How does a phagocyte recognize pathogens and infected cells? Recall from Chapter 3 that channel, receptor, and marker proteins occur in the membrane of a cell. Certain marker proteins serve as identification tags that enable white blood cells to distinguish your cells from foreign cells. As shown in **Figure 32.3**, each cell in your body carries "self" marker proteins that identify it as your cell. Your marker proteins have a unique amino acid sequence and therefore a unique shape. The shape of your marker proteins differs from the shape of marker proteins on cells of other humans and on the cells of other species, such as pathogenic bacteria. Your marker proteins are also different from viral surface proteins. White blood cells have receptor proteins in their membranes that enable them to recognize proteins on the surface of a cell or virus. Your white blood cells recognize proteins shaped differently from your own as foreign, or "non-self." Molecules that can be recognized by white blood cells and that can trigger a defensive response are known as **antigens**.

Section Review

❶ Describe the chemical defenses of skin.

❷ Explain two ways that your mucous membranes protect you against pathogenic bacteria.

❸ Describe the events that occur after a splinter enters your finger.

❹ Explain how white blood cells can distinguish your cells from those of invading pathogens.

THE PAIN OF A SPLINTER DOES NOT COMPARE TO THE SUFFERING YOU
ENDURE WHEN YOU HAVE THE FLU. THE SYMPTOMS OF FLU ARE MORE
SEVERE BECAUSE INFLUENZA VIRUSES BREAK THROUGH YOUR FIRST LINE
OF DEFENSE. AS YOU WILL SEE IN THIS SECTION, YOUR BODY ACTIVATES
ANOTHER SET OF DEFENSES, CALLED THE IMMUNE RESPONSE, THAT
COMBATS AND DEFEATS THE FLU VIRUSES.

32.2   *The Immune Response*

Objectives

① **Identify the functions of helper T cells, killer T cells, and B cells.**

② **Explain how fever helps to defeat pathogens.**

③ **Describe how the immune system protects against a pathogen's second attack.**

④ **Relate vaccines to the functioning of the immune system.**

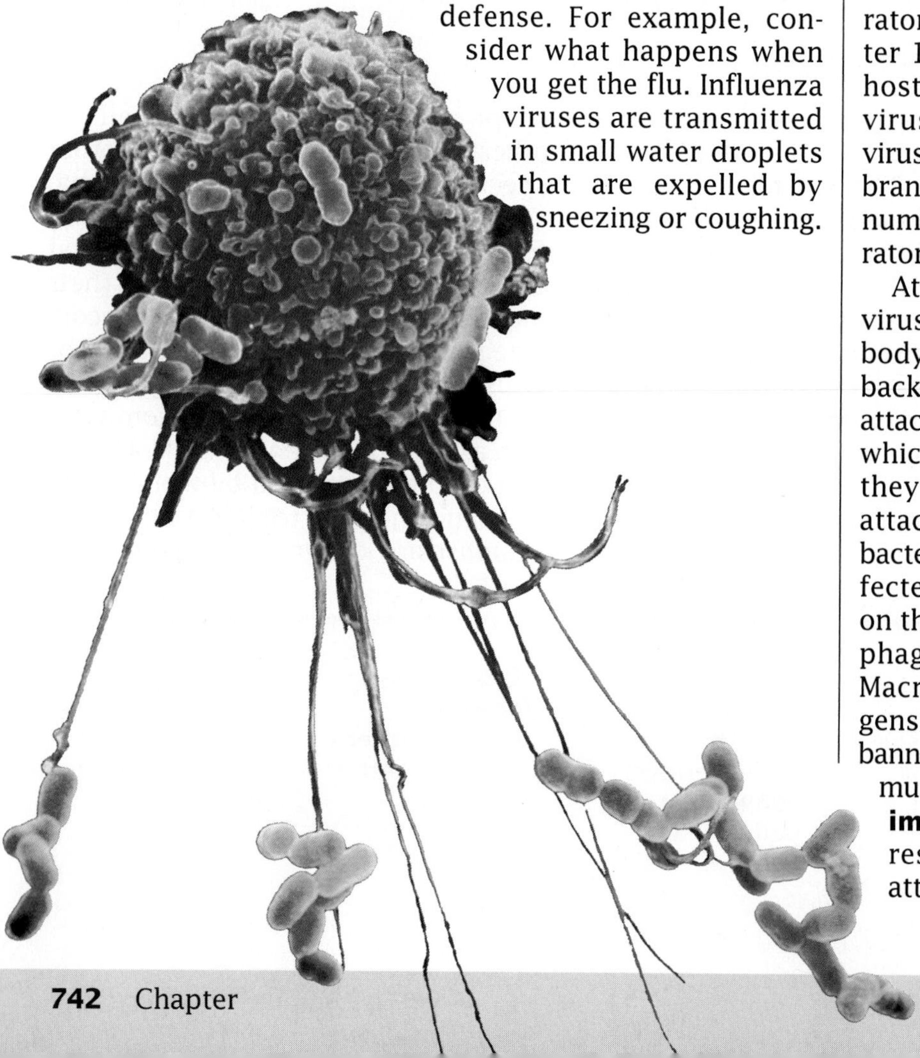

**Figure 32.4
Macrophages
attack foreign
pathogens, such
as viruses or the
E. coli bacteria
shown here.**

Second Line of Defense

The immune system is the defenses your body uses to attack pathogens that get past your first line of defense. For example, consider what happens when you get the flu. Influenza viruses are transmitted in small water droplets that are expelled by sneezing or coughing.

If you inhale some of these droplets, influenza viruses can enter cells of the mucous membrane lining your respiratory tract. As you recall from Chapter 16, viruses seize control of their host cells and transform them into virus-producing factories. Influenza viruses take over and kill mucous membrane cells. You feel sick because large numbers of the cells lining your respiratory tract are dying.

At this point during a case of flu, the viruses have the upper hand, but your body's defenses are beginning to fight back. The first stages of this counter-attack are carried out by macrophages, which consume any influenza viruses they encounter. Macrophages also attack other invaders, including the bacteria shown in **Figure 32.4**. Cells infected by viruses have viral antigens on their membranes. Patrolling macrophages consume these cells as well. Macrophages then display viral antigens on their cell surface, like victory banners. This display stimulates the immune system to carry out a full-blown **immune response**. The immune response is the immune system's attack on a specific pathogen.

T Cells: Command and Attack

White blood cells called **T cells** control the immune response and attack infected cells. T cells are so named because they mature in the thymus, a small gland above the heart. T cells circulate in your blood and lymph, and occur in your spleen and lymph nodes. The three important classes of T cells are helper T cells, killer T cells, and suppressor T cells. Each class carries out a different task during the immune response.

An individual T cell carries specifically shaped receptor proteins on its cell membrane. These receptor proteins enable the T cell to recognize and bind to one particular antigen. Your body can respond to millions of different antigens because it manufactures millions of different types of T cells, each type bearing uniquely shaped proteins. When you have the flu, T cells with receptors that match the antigens of the influenza viruses will be "called up" to fight the viruses.

Helper T cells command the immune response

Figure 32.5 describes the roles of T cells in the immune response. In order for T cells to begin combating influenza viruses, helper T cells, the commanders of the immune response, must be activated. Activation occurs when a helper T cell with a receptor matching the influenza antigen meets a macrophage displaying this antigen (as a "trophy" of its recent encounter with the virus) on its cell membrane.

Figure 32.5
The immune response is activated when a macrophage that has consumed a pathogen, such as an influenza virus, comes into contact with a helper T cell.

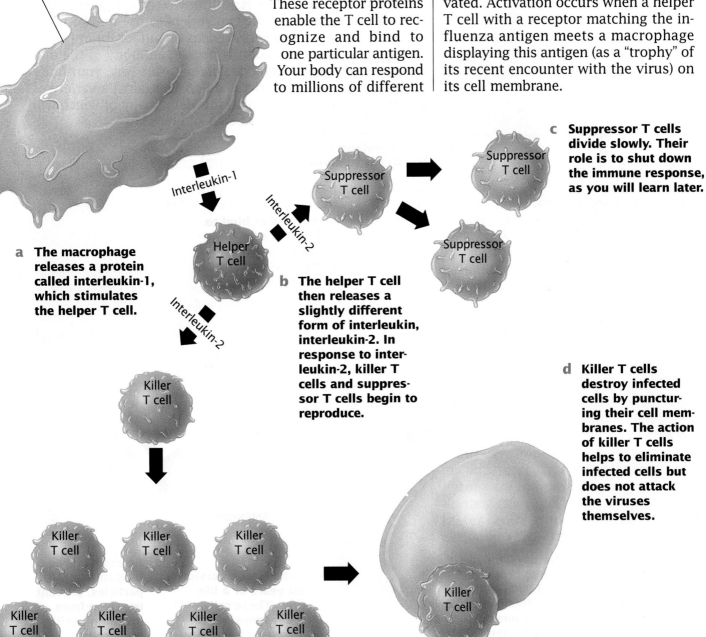

Macrophage

Interleukin-1

Interleukin-2

Helper T cell

Suppressor T cell

Suppressor T cell

Suppressor T cell

Interleukin-2

Killer T cell

Killer T cell

Killer T cell

Killer T cell

Killer T cell

Killer T cell

Killer T cell

Killer T cell

Killer T cell

a The macrophage releases a protein called interleukin-1, which stimulates the helper T cell.

b The helper T cell then releases a slightly different form of interleukin, interleukin-2. In response to interleukin-2, killer T cells and suppressor T cells begin to reproduce.

c Suppressor T cells divide slowly. Their role is to shut down the immune response, as you will learn later.

d Killer T cells destroy infected cells by puncturing their cell membranes. The action of killer T cells helps to eliminate infected cells but does not attack the viruses themselves.

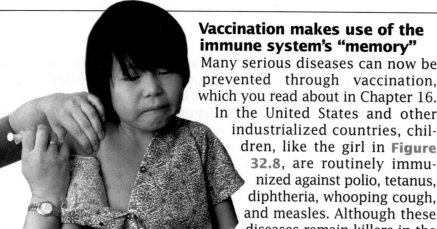

Vaccination makes use of the immune system's "memory"

Many serious diseases can now be prevented through vaccination, which you read about in Chapter 16. In the United States and other industrialized countries, children, like the girl in **Figure 32.8**, are routinely immunized against polio, tetanus, diphtheria, whooping cough, and measles. Although these diseases remain killers in the less industrialized countries, groups like the World Health Organization have begun mass immunization programs in these nations. **Figure 32.9** shows a volunteer telling an Ethiopian woman about the vaccination procedure.

Vaccination triggers an immune response against a particular pathogen without causing the disease itself.

Figure 32.8
This Vietnamese-American child is receiving the vaccinations required before entering school.

Recall that the immune response is triggered by antigens. Vaccines are effective because they contain antigens that have been stripped of their disease-causing abilities. For instance, vaccines for bacterial diseases usually contain bacteria killed by heat or chemical treatment. B cells, T cells, and macrophages respond to the antigens on the dead bacteria as if encountering live, harmful bacteria. Similarly, viral vaccines stimulate the immune response with viruses made harmless by chemicals or genetic engineering. As you learned in Chapter 8, scientists can now produce "piggyback" vaccines through genetic engineering. These vaccines contain harmless viruses that have been altered to express the surface proteins of a pathogen. When injected into the body, the harmless viruses serve as antigens, stimulating production of antibodies and memory cells against the pathogen.

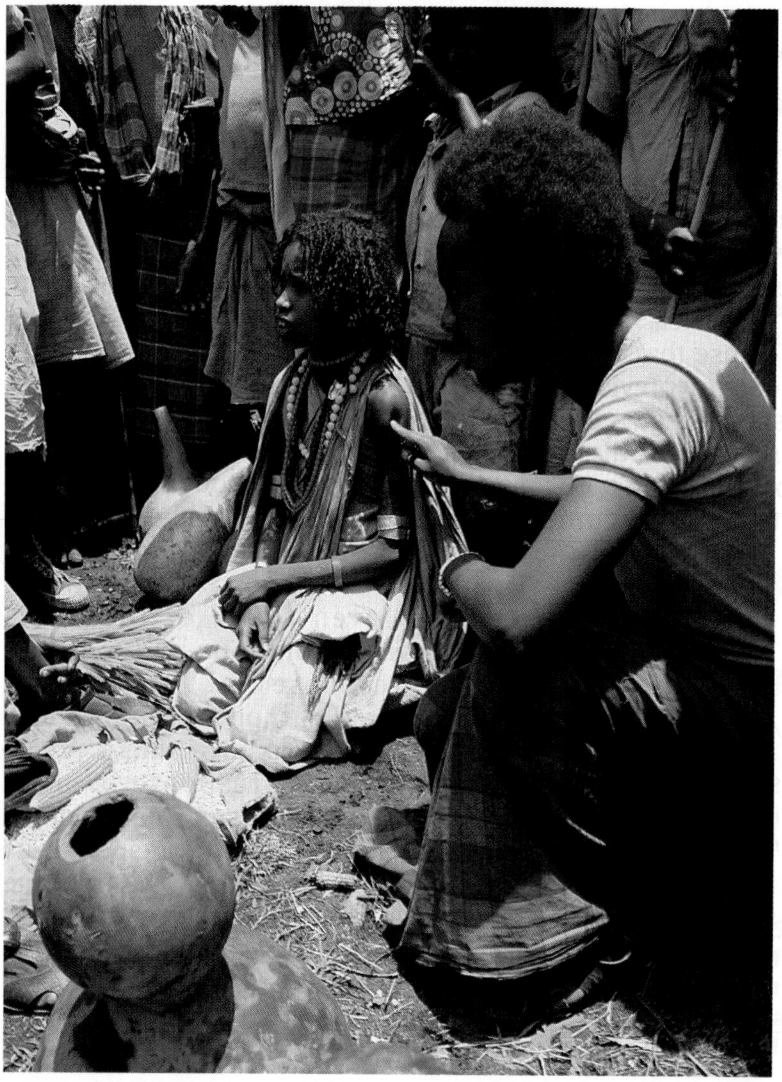

Vaccines aren't effective against rapidly evolving viruses

You can catch the flu more than once, unlike measles or mumps. Furthermore, although you can be immunized against influenza, vaccination does not provide long-term protection against infection. The viruses that cause flu have evolved a way to evade the immune system. Genes coding for the surface proteins of flu viruses mutate, or change, rapidly. Thus, the shapes of these surface proteins alter swiftly. Your body does not recognize viruses with altered surface proteins as the same viruses it has already successfully defeated or been vaccinated against. When these viruses invade your body, they provoke a new primary immune response and you get sick again. HIV (the AIDS virus) also mutates rapidly and has evaded scientists' attempts to produce successful vaccines against it.

Figure 32.9
This rural doctor is explaining the vaccination procedure and its benefits to a woman at an Ethiopian street market.

Antigens and Blood Types

Figure 32.10
Some blood types are not compatible. When incompatible blood types are mixed, they form clumps (shown below) that prevent blood from circulating properly.

You learned about the different blood types in Chapter 31. Transfusions between people of different blood types usually are not compatible because the immune system of the recipient attacks the transfused blood. People with type A blood have a marker protein known as the A antigen on the surface of their red blood cells. Red blood cells from individuals with type B blood have a slightly different marker protein called the B antigen. People with type AB blood have both A and B antigens on their red blood cells. Individuals with type O blood have neither A nor B antigens on their red blood cells. Individuals produce antibodies to the marker proteins not found on their own cells. For instance, individuals with type A blood produce antibodies against the B antigen, even if they have never been exposed to it. **Table 32.1** summarizes the antigens and antibodies found in people of each blood type.

If type A blood is transfused into a person with type B or type O blood, antibodies against the A antigen will attack the foreign red blood cells, causing them to clump together, as shown in **Figure 32.10**. Clumps of red blood cells can block capillaries and cut off blood flow, which can be fatal. Clumping also occurs if type B blood is transfused into individuals with type A or type O blood, or if type AB blood is given to people with type O blood.

Clumping of incompatible blood types

Table 32.1 Antibodies Produced in People of Various Blood Types

Blood type	Antigen present	Antibodies made
A	A	Anti-B
B	B	Anti-A
AB	A,B	Neither
O	None	Anti-A, Anti-B

Section Review

❶ **Explain how destruction of helper T cells would affect the immune response.**

❷ **Contrast the roles of B cells and T cells in the immune response.**

❸ **Explain the role of memory cells in the immune system.**

❹ **What can you conclude about the surface proteins of viruses for which effective vaccines exist?**

TYPE 1 DIABETES IS A DISEASE IN WHICH CELLS ARE UNABLE TO TAKE IN GLUCOSE BECAUSE THE PANCREAS FAILS TO PRODUCE INSULIN. TYPE 1 DIABETES IS THOUGHT TO BE THE RESULT OF AN IMMUNE SYSTEM DEFECT. INSTEAD OF ATTACKING INVADING PATHOGENS, THE IMMUNE SYSTEM ATTACKS INSULIN-MANUFACTURING CELLS OF THE PANCREAS. TYPE 1 DIABETES IS ONE EXAMPLE OF AN IMMUNE SYSTEM FAILURE.

32.3 Immune System Failure

Objectives

❶ Describe the events of an allergic reaction.

❷ Contrast cancer cells with normal cells.

❸ Relate the symptoms of AIDS to the action of HIV.

❹ Describe how HIV is transmitted.

Immune Overreaction

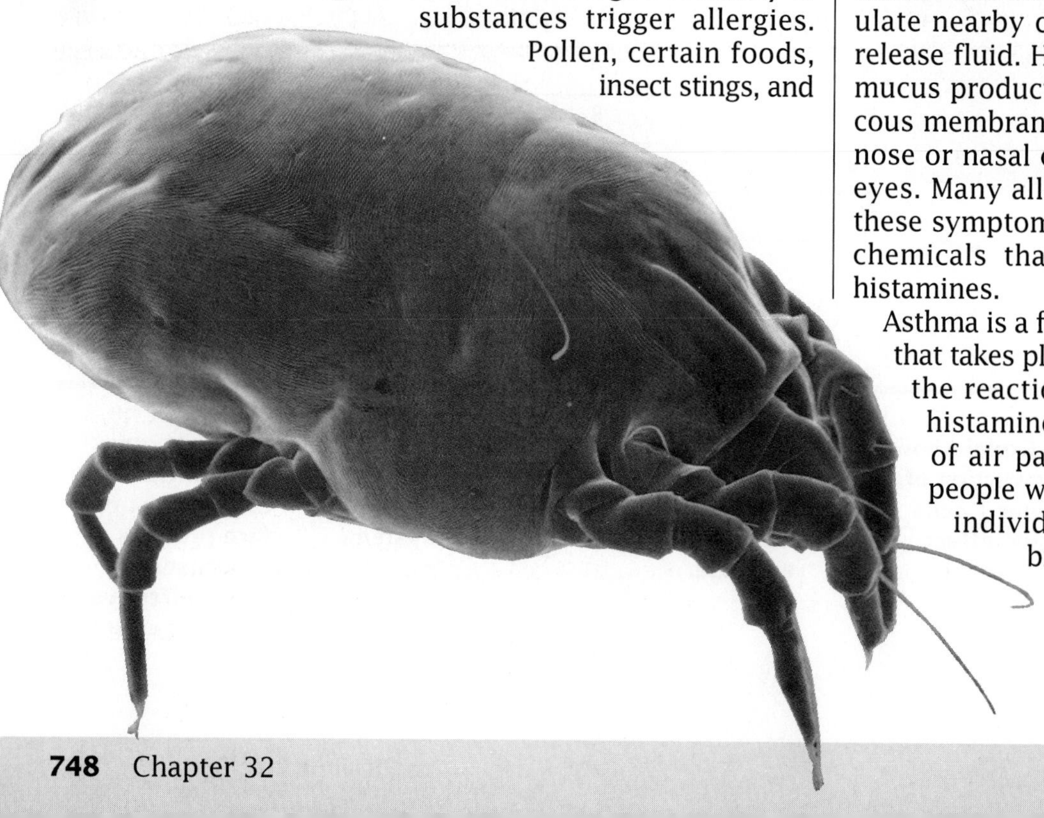

Figure 32.11
People with allergies to dust are actually allergic to the feces of the house-dust mite, which lives on dust particles. This photograph is magnified 300 times.

Petting a cat, smelling some flowers, or walking into a dusty room can be miserable experiences for some people. In these individuals, such activities rapidly lead to sneezing, itchy nose and eyes, nasal congestion, and even difficulty in breathing. These are some of the symptoms of an **allergy**. An allergy is an immune system response against a harmless antigen. A variety of substances trigger allergies. Pollen, certain foods, insect stings, and dust can all cause allergies. Allergies to dust are responses to proteins in the feces of tiny mites, such as the one in **Figure 32.11**, that live on dust particles.

If you inhale pollen you are allergic to, cells in the nasal passages release a set of chemical messengers that includes **histamine** (*HIHS tuh meen*), as shown in **Figure 32.12** on page 749. Histamine and the other messengers stimulate nearby capillaries to swell and release fluid. Histamines also increase mucus production by cells of the mucous membranes, resulting in a runny nose or nasal congestion, and watery eyes. Many allergy medicines relieve these symptoms with **antihistamines**, chemicals that block the action of histamines.

Asthma is a form of allergic response that takes place in the lungs. Besides the reactions already described, histamines cause the narrowing of air passages in the lungs of people who have asthma. These individuals will have trouble breathing when exposed to antigens to which they are allergic.

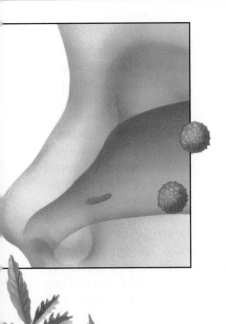

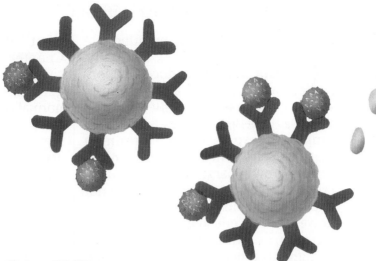

Figure 32.12

Phase 1
When pollen from a plant enters the nose of a person who is allergic to it, pollen antigens attach to antibodies on the surfaces of the cells lining the nasal passages.

Phase 2
Binding of antibodies stimulates cells to release histamine and other chemicals that cause symptoms including sneezing, runny nose, and itchy eyes.

Autoimmune diseases result when the body manufactures "anti-self" antibodies

Distinguishing self from non-self is the key ability of the cells of the immune system. In certain diseases, this ability breaks down, and the body attacks its own tissues. Such diseases are called **autoimmune diseases**. Multiple sclerosis is an autoimmune disease that usually strikes people between the ages of 20 and 40. In multiple sclerosis, the immune system attacks and destroys the insulating myelin sheath that covers motor nerves. Degeneration of the myelin sheath interferes with transmission of nerve impulses, which eventually cannot travel at all. Voluntary functions, such as movement of the limbs, and involuntary functions, such as bladder control, are lost. Multiple sclerosis usually leads to paralysis and death. Scientists do not know what stimulates the immune system to attack myelin. Some other autoimmune diseases, the tissues or organs they affect, and their symptoms are listed in **Table 32.2.**

Table 32.2 Some Autoimmune Diseases

Disease	Areas affected	Symptoms
Systemic lupus erythematosus	Connective tissue, joints, kidney	Facial skin rash, painful joints, fever, fatigue, kidney problems, weight loss
Type 1 diabetes	Insulin-producing cells in pancreas	Excessive urine production, blurred vision, weight loss, fatigue, irritability
Graves' disease	Thyroid	Weakness, irritability, heat intolerance, increased sweating, weight loss, insomnia
Rheumatoid arthritis	Joints	Crippling inflammation of the joints

Cancer: Unrestrained Cell Division

Evolution

The protists that cause African sleeping sickness can rapidly change the antigens on their cell surfaces. Why is this an effective way to evade the immune system?

A major function of your immune system is to ward off cancer. Normally, cells in your body reproduce at a controlled rate. Cancer is a condition in which cells lose the ability to stop dividing. In benign cancers, cells divide rapidly but spread little, forming a mass of cells called a **tumor**. Unless tumors grow so large that they actually crush the internal organs, they are not usually fatal. In malignant cancers, however, the cells aggressively penetrate surrounding tissues. Cells of malignant cancers can even spread throughout the body via the lymph system and bloodstream. Because of the damage they cause to tissues they invade, malignant cancers are often fatal unless detected early.

What causes normal cells to become cancer cells? Scientists think that mutation of growth-regulating genes is what sets off cancer. As you read in Chapter 14, a mutagenic agent that causes cancer is known as a carcinogen. Carcinogens that can transform normal cells into cancer cells are found in cigarette smoke (smoking is a leading cause of lung cancer) and in a variety of industrial chemicals. Some viruses can also cause cells to become cancerous. Mutation also occurs spontaneously in your body all the time, so some cancer cells are being produced "naturally" each day. The immune system usually identifies and destroys these occasional cancer cells before they spread. **Figure 32.13** shows a T cell killing a cancer cell. Surveillance by the immune system is your body's primary defense against cancer. Among AIDS patients, who lack effective immune systems, cancer is the leading cause of death.

Figure 32.13

a When a killer T cell comes into contact with the cells of a tumor, it is able to recognize the tumor cells as enemies. Below you can see a killer T cell (upper right) that has bound to a tumor cell.

b The killer T cell releases proteins that disrupt the tumor cell's membrane. The tumor cell tries to defend itself by forming blisters. Eventually, the tumor cell's membrane breaks and the integrity of the cell is lost.

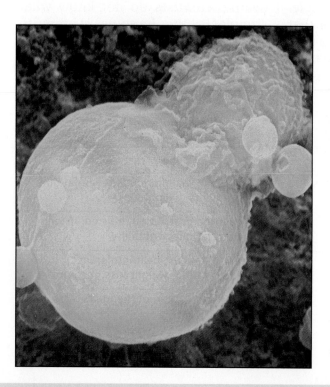

AIDS: Immune System Collapse

When this book was published, more than 150,000 Americans had died of AIDS (Acquired Immune Deficiency Syndrome) and more than 250,000 had the disease. More than 1.5 million Americans were thought to be infected by HIV (Human Immunodeficiency Virus), the virus that causes AIDS.

HIV attacks and cripples the immune system. It invades macrophages and helper T cells, like the one shown in **Figure 32.14**. HIV transforms these cells into virus factories, killing large numbers of helper T cells. When the number of helper T cells in the blood has fallen to very low levels, a person is said to have AIDS. Without helper T cells to stimulate and direct B cells and killer T cells, the immune response cannot occur. The body is overwhelmed by pathogens and cancers that it normally would defeat. Scientists think that practically everyone infected with HIV will eventually develop AIDS. The time between infection and onset of AIDS can be 10 years or longer. During this time, an infected person can still transmit the virus to others.

HIV is transmitted in body fluids

There is no cure for AIDS. You can protect yourself from AIDS by avoiding exposure to HIV. HIV is a fragile virus that cannot exist for long outside cells. You can get HIV only by contacting the HIV-infected blood cells or body fluids of infected individuals. Because HIV is found in both semen and vaginal secretions, you can contract HIV through sexual intercourse with an infected person; worldwide, most HIV infections are spread this way. The virus can be transmitted in either direction during intercourse. Use of a condom during sex greatly reduces but does not eliminate the risk of getting HIV.

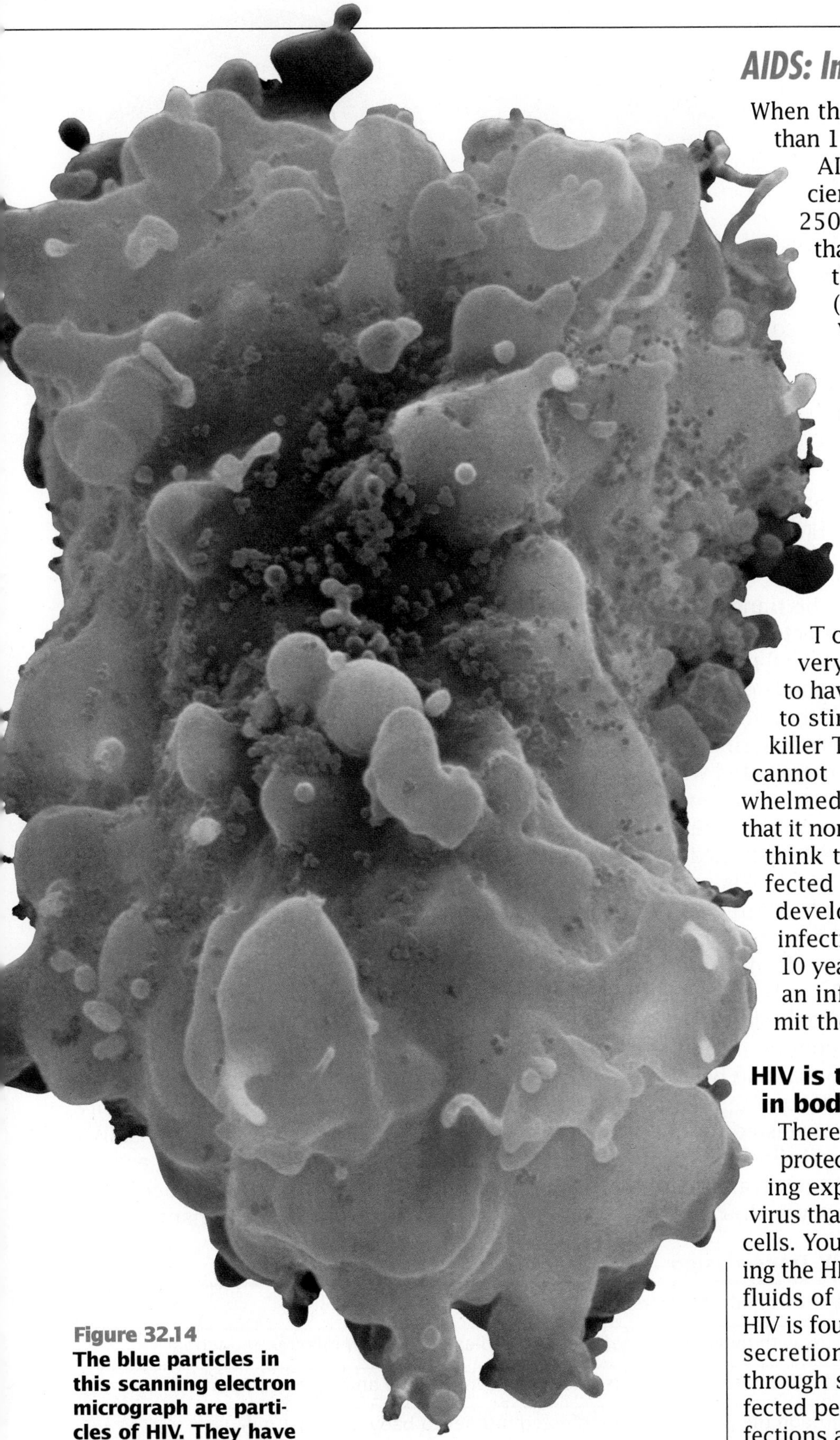

Figure 32.14
The blue particles in this scanning electron micrograph are particles of HIV. They have been produced by this helper T cell and are leaving the cell to infect other cells.

Figure 32.15
AIDS is an equal opportunity disease. Anyone can be infected with HIV, regardless of age, sex, or ethnic background.

Ryan White was infected with HIV by blood clotting factor. He died of AIDS at the age of 19.

Arthur Ashe, a professional tennis player, was infected with HIV through a blood transfusion.

HIV can be transmitted when blood from an infected person is transferred to an uninfected person. People who inject intravenous drugs can be infected with HIV if they share or reuse needles or hypodermic syringes, both of which can be contaminated with blood carrying the virus. The majority of HIV infections in the United States in late 1980s were transmitted in this manner.

In the late 1970s and early 1980s, many people contracted HIV by receiving blood transfusions from infected individuals. Many hemophiliacs, including Ryan White (one of the HIV-infected people shown in **Figure 32.15**), were infected by receiving blood clotting factor that had been isolated from blood of infected individuals. Donated blood is now tested for the presence of HIV, so the likelihood of contracting HIV through blood transfusions or blood products is very low.

If a mother is infected, her child has about a one-in-three chance of being infected during pregnancy. HIV can also be transmitted from mother to child in breast milk. **Table 32.3** summarizes the ways that HIV can be transmitted.

HIV is not transmitted through the air or on toilet seats. It cannot be contracted through casual contact, such as shaking hands, sharing food, or drinking from the same water fountain as an infected person. Although HIV is found in saliva, tears, and urine, it occurs there in very low concentrations. One drop of infected blood contains as many viruses as one quart of infected saliva. Scientists think that a large quantity of virus must be received for infection to occur. You cannot catch HIV from the small amount of saliva exchanged when kissing. Biting arthropods, such as mosquitoes, bedbugs, fleas, and ticks, do not transmit HIV.

Table 32.3 Known Routes of HIV Transmission

- Vaginal, oral, or anal intercourse with an infected person
- Injecting drugs or other substances with hypodermic syringes or needles used by an infected individual
- Use of skin piercing equipment, such as tatooing needles, that has been used by an infected person
- From infected mother to fetus through the placenta; from infected mother to baby in breast milk
- Transfusions or injections of blood or blood products drawn from an infected person (Transmission no longer occurs by this route in the United States and other developed nations because blood is tested for the presence of HIV. It is still a transmission route in the less developed countries, where such blood tests are often unavailable.)

Elizabeth Glaser, who is infected with HIV, spoke at the 1992 Democratic National Convention.

Actor Rock Hudson died of AIDS on October 2, 1985.

AIDS Is a Worldwide Disease

AIDS was first recognized as a disease in 1981. Where did this fatal disease come from? HIV probably evolved from a very similar virus, simian immunodeficiency virus (SIV), which occurs in monkeys and apes in Africa. SIV causes an AIDS-like disease in some primates. Scientists are not sure when HIV evolved from SIV or what events allowed the virus to spread from its point of origin.

AIDS is now a worldwide disease, as you can see by looking at the map in **Figure 32.16**. The World Health Organization estimates that 40 million people will be infected with HIV by the year 2000. Most of these cases will occur in the less-developed countries, where AIDS is now spreading rapidly. Indeed, childbirth and AIDS are currently the biggest killers of women in Africa.

Figure 32.16
This map shows the estimated numbers of HIV-infected people throughout the world.

500,000

20,000

20,000

1 Million +

50,000

1 Million +

6.5 Million +

1 Million +

30,000

The number of AIDS cases is also increasing in the industrialized countries. As you can see in **Figure 32.17**, the number of AIDS cases and the number of deaths from AIDS in the United States have risen dramatically each year since 1982. American teenagers, such as those in **Figure 32.18**, are also increasingly at risk of contracting HIV, as shown in **Figure 32.19**. More than 20 percent of all reported AIDS cases occur among people in their 20s. Given the average 10-year period before symptoms appear, the majority of these people were probably infected during their teenage years. In 1992, AIDS was the sixth leading cause of death among Americans between the ages of 15 and 24.

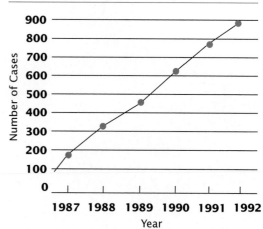

Figure 32.18
Anyone can get AIDS—even you.

Figure 32.17
This graph shows the cumulative numbers of AIDS cases and deaths in the United States.

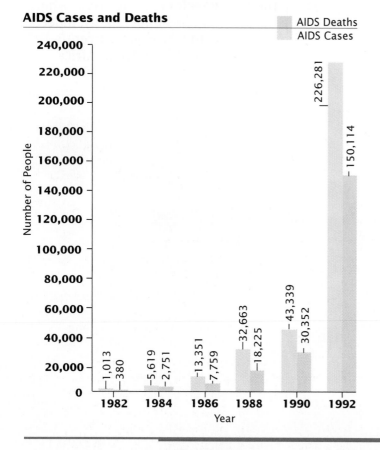

AIDS Cases and Deaths

AIDS Deaths
AIDS Cases

Number of People

240,000
220,000
200,000
180,000
160,000
140,000
120,000
100,000
80,000
60,000
40,000
20,000
0

1982 — 1,013 — 380
1984 — 5,619 — 2,751
1986 — 13,351 — 7,759
1988 — 32,663 — 18,225
1990 — 43,339 — 30,352
1992 — 226,281 — 150,114

Year

AIDS Cases Among 13- to 19-year-olds

Number of Cases

900
800
700
600
500
400
300
200
100
0

1987 1988 1989 1990 1991 1992
Year

Figure 32.19
The incidence of AIDS among 13- to 19-year-olds is rising.

Section Review

❶ **Describe what happens when you are exposed to something to which you are allergic.**

❷ **How are cancer cells different from normal cells?**

❸ **Explain why AIDS patients are unable to resist infections.**

❹ **List four ways HIV is transmitted.**

Killer T cells protect the body by destroying tumors, like the one at the left of the photograph.

Key Terms

Summary

32.1 First Line of Defense

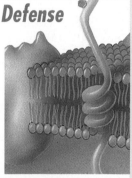

The immune system depends on marker proteins to distinguish self from foreign invaders.

mucous membrane (p. 739)

inflammatory response (p. 740)

phagocyte (p. 740)

macrophage (p. 740)

antigen (p. 741)

- When pathogens enter the body through a wound they trigger an inflammatory response.

- White blood cells called phago-cytes are able to recognize pathogens because the proteins of pathogens differ from the marker proteins of body cells.

- A macrophage is a type of phagocyte.

32.2 The Immune Response

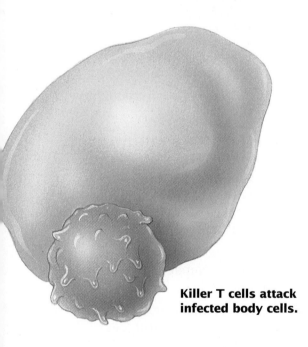

Killer T cells attack infected body cells.

immune response (p. 742)

T cell (p. 743)

B cell (p. 744)

antibody (p. 744)

memory cell (p. 745)

secondary immune response (p. 745)

primary immune response (p. 745)

- Macrophages carrying antigens stimulate an immune response. Defensive white blood cells called T cells begin to divide. Helper T cells stimulate division of killer T cells.

- Helper T cells stimulate B cells to divide. B cells are white blood cells that produce antibodies. Macrophages destroy pathogens marked by antibodies.

- Some B cells become memory cells. If a pathogen that has already been defeated is encoun-tered again, memory cells produce antibodies against it.

- Vaccination stimulates the produc-tion of memory cells by exposing B cells to antigens stripped of their ability to cause disease.

- Viruses that cause colds, flu, and AIDS mutate rapidly. Your immune system fails to recognize the mutated viruses.

32.3 Immune System Failure

This house-dust mite is a source of misery for thousands of Americans aller-gic to dust.

allergy (p. 748)

histamine (p. 748)

antihistamine (p. 748)

autoimmune disease (p. 749)

tumor (p. 750)

- An allergy is a response to a harmless antigen.

- Cancer is uncontrolled cell divi-sion. The immune system normally destroys cancer cells before they spread.

- HIV causes the immune system to fail by invading helper T cells and macrophages.

Understanding Vocabulary

1. For each pair of terms, explain the differences in their meanings.
   a. T cells, B cells
   b. antigens, antibodies
   c. HIV, AIDS
   d. inflammatory response, immune response
   e. antihistamine, histamine

Relating Concepts

2. Copy the unfinished concept map below onto a sheet of paper. Then complete the concept map by writing the correct word or phrase in each oval containing a question mark.

Understanding Concepts

Multiple Choice

3. The skin repels pathogens by
   a. launching phagocytes.
   b. B cell production.
   c. mucous membranes.
   d. secreting oils and sweat.

4. What enables phagocytes to distinguish between pathogens and normal cells?
   a. macrophages
   b. marker proteins
   c. lymphatic system
   d. enzymes

5. What cells display antigens on their cell membranes but are not attacked by killer T cells?
   a. macrophages
   b. helper T cells
   c. B cells
   d. pathogens

6. B cells manufacture
   a. antigens.
   b. macrophages.
   c. killer T cells.
   d. antibodies.

7. A successful, long-term vaccine for influenza has not been produced because
   a. the virus that causes influenza has not been isolated.
   b. production of the vaccine is too expensive.
   c. the genetic code for the viral surface protein mutates often.
   d. it is caused by the same virus as AIDS.

8. Cancer cells are different from normal cells in that
   a. the growth regulating genes of cancer cells have mutated.
   b. cancer cells occur in adults, but not in children.
   c. cancer cells are found in the lymph; normal cells are not.
   d. carcinogens change cancer cells to normal cells.

9. The immune cells most affected by HIV are
   a. B cells.
   b. killer T cells.
   c. helper T cells.
   d. neutrophils.

10. Sneezing, nasal congestion, and itchy nose and eyes are symptoms of
    a. an allergy.
    b. AIDS.
    c. multiple sclerosis.
    d. an autoimmune disease.

11. AIDS may be contracted by
    a. kissing.
    b. sharing food.
    c. sexual intercourse.
    d. shaking hands.

Completion

12. A(n) _____ is an injection of a weakened form of a pathogen to produce immunity.

13. The immune system _____ the body against threats from pathogens.

14. A secondary immune response is _____ and produces _____ antibodies than a primary immune response.

15. After invading pathogens are defeated, _____ cells shut down the immune response.

Short Answer

16. What is the function of antibodies in the immune response?

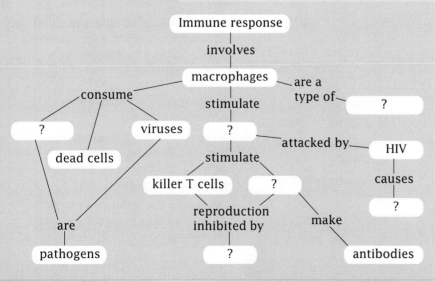

17. How do macrophages contribute to the defense of the body?

18. What measures should you take to prevent HIV infection?

19. Why is the risk of cancer greater among people infected with HIV than among people not infected?

20. What happens to the immune system's ability to distinguish self from non-self in an autoimmune disease like multiple sclerosis?

Interpreting Graphics

21. Study the diagram below.

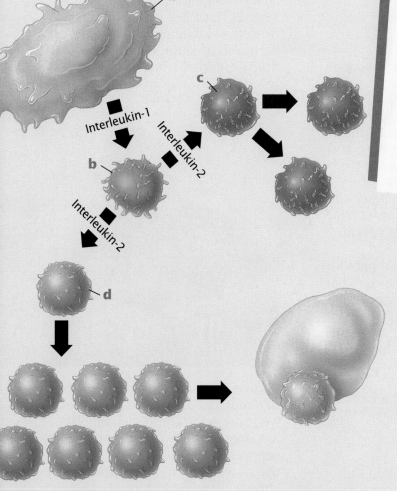

- Identify the cell labeled **a**. What is its function in the immune response?

- What is the function of the cells labeled **d**?

- Describe how HIV infection affects the process illustrated above.

Reviewing Themes

22. *Patterns of Change*
What symptoms are associated with an asthma attack? What effect would taking medicines that contain antihistamines have on the symptoms?

23. *Interacting Systems*
Long term smoking causes paralysis of cilia lining the treachea. How would this affect the body's ability to repel pathogens?

24. *Scale and Structure*
Relate the symptoms of an inflammatory response to the events occurring at the cellular level.

Thinking Critically

25. *Inferring Conclusions*
What would be your chances of contracting German measles for a second time if memory cells lived only three months?

26. *Building on What You Have Learned*
In Chapter 28 you learned about the function of myelin in the transmission of nerve impulses. How is myelin affected by the disease multiple sclerosis?

Cross-Discipline Connection

27. *Biology and Geography*
According to the World Health Organization (WHO), the majority of future AIDS cases will occur in Africa, Asia, and South America. Look for information in your library about living conditions and education in these areas, and about the projected population growth. Do you agree with the WHO's prediction about the spread of AIDS?

Discovering Through Reading

28. Read the article "AIDS or Chronic Fatigue?" in *Newsweek*, September 7, 1992, pages 66 and 69. What is ICL and what are its symptoms? How do doctors plan to treat ICL patients like Rosemary Stevens in the future?

Investigation

How Do Antibody-Antigen Reactions Work?

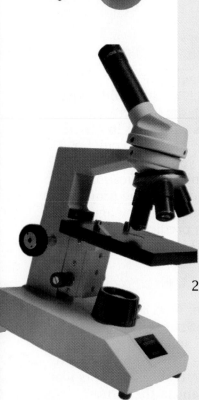

Objectives

In this investigation you will:
- *observe* evidence of antibody-antigen reactions
- *relate* these reactions to blood type

Materials

- blood typing slides
- wax pencil
- vials of simulated blood (blood type unknown)
- simulated anti-A blood typing serum
- simulated anti-B blood typing serum
- toothpicks
- microscope slides
- coverslips
- compound light microscope

Prelab Preparation

1. Review what you have learned about antigens, antibodies, and blood types by answering the following questions.
   - What antigens does a person with type A blood have? What antibodies does this person produce?
   - What antigens does a person with type B blood have? What antibodies does this person produce?
   - What antigens does a person with type AB blood have? What antibodies does this person produce?
   - What antigens does a person with type O blood have? What antibodies does this person produce?
2. Review the procedures in the Appendix for using a compound microscope.

Procedure: Antigen-Antibody Reactions

1. Form a cooperative team with another student to complete steps 2–12.
2. Make a table like the one shown at the right.
3. Use a wax pencil to label four blood typing slides "1," "2," "3," and "4."
4. Thoroughly shake each of the four vials of blood. Place 3–4 drops of blood from the vial labeled "Mr. Green" in each well on Slide 1. Place 3–4 drops of blood from the vial labeled "Mr. Smith" in each well on Slide 2. Place 3–4 drops of blood from the vial labeled "Ms. Jones" in each well of slide 3, and place 3–4 drops of blood from the vial labeled "Ms. Brown" in each well on Slide 4.

5. Choose one slide. Place 3–4 drops of simulated anti-A serum in the well labeled "A" on the slide. Place 3–4 drops of simulated anti-B serum in the well labeled "B" on the slide. Use a new toothpick to stir each sample of blood and serum. Dispose of the toothpicks as directed by your teacher.

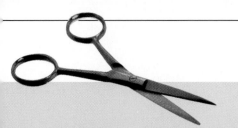

Person	Reaction to Anti-A Serum	Reaction to Anti-B Serum	Blood Type
Mr. Green			
Ms. Jones			
Ms. Brown			
Mr. Smith			

6. Observe each well for two minutes. Look for evidence of clumping. Clumping is evidence that the antibodies in the serum are reacting with the antigens on the blood cells. Record your observations in your table.

7. Repeat steps 5 and 6 for each of the three remaining slides.

8. Thoroughly shake one of the vials of simulated blood. Place a small drop of the blood on a microscope slide. Cover the blood with a coverslip, trying not to trap any air bubbles beneath it.

9. Observe the slide at high power under the microscope. Red blood cells will appear red, while white blood cells will appear blue. *Which kind of blood cell is more common?*

10. Select one vial of blood and place a small drop of blood on a second microscope slide. Now add a drop of blood typing serum that will react to this blood, and mix it with a clean toothpick. Place a coverslip over the blood-serum mixture.

11. Observe the slide at high power under the microscope. *What kind of blood cell has clumped?*

12. Clean up your materials and wash your hands before leaving the lab.

Analysis

1. *Inferring Relationships* What antigens are found in Mr. Green's blood? What evidence supports your conclusion?

2. *Inferring Relationships* What antibodies would be found in Ms. Brown's blood? What evidence supports your conclusion?

3. *Making Predictions* During surgery Ms. Jones is given type O blood. Will her immune system react to the transfusion?

4. *Making Inferences* Do white blood cells have A or B antigens? What evidence supports your conclusion?

5. *Making Inferences* What might cause a patient to have fewer white blood cells than normal?

Thinking Critically

A wounded soldier with type O blood needs an emergency transfusion, but no type O blood is available. Could this soldier safely receive any other type of blood? Explain your answer.

In a Los Angeles suburb, a couple decides to conceive a child who can serve as a bone marrow donor for their daughter. In a Turkish city, a man sells one of his kidneys so that his daughter can have the operation that will save her life. In an Illinois medical center, a physician asks an accident victim's grieving parents for permission to remove their child's liver so that it can be transplanted to a youngster suffering from liver disease.

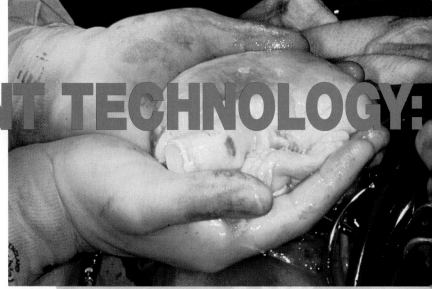

This human heart has just been removed from a donor.

TRANSPLANT TECHNOLOGY:

In a Texas hospital, physicians tell a man who needs a heart transplant that 60 other people are on their list waiting to be matched to a suitable heart donor.

In each of the situations described above, transplant technology provides a way to save a life. Each case also involves decisions that raise legal issues and present ethical dilemmas. The case of the Los Angeles couple, for example, raises a legal question about the rights of the infant. Should a legal guardian be appointed to speak for the infant's rights? This case also raises a serious ethical question: Should parents conceive a child for the purpose of saving another child's life? Some people point out that this kind of thinking might lead to treating offspring as medically useful objects instead of as children.

The buying and selling of kidneys, livers, and hearts raises other serious issues. In the United States and most other industrialized nations, the buying and selling of these organs is illegal and subject to heavy penalties. In these countries, organs for transplant are donated.

So far, the supply of donated organs has not kept up with the demand for them. For example, the number of patients on the national waiting list for transplants rose from about 12,000 in 1987 to more than 20,000 in 1989. During that time, the number of donors has remained around 4,000. Many people who are on hospital waiting lists do not survive long enough for a donated organ to become available to them. As the demand for organs continues to exceed the supply, difficult decisions must be made. Who will receive organs for transplant and who will not? How will these decisions be made and who will make them?

The legal aspects of these questions have not yet been resolved in many states. Many times, the guidelines for arriving at decisions are established by the individual hospitals and medical facilities that perform transplant operations. These guidelines often vary from place to place.

The demand for organs has led some people to suggest that the sale of organs should be legalized in the United States. The people who promote this view argue that legalizing the sale of organs would increase the supply of otherwise rare organs needed for transplants. They suggest that citizens who cannot obtain donated organs in the United States may seek them in countries where the buying and selling of organs is legal. They also point out that if blood, semen, and eggs can be sold—then why not kidneys, hearts, livers, or other organs?

This human liver is about to be transplanted into a recipient.

Other people resist legalization of organ sales under any circumstances. These people suggest that it would be impossible to regulate organ sales. They argue that questions about ownership of an organ are difficult to resolve. An adult, for example, may be able to make a decision about whether to sell one of his or her kidneys. But who makes the decision in the case of a child or the victim of a fatal accident? The issues raised by the increasing demand for tissues and organs for transplant present challenges that could not have been imagined a few decades ago.

SAVING LIVES AND FACING DILEMMAS

Thinking Critically

❶ Do parents have the right to create a life for the purpose of saving another life? In the case of the Los Angeles couple, who owns the infant's bone marrow? Support your view.

❷ What are the similarities and differences between a person selling his or her blood and a person selling one of his or her kidneys? What problems might arise if organs such as kidneys, hearts, and livers could be legally bought and sold in the United States?

❸ When demand for an organ exceeds the supply, how should decisions be made about who receives a transplant? Who should make these decisions?

Acting on the Issue

❶ Contact a local hospital, clinic, or blood bank and find out the criteria for becoming an organ donor.

❷ Write a set of guidelines that prioritize criteria for recipients of scarce organs.

❸ Propose possible ways for organizations to encourage organ donation.

❹ Find out how your state allows a licensed driver to indicate his or her willingness to be an organ donor. What other options are available to people who

want to indicate their willingness to be an organ donor?

❺ Do library research to find information about medical cases in which the organ of an animal was transplanted into the body of a human.

Review

- **organic molecules (Section 2.3)**
- **enzymes (Section 5.1)**
- **water conservation (Section 22.2)**

Digestion and Excretion

33.1 Nutrition: What You Eat and Why

- **Carbohydrates, Proteins, and Lipids**
- **Vitamins and Minerals**
- **Nutrition and Health**

33.2 The Digestive System

- **Digestion**
- **Activity in the Intestines**

33.3 The Excretory System

- **Kidney Form and Function**
- **Urine Formation**
- **Kidney Disorders and Treatment**

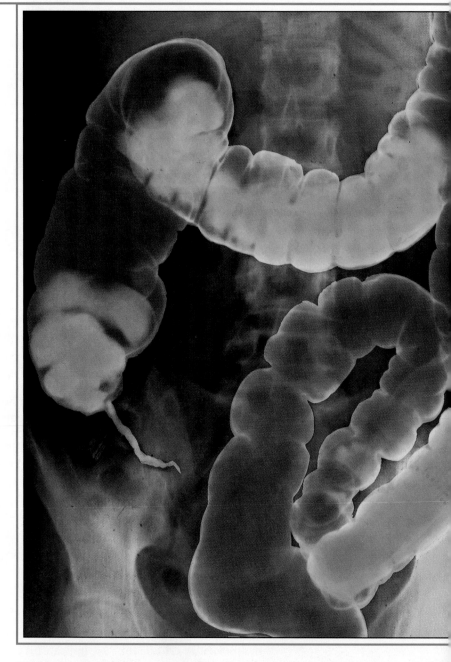

This is an X-ray photo of a human large intestine. The large intestine is an important last stop for food in the digestive system.

ALL ANIMALS MUST GET THE NUTRIENTS THEY NEED FROM A SOURCE OTHER THAN THE SUN. ANIMALS CANNOT MAKE THEIR OWN FOOD AS PLANTS DO, SO THEY MUST EAT OTHER ORGANISMS. SOME ANIMALS, CALLED CARNIVORES, EAT ONLY OTHER ANIMALS; OTHERS, CALLED HERBIVORES, EAT ONLY PLANTS. BUT HUMANS, LIKE BEARS, ARE OMNIVORES, EATING BOTH PLANTS AND ANIMALS.

33.1 Nutrition: What You Eat and Why

Objectives

❶ List the five nutrients required to maintain good health.

❷ List two ways the body uses lipids, proteins, carbohydrates, vitamins, and minerals.

❸ Discuss two diseases that result from nutrient deficiencies.

❹ Discuss the effects of the eating disorders anorexia nervosa and bulimia.

Carbohydrates, Proteins, and Lipids

Figure 33.1
The USDA food pyramid recommends the foods at the top of the pyramid be eaten sparingly, while the other foods may be eaten more often.

Fats, oils, and sweets

Milk products, meat, poultry, fish, beans, eggs, and nuts (2–3 servings daily)

Vegetables (3–5 servings daily) and fruit (2–4 servings daily)

Bread, cereal, rice, and pasta (6–11 servings daily)

Food contains nutrients, the substances that provide your body with the energy and materials it needs for growth, maintenance, and repair. You need several kinds of nutrients in your diet. These include the carbohydrates, lipids, and proteins that you learned about in Chapter 2, as well as vitamins, minerals, and water.

Carbohydrates, lipids, and proteins make up the bulk of what you eat. Vitamins and minerals, while crucial for health, are required in smaller amounts. Water makes up about two-thirds of your body weight. As shown in **Figure 33.1**, most of the food you eat provides a combination of these nutrients.

Carbohydrates are the body's main source of fuel

Most of your body's energy needs come from carbohydrates, also known as sugars and starches. Carbohydrates have other roles as well. One type of sugar, deoxyribose, is a building block of DNA. Another sugar, glucose, is assembled into long chains called cellulose, the main structural component of plant tissues. Cellulose is not digested by humans but provides roughage, or fiber, in our diets. Fiber aids the passage of food through the digestive system.

The carbohydrate molecule ultimately used by your body cells is also glucose. It is used to make ATP. Brain cells and red blood cells rely almost entirely on glucose to supply their energy needs. Even a temporary shortage of blood glucose can severely depress brain function and lead to the death of neurons. When blood glucose is present in excess amounts, it is often converted to fat and stored.

ENERGY SUPPLIES AND BUILDING MATERIALS FOR THE BODY EXIST ONLY IN POTENTIAL FORMS IN FOOD. WHATEVER WE EAT MUST BE PROCESSED INTO SMALLER PIECES BEFORE IT CAN BE USED BY THE BODY. FOOD UNDERGOES THIS TRANSFORMATION IN THE DIGESTIVE SYSTEM, A HOLLOW MUSCULAR TUBE THAT BREAKS DOWN FOOD AND MOVES IT THROUGH YOUR BODY.

33.2 The Digestive System

Objectives

1 List the main organs of the digestive system.

2 Identify the sites of digestion for each of the three nutrients.

3 Describe the absorption of food from the digestive tract.

4 Explain the roles of the pancreas and liver in the digestive system.

Digestion

Suppose you eat a taco for lunch. The taco is rich in the three kinds of large molecules found in food. Proteins occur in the meat, cheese, and beans. Carbohydrates make up the tortilla and beans. Lipids are found in the meat and cheese. The taco also contains essential vitamins and minerals. When you eat the taco, however, the nutrients are in forms that your body cannot absorb and are combined with molecules that your body cannot use. As the taco travels through your digestive system, a journey of more than 8 m (26.24 ft.), it is crushed and churned, and is assaulted by various chemicals in order to release its store of nutrients. In short, it is digested.

Digestion begins in the mouth

Taking a bite of the taco, as shown in **Figure 33.4**, begins the process of digestion. As the food is chewed, it is broken apart so that it will be accessible to digestive enzymes. You have probably noticed that your mouth "waters" when you are hungry and smell or see food (or sometimes just think of food). Salivary glands lying in and near the mouth increase their production of saliva before you eat, and saliva plays an important role in digestion. It moistens and lubricates food, making it easier to swallow. Saliva also contains enzymes that begin to break down starches and other complex carbohydrates present in the food.

Carbohydrates

Lipids

Protein

Figure 33.4
Biting into a taco and chewing it begins its physical breakdown. Chewing not only makes the taco easier to swallow, but also increases the surface area of the food so that digestive enzymes can come into contact with more of the food.

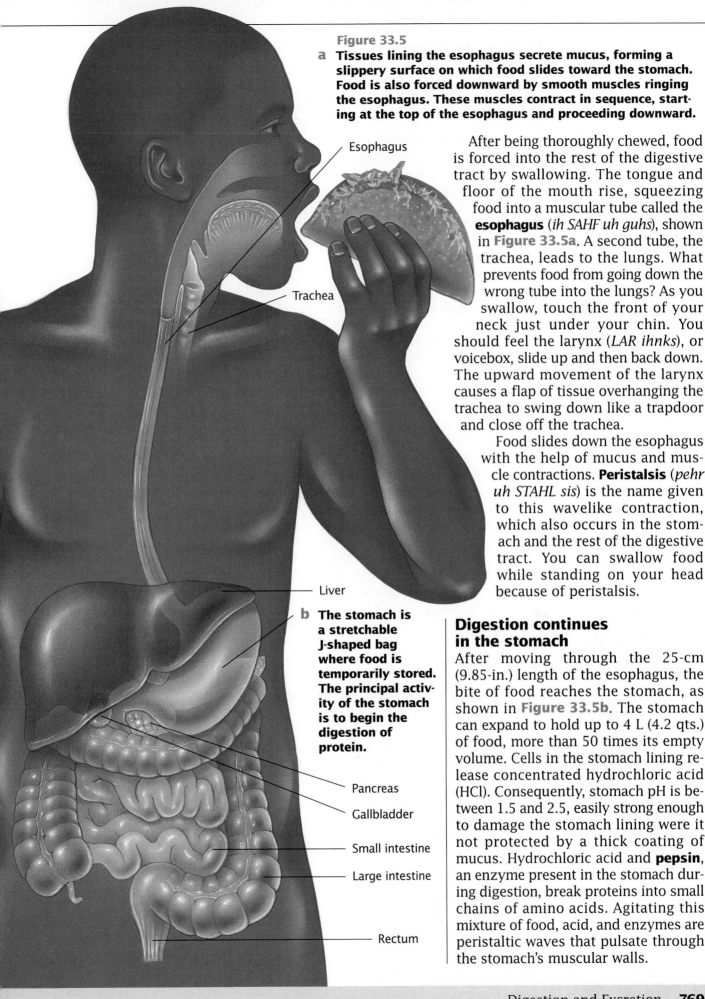

Figure 33.5

a Tissues lining the esophagus secrete mucus, forming a slippery surface on which food slides toward the stomach. Food is also forced downward by smooth muscles ringing the esophagus. These muscles contract in sequence, starting at the top of the esophagus and proceeding downward.

Esophagus

Trachea

Liver

b The stomach is a stretchable J-shaped bag where food is temporarily stored. The principal activity of the stomach is to begin the digestion of protein.

Pancreas

Gallbladder

Small intestine

Large intestine

Rectum

After being thoroughly chewed, food is forced into the rest of the digestive tract by swallowing. The tongue and floor of the mouth rise, squeezing food into a muscular tube called the **esophagus** (*ih SAHF uh guhs*), shown in **Figure 33.5a**. A second tube, the trachea, leads to the lungs. What prevents food from going down the wrong tube into the lungs? As you swallow, touch the front of your neck just under your chin. You should feel the larynx (*LAR ihnks*), or voicebox, slide up and then back down. The upward movement of the larynx causes a flap of tissue overhanging the trachea to swing down like a trapdoor and close off the trachea.

Food slides down the esophagus with the help of mucus and muscle contractions. **Peristalsis** (*pehr uh STAHL sis*) is the name given to this wavelike contraction, which also occurs in the stomach and the rest of the digestive tract. You can swallow food while standing on your head because of peristalsis.

Digestion continues in the stomach

After moving through the 25-cm (9.85-in.) length of the esophagus, the bite of food reaches the stomach, as shown in **Figure 33.5b**. The stomach can expand to hold up to 4 L (4.2 qts.) of food, more than 50 times its empty volume. Cells in the stomach lining release concentrated hydrochloric acid (HCl). Consequently, stomach pH is between 1.5 and 2.5, easily strong enough to damage the stomach lining were it not protected by a thick coating of mucus. Hydrochloric acid and **pepsin**, an enzyme present in the stomach during digestion, break proteins into small chains of amino acids. Agitating this mixture of food, acid, and enzymes are peristaltic waves that pulsate through the stomach's muscular walls.

Activity in the Intestines

The swallowed bites of taco spend up to four hours in the stomach, where the food is reduced to a thin liquid. Most of the proteins and carbohydrates in the taco are broken down in the stomach. Squeezed out of the stomach by peristalsis, the food then moves into the small intestine. Six meters (20 ft.) of small intestine are tightly coiled within your abdomen.

The small intestine absorbs nutrients

The rest of digestion occurs in the entrance to the **small intestine**. The first 25 cm (9.85 in.) of the small intestine is known as the duodenum (*doo uh DEE nuhm*), and it is here that the remaining carbohydrates and proteins are broken down. In adults, almost all lipids are also digested in the small intestine. As in the stomach, peristalsis in the duodenum churns its contents.

The remaining 90 percent of the small intestine has a critical function: it absorbs the nutrients released by digestion and transfers them to the blood. Water is also absorbed. The small intestine is well adapted for absorption. In a cutaway view, the intestine looks as if it were lined with shag carpeting, as shown in **Figure 33.6c**.

The numerous projections of the intestinal lining are **villi** (singular, villus). Each villus is coated with hairlike projections known as **microvilli**. With villi and microvilli, the total surface area available for absorption in the small intestine is about 300 m² (3,229 sq. ft.) —an area larger than a tennis court.

The pancreas and liver secrete digestive enzymes

The products of digestion are absorbed into the bloodstream during the three to six hours food spends in the small intestine. Enzymes and chemicals secreted by the upper end of the small intestine cause additional food breakdown. Most of these secretions come from two organs, the **liver** and the **pancreas**. In Chapter 29 you studied the role of the pancreas in regulating blood sugar. The pancreas also secretes a variety of digestive enzymes. These enzymes break down carbohydrates into simple sugars. They also split proteins into amino acids, and fats into short chains of carbon and hydrogen molecules. In addition, the pancreas secretes bicarbonate (the chemical found in antacids and baking soda), which helps neutralize the stomach acids so that they do not digest the wall of the

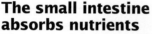

Figure 33.6

a The small intestine is located in the lower abdomen.

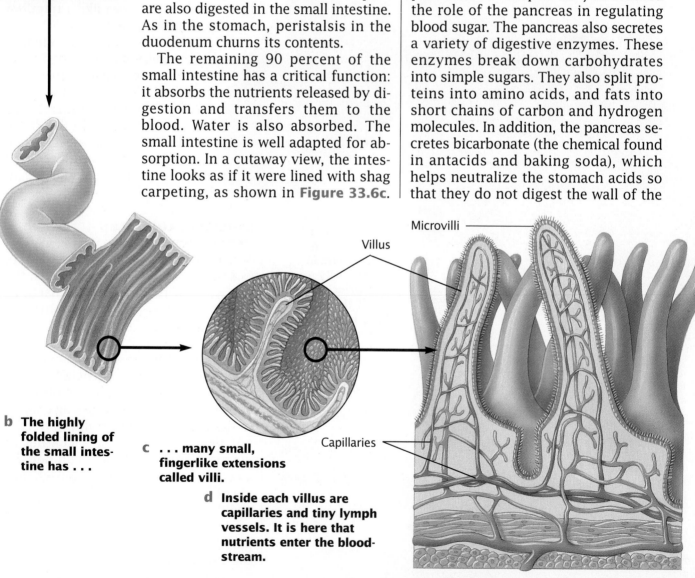

b The highly folded lining of the small intestine has . . .

c . . . many small, fingerlike extensions called villi.

d Inside each villus are capillaries and tiny lymph vessels. It is here that nutrients enter the bloodstream.

Villus

Microvilli

Capillaries

Figure 33.7

a The liver, a football-sized organ, is located above and to the right of the stomach, just below the diaphragm.

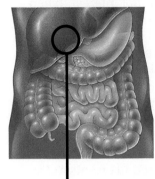

small intestine. When too much stomach acid is produced, the pancreas cannot form enough bicarbonate to compensate and acid begins to eat away the lining of the small intestine or stomach. This painful condition is known as an **ulcer**. Stress and certain foods or medications (alcohol, coffee, aspirin) can cause the stomach to secrete excess acid.

Recall from Chapter 3 that fats are not soluble in water. Fat molecules avoid water and form small clusters or globules in the stomach. Your liver secretes bile, a greenish fluid that emulsifies fat globules so they can be broken down and absorbed. The gallbladder is a green muscular sac attached to the liver that stores bile until it is needed in the small intestine.

Nutrients are collected in the liver

After digested food molecules are absorbed into the bloodstream, they go first to the liver. The liver, shown in **Figure 33.7**, then sends these molecules into the rest of the bloodstream as needed. For example, when you eat a meal, the liver removes excess glucose from the blood and stores it. When the level of glucose in the blood falls, a hormone causes the liver to release some of this glucose back into the blood. The liver also detoxifies many substances. For instance, it produces enzymes that break down drugs and alcohol.

The large intestine compacts wastes

Your lunch-time taco now is largely undigestible material such as cellulose. From the small intestine, what is left of the food now moves into the **large intestine**. The large intestine's role is to store and compact this undigestible material, and then eliminate it. In addition, some water, sodium, and vitamin K are absorbed there. This vitamin K is produced by the billions of *Escherichia coli* bacteria living in the large intestine.

After traveling the 1.5-m (5-ft.) length of the large intestine, what remains of the bite of taco passes out through the rectum and anus as feces. Feces contain undigested food such as cellulose and bacteria.

Gallbladder

b **The liver produces bile, stores nutrients, regulates the level of sugar in the blood, and breaks down drugs and alcohol. Blood going into the liver gets cleansed of its dead or damaged red blood cells, debris, and pathogens. Blood exiting the liver contains plasma proteins and nutrients.**

Section Review

❶ What organs are needed to digest food?

❷ Which nutrient is digested mainly in the stomach? In the small intestine?

❸ How is the small intestine specialized for absorption?

❹ How would digestion be affected if the pancreas or liver were damaged?

Cheryl Coldwater: Pediatrician

How I Became Interested in Pediatrics

"I was born with a birth defect; a broken leg that had healed itself in the womb caused it to be shorter. I got a brace at nine months of age in order to learn how to walk. I wore the brace until I was six, at which point my leg was amputated. That way I could wear a prosthesis and move around as much as possible. As a child I had to spend quite a bit of time in hospitals. And so I was exposed to the medical profession at a very young age.

"In high school I worked as a volunteer both in a clinic and in a hospital. I had a lot of fun and got to see what people did. I worked mostly in the pediatric clinic measuring and weighing the babies and helping the nurses.

"My parents knew before I did that I would go into medicine, because I always had that interest. When I told my mother, she said, 'Well, I thought you would.' They had already figured it out but were going to let me make my own decision. They were very supportive.

"In college I majored in psychology and minored in pre-med. At my college, you couldn't actually major in pre-med. You had to major in something else and fulfill your pre-med requirements at the same time.

"Medical school was very hard—a lot of new things all at once. I didn't pass everything the first time. I had to take two subjects over the first year. It was a lot of work, every day, all day. Some other students had more trouble when it came to working with patients. That was not a big problem for me. That's why I like my specialty so much—you get to deal with people."

Dr. Coldwater is married and has a three-year-old son named Devon. "When I'm not on call, I'm really on my own. That's a new trend in medicine today, particularly with more women going into the field. "

Name:	**Cheryl Coldwater**
Home:	**Austin, Texas**
Employer:	**Austin Regional Clinic**
Personal Traits:	▪ **Perseverance**
	▪ **Good writing skills**
	▪ **Caring**
	▪ **Patience**
	▪ **Sense of humor**

The Satisfaction of Pediatrics

"As a pediatrician, I get to do a lot of preventive medicine. I can help kids with decisions before health becomes a problem. I like to talk to them about good nutrition beginning when they're four years old. I make sure they're eating a fairly balanced diet and keeping junk food for only special occasions. If they learn to eat healthy from the beginning, it's not a big sacrifice. They won't have to go on a diet later on.

"I like talking to kids, questioning them rather than their parents. I talk to them about drugs and alcohol, too. When they get advice only from their parents, they tend not to listen. But when they hear it from their doctor, they pay more attention.

"I like to talk to kids about safety a lot—wearing seat belts and bicycle helmets. It's nice to see that what you're doing is having some effect. When you talk to a child one year about wearing a bicycle helmet, and he comes back the next year and says, 'Oh yeah, I got one,' that feels good.

"I also talk to them about exercise. Many kids aren't interested in organized sports and may not be getting any exercise at all. We talk about ways of getting some movement into their lives. If it's something they're interested in, it will have a long-lasting effect, even into adulthood.

"For our medical group, I produce many patient-education pamphlets and handouts. It's helpful for people to have something they can take home and read later. That way, they can remember the things we talk about in the office."

One of the many aspects of her job that Dr. Coldwater enjoys is working with children. She sees many of the same children year after year and is happy knowing that she is helping them stay healthy.

Research Focus

Sometimes Dr. Coldwater helps pharmaceutical researchers do short-term studies for new pediatric medications. The researchers talk with Dr. Coldwater and her colleagues about using their patients to answer certain questions: Does the drug have any major side effects? Does the medicine work? Do the patients like or dislike the taste? These are things that cannot be learned in the laboratory.

Dr. Coldwater recently helped test a new kind of eye ointment. Newborns are treated with eye ointment to protect them from eye infections caused by bacteria they encounter during the birth process. If there is an infection, it may not show up for a few days. With the parents' permission, Dr. Coldwater helped the researchers find out which eye ointment was most effective and had the fewest side effects.

REGULATING THE CONCENTRATIONS OF SUBSTANCES IN YOUR BODY'S FLUIDS IS ESSENTIAL. A SLIGHT RISE IN POTASSIUM LEVELS IN THE BLOOD CAN CAUSE HEART FAILURE. BUT THIS RARELY HAPPENS BECAUSE THE KIDNEYS MAINTAIN THE LEVEL OF POTASSIUM AND OTHER SUBSTANCES WITHIN NARROW LIMITS. KIDNEYS ALSO SERVE AS FILTERS TO REMOVE NITROGEN-CONTAINING WASTES FROM THE BLOOD.

33.3 The Excretory System

Objectives

❶ Explain how the kidneys determine what leaves the blood and what remains.

❷ Describe the structure of a kidney nephron.

❸ Compare and contrast the two stages of urine formation.

❹ Explain how medical technology has helped people with kidney disorders.

Kidney Form and Function

When the body breaks down excess amino acids, other metabolic wastes—especially nitrogen compounds in the form of ammonia—are released. In Chapter 22 you read that land animals must rid their bodies of ammonia and other nitrogen-containing wastes. The body is able to make ammonia less poisonous by combining it with carbon dioxide in the liver to form a less toxic compound called urea. Urea enters the bloodstream and circulates throughout the body. Some urea is eliminated from the body through the skin as perspiration, which is a mixture of water, minerals, and urea. Most of the urea, however, is eliminated by the excretory system, illustrated in **Figure 33.8**.

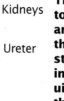

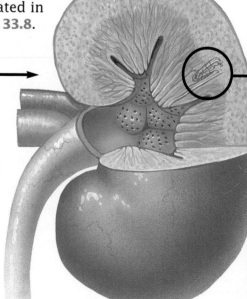

Kidneys

Ureter

Figure 33.8
The kidneys filter out toxins, urea, water, and mineral salts from the blood. These substances are then stored in the bladder as a liquid called urine until they are eliminated.

Bladder

Urethra

The kidneys clean the blood

Your **kidneys** are fist-sized organs located toward the back of your abdomen, about even with the bottom of the rib cage. The kidneys play a vital role in maintaining homeostasis; they remove urea and other wastes, regulate the amount of water in the blood, and adjust the concentrations of various substances in the blood. Because of their crucial function, the kidneys receive large amounts of blood. About one-fourth of the blood your heart pumps every minute travels to the kidneys for cleansing.

Nephrons carry out the processes that form urine

Each kidney contains over 1 million blood-cleaning units called **nephrons** (*NEHF rahns*). A nephron, shown in **Figure 33.9**, is a tiny tube with a cup-shaped capsule at one end. The capsule surrounds a tight ball of capillaries. Substances in the blood inside the capillaries filter into the capsule. The substances then travel through the twists and loops in the nephron, forming urine, an amber-colored fluid that contains nitrogenous waste products and excess amounts of water and solutes.

Figure 33.9

a During filtration, blood pressure in the capillaries forces water, glucose, amino acids, and various salts through small pores in the capillaries and into the capsule of the nephron.

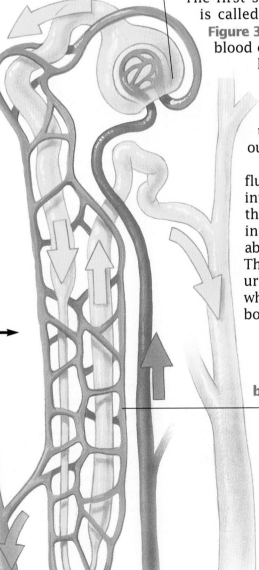

Urine Formation

The first stage of urine formation is called **filtration**, as shown in **Figure 33.9a**. During filtration, the blood cells, proteins, and other large solutes remain in the blood, while smaller solutes (such as glucose, salts, amino acids, and urea) and water are forced out of the blood.

Each day, about 180 L of fluid filter out of the blood into the nephrons. Most of this fluid is absorbed back into the blood, leaving about 1 L of urine every day. The amount and content of urine varies, depending on what has been taken into the body.

b During reabsorption, water and solutes move out of the nephron and into the capillaries.

Reabsorption returns important substances to the blood

From the nephron capsule, the filtered fluid passes into the nephron, beginning **reabsorption**, the second process of urine formation, shown in **Figure 33.9b**. About 99 percent of the water, all of the glucose and amino acids, and many of the salts are reclaimed and sent back into your bloodstream. Water reabsorption is controlled by your hypothalamus. When your body needs to conserve water, the hypothalamus triggers the secretion of a hormone called ADH (antidiuretic hormone). ADH makes the ends of nephrons more permeable to water. As a result, more water is reabsorbed and the urine leaving your body is more concentrated. When your body must get rid of excess water, the secretion of ADH is inhibited, and your urine is more dilute.

After reabsorption, the fluid remaining in the nephron tubes is urine. It is made up of water, urea, and various salts. Urine flows from each kidney into a slender tube called a ureter. It then flows into the urinary bladder, where it is stored. It leaves the body through a tube called the urethra, which leads to the outside of your body.

Kidney Disorders and Treatment

An estimated 13 million people in the United States suffer from kidney disorders. One common disorder is the development of kidney stones. Kidney stones are deposits of uric acid, calcium salts, and other substances that have collected inside the kidney. The stones may become lodged in the urethra, the tube through which urine leaves the body, where they interfere with urine flow and cause pain. Kidney stones usually pass naturally from the body, or they can be eliminated by medical or surgical procedures.

When kidneys are damaged by disease or injury and are unable to function, the blood must be artificially filtered. Filtering of the blood is called **hemodialysis**. The term "dialysis" refers to the use of a semipermeable membrane to separate large particles from smaller ones. The kidney machine shown in **Figure 33.10** can be used twice a week to filter the blood of patients with kidney disorders. A recent development in dialysis uses an abdominal cavity membrane as the dialyzing membrane. A dialyzing solution is fed into the abdomen through a catheter. The fluid is changed four times a day.

a **Blood in a solution is passed through a membrane where the wastes are removed.**

Figure 33.10
The kidney machine is an efficient device for filtering blood.

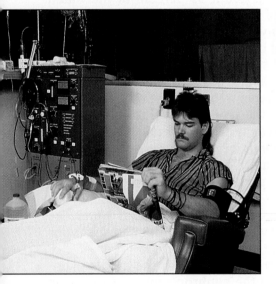

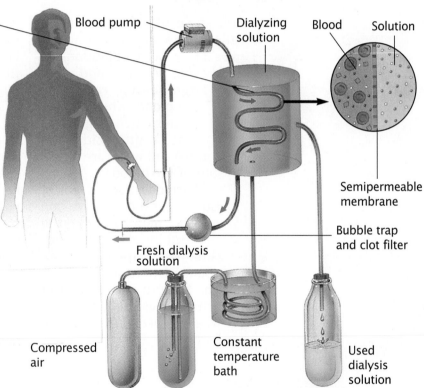

b **The blood is maintained at a constant temperature and pH so that it can be pumped back into the patient after it is filtered.**

Section Review

❶ How do the kidneys help maintain homeostasis?

❷ How are nephrons specialized for urine formation?

❸ Which substances do nephrons filter out of the blood? Which are reabsorbed?

❹ How does the process of hemodialysis work?

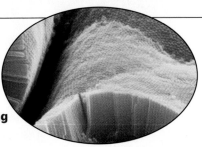

Microvilli can be seen using a scanning electron microscope.

Key Terms	Summary

33.1 Diet: What You Eat and Why

Eating healthful foods is important for maintaining a healthy body.

Key Terms

vitamin (p. 765)

mineral (p. 766)

obesity (p. 766)

anorexia nervosa (p. 767)

bulimia (p. 767)

Summary

- A healthful diet includes carbohydrates and lipids (for fuel), proteins (for building tissues), vitamins, and minerals.

- Diseases such as scurvy and rickets result from diets deficient in certain nutrients.

- Anorexia nervosa and bulimia are eating disorders that can cause serious health problems.

33.2 The Digestive System

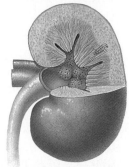

Digestion is the process of breaking down food so that the body can absorb its nutrients.

Key Terms

esophagus (p. 769)

peristalsis (p. 769)

pepsin (p. 769)

small intestine (p. 770)

villus (p. 770)

microvillus (p. 770)

liver (p. 770)

pancreas (p. 770)

ulcer (p. 771)

large intestine (p. 771)

Summary

- In the first stage of digestion, the teeth and saliva break down food.

- In the stomach, proteins are broken down by pepsin.

- Digestion is completed in the small intestine, where nutrients are absorbed.

- The large intestine reabsorbs most of the water from the mass of undigested material left after the removal of nutrients in the small intestine.

- The liver stores and regulates the levels of food molecules in the blood.

33.3 The Excretory System

The kidneys are the filtering organs in the body, removing urea and other wastes from the body.

Key Terms

kidney (p. 775)

nephron (p. 775)

filtration (p. 775)

reabsorption (p. 776)

hemodialysis (p. 776)

Summary

- The functional unit of the kidney is the nephron.

- During filtration, water, glucose, amino acids, salts, and urea move out of the capillary and into the nephron. During reabsorption, water, glucose, amino acids, and many salts move out of the nephron and back into the bloodstream.

- Some kidney disorders can be managed with hemodialysis, a process that uses a semipermeable membrane to simulate the filtering action of the kidney.

Understanding Vocabulary

1. For each set of terms, complete the analogy.
   a. loss of body weight:anorexia nervosa::binge-purge cycle: _____
   b. bicarbonate: pancreas::bile: _____
   c. movement into nephron:filtration::movement out of nephron: _____

Relating Concepts

2. Copy the unfinished concept map below onto a sheet of paper. Then complete the concept map by writing the correct word or phrase in each oval containing a question mark.

Understanding Concepts

Multiple Choice

3. What substances obtained from food regulate body functions but do not supply energy?
   a. proteins
   b. carbohydrates
   c. vitamins
   d. lipids

4. Vegetarians must take greater care in planning their diets than people who eat meat because
   a. vegetables do not provide proteins.
   b. scurvy can result.
   c. their diets may lack some amino acids.
   d. beans and corn have no fiber.

5. Most carbohydrates come from
   a. dairy products.
   b. meat.
   c. plants.
   d. fish and poultry.

6. The tube that carries food from the mouth to the stomach is the
   a. epiglottis.
   b. esophagus.
   c. duodenum.
   d. trachea.

7. Reabsorption of water and sodium is the function of the
   a. stomach.
   b. kidneys.
   c. large intestine.
   d. microvilli.

8. Humans can live for several weeks without eating because the energy needed by the body is stored as
   a. fat.
   b. amino acids.
   c. carbohydrates.
   d. proteins.

9. The functional unit of the kidney is the
   a. nephron.
   b. capillaries.
   c. ureter.
   d. urinary bladder.

10. Emulsifying fats is a function performed by
    a. mucus.
    b. bile.
    c. HCl.
    d. bicarbonate.

11. The fingerlike villi of the small intestine function in
    a. enzyme secretion.
    b. nutrient absorption.
    c. bile production.
    d. ulcer formation.

12. ADH secretion increases when
    a. water intake is less than needed.
    b. a lot of fluids are consumed in a short time.
    c. the nephrons are not permeable to water.
    d. your body needs to get rid of excess water.

Completion

13. The breakdown of food to supply the body with amino acids, sugars, lipids, and other nutrients is called _____ .

14. The first section of the small intestine into which bile and pancreatic juice are secreted is called the _____ .

15. When cells lining the stomach or small intestine are attacked by stomach acids, a(n) _____ may form.

16. When kidneys fail, _____ can be used to maintain a proper balance of solutions in the body.

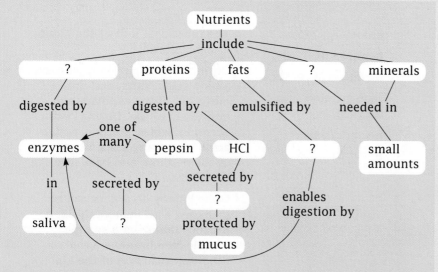

Short Answer

17. Most medical specialists recommend a high-fiber diet. What is fiber? Why is fiber in the diet important?

18. Astronauts in space are often seen eating while upside down. How is it possible for them to swallow and digest food in this inverted position?

19. The pH of the stomach is very acidic. What prevents damage to the stomach lining?

20. Describe two functions of the liver that are related to digestion.

Interpreting Graphics

21. Name the identified parts of the digestive tract and describe the function of each.

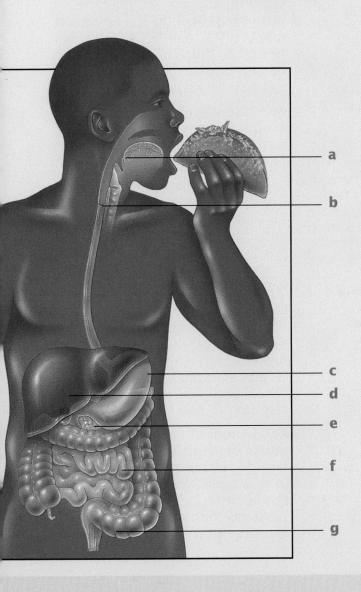

Reviewing Themes

22. *Scale and Structure*
Both chemical and mechanical processes are involved in digestion in the stomach. Identify one chemical and one mechanical process that occur in the stomach and describe how each aids digestion.

23. *Stability*
How do the kidneys function to regulate water balance in the human body?

Thinking Critically

24. *Inferring Conclusions*
A man was diagnosed with cancer of the small intestine, and about a third of his small intestine was removed to save his life. What are the consequences of this life-saving action on the functioning of the small intestine?

25. *Building on What You Have Learned*
In Chapter 22 you learned about the different forms of nitrogen-containing wastes excreted by organisms. Why would it be a problem for humans to excrete ammonia directly rather than excreting urea?

Cross-Discipline Connection

26. *Biology and Health*
The "best if used by" date and the "use by" date on products are of interest to consumers. Do library research to learn why product dates are on the labels of packaged foods.

Discovering Through Reading

27. Read the article "There's Always Room for...," in *Health*, May/June 1992, pages 20 and 24–25. What is the source of most of the gelatin eaten today? Besides some desserts, what are some other foods that contain gelatin?

Investigation

How Do You Test Foods for Nutrients?

Objectives

In this investigation you will:
- *perform* chemical identification tests
- *record, interpret,* and *evaluate* data
- *compare* the nutrient contents of samples of unknown foods

Materials

- safety goggles
- lab apron
- disposable gloves
- test tubes
- test-tube holder and rack
- wax pencil
- glucose solution
- starch solution
- Benedict's solution
- water bath
- hot plate
- Lugol's iodine
- brown paper
- cooking oil
- albumin solution
- Biuret reagent
- unknown food substances

Prelab Preparation

1. Review what you have learned about nutrition by answering the following questions.
   - Why are nutrients important to your health?
   - List the five nutrients that you need in your diet in order to maintain good health.
2. Make a data table like the one shown on the next page.
3. In the first row of the table, briefly describe the nutritional role of the substances in each column.

Procedure: Testing Substances for Nutrients

1. Form a cooperative group of four students. Work with another member of your group to complete steps 2–11.
2. **Caution: Put on safety goggles, a lab apron, and disposable gloves. Leave them on for the entire investigation.**
3. Benedict's Test
   Caution: If you get Benedict's solution on your skin, wash it off immediately.

   Select three clean test tubes. With a wax pencil, label the top of the test tubes "1," "2," and "3." To test tube "1" add 40 drops of glucose. To test tube "2" add 40 drops of starch. To test tube "3" add 40 drops of water.
4. Add 10 drops of Benedict's solution to each test tube. Heat the test tubes in a hot water bath for five minutes.

   In which test tube do you see a positive reaction? In your table, write the name of this identification test under the substance that showed a positive reaction. *What color indicates the presence of this substance?* Record this result in your table.
5. Lugol's Iodine Test
   Select three clean test tubes. With a wax pencil, label the tops of the test tubes "1," "2," and "3." To test tube "1" add 40 drops of glucose. To test tube "2" add 40 drops of starch. To test tube "3" add 40 drops of water.
6. Add 2 drops of Lugol's iodine solution to each test tube. *In which test tube do you see a positive reaction?* In your table, write the name of this identification test under the substance that showed a positive reaction. *What color indicates the*

presence of this substance? Record this result in your table.

7. Brown Paper Test.
Rub a few drops of cooking oil (a lipid) on a small piece of brown paper. On a second piece of brown paper, rub an equal amount of water. After 10 minutes, observe both pieces of paper. *On which piece of paper do you see a reaction?* In your table, write the name of the test that identifies this substance. *How does the appearance of the brown paper change in the presence of this substance?* Record this result in your table.

8. Biuret Reagent Test
Caution: Do not get Biuret solution on your skin. If you do, wash it off immediately.

Select two clean test tubes. With a wax pencil, label the top of the test tubes "1" and "2." To test tube "1" add 40 drops of albumin solution (a protein). To test tube "2" add 40 drops of water.

9. Add 3 drops of Biuret reagent to each test tube. *In which test tube do you see a positive reaction?* In your table, write the name of this identification test under the substance that showed a positive reaction. *What color indicates the presence of this substance?* Record this result in your table.

10. Your team will be assigned two unknown food substances for testing. Conduct each of the above tests to identify the presence of nutrients and other substances. For the Benedict's solution test, the Lugol's iodine test, and the Biuret reagent test, use a 5-mL sample of each unknown food substance. In your table, record whether the nutrient is present.

Substance	Glucose	Starch	Lipid	Protein
Nutritional role				
Identification test				
Positive result				
Unknown food substance A				
Unknown food substance B				

11. Clean up your materials and wash your hands before leaving the lab.

Analysis

1. *Summarizing Methods*
   Summarize the results of each test you conducted.

2. *Communicating Results*
   Compare your results with other members of your group. Offer an explanation for any differences in the test results among the teams.

3. *Analyzing Data*
   Which food contains the greatest variety of nutrients? Which food contains the fewest nutrients?

4. *Applying Methods*
   Describe an experiment for determining whether milk contains glucose.

5. *Making Inferences*
   No color change occurs when table sugar is tested by the Benedict's solution test or the Lugol's iodine test. What conclusion can you draw from these results?

6. *Evaluating Methods*
   What was the purpose of conducting the Benedict's solution test on starch and on water?

Thinking Critically

1. How would tests such as these help a nutritionist plan a balanced diet?

2. What additional information would a nutritionist require about food to plan a balanced diet?

Reproduction and Development

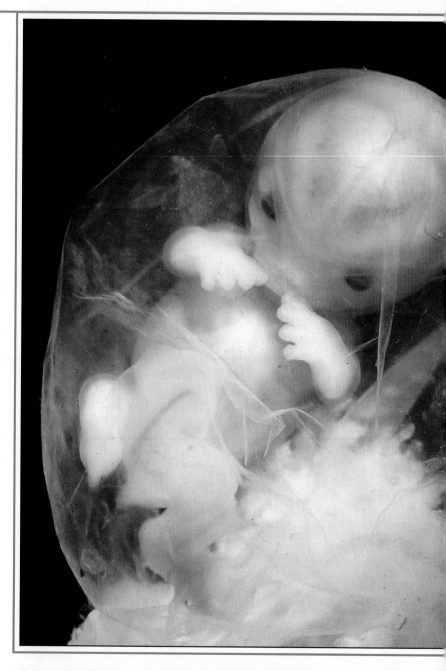

After beginning as a fertilized egg, a human fetus grows and develops inside its mother's uterus for nine months.

BOTH MALES AND FEMALES PRODUCE SPECIALIZED SEX CELLS, CALLED GAMETES, WHICH CONTAIN ONE-HALF OF THE GENETIC INFORMATION NEEDED TO FORM A NEW INDIVIDUAL. THE ROLE OF THE MALE REPRODUCTIVE SYSTEM IS TO PRODUCE SPERM, THE MALE GAMETES, AND DELIVER THEM TO THE FEMALE GAMETE SO THAT FERTILIZATION CAN OCCUR.

34.1 *The Male Reproductive System*

Objectives

❶ Identify two features of sperm cells.

❷ Draw the pathway of sperm from the testes to the outside of the body.

❸ Explain the two functions of the testes.

❹ Describe how the hypothalamus regulates the functioning of the testes.

Structure of Sperm

A male produces millions of sperm each day. How are sperm cells different from other cells of the body? Recall from Chapter 6 that sperm are haploid. They have only one set of chromosomes instead of the two sets found in the body's other cells. The cells that give rise to sperm are initially diploid. These cells undergo meiosis, which reduces the number of chromosomes in the developing sperm cell by half.

After meiosis, a sperm matures for several weeks, becoming a highly specialized DNA delivery cell. As you can see in **Figure 34.1**, a sperm has a long tail and has discarded most of its cytoplasm and organelles. Mitochondria remain in the sperm cell, clustering around the top part of the tail. Mitochondria are "motors" for the sperm, providing energy for the tail to whip back and forth. A cap of digestive enzymes forms at the tip of the sperm's head. These enzymes enable the sperm to penetrate the egg.

Figure 34.1

a A sperm cell consists of a tail used for locomotion and a head that contains DNA.

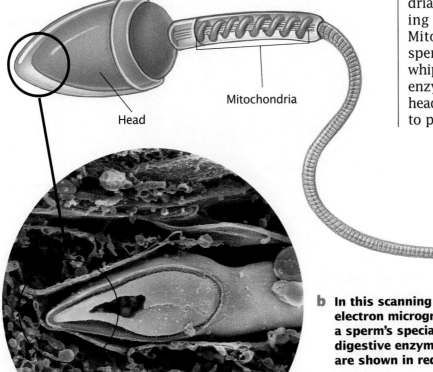

Head

Mitochondria

Tail

b In this scanning electron micrograph, a sperm's special digestive enzymes are shown in red.

The Path Traveled by Sperm

Sperm are produced in two oval-shaped organs called **testes** (*TEHS tees*), or testicles. The testes are located in the scrotum, a loose sac hanging below the base of the penis. After sperm form, they mature and are stored temporarily in the **epididymis** (*ehp uh DIHD ih mihs*), shown in **Figure 34.2**. Some sperm are stored in the lower region of the **vas deferens** (*vas DEHF uh REHNZ*).

Figure 34.2
The male reproductive system is a set of organs that produces and delivers sperm.

a The seminal vesicles produce fluid that nourishes sperm.

b The prostate gland secretes an alkaline fluid that counteracts the acids found in the male urethra and in the vagina of the female.

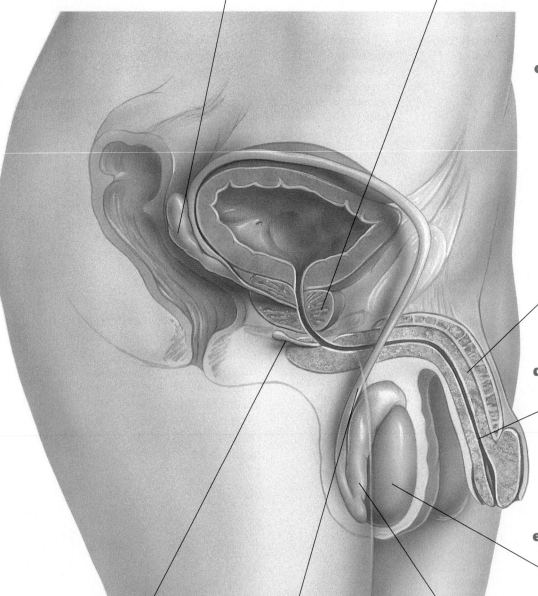

c When a male becomes sexually excited, blood accumulates in the penis, causing it to become firm and erect. The firmness of the erect penis makes it possible for the penis to be inserted into the vagina of the female.

d The urethra is the tube that carries urine during urination and semen during ejaculation.

e Sperm production begins in the testes at puberty and continues throughout life.

h The bulbourethral gland secretes an alkaline fluid that becomes part of the semen.

g The vas deferens passes over the urinary bladder, connecting the epididymis to the urethra.

f The epididymis is a coiled tube attached to the surface of each testis. If uncoiled, an epididymis would be about 6 m (20 ft.) long.

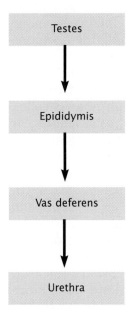

Testes

↓

Epididymis

↓

Vas deferens

↓

Urethra

For fertilization to occur, sperm first must be transferred from a male to a female. The penis is the male reproductive organ that makes it possible for sperm to be delivered to the body of the female. **Ejaculation** (*ee jak yoo LAY shuhn*) is the process by which sperm leave the male's body. The walls of the vas deferens contain a thick layer of smooth muscle. During ejaculation, rhythmic contractions of this smooth muscle propel sperm through the vas deferens. Sperm move from the vas deferens into the urethra. Smooth muscles surrounding the urethra near the opening to the urinary bladder contract during ejaculation. Contraction of these muscles prevents urination during ejaculation. The path that sperm travel from the time they are produced to the time they are expelled during ejaculation is outlined in **Figure 34.3**.

Several glands add fluids to the sperm as they travel through the vas deferens and urethra. The seminal vesicles (*SEHM uh nuhl VEHS ih kuhlz*) are two small glands located near the bladder. The fluid produced by these glands nourishes the sperm. The bulbourethral *(buhl boh yoo REE thruhl)* gland is located just past the prostate (*PRAH stayt*) gland. Both glands secrete a thick, clear, alkaline fluid. These

secretions mix with sperm to form **semen**. Semen is expelled from the penis during ejaculation. **Figure 34.4** shows sperm that have reached the uterus of a female.

If sperm cannot reach the egg, fertilization cannot occur. The surest way to prevent a pregnancy is to not have sexual intercourse. Blocking the path travelled by sperm can prevent fertilization. One way to accomplish this is for a male to wear a latex rubber sheath called a condom during sexual intercourse. If the condom is used properly and no semen leaks out, fertilization cannot occur, because no sperm are deposited in the body of the female. If a male chooses to have no more children, he can have an operation called a vasectomy. During a vasectomy, a small incision is made in the scrotum. Each vas deferens is then cut and the ends are tied. A male who has had a vasectomy is sterile, because the sperm's path to the outside of the body is blocked. Sperm are still produced, but they disintegrate in the epididymis.

Figure 34.4
As many as 350 million sperm, shown here in the uterus, are released into the body of the female during sexual intercourse.

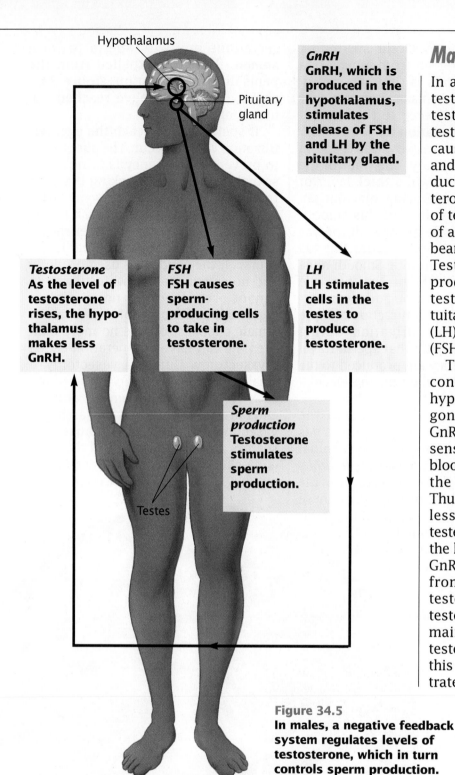

Hypothalamus

Pituitary gland

GnRH
GnRH, which is produced in the hypothalamus, stimulates release of FSH and LH by the pituitary gland.

Testosterone
As the level of testosterone rises, the hypothalamus makes less GnRH.

FSH
FSH causes sperm-producing cells to take in testosterone.

LH
LH stimulates cells in the testes to produce testosterone.

Sperm production
Testosterone stimulates sperm production.

Testes

Male Hormones and Reproduction

In addition to producing sperm, the testes also produce the male hormone testosterone. A male begins to make testosterone before birth. Testosterone causes an embryo to develop a penis and scrotum rather than female reproductive structures. At puberty, testosterone levels rise. These higher levels of testosterone cause the development of adult male characteristics, such as a beard, a low voice, and large muscles. Testosterone also stimulates sperm production. Production of sperm and testosterone are regulated by two pituitary hormones: luteinizing hormone (LH) and follicle-stimulating hormone (FSH).

The release of LH and FSH is controlled by a hormone from the hypothalamus. This hormone is called gonadotropin-releasing hormone, or GnRH for short. The hypothalamus is sensitive to testosterone levels in the blood. If the level of testosterone rises, the hypothalamus makes less GnRH. Thus, less LH and FSH are released, and less testosterone is produced by the testes. As the levels of testosterone fall, the hypothalamus begins to make more GnRH. More LH and FSH are released from the pituitary gland, and more testosterone is then produced by the testes. This negative feedback system maintains a nearly constant level of testosterone in the blood. The steps of this negative feedback system are illustrated in **Figure 34.5.**

Figure 34.5
In males, a negative feedback system regulates levels of testosterone, which in turn controls sperm production.

Section Review

❶ In what ways are sperm different from the body's other cells?

❷ What is the function of the vas deferens in sperm delivery?

❸ Describe two functions of the testes.

❹ What role does the hypothalamus play in sperm production?

THE MALE'S DIRECT CONTRIBUTION TO THE FERTILIZED EGG IS THE DONATION OF ONE SET OF CHROMOSOMES. THIS CONTRIBUTION IS COMPLETE ONCE FERTILIZATION HAS OCCURRED. IN CONTRAST, THE FEMALE CARRIES THE DEVELOPING FETUS FOR ABOUT NINE MONTHS UNTIL BIRTH. THE FEMALE REPRODUCTIVE SYSTEM THUS MUST PLAY TWO ROLES: MAKING GAMETES AND NOURISHING THE FETUS.

34.2   *The Female Reproductive System*

Objectives

❶ **Identify two differences between the male and female reproductive systems.**

❷ **Compare and contrast sperm production with egg production.**

❸ **Relate the events of the ovarian cycle to the levels of LH, FSH, and estrogen.**

❹ **Describe the events of the menstrual cycle.**

Structure of the Female Reproductive System

Female gametes are called **ova** (singular, ovum), or eggs. The **ovaries** are the organs responsible for producing eggs. Ovaries are located in the lower part of the abdomen. Eggs released from the ovaries move into the **Fallopian tubes**, which carry the egg into the uterus, as you can see in **Figure 34.6**. The **uterus** is a hollow, muscular organ about the size of your fist. During pregnancy, the uterus expands to many times its usual size.

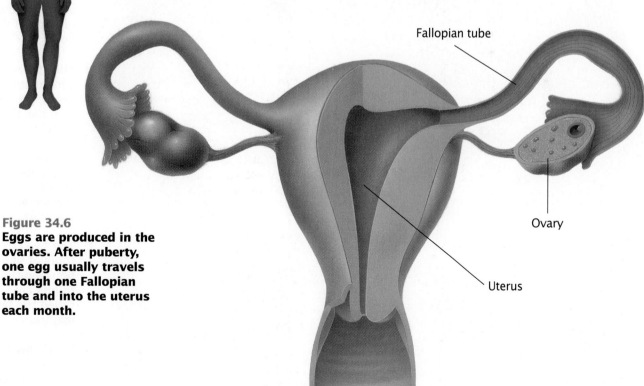

Fallopian tube

Ovary

Uterus

Figure 34.6
Eggs are produced in the ovaries. After puberty, one egg usually travels through one Fallopian tube and into the uterus each month.

The lower part of the uterus is a ring of strong muscles known as the **cervix**. As shown in **Figure 34.7**, the **vagina** (*vuh JEYE nuh*) is a muscular tube that receives sperm ejaculated by a male during intercourse. The lower end of the vagina opens to the outside of the body. During birth, the baby must pass through the cervix and the vagina.

Pregnancy can be prevented by not allowing sperm to enter the opening in the cervix. A diaphragm is a shallow, round cup made of rubber that is placed inside the vagina to block the entrance to the cervix. A cervical cap is similar to a diaphragm but fits tightly over the cervix. A diaphragm or a cervical cap must be obtained from a doctor or health-care professional. Spermicidal creams, spermicidal jellies, spermicide-filled sponges, vaginal suppositories, and spermicidal foams work by killing sperm in the vagina so that no living sperm can enter the opening in the cervix ("spermicidal" means "sperm-killing"). Usually spermicides are used with a diaphragm or a cervical cap. A diaphragm and spermicide used together are much more effective than either used by itself.

Figure 34.7
The female reproductive system releases eggs and provides nourishment to a developing fetus.

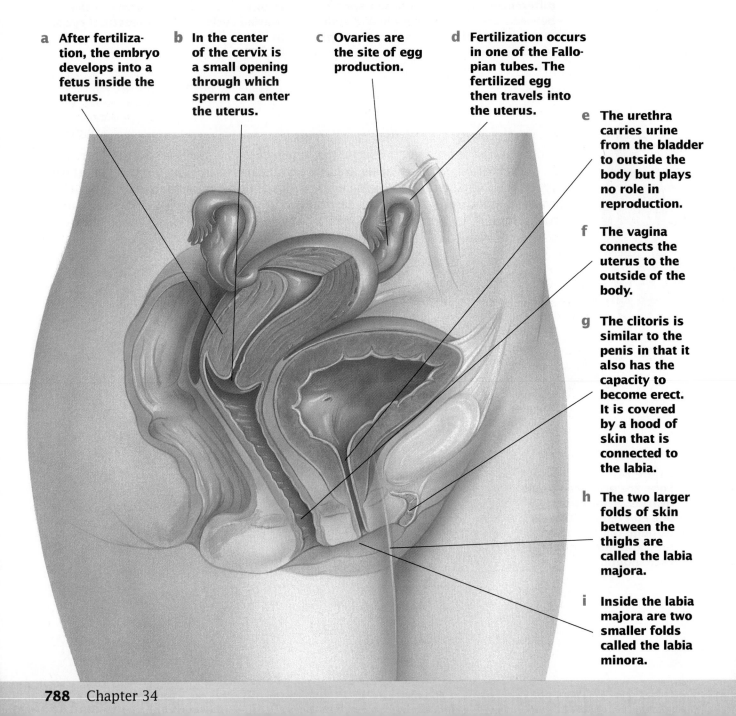

a After fertilization, the embryo develops into a fetus inside the uterus.

b In the center of the cervix is a small opening through which sperm can enter the uterus.

c Ovaries are the site of egg production.

d Fertilization occurs in one of the Fallopian tubes. The fertilized egg then travels into the uterus.

e The urethra carries urine from the bladder to outside the body but plays no role in reproduction.

f The vagina connects the uterus to the outside of the body.

g The clitoris is similar to the penis in that it also has the capacity to become erect. It is covered by a hood of skin that is connected to the labia.

h The two larger folds of skin between the thighs are called the labia majora.

i Inside the labia majora are two smaller folds called the labia minora.

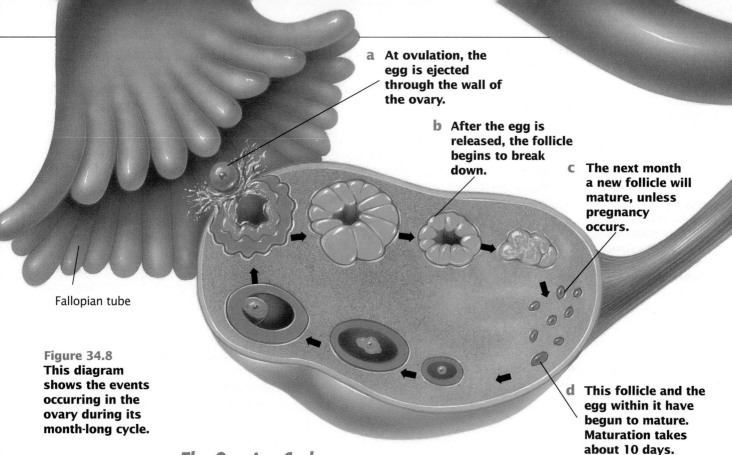

a At ovulation, the egg is ejected through the wall of the ovary.

b After the egg is released, the follicle begins to break down.

c The next month a new follicle will mature, unless pregnancy occurs.

d This follicle and the egg within it have begun to mature. Maturation takes about 10 days.

Fallopian tube

Figure 34.8
This diagram shows the events occurring in the ovary during its month-long cycle.

The Ovarian Cycle

Scale and Structure

How are mature

sperm and eggs

different in size

and structure?

Once a male has reached puberty, his testes continuously produce and release millions of sperm. In a female, however, egg release occurs only at a specific time each month. The release of an egg from the ovary is called **ovulation** (*ahv yoo LAY shun*). A female usually releases only one egg each month. If this egg is not fertilized, the ovaries prepare another egg for release the next month. Thus, the activities of the ovaries occur in a cycle, which is called the ovarian cycle.

Eggs begin to mature in the ovaries of a female while she is still in her mother's uterus. Thousands of her eggs grow and begin to undergo meiosis. Growth and meiosis stop before the eggs are completely mature, however. All the eggs in the ovaries remain in this immature state until puberty.

After puberty, follicle-stimulating hormone (FSH) released by the pituitary signals several eggs to resume the process of maturation each month. As in males, FSH release is stimulated by gonadotropin-releasing hormone (GnRH) produced by the hypothalamus. The maturing eggs become large, highly complex cells, growing to be nearly 200,000 times larger than a sperm. Although many eggs begin to mature each month, usually only one egg completes the process. If two eggs mature, fraternal, or nonidentical, twins may result.

The egg matures within the follicle

Eggs grow inside a fluid-filled chamber called a **follicle** (*FAHL ih kuhl*), which is located in the ovary. The follicle supports growing eggs and produces the hormone estrogen. As you read in Chapter 29, estrogen is the female sex hormone. At puberty, estrogen causes females to develop breasts and wider hips than males. Cells of the follicle produce the greatest amount of estrogen when the egg nears maturity. This high level of estrogen triggers the pituitary to release a burst of luteinizing hormone (LH). LH triggers the egg to resume meiosis. Stimulated by high LH levels, the fluid-filled follicle ruptures, as shown in **Figure 34.8**. After ovulation, currents of fluid sweep the egg into one of the Fallopian tubes. The Fallopian tubes provide a way for an egg to travel from the ovary to the uterus.

If a woman chooses to have no more children, she can undergo tubal ligation, an operation in which the Fallopian tubes are severed or tied.

If the egg is fertilized it is then known as an **embryo**. An egg must be fertilized within 24 hours of its release. After that, the egg begins to break down. Unfertilized eggs dissolve in the uterus.

Hormone levels change during the ovarian cycle

Levels of estrogen and progesterone are high after ovulation. High levels of estrogen and progesterone cause the pituitary gland to stop releasing LH and FSH. If no embryo reaches the uterus, estrogen and progesterone levels fall, and FSH and LH levels rise again. Another egg will mature and be released. During pregnancy, however, estrogen and progesterone levels remain high. High levels of these hormones suppress release of FSH and LH. Because no eggs mature, a woman does not ovulate during pregnancy.

Prescription pills containing synthetic estrogen and progesterone can prevent the egg from maturing. Use of these pills prevents pregnancy by keeping the levels of estrogen and progesterone at a steady level, as shown in **Figure 34.9**. The pituitary responds as if a pregnancy were in progress. It does not release FSH or LH. Since ovulation does not occur, pregnancy cannot take place.

Figure 34.9
The graph below compares a woman's levels of estrogen and progesterone under normal conditions (solid lines) and while she is using pills containing synthetic estrogen and progesterone (dotted lines).

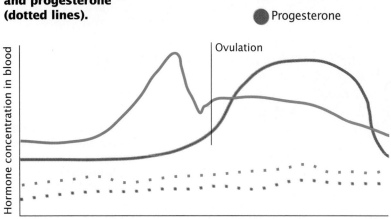

Estrogen

Progesterone

Ovulation

Hormone concentration in blood

Length of one ovarian cycle (about 28 days)

The Menstrual Cycle

Like the ovaries, the uterus follows a monthly cycle. Each month, the uterus prepares to receive and nourish an embryo. The **menstrual** (*MEHN struhl*) **cycle** is the series of changes that occur in the uterus each month.

During the first part of the menstrual cycle, as the egg matures, the lining of the uterus grows thicker. Many tiny blood vessels grow into this thickened lining. The lining will provide a safe environment for an embryo. After ovulation, the follicle begins to produce progesterone. This hormone causes blood vessels in the lining of the uterus to enlarge even more. The lining is prepared to receive the embryo four or five days after the egg is released from the ovary. It takes an embryo about this long to travel to the uterus. An embryo that settles into the rich lining of the uterus releases hormones. One of these hormones causes the uterus to maintain its thickened lining. Most of the

time, no embryo arrives. The follicle begins to break down and produces less and less estrogen and progesterone. As the levels of estrogen and progesterone fall, the blood vessels in the uterine lining begin to close and then to break. The cells of the uterine lining do not receive adequate blood supply and come loose from the inside of the uterus. Blood from the broken blood vessels helps wash these cells out of the uterus. This mixture of blood and the cells that made up the lining of the uterus is called menstrual fluid. The passage of this fluid through the vagina and out of the body is called a menstrual period. It usually lasts from three to seven days. The steps of the menstrual cycle are shown in **Figure 34.10**.

The average menstrual cycle is 28 days long. Almost all women start their menstrual period 14 days after ovulation occurs. The length of the first stage of the cycle, the period when the follicle is growing, differs from woman to woman. For example, women with regular 28-day cycles probably ovulate 14 days after the first day of their last menstrual period. Women with regular 33-day cycles probably ovulate 19 days after the first day of their last period. Women with irregular menstrual cycles have a difficult time determining when ovulation occurs.

Some couples try to prevent pregnancy by abstaining from sexual intercourse around the time of ovulation. This approach, usually called the rhythm method, is not an effective means of preventing pregnancy. Since very few women know exactly when their next period will begin, it is difficult to know when ovulation will occur. Also, sperm can survive in the female reproductive tract for two days or longer, so there are usually three or four days each month when fertilization is possible.

The menstrual and ovarian cycles stop in middle age

Between the ages of 45 and 55, women usually stop having menstrual periods and their ovaries stop releasing eggs. The shutdown of menstrual and ovarian cycles is known as **menopause**. Once a woman has undergone menopause, she is no longer able to have children.

Figure 34.10
This graph shows the changes in uterine lining during the menstrual cycle.

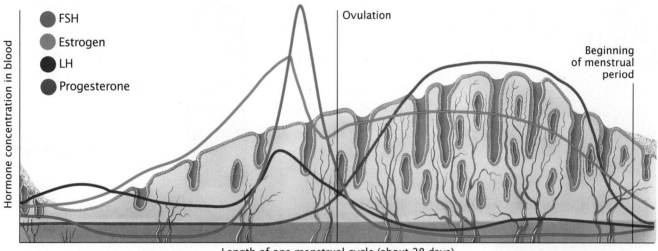

FSH
Estrogen
LH
Progesterone

Hormone concentration in blood

Ovulation

Beginning of menstrual period

Length of one menstrual cycle (about 28 days)

a During the first part of the menstrual cycle, blood vessels grow in the uterine lining as it thickens.

b The lining is ready to receive an embryo four or five days after the egg is released from the ovary.

c If no embryo arrives, the blood vessels in the uterine lining begin to break. Blood and cells from the uterine lining pass through the vagina during the menstrual period. After the menstrual period, menstrual and ovarian cycles begin anew.

Section Review

1 Trace the path of an egg through the female reproductive system.

2 Identify two differences between a sperm and an egg.

3 What role does LH play in the uterine cycle?

4 Explain the role of progesterone in the menstrual cycle.

No Alcohol During Pregnancy

Alcohol and Fetal Development

New signs may already be up or may be going up in restaurants and bars in your state. The signs are similar to the caution labels found on containers for alcoholic beverages. These labels warn women about the dangers of drinking alcohol during pregnancy. Alcohol use by pregnant women is one of the leading causes of birth defects.

When a woman drinks alcohol during pregnancy, the alcohol in her blood travels through the placenta and into the blood of the fetus. What may not seem like much alcohol to the mother can be a harmful dose for the developing fetus.

In the first three months of pregnancy, the major organs and the circulatory system of a fetus are forming. The last six months of a fetus's development centers on growth and maturation. Ingesting alcohol during the first three months of pregnancy can result in severe physical damage—organs can be malformed, the spine can be misshapen, and brain development can be impaired. Drinking later in pregnancy often results in inhibited physical growth, hyperactivity, and behavioral problems after the child is born. Because the brain develops throughout the nine months of pregnancy, brain damage due to alcohol can occur at any time.

The mother of this healthy fetus has decided not to drink alcoholic beverages while she is pregnant.

Warning labels now appear on all alcoholic beverage containers.

GOVERNMENT WARNING: ACCORDING TO THE SURGEON GENERAL WOMEN SHOULD NOT DRINK ALCOHOLIC BEVERAGES DURING PREGNANCY BECAUSE OF THE RISK OF BIRTH DEFECTS.

Substance:	**Alcohol**
Related Illnesses:	**Fetal Alcohol Syndrome Fetal Alcohol Effect**
Symptoms:	**Brain damage, facial deformities, behavioral problems, hyperactivity**
Causes:	**Drinking during pregnancy**

Fetal Alcohol Syndrome

A baby born with severe impairment due to his or her mother's drinking has a condition called fetal alcohol syndrome (FAS). FAS babies often have deformed faces, with improperly formed eyes and lips. They often are mentally retarded and tend to have retarded growth. Most cannot advance beyond the second- to fourth-grade level. These children often have trouble with simple functions such as telling time or counting. The average IQ of children with FAS is 30–40 points below normal.

The damage to these babies is non-treatable and permanent. When FAS children become adults, they cannot take care of themselves and therefore always require care from others.

Some babies may be born with less severe, and less obvious, forms of impairment due to their mother's drinking. These babies have a condition called fetal alcohol effect (FAE). FAE babies usually have fewer physical problems but often have trouble learning and have behavioral problems including a short attention span, consistently poor judgment, and social withdrawal. These behavioral problems, though most obvious during childhood, do seem to continue into adulthood.

This child was born with Fetal Alcohol Syndrome (FAS). The symptoms of FAS include a small head; low, prominent ears; poorly developed cheek bones; and a long, smooth upper lip. In the United States, about 1 in 750 newborns are affected with FAS.

How Much Is Too Much?

The severity of FAS or FAE is directly related to the amount of alcohol the mother consumed during pregnancy—the more alcohol, the worse the impairment. Also, binge drinking, where a large amount of alcohol is consumed at one time, is especially harmful. No one knows if there is any amount of alcohol a pregnant woman can safely drink. For this reason, a "better-safe-than-sorry" approach is recommended. The Surgeon General of the United States, the American Medical Association, and the March of Dimes Birth Defects Foundation all recommend that women consume no alcohol during pregnancy.

Warnings have appeared on every bottle of beer, wine, and liquor sold in the United States since 1989. Despite the warnings, 3,800 to 7,600 FAS babies and an estimated 15,000 to 30,000 FAE babies are born in the United States each year. Getting the word out to women is an important goal of the signs in restaurants and bars.

There is hope for women who drink alcohol. If a woman drinks prior to but not during pregnancy, there is no risk of an FAS or FAE baby. Unfortunately, many women do not realize they are pregnant during the early weeks of pregnancy. If a woman stops drinking as soon as she knows she is pregnant, the risk of FAS and FAE greatly decreases.

The message is clear: Don't drink alcohol while you are pregnant. Give your baby a fair start in life.

How Maternal Consumption of Alcohol May Affect the Fetus and Newborn

Interferes with delivery of maternal nutrients to the fetus
Reduces the supply of oxygen to the fetus
Inhibits protein synthesis and metabolism in the fetus
Causes low birth weight
Leads to alcohol withdrawal in newborns—tremors, abnormal muscle tension, inconsolable crying, reflex abnormalities, restlessness

BEFORE FERTILIZATION, A SPERM AND AN EGG ARE SEPARATE CELLS, EACH CONTAINING ONE-HALF OF THE INFORMATION NEEDED FOR THE DEVELOPMENT OF AN EMBRYO. DURING FERTILIZATION, GAMETES FUSE INTO ONE CELL CONTAINING ALL THE INSTRUCTIONS NEEDED TO FORM A NEW INDIVIDUAL. AFTER FERTILIZATION, THE EMBRYO WILL GROW AND DEVELOP FOR ABOUT NINE MONTHS INSIDE ITS MOTHER'S UTERUS.

34.3 *Fertilization and Development*

Objectives

1 Describe the process of fertilization.

2 Explain the process of implantation.

3 List three substances that can pass through the placenta.

4 Describe three changes that occur as the fertilized egg develops into a new individual.

Fertilization: Fusion of Gametes

During ejaculation, 150 million to 350 million sperm are deposited just a few inches away from the Fallopian tubes, in which fertilization can occur. Now a long and difficult journey begins. To reach the egg, a sperm must first pass through the small opening in the cervix. Then it must cross the uterus and swim up one of the Fallopian tubes. The vast majority of ejaculated sperm never leave the vagina. Of the sperm that find the small opening to the cervix, many will be trapped in mucus that covers this opening. Over the next two days, sperm will gradually free themselves from this mucus and start their journey across the uterus. Even if a mature egg is waiting in one of the Fallopian tubes, half of the sperm will swim up the wrong tube.

Although only one sperm can fuse with the egg, many sperm can arrive at the egg at about the same time. The egg is surrounded by a protective layer of smaller cells. Enzymes in the heads of the sperm cells help loosen this layer of cells. Once a sperm reaches the surface of the egg, as shown in **Figure 34.11**, it leaves its tail behind. The genetic information inside the head of the sperm enters the egg and fuses with the genetic information inside the egg. A protective shield then forms around the egg to prevent other sperm from entering.

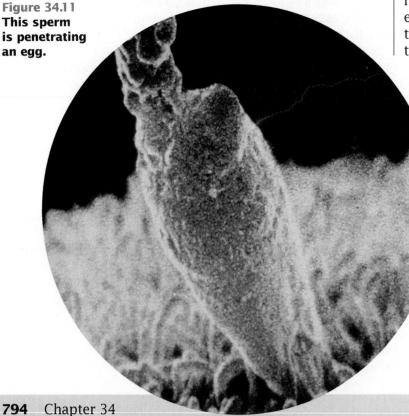

Figure 34.11
This sperm is penetrating an egg.

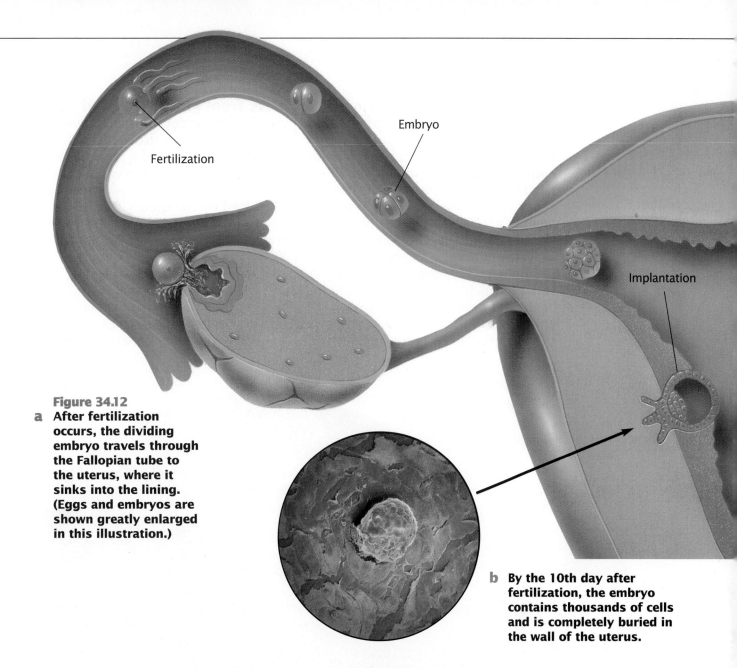

Fertilization

Embryo

Implantation

Figure 34.12

a **After fertilization occurs, the dividing embryo travels through the Fallopian tube to the uterus, where it sinks into the lining. (Eggs and embryos are shown greatly enlarged in this illustration.)**

b **By the 10th day after fertilization, the embryo contains thousands of cells and is completely buried in the wall of the uterus.**

The Embryo Enters the Lining of the Uterus

After the egg is fertilized, it continues to travel through the Fallopian tube toward the uterus, as shown in **Figure 34.12**. After about 30 hours the fertilized egg cell divides into two cells and is called an embryo. The cells of the embryo continue to divide by mitosis. (If the young embryo splits in two, identical twins will result.) From the ninth week of development until birth, it is called a **fetus**.

By the time the embryo reaches the uterus, it is made up of a hundred or more cells. The embryo embeds itself in the soft, thickened uterine lining. The embryo then releases hormones to ensure that the thick lining of the

uterus remains healthy and is not washed away by a menstrual period. The embryo's entry into the uterine wall is known as **implantation**.

There are several ways for a female to determine whether she is pregnant. A missed menstrual period is one possible sign of pregnancy. A more accurate way of determining pregnancy is to test her blood or urine for the presence of the hormone produced by the embryo. By three weeks after fertilization, a pregnant woman will probably have enough of this hormone in her blood and urine to produce a positive test result. Doctors and health clinics are equipped to administer these tests.

The Placenta

As the embryo implants into the wall of the uterus, it is beginning to use up the raw materials that were stored in the egg. The body of the mother begins to transfer nutrients and oxygen to the embryo through the placenta, which you read about in Chapter 22. The placenta is a two-layered, disk-shaped membrane. One layer of the placenta is derived from the chorion (*KAWR ee ahn*), the outermost membrane around the embryo. The other layer of the placenta forms from the thickened lining of the mother's uterus. Cells from the embryo form the umbilical cord, which connects the placenta to the embryo.

Nutrients and oxygen pass from mother to child

As the embryo develops, blood vessels from the embryo grow through the umbilical cord into the inner layer of the placenta. The outer layer of the placenta is rich in blood vessels, which are produced by the uterus. The blood of the mother does not flow into the embryo, however. As shown in **Figure 34.13**, the mother's part of the placenta forms pools of blood that bathe the outside of the embryo's blood vessels. Even though the mother's blood and the embryo's blood come very close to each other, they do not mix in the

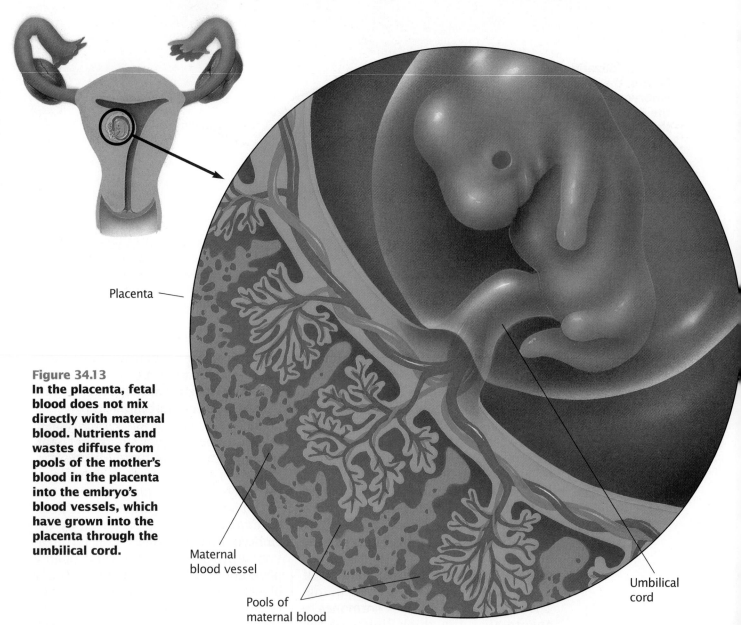

Figure 34.13
In the placenta, fetal blood does not mix directly with maternal blood. Nutrients and wastes diffuse from pools of the mother's blood in the placenta into the embryo's blood vessels, which have grown into the placenta through the umbilical cord.

Placenta

Maternal blood vessel

Pools of maternal blood

Umbilical cord

Drugs
Babies of mothers who use drugs during pregnancy are often born underweight. These babies can face a variety of medical complications. For this reason, no pregnant woman should take any drug without checking with her doctor first.

Nutrition and Exercise
Nutrition and exercise are important during pregnancy. A mother must provide all the nutrients a fetus needs to develop normally. An undernourished mother will give birth to a smaller, less healthy baby.

Smoking
Smoking cigarettes decreases the amount of oxygen available to the growing fetus. Babies born to mothers who smoke heavily tend to be much smaller than the babies of nonsmokers. Babies of heavy smokers also face a much higher risk of complications after birth than the babies of nonsmokers.

Alcohol
Women who drink alcohol during pregnancy may have babies with severe facial deformities and mental retardation. Most doctors think that even a small amount of alcohol may damage a fetus, and recommend that women do not drink at all during pregnancy.

Figure 34.14
This baby was born healthy because his mother was careful during pregnancy. She ate healthy foods, exercised regularly, and avoided drugs and alcohol.

placenta. Oxygen and nutrients from the mother's blood diffuse through the walls of the embryo's blood vessels in the placenta. The oxygen and nutrients then travel through the blood vessels in the umbilical cord to the embryo.

Since the fetus is completely enclosed in the mother's body, it has no way to get rid of waste products on its own. Waste products such as carbon dioxide and urea diffuse through the placenta into the mother's bloodstream and are eliminated by the mother's body.

The behavior of a pregnant woman can affect her child

Although an embryo contains all the genetic information necessary for development, the health and activities of its mother influence whether it develops normally. As **Figure 34.14** indicates, the diet and behavior of a woman during pregnancy can affect the mental and physical development of the

child, both before and after birth. For example, viruses can pass from the mother's blood, through the placenta, into the blood of the fetus. Viruses, such as those that cause rubella (German measles) and herpes, can cause a fetus to develop abnormally.

Drugs can also pass through the placenta. Some drugs can cause birth defects. A growing problem in the United States is the increasing number of babies born to mothers who use crack cocaine during pregnancy. These "crack babies" have a very high risk of mental retardation, learning disabilities, and personality disorders. Other illegal drugs also affect the fetus. Babies of mothers addicted to heroin may themselves be addicted when born.

Alcohol is a legal drug that is also very dangerous when consumed during pregnancy. Nicotine, which is found in cigarettes, is also harmful to babies before birth.

Growth and Development

Figure 34.15
Profound changes take place during the nine months of development.

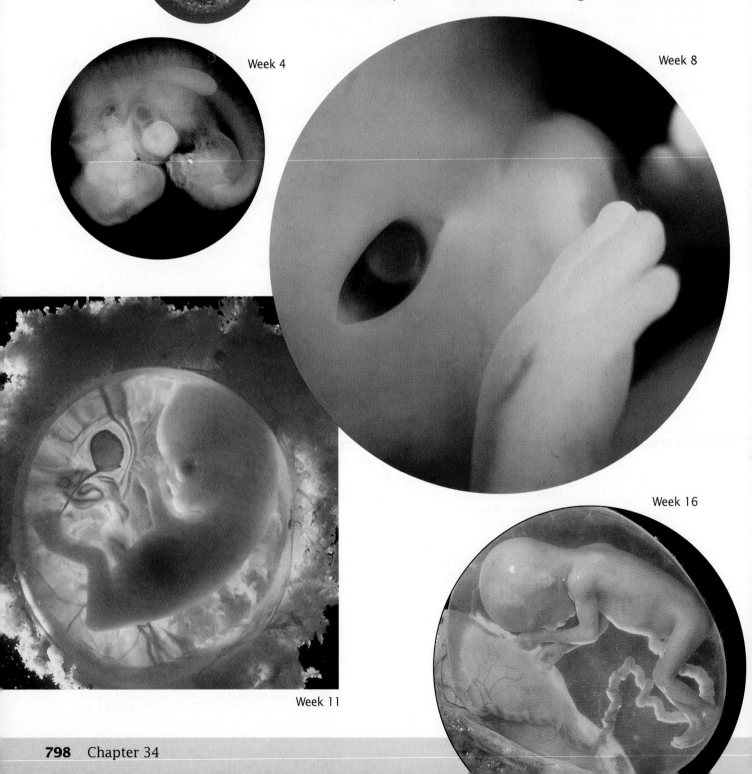

24 hours

Once the embryo has implanted in the uterus, it grows rapidly. A heart and brain begin to form, eyes appear, the face takes shape, small buds on the sides of the body become arms and legs, and internal organs such as the lungs, stomach, and liver develop.

Eight weeks after fertilization, the embryo is only 30 mm (a little over 1 in.) long. However, all the major organ systems of its body have begun to form and are growing, and its heart is beating. The nine months of pregnancy are often divided into three 3-month periods known as **trimesters**. Most of the fetus's development is complete by the end of the second trimester. Follow the changes that occur during development in **Table 34.1** and **Figure 34.15**.

Week 4

Week 8

Week 16

Week 11

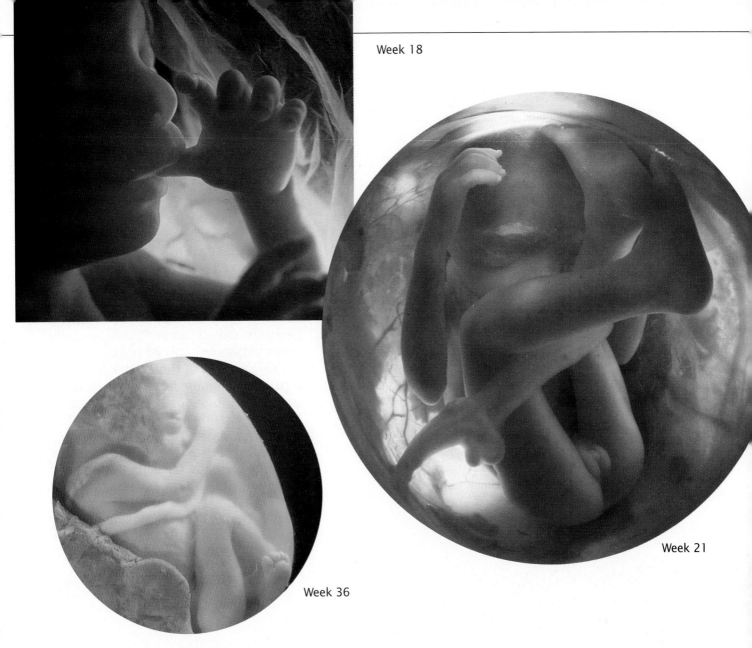

Week 18

Week 21

Week 36

Table 34.1 Stages of Embryonic and Fetal Development

Stage	Major Changes
0–4 weeks	Fertilization occurs; embryo travels through Fallopian tubes and implants in uterine wall; brain, ears, and arms begin to form; heart forms and begins to beat
5–8 weeks	Nostrils, eyelids, nose, hands, fingers, legs, feet, toes, and bones begin to form; females develop ovaries, males develop testes; head is nearly as large as body; cardiovascular system is fully functional; about 22 mm (less than 1 in.) long
9–12 weeks	Embryo becomes a fetus; penis of males is distinct; growth of chin and other facial structures give fetus a human face and profile; head still dominant, but body is lengthening; about 36 mm (1.5 in.) long
13–16 weeks	Blinking of eyes and sucking of lips occurs; body begins to outgrow head; mother can feel muscular activity of fetus; about 140 mm (5.5 in.) long
17–20 weeks	Limbs achieve final proportions; eyelashes and eyebrows are present; about 190 mm (6.5 in.) long
21–30 weeks	Substantial increase in weight; may survive if born at this stage; skin is wrinkled and red; about 280 mm (13 in.)
30–40 weeks	Fingernails and toenails are present; about 360 mm (14.5 in.) long

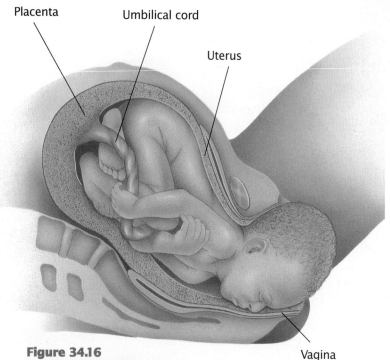

Placenta   Umbilical cord

Uterus

Figure 34.16

a **During birth, the cervix is forced open and the baby is pushed out through the vagina.**

Vagina

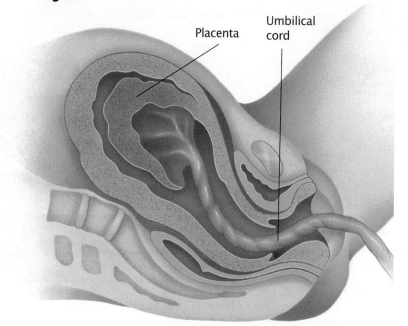

Placenta   Umbilical cord

b **Shortly after a child is born, fluid, blood, and the placenta are expelled from the uterus. This material is the afterbirth.**

Birth

About nine months after fertilization, the fetus is ready to be born. By this time, it has usually moved so that its head is against the cervix. Near the end of pregnancy, hormones cause muscles in the walls of the uterus to begin contracting. Weak, irregular contractions may occur for several weeks before birth. As the time for birth approaches, contractions become much stronger. They also start to come at regular intervals and are closer together. The final stage of pregnancy, when the uterus is contracting strongly enough to push the baby out of the mother's body, is called **labor**. Labor usually takes between 5 and 20 hours.

Contractions of the uterus press the fetus's head against the cervix, forcing it open enough for the baby to pass through, as shown in **Figure 34.16**. The baby then passes through the vagina and out into the world. Within a few seconds the baby takes its first breath. The placenta separates from the wall of the uterus. A few minutes after the baby is born, the placenta and the umbilical cord are pushed out of the uterus. The process of childbirth is complete.

Section Review

❶ **Explain why only one sperm fertilizes an egg.**

❷ **Describe the events of implantation.**

❸ **Why can it be harmful to a fetus for its mother to drink alcohol during pregnancy?**

❹ **Describe two changes that occur in the fertilized egg before it implants in the uterus.**

DISEASE-CAUSING BACTERIA AND VIRUSES TRAVEL FROM ONE HOST TO ANOTHER IN CHARACTERISTIC WAYS. SOME ARE CARRIED IN WATER, AIR, OR BY INSECTS, WHILE OTHERS ARE TRANSMITTED BY SEXUAL CONTACT. DISEASES THAT CAN BE SPREAD FROM ONE PERSON TO ANOTHER BY SEXUAL CONTACT ARE CALLED SEXUALLY TRANSMITTED DISEASES, OR STDs.

34.4 *Sexually Transmitted Diseases*

Objectives

1 **Identify** three ways HIV is transmitted.

2 **Identify** three ways to avoid HIV infection.

3 **List two** sexually transmitted diseases that are caused by bacteria.

4 **Describe** the cause and results of pelvic inflammatory disease.

Sexually Transmitted Viral Diseases

As you learned in Chapter 16, viruses cause many human diseases, including rubella, measles, mumps, and influenza. In addition, viruses cause the sexually transmitted diseases AIDS and genital herpes.

Figure 34.17
This California doctor is examining a patient with AIDS.

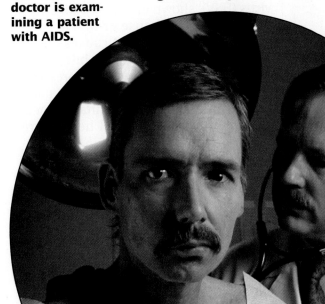

AIDS is a fatal viral disease

As you learned in Chapter 32, Acquired Immune Deficiency Syndrome (AIDS) is caused by HIV, Human Immunodeficiency Virus. HIV destroys the immune system of infected individuals by destroying helper T cells. Helper T cells are essential for the immune system to defend the body against pathogens. AIDS patients, such as the man being examined by his doctor in **Figure 34.17**, die from infections and cancers that the immune system normally defeats.

Researchers now know that HIV is found at high levels within white blood cells called macrophages that are present in blood, semen, and vaginal fluids of infected individuals. HIV and HIV-containing cells can be transmitted from one person to another during sexual intercourse. Anal intercourse, oral sex, and vaginal intercourse are all capable of transmitting HIV, as all permit macrophages to pass from one person to another—*in either direction*. During vaginal intercourse, a woman is at a somewhat higher risk of getting the disease from an infected man than vice versa. However, female-to-male transmission also occurs easily. Basketball star Earvin "Magic" Johnson contracted HIV in this way. In Africa, where AIDS is most widespread, the numbers of HIV-infected males and females are equal.

Sexual intercourse is not the only way in which AIDS is spread. Since HIV is found in blood, it can be transmitted from one person to another by sharing hypodermic syringes and needles used for injecting drugs, and by needles used for tattooing. In a doctor's office or hospital, a new, disposable syringe is used for each injection. Users of illegal drugs sometimes reuse and share syringes and needles, both of which can be contaminated with HIV.

Like the viruses that cause rubella and herpes, HIV can be transmitted from mother to child across the placenta. About one-third of the babies born to HIV-infected mothers are infected with the virus.

Since it takes up to 10 years for HIV to damage the immune system enough to cause illness, a person can be infected with HIV for many years without being aware of it. During this time, the infected person can be spreading the disease through sexual contact or shared needles. Laboratory tests, however, can reveal whether an individual has been infected with HIV.

There is no cure for AIDS. HIV infection can be prevented by avoiding behaviors that allow HIV to be transmitted. Abstaining from sexual intercourse and not sharing needles are the only ways to completely avoid infection with HIV. Using a condom during sexual intercourse greatly reduces but does not eliminate the risk of being infected with HIV.

Genital herpes is also caused by a virus

Genital herpes is a sexually transmitted disease caused by herpes simplex virus (HSV), shown in **Figure 34.18**. Two types of HSV can cause genital herpes: HSV-1 and HSV-2. HSV-2 causes about 80 percent of genital herpes infections. HSV-1 commonly causes cold sores and fever blisters but can also be sexually transmitted and cause genital herpes. The symptoms of genital herpes are painful, recurring blisters in the genital area. Antiviral drugs can temporarily eliminate the blisters, but these drugs cannot eliminate the virus from the body.

HSV can be passed from mother to fetus during pregnancy or birth. A baby infected with HSV can suffer severe damage to its nervous system or even die as a result of the infection.

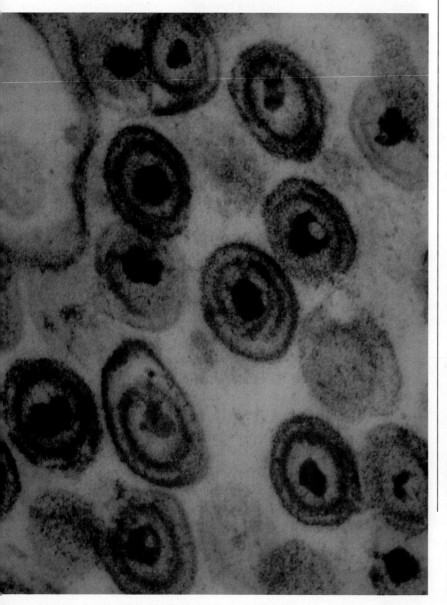

Figure 34.18
Herpes virus 2 causes the majority of genital herpes infections.

Sexually Transmitted Bacterial Diseases

Evolution

Strains of

penicillin-resistant

gonorrhea bacteria

are becoming more

common. How does

penicillin resistance

represent an

adaptation by

these bacteria?

Like viruses, bacteria cause many human diseases. Among these are sexually transmitted diseases such as syphilis, gonorrhea, and chlamydia. These sexually transmitted diseases, unlike those caused by viruses, can be treated with antibiotics.

Syphilis often goes unnoticed

Unlike HIV and herpes, syphilis (*SIHF uh lihs*) is caused by a bacterium. **Figure 34.19** shows the incidence of syphilis among 15- to 19-year-olds. Even though it can be cured with antibiotics, syphilis often is not diagnosed or treated because its early symptoms are mild or go unnoticed. About two to three weeks after infection, syphilis usually causes a small painless ulcer called a chancre (*SHAHNG kuhr*) on the genitals. In males, the chancre usually appears on the penis. Because it is painless, the chancre may not be noticed or may be ignored. In females, the chancre may form inside the vagina or on the cervix, making it even less likely to be detected.

If syphilis is not treated, a few weeks later it may cause a rash, fever, or swollen lymph glands. These symptoms also disappear without treatment. Years later, however, the infection may cause serious damage to the nervous system, blood vessels, or skin.

Untreated syphilis can be transmitted from an infected mother to her fetus. Babies infected in the uterus may be stillborn or have serious complications involving damage to major organ systems.

Gonorrhea can damage the reproductive organs

Gonorrhea (*gahn uh REE uh*) is another sexually transmitted disease that is caused by a bacterium. **Figure 34.19** also shows the incidence of gonorrhea among 15- to 19-year-olds. Gonorrhea can be cured with antibiotics, although some strains of gonorrhea are resistant to the more commonly used antibiotics, such as penicillin. In males, gonorrhea usually causes painful urination and pus discharge from the penis. In females, it sometimes causes vaginal discharge but more often causes no symptoms.

In males, untreated gonorrhea can spread to the vas deferens, epididymis, or testes. In females, it can spread to the Fallopian tubes. Infection of the Fallopian tubes can cause scarring that results in infertility.

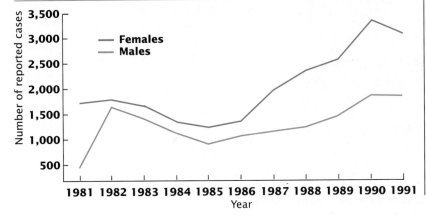

Cases of Syphilis in 15- to 19-year-olds in the United States

Figure 34.19
The incidence of both syphilis and gonorrhea is quite high among 15- to 19-year-olds. As you can see in the graphs, the number of reported cases of syphilis and gonorrhea is greater for females than for males.

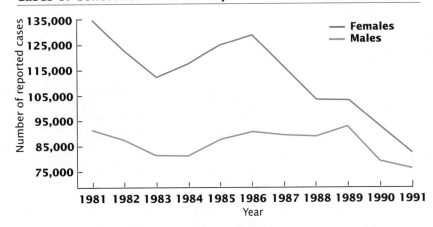

Cases of Gonorrhea in 15- to 19-year-olds in the United States

Chlamydia is the most common sexually transmitted disease

Chlamydia (*kluh MIHD ee ah*) is the most common sexually transmitted disease in the United States. There are between 3 million and 10 million new cases each year. Like gonorrhea, chlamydia is caused by a bacterium and can be cured with antibiotics. The *Chlamydia* bacterium shown in **Figure 34.20** lives within an infected person's cells. The symptoms of chlamydia are similar to those of a mild case of gonorrhea: painful urination in males and vaginal discharge in females. Also like gonorrhea, chlamydia often produces no symptoms. In people with mild symptoms or no symptoms, chlamydia is often not diagnosed. Untreated chlamydia can spread to the Fallopian tubes and can cause infertility. A baby born to a mother with chlamydia can be infected as it passes through the birth canal and may suffer eye or lung damage.

Although gonorrhea and chlamydia often have no symptoms, routine laboratory tests available at hospitals and clinics can detect infection by either bacterium.

Untreated gonorrhea or chlamydia can cause pelvic inflammatory disease

Infection of the uterus, ovaries, pelvic cavity, or Fallopian tubes is called pelvic inflammatory disease, or PID. PID is usually caused by gonorrhea or chlamydia. When the Fallopian tubes are infected, scar tissue may form and close the tubes. If the tubes are closed, an egg can no longer be fertilized. PID is one of the most common causes of infertility in women.

Fallopian tubes that are damaged, but not completely closed, may allow sperm to reach the egg, but not allow the embryo to reach the uterus. This can cause an ectopic pregnancy. An **ectopic pregnancy** occurs when the embryo grows somewhere other than the uterus. An ectopic pregnancy in a Fallopian tube can be life-threatening to a woman. As the embryo grows, it can cause the Fallopian tube to rupture.

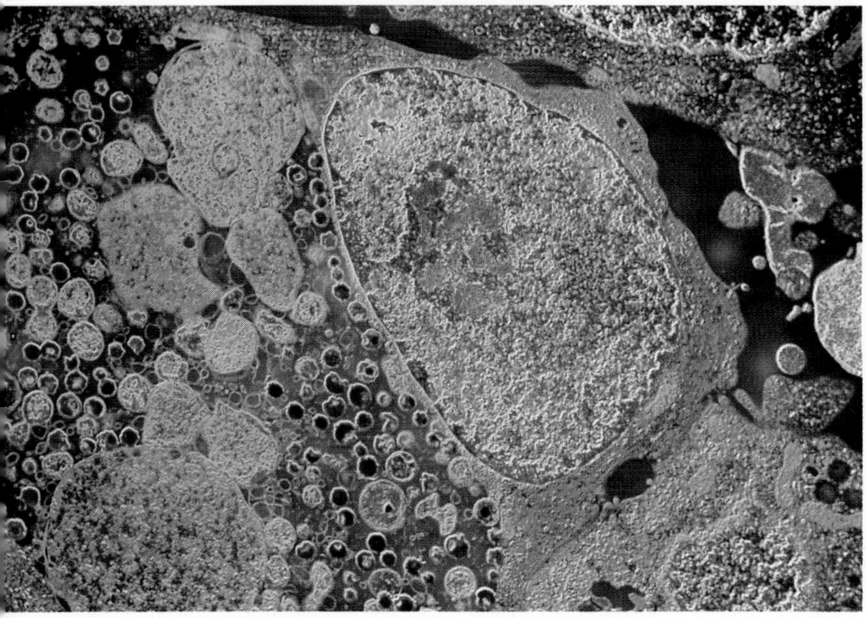

Figure 34.20
The bacterium that causes chlamydia, shown here in brown, grows within cells of its human host.

Section Review

❶ Describe three ways HIV can be spread from one person to another.

❷ List two ways you can avoid HIV infection.

❸ List two sexually transmitted diseases that are caused by bacteria.

❹ What are the possible results of untreated gonorrhea or chlamydia?

This young lady, not yet two years old, has much to learn before she can wear daddy's hat.

	Key Terms	Summary

34.1 The Male Reproductive System

Millions of sperm are expelled by the male during ejaculation.

Key Terms:
testis (p. 784)
epididymis (p. 784)
vas deferens (p. 784)
ejaculation (p. 785)
semen (p. 785)

Summary:
- Sperm are produced in the testes. They are stored in the epididymis and vas deferens, and are expelled during ejaculation.
- Luteinizing hormone stimulates the testes to produce testosterone. Follicle-stimulating hormone stimulates sperm production.

34.2 The Female Reproductive System

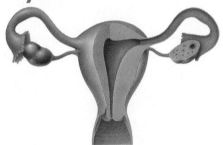

Usually, one egg is released from an ovary each month. If it is fertilized by a sperm, an embryo will develop.

Key Terms:
ovum (p. 787)
ovary (p. 787)
Fallopian tubes (p. 787)
uterus (p. 787)
cervix (p. 788)
vagina (p. 788)
ovulation (p. 789)
follicle (p. 789)
embryo (p. 790)
menstrual cycle (p. 790)
menopause (p. 791)

Summary:
- Eggs are produced in the ovaries. Ovulation occurs about 14 days after the egg begins to develop.
- Follicle-stimulating hormone causes eggs to mature. Luteinizing hormone causes release of eggs.
- Each month, the uterus prepares to receive an embryo. If no embryo arrives, the lining of the uterus is washed out of the body by blood. This bleeding is known as the menstrual period.

34.3 Fertilization and Development

The fetus will grow and develop for about nine months, until it is ready to be born.

Key Terms:
fetus (p. 795)
implantation (p. 795)
trimester (p. 798)
labor (p. 800)

Summary:
- A fertilized egg drifts down the Fallopian tube into the uterus.
- After the embryo settles into the uterine lining, the placenta forms. Nutrients, oxygen, wastes, and carbon dioxide are transferred through the placenta.
- After eight weeks, the embryo is known as a fetus.
- Birth occurs after about nine months of pregnancy.

34.4 Sexually Transmitted Diseases

In the United States, more than 500,000 new cases of genital herpes are reported each year.

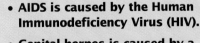

Key Terms:
ectopic pregnancy (p. 804)

Summary:
- AIDS is caused by the Human Immunodeficiency Virus (HIV).
- Genital herpes is caused by a virus; syphilis, gonorrhea, and chlamydia are caused by bacteria.

Understanding Vocabulary

1. For each set of terms, choose the term that does not fit the pattern and explain why it does not fit.
   a. uterus, ovaries, epididymis, cervix
   b. genital herpes, syphilis, gonorrhea, chlamydia
   c. ovulation, fertilization, ejaculation, menstrual period
   d. estrogen, FSH, progesterone, testosterone

Relating Concepts

2. Copy the unfinished concept map below onto a sheet of paper. Then complete the concept map by writing the correct word or phrase in each blank oval.

Understanding Concepts

Multiple Choice

3. Testosterone and sperm are produced by the
   a. testes.
   b. epididymis.
   c. vas deferens.
   d. semen.

4. Sperm are stored in the
   a. urethra.
   b. epididymis.
   c. kidneys.
   d. urinary bladder.

5. Adult male characteristics such as facial hair and a deep voice are caused by
   a. high levels of testosterone.
   b. the onset of sperm production.
   c. low levels of LH and FSH.
   d. ejaculation.

6. The muscular tube in females that serves as a birth canal is the
   a. vagina.
   b. cervix.
   c. ovum.
   d. uterus.

7. In females, FSH and LH are at their lowest levels during
   a. puberty.
   b. ovulation.
   c. menstruation.
   d. sexual intercourse.

8. An egg travels from the ovary to the uterus through the
   a. Fallopian tube.
   b. urethra.
   c. vas deferens.
   d. follicle.

9. Rupture of the fluid-filled follicle, which signals the beginning of ovulation, is stimulated by
   a. fertilization.
   b. the release of FSH.
   c. high LH levels.
   d. high levels of estrogen.

10. Which substance *cannot* be passed from mother to fetus through the placenta?
    a. blood
    b. heroin
    c. oxygen
    d. nutrients

11. Which sexually transmitted disease can be cured with antibiotics?
    a. genital herpes
    b. HIV
    c. gonorrhea
    d. measles

Completion

12. Pelvic inflammatory disease is caused by untreated _____ or _____ and may result in blocked Fallopian tubes or a(n) _____ pregnancy.

13. After eight weeks of development, an embryo is called a(n) _____ .

14. The sequence of events during which one or more eggs mature and are released is known as the _____ cycle.

15. Pregnancy in the human female lasts _____ months. This time is divided into three-month periods called _____ .

Male hypothalmus
produces
GnRH
suppresses
stimulates
?
to release
testes — stimulates— ? — FSH
to produce   in   stimulates
? — ?
causes
male characteristics —including— facial hair / ?

Short Answer

16. How can alcohol consumption by a pregnant woman affect the fetus she carries?

17. Urine is acidic. After urine passes through the urethra, it too becomes acidic. How does the body prevent sperm cells from being killed as they pass through the urethra?

18. List behaviors that enable HIV to be transmitted.

19. Describe how the diaphragm prevents pregnancy.

20. What happens to an egg if it is not fertilized within 24 hours after its release?

Interpreting Graphics

21. Examine the diagram below.

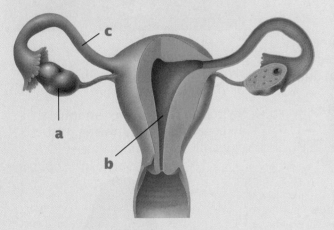

- In which structure does fertilization occur?
- Where does the embryo implant?
- In which structure could an ectopic pregnancy occur? What could cause an ectopic pregnancy?

Reviewing Themes

22. *Patterns of Change*
Describe how the ovarian and menstrual cycles are related.

23. *Interacting Systems*
Why does a woman who has undergone tubal ligation not get pregnant?

24. *Evolution*
What characteristics of sexual reproduction are responsible for the similarities between parents and their children?

Thinking Critically

25. *Applying Concepts*
Why is it important for men who have contracted gonorrhea to inform their female sex partners?

26. *Applying Concepts*
A 48-year-old woman stops having menstrual periods. She believes she is pregnant. What is another possible explanation?

27. *Building on What You Have Learned*
In Chapter 28 you learned about the hypothalamus. Describe how the hypothalamus regulates the levels of testosterone and sperm production in males.

Cross-Discipline Connection

28. *Biology and Health*
The Lamaze method of childbirth stresses muscle relaxation, controlled breathing, and the involvement of a birthing coach. Look for information about the advantages of Lamaze.

Discovering Through Reading

29. Read the article "The Mysterious Origin of AIDS," in *Natural History*, September 1992, pages 24–29. Summarize the three hypotheses to account for the appearance of the AIDS epidemic. Explain the weaknesses in each hypothesis.

Investigation

How Are Sperm and Eggs Different?

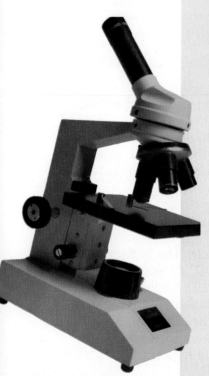

Objectives

In this investigation you will:
- *observe* and *compare* sperm and eggs
- *observe* early stages of development

Materials

- prepared slides of sea-star eggs and sperm
- compound light microscope
- prepared slide of early developmental stages of a sea star

Prelab Preparation

1. Review what you have learned about eggs and sperm by answering the following questions:
   - Where are sperm produced?
   - Where are eggs produced?
   - What purpose do sperm and eggs serve?
   - What process forms both sperm and eggs?
2. Review what you have learned about early development by answering the following question:
   - What are the first changes to occur in an egg after fertilization?

Procedure

1. Form a cooperative group with another student to complete steps 2–7.
2. Place the slide of the sea-star egg on the microscope stage. Observe the egg cell using low power. Make a labeled drawing of the egg cell.

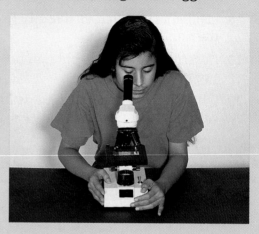

3. Describe the appearance of the egg. *Which region is larger, the nucleus or the cytoplasm?*
4. Switch to high power and make a second labeled drawing of the egg cell.

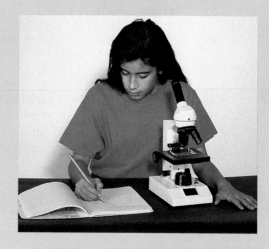

5. Place the slide of sea-star sperm on the microscope stage. Observe the sperm using low power. Make a labeled drawing of the sperm.

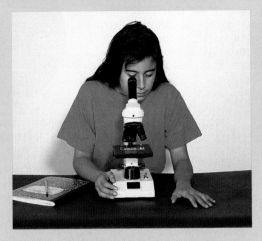

6. Switch to high power and make a second labeled drawing of an individual sperm. *Which region of the sperm is larger, the nucleus or the cytoplasm?*

7. Now examine the slide of early developmental stages of the starfish. Make a drawing of each stage in development. Note the number of cells and the size of the cells at each developmental stage.

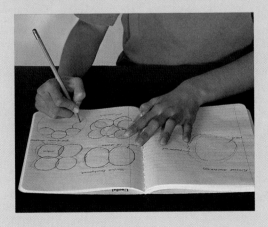

Analysis

1. *Analyzing Observations* How do egg and sperm cells compare in size?

2. *Analyzing Observations* How does the amount of cytoplasm in the sperm compare with the amount in the egg?

3. *Making Inferences* Suggest an explanation for the difference in size between sperm and egg cells.

4. *Analyzing Observations* Why are sperm cells capable of locomotion while egg cells are not?

5. *Relating Structure to Function* How does the shape of a sperm cell reflect its function?

6. *Analyzing Observations* Of the developmental stages of the sea star you observed, which stage was the latest in development? Explain.

7. *Comparing Structures* Which developmental stage of the sea star was the earliest in development? Where would you find a human embryo at the same developmental stage?

Thinking Critically

A sperm and an egg each contribute one-half of an embryo's genes. Yet there are some genes that can only be inherited from the mother. Where in the genome are these genes located?

UNCOVERING THE SECRETS OF THE BODY

1625

1628 British physician **William Harvey** shows that blood circulates around the body through blood vessels. Harvey bases his conclusions on his own observations and experiments, not the untested ideas of authority. The experimental method used by Harvey provides a model that continues to be used by scientists.

1822 and years following American physician **William Beaumont** begins treating a patient whose injury leaves an opening through the body wall to the stomach. Beaumont discovers much about the digestive process by observing the action of the patient's stomach directly through the wound.

William Beaumont

1796 After developing a vaccine for smallpox, English physician **Edward Jenner** gives the first vaccination.

Edward Jenner

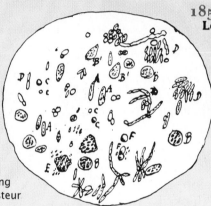
Drawing by Pasteur

1850 and years following **Louis Pasteur** performs experiments showing that bacteria cause fermentation and decay. Pasteur's work leads to the Germ Theory of Disease, which states that bacteria, not "bad humors," cause diseases.

1900

Emil Behring

1901 German scientist **Emil Behring** shows that immunity to a disease can be provided by inoculating animals with the blood serum of other animals that have recovered from the disease.

1921 Canadians **Frederick Banting**, **Charles Best**, and **J.J.R. MacCleod** discover the hormone insulin. Their work links inadequate insulin production with the disease diabetes and shows that insulin injections can control this disease.

1928 British bacteriologist **Alexander Fleming** notices that the mold *Penicillium notatum* prevents the growth of other microorganisms. Fleming's observation leads to the development of the first antibiotic, penicillin, which proves remarkably successful against bacterial infections.

Penicillin mold

Jonas Salk giving polio vaccine

1954 American physician **Jonas Salk** develops an effective vaccine against the virus that causes polio, a crippling disease of children and young adults. The Salk vaccine reduces the number of polio cases in the United States by 90 percent between 1954 and 1962.

Anopheles mosquito

1900

1858 German physician **Rudolf Virchow's** work shows that different kinds of cells in the body perform different tasks that contribute to the smooth functioning of the body.

Human body cells

1877–1897 Scottish physician **Sir Patrick Manson** shows how insects may carry diseases. Manson's work leads other scientists to search for insect carriers of human diseases. English physician **Sir Ronald Ross** finds the malaria parasite in the stomach of the *Anopheles* mosquito. Observations made by American physician **Carlos Finlay** in Cuba help Finlay hypothesize that the *Aedes* mosquito transmits yellow fever to humans.

1900 Austrian-born American scientist **Karl Landsteiner** defines four major human blood groups— A, B, AB, and O— based on antigens found on red blood cell surfaces.

Red blood cells

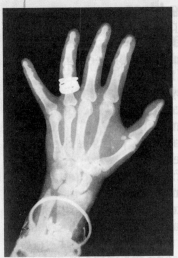

X ray of a hand

1865 Armed with his knowledge of Pasteur's work, English surgeon **Joseph Lister** introduces a system of antiseptic surgery. Lister's methods reduce the number of fatal infections resulting from surgery.

1895 German physicist **Wilhelm Roentgen** discovers X rays. X rays revolutionize medicine by allowing physicians to "see" into the body without performing surgery.

1990

1967 South African surgeon **Christiaan Barnard** performs the first successful human heart transplant.

Christiaan Barnard

1981 The United States Centers for Disease Control officially recognizes **AIDS (Acquired Immune Deficiency Syndrome)** as a disease.

1990 Researchers use **gene therapy** on a human for the first time in an attempt to cure a patient of severe combined immune deficiency, a fatal hereditary disease of the immune system. Gene therapy is a technique in which healthy genes are introduced into a patient in order to cure or control an inborn error of metabolism.

DNA sequencing

1972 The **CAT (Computerized Axial Tomography) scanner** is developed in Britain. The CAT scanner takes a series of X rays through an organ. When viewed as a set, these X-ray images allow physicians to pinpoint the exact size and location of diseased tissue in the body.

1982 The first commercial product of **genetic engineering** is approved for sale by the United States Food and Drug Administration. The product, humulin, is human insulin produced by bacteria that have had the human gene for insulin production spliced into their DNA.

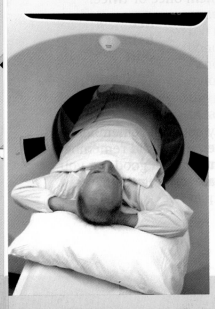

CAT scanner

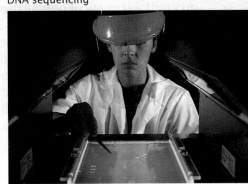

Laboratory Skills

Using a Compound Light Microscope

Parts of the Compound Light Microscope

- The *eyepiece* magnifies the image, usually 10X.

- The *low-power objective* magnifies the image more, often 10X.

- The *high-power objective* magnifies the image even more, such as 43X.

- The *revolving nosepiece* holds the objectives and can be turned to change from one magnification to another.

- The *body tube* maintains the correct distance between eyepiece and objectives. This is usually about 25 cm (10 in.) the normal distance for reading and viewing objects with the naked eye.

- The *coarse adjustment* moves the body tube up and down in large increments to allow gross positioning and focusing of the objective lens.

- The *fine adjustment* moves the body tube slightly to bring the image into sharp focus.

- The *stage* supports a slide.

- *Stage clips* secure the slide in position for viewing.

- The *diaphragm* (or *iris diaphragm*) controls the amount of light allowed to pass through the object being viewed.

- The *light source* provides light for viewing the image. It can either be a light reflected with a mirror or incandescent light from a small lamp.

  NEVER use reflected direct sunlight as a light source.

- The *arm* supports the body tube.

- The *base* supports the microscope.

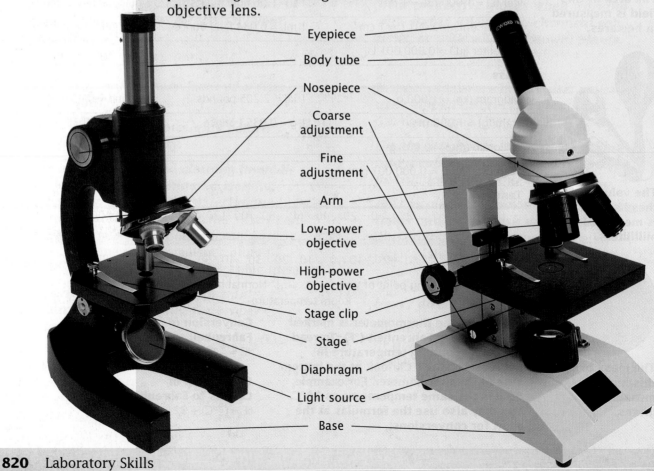

Eyepiece

Body tube

Nosepiece

Coarse adjustment

Fine adjustment

Arm

Low-power objective

High-power objective

Stage clip

Stage

Diaphragm

Light source

Base

Credits

List of Abbreviations

All photographs attributed to SP/Foca were taken by Sergio Purtell/Foca Co., NY, NY.

AA = Animals Animals; AP = Archive Photos; AP/WW = AP/Wide World; AS = AllStock; AT = Alexander & Turner; BA = rep. Bernstein & Andriulli Inc.; BPA = Biophoto Associates; BPS = BPS & Biological Photo Service; CBMI = Claire Booth Medical Illustration; CM = Custom Medical; CP = Culver Pictures; CS = Comstock; HS = Highlight Studios; ES = Earth Scenes; FPG = FPG International, Corp.; HPS = Historical Pictures Service; KBI = Karapelou BioMedical Illustrations; KC= Kip Carter, M.S. certified medical illustrator; LM = Les Mintz/Incandescent Ink, Inc.; LPI = Lynne Prentice Illustration; MA = rep. Mattleson Associates, Ltd.; MC = MediChrome; MI = Magnum; MIS = Medical Illustration Studio; MMA = Margulies Medical Art; MT = rep. Melissa Turk; MHI = Mark Hallett Illustrations; NSI = Natural Science Illustration; OSF = Oxford Scientific Films; PA = Peter Arnold, Inc.; PR = Photo Researchers; RMF = rep. Rita Marie & Friends; RWS = rep. Richard W. Salzman S.F., CA; SPL = Science Photo Library; SS = Studio Sloth; TBA = The Bettmann Archive; TM = Teri J. McDermott, M.A., certified medical illustrator; TMMK = Teri J. McDermott, M.A., certified medical illustrator/Michael Kress-Russick, M.A., M.S.; TSA = Tom Stack & Associates; TSM = The Stock Market; TWC = The Wildlife Collection; VU = Visuals Unlimited; WCA = Woodfin Camp & Assoc.

All Safety Symbols were designed by Michael Helfenbein/Evolution Design.

Cover (sparrow) Daniel Kirk/BA; (pea pod,sunflower,dancer) SP/Foca; (pea flowers) Karen Kluglein/MA; (soil) Lynne Prentice/LPI; (dinosaur) Mark Hallet/MHI; (DNA) R Margulies/Virginia Ferrante/MMA; (cell) R Margulies/MMA; (shark) © 1987 J Stafford-Deitsch/SHARKS/Eddison/Sadd Editions; (chromosomes) © BPA/PR; (Earth) © Frank Rossotto/TSM; (fly) © Herman Eisenbeiss/PR; (eagle) © Jim Simmen/AS; (flowers) © Richard Parker/PR; **i** (l) C Booth/CBMI; (cr) David Beck/RMF; (t) SP/Foca; (br) John W. Karapelou/KBI; **ii** (tl,bl) SP/Foca; (br) Susan Johnston Carlson/MT; **iii** (bl,br) SP/Foca; **iv** Courtesy of George Johnson; **v** (ct, cb) C Booth/CBMI; (t,bl) SP/Foca; (br) T Narashima/SS; **vi** (tl,bl) SP/Foca; (ct) Karen Kluglein/MA; (c) R Margulies/Christine Schaar/MMA; (br) © Glen Nelson; **vii** (bl,br) SP/Foca; (ct) John Daugherty/HS; (c) Lynne Prentice/LPI; (t) Yemi; **viii** (c) C Booth/CBMI; (t) David Beck/RMF; (bl,br) SP/Foca; (ct) © M.I. Walker/PR; **ix** (tr) C Turner/AT; (tl,bl,br) SP/Foca; (c) Peg Gerrity/HS; **x** (t,ct) SP/Foca; (b,bc) Peg Gerrity/HS; **xi** (bl) SP/Foca; (t) Kip Carter/KC; (br) Peg Gerrity/HS; (ct) T McDermott/Michael Kress-Russick/TMMK; (c) Walter Stuart/RWS; **xii** SP/Foca; **xiii** (br) Daniel Kirk/BA; (t) Laurie O'Keefe/HS; (bl) Lynne Prentice/LPI; (c) Walter Stuart/RWS; **xiv** (t) Luiz C. Marigo/PA; (b) © Harvey Lloyd/PA; **xv** (l) See credits for p. 112; (br) See credits for p. 315 ; **2–3** (girl) SP/Foca; **2** (flower) SP/Foca; (tern) © Dean Lee/TWC; (bee) © OSF/AA; **3** (frog) © Lynda Richardson/PA; **4** © Stephen Dalton/PR; **5** Published w/ the permission of Discover Magazine; **6** (t,bl,cl) SP/Foca; (cr) John W. Karapelou/KBI; **7** (bl) (C) Michael Minardi/PA; (t) PEOPLE Weekly ©1990 Thomas S. England; (r) © Tom Raymond/MC; **8** (c) Jacques Jangoux/PA; (bl) SP/Foca; (cr) © 1984 William J. Jahoda /PR; (tl) © 1985 Patrick Grace/PR; (tr) © European Space Agency/SPL/PR; (cl) © Ray Pfortner/PA; **9** (c) © 1985 Thomas S. England/PR; (r) © Bruce Ando/Index Stock; **10** SP/Foca; **11** © Sue Ford/SPL/CM; **12** (l) TBA; (c) UPI/Bettmann; (r) © OSF/AA; **13** (r) Courtesy New York Academy of Medicine; (c) UPI/Bettmann; (l) © BPA/Science Source/PR; **14** © 1989 by Sidney Harris-'Einstein Simplified', Rutgers Univ. Press; **15** Lynne Prentice/LPI; **16** SP/Foca; **17** (l) SP/Foca; (r) John W. Karapelou/KBI; (c) © Michael Tamborrino/MC; **18** © Stephen Dalton/PR; **19** (tl, br) John W. Karapelou/KBI; (tr) © 1978 Carlton Ray/PR; (bl) © Michael Tamborrino/MC; (cl) © OSF/AA; **21** © Matt Meadows/PA; **22** SP/Foca; **24** SP/Foca; **25** Cabisco/Visuals Unlimited; **26** Cabisco/Visuals Unlimited; **27** Cabisco/Visuals Unlimited; **28** © BPA/ Science Source/PR; **29** (c) © BPA/Science Source/PR; (t) © S. Flegler/VU; (b) © Science VU/Lawrence Livermore Laboratory/VU; **30** C Turner/AT; **31** (l) C Booth/CBMI; (r) SP/Foca; **32** (tr,b) C Booth/CBMI; (tl) © W. Kaufmann/JPL/Science Source/PR; **33** (l) SP/Foca; (r) © Jeff Rotman/PA; **34** SP/Foca; **35** C Booth/ CBMI; **36** E Alexander/AT; **37** R Margulies/MMA; **38** (l) C Booth/CBMI; (hair) SP/Foca; (chromosomes) © BPA/Science Source/PR; (cells) © Manfred Kage/PA; (starch) © Omikron/ Science Source/PR; **39** (cl) C Turner/AT; (tr) SP/Foca; (tl) © B. Beatty/VU; (bl) © Jeff Rotman/PA; **41** (t) C Turner/AT; (b) SP/Foca; **42** SP/Foca; **43** SP/Foca; **44** © Dr. Tony Brain & David Parker/SPL/PR; **45** © Eric V. Grave/PR; **46** (tr,br) C Booth/CBMI; (tl,bl) SP/Foca; **47** C Booth/CBMI; **48** (tl,bc,br) C Booth/CBMI; (bl) SP/Foca; **49** (bl) C Booth/CBMI; **49** (r) SP/Foca; **50** Catherine Twomey/Craven Design Studios Inc.; **51** R Margulies/MMA; **52** R Margulies/MMA; **53** R Margulies/MMA; **54** © Chris Wang; **55** (l) © Chris Wang; (r) © Science/VU; **56** (l) © A.B. Dowsett/SPL/PR; (r) © M.I. Walker/PR; **57** R Margulies/MMA; **58** (tl,br) R Margulies/MMA; (tr) © BPA/PR; (bl) © David M. Phillips/The Population Council/ Science Source/PR; **59** (tl,cl,bl) R Margulies/MMA; (br) © B. Bowers/PR; (t) © D.W. Fawcett, S. Ito/PR; **60** R Margulies/MMA; **61** (tr) SP/Foca; (c,b) R Margulies/MMA; (tl) © Eric V. Grave/PR; **63** (t) Catherine Twomey/Craven Design Studios Inc.; (c) R Margulies/MMA; (b) © A.B. Dowsett/SPL/PR; **64** SP/Foca; **66–67** (bkrnd) © Harvey Lloyd/PA; **66** © Andrew McClenaghan/SPL/PR; **67** © P. Shambroom/ PR; **68** © Lennart Nilsson/THE INCREDIBLE MACHINE/National Geographic Society/Boehringer Ingelheim Int'l GmbH; **69** (l,c) C Booth, CBMI; (br) R Margulies/MMA; **70** (l) SP/Foca; (c,r) R Margulies/MMA; **71** (br) R Margulies/MMA; (tr) T Narashima/SS; (tl) © 1989 Tom Tracy/TSM; (bl) © Michael Abbey/Science Source/PR; **72** (br) © Michael English, M.D./CM ; (bl) © Peter Berndt/CM ; (t) © SPL/CM ; **73** (t) Yemi; (b) © 1991 Kay Chernush for Howard Hughes Medical Institute; **74** SP/Foca; **75** (flowers) SP/Foca; (cells) T Narashima/SS; **76** (l) SP/Foca; (r) R Margulies/MMA; **77** R Margulies/ MMA; **78** T Narashima/SS; **79** T Narashima/SS; **80** (tl) © BPA/PR; (bl,bc,br) © Lennart Nilsson/A CHILD IS BORN/Dell Publishing Co.; **81** T Narashima/SS; **82** (c) Yemi; (l,r) © Moredun Animal Health LTD/SPL/CM ; **83** (tr) SP/Foca; (tl) R Margulies/MMA; (b,c) T Narashima/SS; **85** (cells) T Narashima/SS; (membrane) Yemi; **88** © Manny Millan/Sports Illustrated; **89** SP/Foca; **90** SP/Foca; **91** (b) © Walter H. Hodge/PA; **92** R Margulies/MMA; **93** (l) SP/Foca; (r) © Fred Breummer/PA; **94** © A.B. Dowsett/SPL/PR; **95** C Booth/CBMI; **96** (l) NASA; (cl) © 1991 CS; (cr) © David C. Fritts/AA; **96** (r) © Mark Stouffer; **97** © 1992 Bruce Fritz; **98–99** (holly) SP/Foca; **98** (l) SP/Foca; **99** (br) E Alexander/AT; (tr) © 1977 Walker England/PR ; (tl) © Walker England/PR; **100** E Alexander/AT; **101** C Booth/CBMI; **102** © Paul W. Johnson/Biological Photo Service; **103** (c) C Booth/CBMI; (tl,tr) SP/Foca; (bl) © Link/VU; **104** (molecules) C Booth/CBMI; (tr,br) SP/Foca; **105** C Booth/CBMI; **106** R Margulies/MMA; **107** (chloroplast) E Alexander/AT; (person) SP/Foca; (tr) © B. Taurus/OSF/AA; (deer) © David C. Fritts/AA; (enzyme) © Walter H. Hodge/PA; **109** © Vandystadt/ PR; **110** SP/Foca; **112–113** (jars) John Daugherty/Peg Gerrity/HS; **112** (tl) SP/Foca; (tr) TBA; (br) UPI/Bettmann; (bl) © 1989 Kent Wood/PR; © Carolina Biological ; **113** (tr) Archive Photos; (br) NASA; (bl) NASA/Science Source/PR; (tc) SPL/PR; **114–115** © 1990 Peter Menzel; **114** (t) © 1991 Penny Gentieu/Black Star; (c) © Dan Guravich/PR; **115** (b) © Tom Casaline/Sharp Shooters; **116** © A.T. Willett/The Image Bank; **117** (t) CP Pictures; (b) Karen Kluglein/MA; **118** Karen Kluglein/MA; **119** Karen Kluglein/MA; **120** Karen Kluglein/MA; **121** Karen Kluglein/MA; **122** Courtesy Moravské Muzeum-Mendelianum; **123** (l) TBA; (r) © Herman Eisenbeiss/PR; **124** (l) SP/Foca; (c) © 1985 Howard Sochurek/MC; (r) © BPA/PR; **125** T Narashima/SS; **126** (b) SP/Foca; (l) T Narashima/SS; **127** © David York/MC; **128–129** T McDermott/TM; **128** (r) © 1987 Ruseman/CM; (l) © Simon Fraser/SPL/CM; **129** (b) Courtesy The Rice Family; (bl) T McDermott/TM; (cl) © Richard Hutchings/PR; **130** Courtesy The Davidson Family; **131** (t) SP/Foca; (tl) Karen Kluglein/MA; (cl) © BPA/PR; (bl) © David York/MC; **134** SP/Foca; **136** © Johnny Johnson/AS; **137** (t) C Booth/CBMI; (b) Susan Johnston Carlson/MT; **138–139** (b) R Margulies/Virginia Ferrante/MMA; **138** (t,bl) C Booth/CBMI; (cl,b) © Lee D. Simon/PR; **139** (t) © PA; **140** (t) SP/Foca; (b) R Margulies/MMA; **141** R Margulies/MMA; **142–143** (t) SP/Foca; **142** (b) R Margulies/Christine Schaar/MMA; **143** (tr,tc,cr,br) SP/Foca; (bl) © Walker/PR; **144–145** E Alexander/AT; C Turner/AT; **146** (r) Diana Punales-Morejon; (l) SP/Foca; **147** (r) SP/Foca; (l,br) © BPA/Science Source/PR; **148–149** R Margulies/Christine Schaar/MMA; **150** (l) SP/Foca; (r) © D. Goldberg/Sygma; **151** (cl) E Alexander/AT; (tl,bl) R Margulies/MMA; (tr) TBA; **153** E Alexander/AT; **154** SP/Foca; **155** SP/Foca; **156** © Sinclair Stammers/SPL/PR; **157** (l) SP/Foca; (c) R Margulies/MMA; (r) © R. Kessel & G. Shih/VU; **158** (tl,br) E Alexander/AT; (bl) SP/Foca; (tr) R Margulies/MMA; **159** (c) E Alexander/AT; (tl) © David M. Phillips/VU; (tr) © R. Kessel & G. Shih/VU; **160** E Alexander/AT; **161** (l) E Alexander/AT; (r) © Leonard Lessin/PA; **162** (bl) SP/Foca; (br) © E.H. Newcomb & S.R. Tandon, Univ. of Wisconsin/BPS; **163** B Hansen/NSI; **164** © Bruce Fritz; **165** (l) E Alexander/AT; (r,inset) SP/Foca; (t) © BPA/PR; (c) © R. Kessel & G. Shih/VU; **166** (l,cl,cr) E Alexander/AT; (r) SP/Foca; **167** © Dr. A. Liepins/Sciene Photo Library/Science Source/PR; **168** © Matt Meadows/PA; **169** (tl) E Alexander/AT; (cl) © Bruce Fritz; (bl) © Dr. A. Liepins/SPL/PR; (tr) © SPL/CM; **171** (tl,bl) E Alexander/AT; (br) R Margulies/MMA; (tr) © R. Kessel & G. Shih/VU; **174–175** (bkrnd) © Omikron-Science Source/PR; **174** © Omikron-Science Source/PR; **176** © Tom McHugh/California Academy of Sciences/PR; **177** (map) Lazlo Kubinyi/Rep. Gerald & Cullen Rapp, Inc.; (inset) © Bridgeman, London/Art Resource, NYC; **178** (t) © John S. Dunning/PR; (br) © Leonard Lee Rue III/AA; (bl,bc) © M.A. Chappell/AA; **179** (tr) Courtesy of the Department Library Services American Museum Of Natural History; (tl,bl) TBA; **180** SP/Foca; **181** B Hansen/NSI; **182** T McDermott/Michael Kress-Russick/TMMK; **183** T McDermott/Michael Kress-Russick/TMMK; **184** (r) © Doug Wechsler/AA; (l) © Manfred Danegger/PA; **185** (r) © Don W. Fawcett/VU; (bl) © E.R. Degginger /AA; (tl) © Mark Hamblin/OSF/AA; **186** © Glen Nelson; **187** Daniel Kirk/BA; **188** © Breck P. Kent/AA ; **189** C Turner/AT; **190–191** (birds) Daniel Kirk/BA; **190** (map) Randall Zwingler/LM; **192** E Alexander/AT; **193** (cl) B Hansen/NSI; (tr,bl) © Breck P. Kent/AA; (tl) © Bridgeman, London/Art Resource, NYC; **195** B Hansen/NSI; **196** SP/Foca; **197** SP/Foca; **198** © Breck P. Kent/ES; **199** T Narashima/SS; **200** Yemi; **202** © Sidney Fox/VU; **203** © F. Widdel/VU; **204** (tr,br) SP/Foca; (l) © CNRI/SPL/PR; (c) © D. Gotshall/VU; (cr) © Richard Thom/VU; **205** Lynne Prentice/LPI; **206** (l) C Booth/CBMI; (r) © NASA/Science Source/PR; **207** (l) SP/Foca; (r) © R.L. Peterson,Univ. of Guelph/BPS; **208–209** (bkrnd) © Tom & Pat Leeson/PR; **208** (tl,tr,bl,br) Courtesy The George C. Page Museum/John C. Dawson; **209** (Silurian, Odovician) David Beck/RMF; (Jurassic) Mark Hallet/MHI; (Cretaceous, Triassic) Pat Ortega; (Devonian) Peg Gerrity/HS; (bacteria) © M.D. Maser/VU; (Cambrian) © Sinclair Stammers/SPL/PR; (Quarternary) © VU; **210** (c) Courtesy of the Peabody Museum of Natural History, Yale Univ. Painted by Rudolph F. Zallinger; (inset) Lynne Prentice/LPI; **211** (r) David Beck/RMF; (l) © 1988 Hans Reinhard/Okapia/PR; **212** © Stephen Dalton/AA; **213** (bkrnd) Laurie O'Keefe/HS; (c) Mark Hallet/MHI; **214** (c) © 1988 Steve Kaufman/AS; (l) © AA; (r) © Robert A. Tyrell/OSF/AA; **215** (Earth) NASA/Science Source/PR; (ostrich) © 1988 Steve Kaufman/AS; (space) © Dr. Fred Espenak/SPL/PR; (bacteria) © F. Widdel/VU; (lipid) © Sidney Fox/VU; **217** Yemi; **218** SP/Foca; **219** SP/Foca; **220** © Erich Lessing/Musée de L'Homme, Paris/Art Resource, NYC; **221** Todd Buck/HS; **222** © 1973 E.H. Rao/PR; **223** (r) © Leonard Lee Rue III/AS; (l) © Patti Murray/AA; **224** © George Holton/PR; **225** (l to r) © E.R. Degginger/AA; © 1986 Christopher Arnesen/AS; © Pat Crowe/AA; © Michael Dick/AA; SP/Foca; SP/Foca; **226** © Paul Fusco/MI ; **227** (jaws) Dana Geraths; (bl,br) SP/Foca; (tl) © Ian Berry/MI ; **228** Mark Hallet/MHI; **229** (jaws) Dana Geraths; (skulls) © Institute of Human Origins; (b) © John Reader/ SPL/PR; **230** (r) © Fleishman-Hillard/The Living World/St. Louis Zoological Park; (l) © John Reader/SPL/PR; **231** Mark Hallet/MHI; **232–233** (map) Randall Zwingler/LM; **232** (skulls) Mark Hallet/MHI; **233** (skulls) Mark Hallet/MHI; **234** (t) SP/Foca; (b) Randall Zwingler/LM; **235** Courtesy American Museum of Natural History; **236** (b) Art Resources, NYC; (t) Courtesy American Museum of Natural History; **237** (bl) Courtesy American Museum of Natural History; (jaw) Dana Geraths; (tr) Reprinted Courtesy OMNI Magazine © 1991; (skull) © Institute of Human Origins; (tl) © Leonard Lee Rue III/AS; **239** Mark Hallet/MHI; **240** SP/Foca; **241** Dana Geraths; **242–243** (B/W chromosomes) C Booth/CBMI; **242** (bc) PR; (br) R Margulies/MMA; (tc) UPI/Bettmann; (tl) © Ray Glover/PR; (bl) © Science Source/PR; (tr) © SIU/Photo Reseachers; **243** (br) Courtesy Agnes Stroud-Lee; (bl) E Alexander/AT; (tr) SP/Foca; (tcl,tcr) UPI/Bettmann; (bc) © 1985 Sygma; (tl) © Gennaro & Grillone/Science Source/PR; **244–245** (b) © Christine M. Douglas; **244** (t) © K. Straiton/PR; **245** SP/Foca; **246** © Terry Domico/Earth Images; **247** © 1989 Kevin Schafer/PA; **248** Lynne Prentice/LPI; **249** Lynne Prentice/LPI; **250** (l) David Beck/RMF; (br) © Ed Reschke/PA; **251** David Beck/RMF; **252** Laurie O'Keefe/HS; **253** © James Hornbeck/US Forest Service; **254** (molecules) C Booth/CBMI; (plants,deer) Lynne Prentice/LPI; **255** B Hansen/NSI; **256** (molecules) C Booth/CBMI; (b) Lynne Prentice/LPI; **257** SP/Foca; **258** (br) © Bill Bachman/PR; (bl) © Gregory Dimijian/PR; (tl) © Klaus Uhlenhut/ES; **259** (tr) © 1992 Stephen J.Krasemann/ PR; (tl) © Mickey Gibson/AA; (bl,br) © Stephen J. Krasemann/PA; **260** David Beck/RMF; **261** (tl) © Ed Reschke/PA ; (c) © James Hornbeck/US Forest Service; (tr) © Raymond Mendez/AA; (b) © Stephen J. Krasemann/PA; **263** David Beck/RMF; **264** SP/Foca; **266** © Gregory K. Scott/PR; **267** © Ken Brate/PR; **268** (flies) SP/Foca; (bl) © 1984 Fletcher W.K./PR; **269** SP/Foca; **270** (bl) © 1985 L. West/PR; (inset) © V. Ahmadjian/VU; **271** (b) © 1983 F. Gohier/PR; (t) © Fred Bavendam; **272** SP/Foca; **273** Wendy Smith-Griswald/MT; **274** (t) © Francis, Donna Caldwell/Affordable Photo Stock; (c,b) © Gary Braasch/WCA; **275** (br) SP/Foca; **275** (tl) © Adrienne T. Gibson/AA; **276–277** Laurie O'Keefe/HS; **278** © Zig Leszczynski/AA; **279** (bl) © 1981 Patrick Montagne/PR; (r) © Julian Baum & David Angus/SPL/PR; (tl) © MCMXXCI Tom McHugh/PR; (cl) © Milton H Tierney Jr./VU; **280** Courtesy Carmen R. Cid-Benevento; **281** Courtesy Carmen R. Cid-Benevento; **282** © 1991 Catherine Karnow; **283** © Harry Hartman/Bruce Coleman, Inc.; **284** (r) © Jack Dermid/PR; (l) © PA; (c) © Walt Anderson/VU; **285** (tr) AP/WW; (cl) © Adrienne T. Gibson/AA; (tl) © Fred Bavendam; (bl) © Harry Hartman/Bruce Coleman, Inc.; **287** (tr,bl) SP/Foca; (cr) © G.I. Bernard/OSF/AA; (tl) © John Gerlach/VU; (br) © Patti Murray/AA; (cl) © W.K. Fletcher/PR; **288** SP/Foca; **289** SP/Foca; **290–291** (bkrnd) © George Loun/VU; **290** (t) © Luiz C. Marigo/PA; (b) © Martin Wendler/PA; **291** © Tom & Pat Leeson/PR; **292** © 1991 Photri/TSM; **293** ©1985 Earl Roberge/PR; **294** (b) AP/WW; (t) © IFA/PA; **295** (c) Randall Zwingler/LM; (l) © 1989 D. A. Glawe, Univ. of Ill./BPS; (tr) © 1991 Ray Pfortner/PA; (br) © Master Communications; **296–297** (tr) NASA/Goddard Institute For Space Studies/SPL/PR; **296** NASA ; **298** SP/Foca; **299** SP/Foca; **300** (l,r) Gene Alexander/Soil Conservation Service; (c) Tim McCabe/ Soil Conservation Service; **301** (tl) Jim Williams/U.S. Fish & Wildlife Service; (tr) Rich Krueger/US Fish & Wildlife Service; (bl) © Ken Cole/AA; **302–303** Patricia J. Wynne; **304** SP/Foca; **305** (t) SP/Foca; (b) © Renee Lynn/PR; **306** © Mickey Gibson/AA; **307** SP/Foca; **308** SP/Foca; **309** (b) SP/Foca; (c) Rick Krueger/U.S. Fish & Wildlife Service; (tl) © IFA/PA; (tr) © James H. Karales/PA; **312** SP/Foca; **313** SP/Foca; **314** (tl,bl) TBA; (br) © Erich Hartmann/MI; (tr) © Erich Lessing/Art Resource, NYC; **315** (bl) Courtesy Dr. David E. Radcliffe; (br) SP/Foca; (tr) Patricia J. Wynne; (tc,bl) TBA; (cl) © M. Wendler/Okapai/PR; **316-317** (mushrooms) © Phil Dotson/PR, Inc.; **316** (flower) © Nawrocki Stock Photo, Inc.; (forest) C. Sims/SuperStock; (paramecium) © Phil Dotson/PR, Inc.; **317** (kelp) © Bob Evans/PA; (polio) © Science/CDC/VU; **318** Courtesy The Audubon Society & General Electric ; **319** © Charles W. Mann/PR; **320** SP/Foca; **321** (bl) Mark Hallet/MHI; (tc) Randall Zwingler/LM; (br) © Bridgeman, London/Art Resource, NYC; (tr) © Dan Sudia/PR; (bc) © M. Fogden/OSF/AA; (tl) © Ray Coleman/PR ; **322** (tc) Courtesy Alma Solis/SEL/USDA; (bc) Courtesy Lucy Bunkley-Williams, Ernest H. Williams/Univ. of Puerto Rico at Mayaguez; (bl) Courtesy Paul Melançon/Univ. of Colorado, Boulder; (tl) Courtesy Yehoshua Annixter, Univ. of Tel Aviv; (br) SP/Foca; **323** © The Pierpont Morgan Library 1992; **325** David Beck/RMF; **326** (tr,bl) E Alexander/AT ; (br) © Chuck Davis Photography; (tl) © G. Soury/Jacana/PR; **327** (fern,flowers,seeds) SP/Foca; (moss) © 1989 K.G. Vock/Okapia/PR; (tree) © 1991 Tom Bean/AS; (system) © P. Dayanandan/PR; **328** (r) R Margulies/MMA; (l) Susan Johnston Carlson/MT; **329** © Mark Boulrow/PR; **330** © 1989 Neil G. McDaniel/PR; **331** © Manfred Kage/PA; **332** (bl) SP/Foca; (r) © Jeremy Stafford- Deitsch/Eddison/Sadd Editions Limited/SHARKS; (tl) © Robert Lee/PR; **333** (tl) SP/Foca; (tr) © Bob Orsillo; (bl) © Neil G. McDaniel/PR; (lc) © The Pierpont Morgan Library 1992; **335** (fern,flowers,seeds) SP/Foca; (moss) © 1989 K.G. Vock/Okapia/PR; (tree) © 1991 Tom Bean/AS; (system) © P. Dayanandan/PR; **336** SP/Foca; **338** © BPA/Science Source/PR; **339** (l) © Bill Longcore/Science Source/PR; (cr) © CNRI/SPL/PR; (cl) © Manfred Kage/PA; (r) © Omikron/Science Source/PR; **340** (tr,br) T Narashima/SS; (bl) © 1988 Martin M. Rotker/PR; (tc) © CNRI/SPL/PR; (tl,bl) © Manfred Kage/PA; **341** (bkrnd) R Margulies/MMA; (c) R Margulies/Virginia Ferrante/MMA; (b) © David M. Philips/VU; **342** (r) © M.I. Walker/PR; (l) © Pierre Berger/PR; **343** (l) © 1990 Dale E. Boyer/PR; (r) © F. Widdel/VU; **344** (r) © 1992 David York/MC; (l) © Charles W. Stratton/VU; (c) © L. O'Shaughnessy/StockShop; **345** © Cabisco/VU; **346** (r) SP/Foca; (l) © Norm Thomas/PR; **347** SP/Foca; **348** (b) © Ed Reschke/PA; (t) © John D. Cunningham/VU; (c) © SPL/PR; **349** (t) SP/Foca; (bl) © 1990 Kent Wood/PR; (br) © Manfred Kage/PA; **350** (t) Reuters/Bettmann ; (br)